Methods in Cell Biology
Centrosome & Centriole

Volume 129

Series Editors

Methods in Cell Biology

Centrosome & Centriole

Volume 129

Edited by

Renata Basto
Cell Biology Department, CNRS, Institut Curie, France

Karen Oegema
Affiliation Ludwig Institute for Cancer Research, University of California - San Diego, USA

ELSEVIER

AMSTERDAM • BOSTON • HEIDELBERG • LONDON
NEW YORK • OXFORD • PARIS • SAN DIEGO
SAN FRANCISCO • SINGAPORE • SYDNEY • TOKYO

Academic Press is an imprint of Elsevier

AP

Academic Press is an imprint of Elsevier
225 Wyman Street, Waltham, MA 02451, USA
525 B Street, Suite 1800, San Diego, CA 92101-4495, USA
125 London Wall, London EC2Y 5AS, UK
The Boulevard, Langford Lane, Kidlington, Oxford OX5 1GB, UK

First edition 2015

ISBN: 978-0-12-802449-2
ISSN: 0091-679X

For information on all Academic Press publications
visit our website at http://store.elsevier.com

Contents

Contributors xi
Preface xvii

CHAPTER 1 Correlative Light and Electron Microscopy Analysis of the Centrosome: a Step-by-step Protocol 1
Dong Kong, Jadranka Loncarek
Introduction 2
1. Experimental Procedure 3
Acknowledgments 16
Supplementary Data 17
References 17

CHAPTER 2 Generation of a Conditional Analog-sensitive Kinase in Human Cells Using CRISPR/Cas9-mediated Genome Engineering 19
Tyler C. Moyer, Andrew J. Holland
Introduction 20
1. CRISPR/Cas System 21
2. Analog-Sensitive Kinases 21
3. Methods 22
4. Functional Analysis 33
Conclusion 34
References 34

CHAPTER 3 Studying Centrosome Function Using Three-dimensional Cell Cultures 37
Teresa Arnandis, Susana A. Godinho
Introduction 38
1. Methods 41
Conclusion 49
Acknowledgments 49
References 49

CHAPTER 4 Analysis of Centrosomes in Human Cancer 51
Stefan Duensing
Introduction 52
Centrosomes and Cancer 52
Centrosome Aberrations and Mitotic Outcomes 53
Not All Centrosome Aberrations are Created Equal 53

1. Methods ... 54
Conclusions ... 58
Acknowledgments ... 58
References ... 58

CHAPTER 5 In Vivo Examination of the Cortical Cytoskeleton in Multiciliated Cells Using Electron Tomography ... 61
Daniel K. Clare, Maud Dumoux, Delphine Delacour
Introduction ... 62
1. Rationale ... 63
2. Methods ... 65
3. Materials ... 78
Conclusion ... 79
List of Abbreviations ... 80
Acknowledgments ... 80
References ... 80

CHAPTER 6 Tackling Centrosome Biology Through Gene Targeting in Chicken B Cells ... 83
Pavithra L. Chavali, Fanni Gergely
Introduction ... 84
1. Methods ... 86
2. Analysis of DT40 Cells ... 93
Conclusion ... 100
Acknowledgments ... 100
References ... 100

CHAPTER 7 Centriole Biogenesis and Function in Multiciliated Cells ... 103
Siwei Zhang, Brian J. Mitchell
Introduction ... 104
1. Visualization of Centrioles and Related Structures in *Xenopus* MCCs ... 106
2. Driving Specific Expression in *Xenopus* MCCs ... 112
3. Generating Mosaic *Xenopus* Embryos to Circumvent Embryo Lethality ... 113
4. Ciliated Epithelial Spheroid Culture ... 114
5. Manipulating MCC Formation ... 115
6. Fully-Automatic, Program-Assisted Counting of Centrioles in Confocal Images Using ImageJ ... 118

Conclusion ... 123
References ... 124

CHAPTER 8 **Subdiffraction Resolution Microscopy Methods for Analyzing Centrosomes Organization** ... 129
Vito Mennella, Rachel Hanna, Moshe Kim
Introduction ... 130
1. Methods ... 133
Summary and Outlook ... 146
Acknowledgments ... 146
References ... 146

CHAPTER 9 **Probing Mammalian Centrosome Structure Using BioID Proximity-dependent Biotinylation** ... 153
Elif Nur Firat-Karalar, Tim Stearns
Introduction ... 154
1. Materials ... 157
2. Methods ... 159
3. Large-Scale BioID Pulldown of BirA*-Fusion Proteins ... 163
4. Data Analysis ... 166
Conclusion ... 167
Acknowledgments ... 168
References ... 168

CHAPTER 10 **Purification of Centrosomes from Mammalian Cell Lines** ... 171
Delphine Gogendeau, Paul Guichard, Anne-Marie Tassin
Introduction ... 172
1. Methods ... 175
Conclusions ... 184
Acknowledgments ... 184
References ... 184

CHAPTER 11 **Isolation, Cryotomography, and Three-dimensional Reconstruction of Centrioles** ... 191
Paul Guichard, Virginie Hamel, Aitana Neves, Pierre Gönczy
Introduction ... 192
1. Methods ... 194
Conclusions ... 206
Acknowledgments ... 207
References ... 207

CHAPTER 12 New Insights into Centrosome Imaging in *Drosophila* and Mouse Neuroepithelial Tissues 211
Maria A. Rujano, Renata Basto, Véronique Marthiens
Introduction 212
1. Imaging Centrosomes in *Drosophila* Neuroepithelial Cells 214
2. High Spatial Resolution Imaging of Centrosomes in Neural Stem Cells of the Mouse Neocortex 220
Conclusion 226
Acknowledgments 226
References 227

CHAPTER 13 Microinjection Techniques for Studying Centrosome Function in *Drosophila Melanogaster* Syncytial Embryos 229
Paul T. Conduit, Daniel Hayward, James G. Wakefield
Introduction 230
1. Protocol 231
2. Collection and Preparation of *Drosophila* Embryos 232
3. Preparing Samples for Microinjection 233
4. Mounting Embryos for Spinning Disc Confocal Microscopy .. 235
5. Microinjection of Embryos 238
6. Perturbation via Cold Treatment 241
7. Imaging and Data Analysis 242
References 247

CHAPTER 14 A Yeast Two-hybrid Approach for Probing Protein–protein Interactions at the Centrosome 251
Brian J. Galletta, Nasser M. Rusan
Introduction 252
1. Identifying Direct Protein Interaction Have Had a Profound Impact on Understanding Centrosome Function 253
2. Challenges to Understanding Protein–protein Interactions in the Centrosome 255
3. Dissection of Complex, Multicomponent Protein Machines Using Y2H 256
4. Performing an Array Based Screen for Centrosome Protein Interactions 259
5. Generating Specific, Separation of Function Mutations by Reverse Y2H 267
Summary 272

Acknowledgments.. 273
References.. 273

CHAPTER 15 Genome-wide RNAi Screens in S2 Cells to Identify Centrosome Components........................279
Jeroen Dobbelaere
Introduction.. 280
Centrosome Function and Structure 280
Early Characterization of the Molecular Composition of Centrosomes 281
Genetic Identification of Centriole Components in *Drosophila* and *C. elegans* and Other Model Systems 281
The First Genome-Wide RNA Interference Screen for Centrosomes 282
Identification of the Complete Proteome of Purified Centrosomes.. 282
Use of RNAi in *Drosophila* Cells to Study Genes Involved in MT Formation and Centrosome Function........................ 283
1. Methods 284
2. Genome-Wide Screen 288
3. Image Acquisition.. 291
4. Scoring Phenotypes and Image Analysis 292
5. Validation of Potential Hits.. 294
Discussion.. 294
Acknowledgments.. 296
References.. 297

CHAPTER 16 Time-lapse Recording of Centrosomes and Other Organelles in *Drosophila* Neuroblasts 301
Judit Pampalona, Jens Januschke, Paula Sampaio, Cayetano Gonzalez
Introduction.. 302
1. Primary Cultures.. 303
2. Ex Vivo Whole Mount.. 309
3. Fluorescent Reporters 311
4. Remarks and Troubleshooting 311
Acknowledgments.. 313
Supplementary Data.. 313
References.. 313

CHAPTER 17 Centrosomes and Spindles in Ascidian Embryos and Eggs 317
Alex McDougall, Janet Chenevert, Gerard Pruliere, Vlad Costache, Celine Hebras, Gregory Salez, Remi Dumollard
Introduction 317
1. Methods 318
References 337

CHAPTER 18 Ultrastructural Analysis of *Caenorhabditis elegans* cilia 341
Daniel Serwas, Alexander Dammermann
Introduction 342
1. Methods 344
Discussion 363
Acknowledgments 364
References 364

CHAPTER 19 Method: In Vitro Analysis of Pericentriolar Material Assembly 369
Jeffrey B. Woodruff, Anthony A. Hyman
Introduction 370
1. Methods 371
Conclusion 381
Acknowledgments 381
References 381

CHAPTER 20 Monitoring SPB Biogenesis in Fission Yeast with High Resolution and Quantitative Fluorescent Microscopy 383
Imène B. Bouhlel, Kathleen Scheffler, Phong T. Tran, Anne Paoletti
Introduction 384
1. Monitoring SPB Duplication in Fixed SPB-Labeled Strains 384
2. Quantitative Analysis of SPB Biogenesis in Live Cells 387
Conclusion 390
Acknowledgments 390
References 390

Volumes in Series 393
Index 407

Contributors

Teresa Arnandis
Barts Cancer Institute, Queen Mary University of London, London, UK

Renata Basto
Institut Curie, CNRS UMR144, Paris, France

Imène B. Bouhlel
Centre de Recherche, Institut Curie, Paris, France; CNRS-UMR144, Paris, France

Pavithra L. Chavali
Li Ka Shing Centre, Cancer Research UK Cambridge Research Institute, Cambridge, UK

Janet Chenevert
Sorbonne Universités, UPMC Univ Paris 06, and CNRS, Laboratoire de Biologie du Développement de Villefranche-sur-mer, Observatoire Océanographique, Villefranche-sur-Mer, France

Daniel K. Clare
Institute of Structural and Molecular Biology, Birkbeck College and University College of London, London, UK

Paul T. Conduit
Department of Zoology, University of Cambridge, Cambridge, UK

Vlad Costache
Sorbonne Universités, UPMC Univ Paris 06, and CNRS, Laboratoire de Biologie du Développement de Villefranche-sur-mer, Observatoire Océanographique, Villefranche-sur-Mer, France

Alexander Dammermann
Max F. Perutz Laboratories, University of Vienna, Vienna Biocenter (VBC), Vienna, Austria

Delphine Delacour
Cell Adhesion and Mechanics Group, Jacques Monod Institute, CNRS-UMR7592, Paris Diderot University, Paris Cedex, France

Jeroen Dobbelaere
Max F. Perutz Laboratories, University of Vienna, Vienna Biocenter (VBC), Vienna, Austria

Stefan Duensing
Division of Molecular Urooncology, Department of Urology, University of Heidelberg School of Medicine, Heidelberg, Germany

Remi Dumollard
Sorbonne Universités, UPMC Univ Paris 06, and CNRS, Laboratoire de Biologie du Développement de Villefranche-sur-mer, Observatoire Océanographique, Villefranche-sur-Mer, France

Maud Dumoux
Institute of Structural and Molecular Biology, Birkbeck College and University College of London, London, UK

Elif Nur Firat-Karalar
Department of Molecular Biology and Genetics, Koç University, Istanbul, Turkey

Brian J. Galletta
Cell Biology and Physiology Center, National Heart Lung and Blood Institute, National Institutes of Health, Bethesda, MD, USA

Fanni Gergely
Li Ka Shing Centre, Cancer Research UK Cambridge Research Institute, Cambridge, UK

Susana A. Godinho
Barts Cancer Institute, Queen Mary University of London, London, UK

Delphine Gogendeau
Institut Curie, CNRS UMR144, Paris, France; Institut Curie, Orsay, France

Pierre Gönczy
Swiss Institute for Experimental Cancer Research (ISREC), School of Life Sciences, Swiss Federal Institute of Technology (EPFL), Lausanne, Switzerland

Cayetano Gonzalez
Institute for Research in Biomedicine (IRB-Barcelona), Barcelona, Spain; Institució Catalana de Recerca i Estudis Avançats (ICREA), Barcelona, Spain

Paul Guichard
Swiss Institute for Experimental Cancer Research (ISREC), School of Life Sciences, Swiss Federal Institute of Technology (EPFL), Lausanne, Switzerland

Virginie Hamel
Swiss Institute for Experimental Cancer Research (ISREC), School of Life Sciences, Swiss Federal Institute of Technology (EPFL), Lausanne, Switzerland

Rachel Hanna
Department of Biochemistry, University of Toronto, Toronto, ON, Canada; Cell Biology Program, The Hospital for Sick Children, Toronto, ON, Canada

Daniel Hayward
Biosciences, College of Life and Environmental Sciences, University of Exeter, Exeter, UK

Celine Hebras
Sorbonne Universités, UPMC Univ Paris 06, and CNRS, Laboratoire de Biologie du Développement de Villefranche-sur-mer, Observatoire Océanographique, Villefranche-sur-Mer, France

Andrew J. Holland
Department of Molecular Biology and Genetics, Johns Hopkins University School of Medicine, Baltimore, MD, USA

Anthony A. Hyman
Max Planck Institute of Molecular Cell Biology and Genetics, Dresden, Germany

Jens Januschke
Division of Cell and Developmental Biology, College of Life Sciences, University of Dundee, Dundee, UK

Moshe Kim
Department of Biochemistry, University of Toronto, Toronto, ON, Canada; Cell Biology Program, The Hospital for Sick Children, Toronto, ON, Canada

Dong Kong
Laboratory of Protein Dynamics and Signaling, NIH/NCI/CCR-Frederick, Frederick, MD, USA

Jadranka Loncarek
Laboratory of Protein Dynamics and Signaling, NIH/NCI/CCR-Frederick, Frederick, MD, USA

Véronique Marthiens
Institut Curie, CNRS UMR144, Paris, France

Alex McDougall
Sorbonne Universités, UPMC Univ Paris 06, and CNRS, Laboratoire de Biologie du Développement de Villefranche-sur-mer, Observatoire Océanographique, Villefranche-sur-Mer, France

Vito Mennella
Department of Biochemistry, University of Toronto, Toronto, ON, Canada; Cell Biology Program, The Hospital for Sick Children, Toronto, ON, Canada; Peter Gilgan Centre for Research and Learning, Toronto, ON, Canada

Brian J. Mitchell
Department of Cell and Molecular Biology, Feinberg School of Medicine, Northwestern University, Chicago, IL, USA

Tyler C. Moyer
Department of Molecular Biology and Genetics, Johns Hopkins University School of Medicine, Baltimore, MD, USA

Aitana Neves
Swiss Institute for Experimental Cancer Research (ISREC), School of Life Sciences, Swiss Federal Institute of Technology (EPFL), Lausanne, Switzerland

Judit Pampalona
Institute for Research in Biomedicine (IRB-Barcelona), Barcelona, Spain

Anne Paoletti
Centre de Recherche, Institut Curie, Paris, France; CNRS-UMR144, Paris, France

Gerard Pruliere
Sorbonne Universités, UPMC Univ Paris 06, and CNRS, Laboratoire de Biologie du Développement de Villefranche-sur-mer, Observatoire Océanographique, Villefranche-sur-Mer, France

Maria A. Rujano
Institut Curie, CNRS UMR144, Paris, France; Imagine Institute, Paris, France

Nasser M. Rusan
Cell Biology and Physiology Center, National Heart Lung and Blood Institute, National Institutes of Health, Bethesda, MD, USA

Gregory Salez
Sorbonne Universités, UPMC Univ Paris 06, and CNRS, Laboratoire de Biologie du Développement de Villefranche-sur-mer, Observatoire Océanographique, Villefranche-sur-Mer, France

Paula Sampaio
Instituto de Investigação e Inovação em Saúde, Universidade do Porto, Portugal; IBMC—Instituto de Biologia Molecular e Celular, Universidade do Porto, Portugal

Kathleen Scheffler
Centre de Recherche, Institut Curie, Paris, France; CNRS-UMR144, Paris, France

Daniel Serwas
Max F. Perutz Laboratories, University of Vienna, Vienna Biocenter (VBC), Vienna, Austria

Tim Stearns
Department of Biology and Department of Genetics, Stanford University, Stanford, CA, USA

Anne-Marie Tassin
Institute for Integrative Biology of the Cell (I2BC), CEA, CNRS, Université Paris Sud, Gif sur Yvette, France

Phong T. Tran
Centre de Recherche, Institut Curie, Paris, France; CNRS-UMR144, Paris, France

James G. Wakefield
Biosciences, College of Life and Environmental Sciences, University of Exeter, Exeter, UK

Jeffrey B. Woodruff
Max Planck Institute of Molecular Cell Biology and Genetics, Dresden, Germany

Siwei Zhang
Department of Cell and Molecular Biology, Feinberg School of Medicine, Northwestern University, Chicago, IL, USA

Preface

Centrosomes are cytoplasmic organelles composed of two centrioles that recruit and organize a network of proteins called the pericentriolar material or PCM. Centrosomes are the major microtubule organizing center of most animal cells, and as such participate in a variety of cellular processes. In yeasts, the spindle pole body (SPB), which lies embedded in the nuclear membrane, is the functional homologue of the centrosome.

Since the initial studies of Edouard Van Beneden, Theodor Boveri, and Walther Fleming during the nineteenth century, the positioning of the centrosome at the center of the cell has contributed to its recognition as an important organelle in cellular organization. Much of the attention given to this organelle during the twentieth century has been related cell division. Upon mitotic entry, the PCM, which is the site of microtubule nucleation, grows in size causing an increase in the microtubule nucleation capacity of the centrosome. Large mitotic embryos containing large microtubule asters emanating from the centrosome, beautifully illustrated by T. Boveri, clearly sustained the view that this organelle participated in cell division.

The observations that plant cells divide without centrosomes combined with fact that most female meiotic divisions across the animal kingdom occur in the absence of centrosomes showed, however, that in certain cell types these organelles are dispensable for cell division. This is also true in the flat worm planaria, which even lacks genes encoding PCM components, and in mice at certain developmental stages. Moreover, in flies, mutations that affect centrosome assembly do not impair mitotic divisions during development from late embryogenesis till adulthood.

The fact that we found that centrosomes were not required for all cell divisions, suggested that maybe other important yet unidentified functions were associated with this organelle. Indeed, centrosomes are required for a variety of other cellular processes. The presence of centrosomes at opposite poles of the mitotic spindle allow, for instance, the correct segregation of centrioles during mitosis into daughter cells. This might be essential if these cells need to assemble a cilium in the following cell cycle. Centrosomes, through aster microtubule nucleation respond to polarity cues and forces exerted by molecular motors to orient the mitotic spindle, which is essential for tissue morphogenesis in a diversity of developmental contexts. Centrosome positioning is also thought to contribute in certain cell types for polarity establishment or maintenance, cell migration, and unexpectedly to generate an efficient immune response.

The last 15 years have remarkably changed our view of the centrosome. The loss-of-function screens first performed in *Caenorhabditis elegans* identified the molecular components responsible for duplicating centrioles and for PCM assembly. Proteomics approaches were also essential to build a part list of the molecules that contribute to centrosome biogenesis. These initial studies were further complemented by other studies performed in other model organisms, which allowed us to obtain a bigger picture of the molecules and pathways involved in centrosome assembly. But, more important than providing with a detailed list of centrosome molecules, all these studies allowed us to be able to manipulate the centrosome assembly machinery to start to unravel how defects in centrosome assembly can lead to loss of cell and organism homeostasis, leading to pathological conditions.

The presence of more than two centrosomes in a cell, centrosome amplification has since the dispermic experiments performed by T. Boveri, been correlated with cancer. Indeed, more than 80% of human solid tumors contain a percentage of cells with extra centrosomes. More recently, centrosome numerical and structural abnormalities have also been described in growth-related disorders such as microcephaly, Seckel syndrome, and primordial dwarfism. The etiology underlying these centrosome alterations and human diseases are still under investigation, but several factors such as defects in mitosis, spindle positioning, ciliogenesis, and cell migration, to name a few seem to be implicated.

The centrosome field has also profited from important technological improvement notably in microscopy. The recent use of Cryo-EM tomography or high resolution techniques such as structural illumination, allowed us to go deeper into the structure of centrioles and PCM.

In these times of translational driven research and applications at the expense of basic research, the centrosome field is a good example of how knowledge accumulated in the past 50 years using basic cell and molecular biology has contributed enormously to our current understanding of cellular alterations implicated in human health. In this method series, we brought together both different model systems and approaches with the aim of providing researchers interested in centrosomes or SPBs with methodologies that can be applied to their questions. These include different types of electron microscopy related techniques (**D. Kong & J. Loncarek, pp. 1–18; D.K. Clare et al., pp. 61–82; P. Guichard & P. Gonczy, pp. 191–210; D. Serwas & A. Dammermann, pp. 341–368**), structured illumination microscopy (**V. Mennella et al., pp. 129–152**), methods for centrosome purification (**Gogendeau et al., pp. 171–190**), methods for in vitro analysis of PCM assembly (**J. Woodruff & A. Hyman, pp. 369–382**), methods for yeast two hybrid screens (**B. Galletta & N. Rusan, pp. 251–278**), and methods for performing genome-wide screens in S2 cells (**J. Dobbelaere, pp. 279–300**). Furthermore, methods for genome engineering in human cells (**T. Moyer & A. J. Holland, pp. 19–36**), methods for centrosome analysis in organotypic cultures (**T. Arnandis & S.A. Godinho, pp. 37–50**), in human cancers (**S. Duensing, pp. 51–60**), in chicken DT40 cells (**P.L. Chavali & F. Gergely, pp. 83–102**), in multiciliated cells (**S. Zhang & B.J. Mitchell, pp. 103–128**), in *Drosophila* embryos (**P. Conduit et al., pp. 229–250**), in Ascidian embryos (**A. McDougall et al., pp. 317–340**), in the *Drosophila* and mouse neuroepithelium (**M. Rujano et al., pp. 211–228**) and in *Drosophila* neuroblasts (**J. Pampalona et al., pp. 301–316**). Finally, a chapter dedicated to SPB biogenesis in fission yeast completes this volume (**I.B. Bouhlel et al., pp. 383–392**).

It is our hope that by bringing together methods for studying centrosomes using a wide range of model systems and techniques, this volume might spark researchers to explore other model systems with unique potential, but also to bring researchers together who work on different models to explore areas of commonality.

Renata Basto
Karen Oegema

CHAPTER

1

Correlative light and electron microscopy analysis of the centrosome: a step-by-step protocol

Dong Kong, Jadranka Loncarek[1]

Laboratory of Protein Dynamics and Signaling, NIH/NCI/CCR-Frederick, Frederick, MD, USA

[1]*Corresponding author: E-mail: jadranka.loncarek@nih.gov*

CHAPTER OUTLINE

Introduction **2**

1. Experimental Procedure **3**

1.1 Cell Culture and Preparation of the Cells for Microscopy 3

1.1.1 Cell culture 3

1.1.2 Preparation of the cells for microscopy 4

1.1.3 Light microscopy 5

1.2 Cell Fixation and Postfixation Recording of Cell/Centriole Position 5

1.3 Prestaining, Dehydration, and Embedding 6

1.4 Marking the Position of the Target Cell on the Polymerized Resin 7

1.5 Removal (Dissolving) of the Glass Coverslip 8

1.6 Trimming, Ultrathin Serial Sectioning, and the Pickup of the Serial Sections 8

1.6.1 Trimming 8

1.6.2 Ultrathin serial sectioning 10

1.6.3 Picking up the serial sections 12

1.6.4 Staining of the sections 12

1.6.5 Electron microscopy 12

1.7 Preparation of Formvar-Coated Slot Grids 13

1.8 Chemicals, Buffers, and Media 15

1.9 Instrumentation 16

Acknowledgments **16**

Supplementary Data **17**

References **17**

Methods in Cell Biology, Volume 129, ISSN 0091-679X, http://dx.doi.org/10.1016/bs.mcb.2015.03.013

Abstract

Correlative light and electron microscopy harnesses the best from each of the two modalities of microscopy it utilizes; while light microscopy provides information about the dynamic properties of the cellular structure or fluorescently labeled protein, electron microscopy provides ultrastructural information in an unsurpassed resolution. However, tracing a particular cell and its rare and small structures such as centrosomes throughout numerous steps of the experiment is not a trivial task. In this chapter, we present the experimental workflow for combining live-cell fluorescence microscopy analysis with classical transmission electron microscopy, adapted for the studies of the centrosomes and basal bodies. We describe, in a step-by-step manner, an approach that can be affordably and successfully employed in any typical cell biology laboratory. The article details all key phases of the analysis starting from cell culture, live-cell microscopy, and sample fixation, through the steps of sample preparation for electron microscopy, to the identification of the target cell on the electron microscope.

INTRODUCTION

Centrosome structure is precisely defined and conserved among various eukaryotic organisms (Carvalho-Santos, Azimzadeh, Pereira-Leal, & Bettencourt-Dias, 2011). The centrosome consists of an unduplicated or a duplicated centriole (Alvey, 1985; Rattner & Phillips, 1973; Vorobjev & Chentsov, 1982), which organizes a proteinaceous pericentriolar material (PCM) in a highly hierarchical and ordered fashion (Lawo, Hasegan, Gupta, & Pelletier, 2012; Mennella, Agard, Huang, & Pelletier, 2014; Mennella et al., 2012; Sonnen, Schermelleh, Leonhardt, & Nigg, 2012). Centrosome function and its ultrastructural features are intimately linked, so the findings discerned through biochemistry and light microscopy should be correlated with analysis at the ultrastructural level whenever possible. Centrioles are ninefold symmetrical microtubule-based structures, easily detectable by electron microscopy (EM). The PCM component of the centrosome is, on the other hand, not electron dense. The PCM components can be visualized by fluorescence microscopy, and their structural organization examined using recently developed super-resolution microscopy techniques (Leung & Chou, 2011; Yamanaka, Smith, & Fujita, 2014) predicts the localization of the florescence signals beyond the resolution limit of classical light microscopy (Keller et al., 2014; Lau, Lee, Sahl, Stearns, & Moerner, 2012; Mennella et al., 2012; Sillibourne et al., 2011). A centriole's ultrastructural features are less reliably predicted by light microscopy due to their small size (centrioles are, depending on the species, ~120–200 nm in diameter and ~200–500 nm in length).

Correlative light and electron microscopy (CLEM) is an imaging approach that combines various modalities of light microscopy and EM in one experiment (Rieder & Bowser, 1985; Rizzo, Parashuraman, & Luini, 2014; Spiegelhalter, Laporte, & Schwab, 2014). It has been used to analyze centrosome-associated processes during cell cycle progression or after various genetically or chemically induced treatments (Kong et al., 2014; Lončarek, Hergert, & Khodjakov, 2010; Loncarek, Hergert, Magidson, & Khodjakov, 2008; Rieder & Bowser, 1985; Tsou et al., 2009). A combination of live- or fixed-cell light microscopy and EM allows the investigator to capitalize on the strengths of both individual techniques. However, it also brings a new layer of complexity to the experiment, as it requires an expertise in both modalities of microscopy.

One of the challenges of CLEM is to trace down the target cell previously analyzed by light microscopy through multiple steps of sample preparation, down to the imaging on the electron microscope. Finding the right cell among hundreds of surrounding cells might seem impossible, but various strategies can be employed to accomplish this task. In this chapter, we describe the strategy utilized in our laboratory for routine study of centrosomes by CLEM. This strategy allows us to first analyze centrosomes by light microscopy, and to reproducibly follow the same cell through embedding, trimming, and sectioning, and to image the same centrosomes on the electron microscope. Due to the complexity of the technique, many researchers interested in centrosome biology may hesitate to employ CLEM and therefore not benefit from the insight provided by ultrastructural analysis. We hope that detailing our approach to CLEM will inspire some to introduce this highly rewarding technique into their daily experimental practice. The strategy we describe can also be employed for the analysis of other cellular organelles with minimal adaptations to the described protocol.

1. EXPERIMENTAL PROCEDURE

1.1 CELL CULTURE AND PREPARATION OF THE CELLS FOR MICROSCOPY

1.1.1 Cell culture

1. Cells are plated on sterile 25-mm round, 0.17-mm thick coverslips previously washed in deionized H_2O, followed by 70% ethanol, then absolute ethanol, and individually dried in a sterile cell culture hood. Most cells will adhere and proliferate on these clean glass coverslips; however, some cell types will require coverslips to be coated with poly-L-lysine for better adherence.
2. To visualize centrioles by fluorescent light microscopy, we express Centrin1-green fluorescent protein (GFP) (or another fluorescent-tagged centrosomal protein) and select for the population of cells with optimal signal-to-noise ratio for imaging (see Figure 5) using cell sorting.
3. For easier identification of the target cell after embedding, the cells should be less than 60% confluent at the time of analysis. This allows us to use the shape of neighboring cells as identification landmarks during trimming and sectioning. In confluent cultures, identification of the target cell throughout the experiment is possible but the incidence for misidentification increases.

1.1.2 Preparation of the cells for microscopy

To image the cells we use homemade reusable Rose chambers (see Figure 1) which are a proven, affordable alternative to glass bottom Petri dishes (Pereira, Matos, Lince-Faria, & Maiato, 2009; Stout, Rizk, & Walczak, 2009). Rose chambers are ideally suited for long-term live-cell imaging at high resolution. Most cell types will grow in a Rose chamber without a change in viability for days. In addition, their square shape allows for an easy orientation and quick repositioning of the sample on the microscope during fixation, marking the target cells, and switching between the high- and low-magnification objectives (as described in the next chapter). Before imaging, assemble a Rose chamber in a sterile hood as follows:

1. Put an empty coverslip on the flat surface of the lower metal plate.
2. Place a sterile silicone gasket on the top followed by a coverslip with the cells facing down. For sterilization, wash the gasket in 70% ethanol and dry in a sterile hood, or sterilize the spacers by autoclaving.
3. Put the upper metal plate on the rubber gasket. Be sure to assemble the chamber fast enough to prevent the cells from drying.
4. Press lightly on the assembly to hold it in place while sealing the chamber with four bolts.
5. Perfuse the chamber with complete, warm, CO_2-independent medium (which will maintain pH during imaging) using a 21–22 gauge needle and a 3 mL syringe. Punch through the rubber gasket, and slowly inject the medium. Be sure to insert another needle to serve as a vent on the opposite side of the rubber gasket before injecting the medium into the chamber.

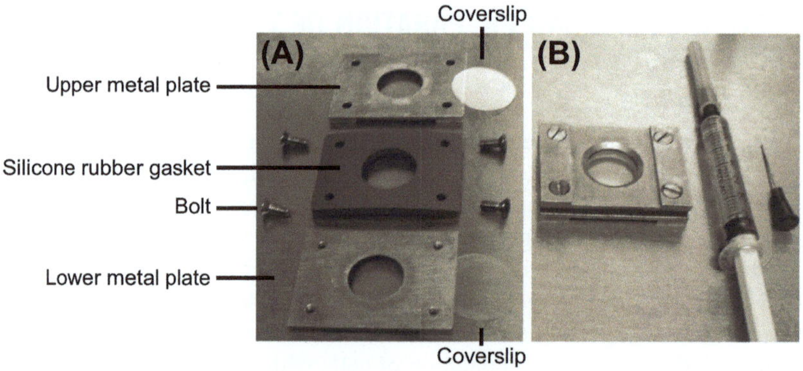

FIGURE 1 Rose chamber for live-cell imaging.

(A) A disassembled Rose chamber to illustrate its components. A Rose chamber contains two metal plates, one 1–3 mm thick silicon rubber gasket, and two coverslips (one that carries the cells and another one that is empty). All components are held together by four bolts. (B) An assembled Rose chamber and a syringe filled with medium. The side of the Rose chamber facing up will be facing the objective during imaging.

6. Remove all traces of media and cell debris from the coverslip. The Rose chamber is now ready for imaging.

Many variations of the original Rose chamber have been developed since its first description in the 1950s (Rose, Pomerat, Shindler, & Trunnell, 1958). We designed a modified version of the chamber to account for our specific microscope setting and objective features. In designing a Rose chamber, be sure that the silicone rubber used for the spacer is made of high quality, FDA approved, nontoxic material as it will be in direct contact with the medium. An alternative to using a Rose chamber is to culture cells in a culture dish with a glass bottom (also available with a gridded bottom, e.g., from MatTek).

1.1.3 Light microscopy

For live-cell microscopy use a research-grade inverted microscope equipped with 20×, 60×, and 100× objectives, a spinning disc confocal, a sensitive CCD camera, bright field illumination, and an environmental enclosure set to 37 °C. To resolve individual centrin-GFP labeled centrioles, cells must be imaged using 60× or 100× objective with a high numerical aperture (NA = 1.4 or higher). The exposure to illuminating fluorescence light should be minimized during imaging by using automated shutters to avoid photo damage. Photo damage may induce a nonphysiological response of the cells during imaging. This is especially important if the cells are imaged for a long period of time. The cell cycle and the centriole cycle can both be perturbed or even completely stalled in cell cultures stressed by excitation light.

1.2 CELL FIXATION AND POSTFIXATION RECORDING OF CELL/CENTRIOLE POSITION

Centrosomes occupy only a small fraction of a cell's volume and can be easily missed during EM analysis. In addition, centrioles in live cells continuously change their position and orientation with respect to the coverslip. The average thickness of a section for transmission EM analysis is 60–80 nm, meaning the centriole(s) belonging to one centrosome will, depending on their orientation, be present in no more than three to six serial sections (out of ~100 needed to span entire volume of a typical adherent interphase cell). Thus, it is important to record the position of the centrosome/centriole with respect to the coverslip and other cellular landmarks as accurately as possible on the light microscope after fixation. This is crucial for overall success of the experiment.

1. After and analysis of the target cell by light microscopy, fix the cells by perfusing the Rose chamber with freshly prepared and prewarmed fixative and return the chamber to the microscope in the same position and orientation as prior to fixation.
2. Within 1–2 min of fixative addition, intracellular movements will cease. A soon as possible after fixation, record a Z-stack (200-nm z step) spanning the entire cell to register the position and orientation of the centrioles.

3. In parallel, record a Z-stack in differential interference contrast (DIC) (or phase) at the same magnification to facilitate subsequent localization of the centrioles within the cell using morphological clues of the cell (the shape, the position of the nucleus, etc.) while searching on the electron microscope.
4. The position of the target cell should then be permanently marked on the coverslip. For this purpose, we use an objective diamond scriber placed on the microscope objective turret. If you have such a scriber, scratch a circular mark on the glass surface around the cell of interest (see Figure 2(A)). Alternatively, the position of the target cell can be marked by an objective slide marker (available from Nikon), and subsequently reinforced using a glass writing diamond pen (many versions are available on the market). The bottom of the glass coverslip can also be prescratched using a diamond pen before plating the cells. In addition, cells can be cultured on "grid" glass bottom dishes (MatTek) or on gridded coverslips (which is a more expensive alternative).
5. Clean immersion oil from the coverslip and use a low-magnification objective (for instance 10× or 20×) to record DIC images of the targeted cell and its surroundings (see Figure 5). Neighboring cells will be helpful landmarks when identifying the cell of interest during further steps, such as trimming of the embedded sample. We recommend taking several images of the area surrounding the target cell.
6. When the position of the target cell is clearly marked on the surface of the coverslip, disassemble the Rose chamber and place the coverslip (cell side facing up) into 5 mL of the fixative in a 60-mm Petri dish, seal the dish with parafilm, and store it at 4 °C until embedding.

1.3 PRESTAINING, DEHYDRATION, AND EMBEDDING

After fixation and marking the position of the target cell on the light microscope, prestaining, dehydration, and embedding, with many possible variations in the

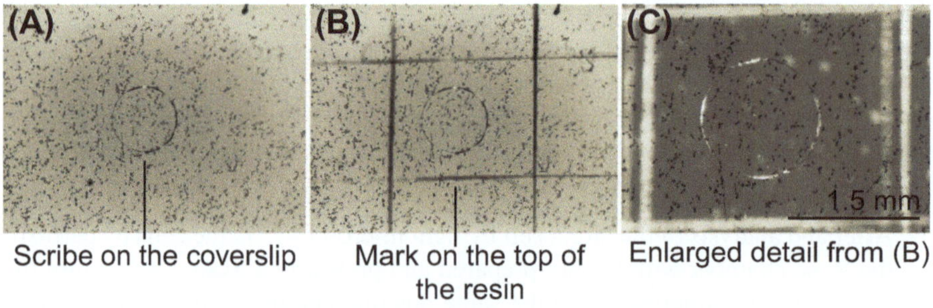

FIGURE 2 Marking the position of the target cell on polymerized resin.

(A) The scribe on the coverslip encircling the target cell is easily visible after embedding. (B) The position of the target cell is marked on the resin with a razor blade. (C) Enlarged detail from (B).

protocol, can be performed in any EM laboratory by a technician skilled in the preparation of EM samples. We will delineate here the protocol used in our laboratory.

1. Transfer the coverslips with the fixed cells into a 35-mm Petri dish and carry out all the following steps in Petri dish at room temperature. Wash the cells with 1X phosphate buffered saline (PBS) (pH 7.2) three times, 10 min/each wash, to remove the fixative.
2. Incubate the cells in 0.15% Tannic acid in 1X-PBS for 1 min. Tannic acid allows for better contrasting of microtubule-based structures such as centrioles. Wash the cells with 1X-PBS for 5 min.
3. Postfix the cells in 2% Osmium tetroxide (OsO_4) diluted in distilled water (hereafter dH_2O) for 1 h at 4 °C in a Petri dish sealed with parafilm. Wash the cells in dH_2O three times, 5 min/each wash.
4. Prestain the cells with 1% uranyl acetate diluted in dH_2O for 1 h at 4 °C in a dish sealed with parafilm. Wash the cells in dH_2O two or three times, 5 min/each wash.
5. Dehydrate the sample by sequential exchange of 20%, 30%, 40%, 60%, 80%, and 95% ethanol solutions, each exchange being 5 min. Finally, dehydrate the sample in three exchanges of 100% ethanol, 10 min each. During this dehydration procedure, be sure that the cells are not allowed to dry out between ethanol exchanges.
6. Incubate the cells in a mixture of 100% ethanol and Embed 812 resin at 1:1 ratio overnight. In the morning remove the old resin and replace with pure Embed 812 resin, and leave for 1 h.
7. Tilt the Petri dish for a few seconds to let the resin flow down, and remove it using a pipette. Keep the Petri dish tilted, apply fresh pure Embed 812 resin to the cells from the top of the coverslip and leave for 1 h. Repeat this procedure two times to completely remove all traces of ethanol.
8. Remove the resin from the Petri dish as described in Step 7. Take the coverslip out of the Petri dish and clean residual resin from the bottom of the coverslip with a Kimwipe tissue. Place the coverslip (with the cells up) on some sort of the holder resistant to high temperature. Apply ~1 mL of fresh Embed 812 resin to the center of the coverslip, and the resin will slowly spread out to cover the whole coverslip. Put the coverslip with resin in an oven at 60 °C for 48 h to polymerize. Be sure that the holder with the coverslips is in horizontal orientation to avoid leaking of the resin from the coverslip. As a holder we use a pipette tips box, with some tips left in the box to support the coverslip from below.

1.4 MARKING THE POSITION OF THE TARGET CELL ON THE POLYMERIZED RESIN

After embedding, the cells will be localized within a thin layer of polymerized resin adjacent to the glass coverslip. The mark on the glass, indicating the position of the target cell, must be transferred to the top of the resin before the glass coverslip is removed. After embedding, the bottom of the embedded sample is usually covered with a thin layer of polymerized resin which makes the mark on the coverslip invisible. To observe the mark, scrape the resin off the glass using a razor blade. Place the sample on the dissecting microscope and transfer the mark from the glass to the top of the resin with a fine needle, diamond pen, or razor blade, as illustrated in Figure 2.

1.5 REMOVAL (DISSOLVING) OF THE GLASS COVERSLIP

This step potentially exposes a researcher to the harmful effects of hydrofluoric acid. Therefore, all steps described in this section should be performed in a vented chemical hood in agreement with required safety procedures.

1. Place embedded sample in a small Petri dish with the glass coverslip facing upwards. Add enough concentrated hydrofluoric acid to overlay the coverslip. Incubate for 1 h at room temperature. Hydrofluoric acid will gradually dissolve the glass coverslip.
2. Check if the glass is completely dissolved; if not, continue incubation until it is.
3. Using tweezers transfer the sample into a large beaker containing glass beads at the bottom and wash the sample under a gentle stream of running water for 1–2 h. The resin will appear turbid. Let the resin thoroughly dry until it becomes clear again. Alternatively, fill up the beaker with water, and exchange water in 20 min intervals.
4. If the cells were cultured on a plastic surface, separate the plastic from the resin using two pliers. Note that this step may damage the tissue close to the Petri dish. Alternatively, immerse the sample into liquid nitrogen to break the plastic.

1.6 TRIMMING, ULTRATHIN SERIAL SECTIONING, AND THE PICKUP OF THE SERIAL SECTIONS

Until this point, all described steps of the sample preparation can be accomplished in a typical laboratory setting after appropriate training. However, trimming, ultrathin serial sectioning, and the pickup of the serial sections is the most delicate part of CLEM, and it is not expected to be routinely preformed in the laboratory. Extensive experience and great skill are required to perform this part of experiment. The assistance of a specialist will usually be required.

1.6.1 Trimming

1. Roughly cut out the region of the resin where your cell of interest is located using pliers and glue it onto the flat bottom of a premade resin block (see Figure 3(A)). Before trimming, make sure that the sample tightly attaches to the resin block. Note that after removal of the glass or the plastic, the thin layer of the resin that contains the cells is unprotected and vulnerable to damage. It is important to keep this surface of the resin unscratched during cutting and trimming, as the

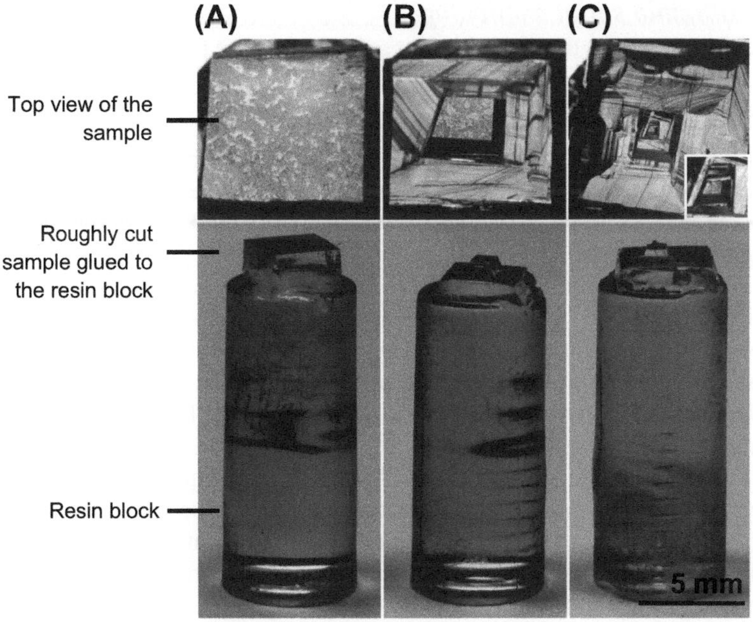

FIGURE 3 Trimming.

(A) Roughly cut the part of the resin containing the cell of interest. A layer of cells is visible under the ultramicrotome stereomicroscope (top view). (B) Shaping of a trapezoidal prism with the target cell in the center of the trapezoid. (C) Final size of the trapezoidal prism ready for sectioning. The inset is showing enlarged prism (top view).

scratches may damage the cells or make it impossible to identify the target cell later on.

2. Place the resin block into the sample holder of an ultramicrotome and secure it tightly. Place the holder into the trimming adaptor that is locked on the ultramicrotome stage.
3. Use previously recorded low-magnification DIC images of the target cell and the neighboring cells to identify the cell of interest under the ultramicrotome stereomicroscope. Trim the sample with a sharp razor blade to the shape of a trapezoidal prism. The trapezoid should have one side slightly slanted, and another side straight (see Figure 3(B)).
4. Center the prism on the target cell, and then carefully continue fine trimming until you form a pyramid with the small trapezoid on the top. The final top surface of the trapezoid should be about 0.5 mm^2 containing only a few neighboring cells, with the target cell in the center (see Figure 3(C)). To get a straight ribbon of serial sections during future sectioning, it is important that the bases of the trapezoid are smooth and parallel to each other. To obtain smooth sides of the pyramid during final trimming steps, it is the best to use only the sharpest and previously unused razor blades.

1.6.2 Ultrathin serial sectioning

This section describes the procedure for sectioning using Leica EM UC7 ultramicrotome (Leica Microsystems). This procedure might have to be adapted for the specific features of other instruments.

1. After fine trimming, place the sample holder into the specimen arm and clamp it with the knurled screw. Set the diamond knife into the knife holder with its clearance angle to 6°, as recommended, and lock the knife in this setting (see Figure 4(A)).
2. Turn off all the overhead lights to reduce distraction and only leave the bottom light on. Go to the highest magnification of the stereomicroscope and slowly advance the diamond knife to the trapezoid, bringing them as close together as possible (within 1 mm) until you can see a bright narrow gap between them. Align the bottom base of the trapezoid to the diamond knife edge to make them parallel to each other.
3. Adjust the orientation of the trapezoid to make the gap between its surface and the diamond knife edge identical for both trapezoid bases (until the width of the bright gap does not change during the up and down movement of the trapezoid). This alignment is important to get the section plane parallel to the sample plane.
4. Retract the diamond knife 1–2 mm from the trapezoid. Take great care not to damage the surface of the sample during the whole procedure.
5. Turn on all overhead lights and set the cutting window with buttons "START" and "END" on the touch screen control panel following the manufacturer's instructions.
6. Fill the boat of the knife with dH_2O (see Figure 4(B)). Water should be leveled with the diamond knife edge until you can see a silver reflection of water surface on the surface of the water. Set the cutting speed to 1 mm/s, the section thickness to 70–80 nm, and start sectioning. One long ribbon of silver-to-light yellow sections should appear floating on the surface of the water (see Figure 4(C) and (D), and Video 1). Usually, one can see the cells within the sections through binoculars.
7. Stop the cutting after the ribbon is 30–40 sections long and gently pull the ribbon toward the water boat to make it float on the surface of the water. Retract the diamond knife ~1 mm away before you are ready to pick up the sections. Separate the long ribbon into shorter segments (7–10 sections) using an eyelash glued to the end of a wooden toothpick (see Figure 4(E)). The eyelash tip should be wet, as a dry eyelash tends to stick to the sections. The eyelash should be precleaned in acetone.
8. Do not cut too many sections at a time, as this can make it harder to distinguish the correct order of those separated shorter ribbons. If the ribbon breaks into too many segments during sectioning, it could mean that the bases of the trapezoid were poorly made, or that the alignment between the trapezoid and the knife was not correct. Try your best to track the correct order of those segments.

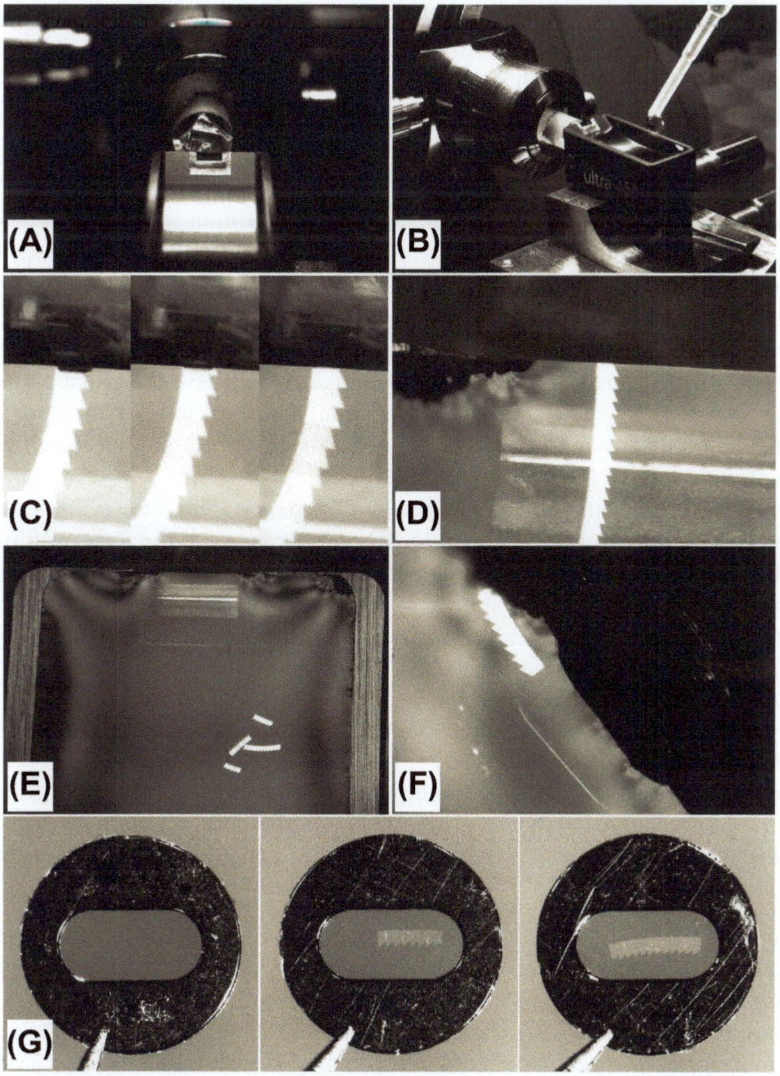

FIGURE 4 Serial sectioning and picking up the sections.

(A) The sample in a position ready for sectioning. (B) A knife boat is filled with water before sectioning. (C) Stills from Video 1 illustrating the movement of the trapezoidal prism during sectioning and the formation of the ribbon of sections. (D) The ribbon floating on the water with the last section still attached to the edge of the knife. (E) The long ribbon is separated in four shorter ribbons before the pickup. The ribbons are free-floating in a knife boat filled with water. (F) The coated grid partially submerged in the water in preparation for the ribbon pickup. (G) Examples of three formvar coated grids; the left one is empty, and the middle and the right one carry a ribbon of serial sections.

1.6.3 Picking up the serial sections

The ribbons of serial sections are picked up on the formvar-coated slot grid which must be prepared at least 24 h in advance prior to sectioning (how to prepare coated slot grids is described in the following section). Great care must be taken not to cause wrinkles in the sections during the picking up procedure. To avoid wrinkles, a ribbon of serial sections should be picked from underneath the water as follows:

1. Hold the formvar-coated grid at the four o'clock position and inspect it under the binoculars. Use only the best coated grids with no cracks, wrinkles, or any other irregularity on the formvar.
2. Paddle the water around the ribbon to bring the ribbon to an isolated area in the water boat. Submerge approximately three-fourth of the grid in the water, keeping the top area of the film dry (see Figure 4(F)). Tilt the grid a bit toward the ribbon, slowly advancing it to the ribbon, then waft the ribbon closer to the grid by gently moving the grid back and forth, or by paddling the water around the ribbon toward the grid using the eyelash.
3. Once one edge of the ribbon attaches to the top, dry area of the film, slowly pull the grid out of the water. The entire ribbon will attach to the formvar film (see Figure 4(G)). Touch the lower edge of the grid with a small piece of filter paper to absorb water, air dry the grid for a few seconds, and put it into a well of the grid storage box. Pick up all ribbons in this way and keep them in the grid storage box in a systematic order.
4. Continue sectioning and picking up the ribbons.

1.6.4 Staining of the sections

Before they are used for imaging on the transmission electron microscope, the sections have to be stained with uranyl acetate and lead citrate. We stain the grids using a grid staining matrix system, which allows for fast and uniform staining of multiple grids.

1. Put the grids into the wells of the matrix body in a sequential order during this procedure (for the analysis of serial sections, it is very important to preserve the order of the grids).
2. Stain the grids in 2% uranyl acetate diluted in dH_2O for 20–30 min at room temperature, followed by three washes in dH_2O, each for 1–2 min. Dry the grids, or immediately stain them in lead citrate solution for 1 min, followed by three 1–2 min washes in dH_2O that has been previously boiled and cooled down to room temperature.
3. Return the grids to a grid storage box and air dry the grids at room temperature for 24 h (do not close the box). Samples are now ready to be analyzed on the transmission electron microscope.

1.6.5 Electron microscopy

1. To facilitate the identification of the targeted cell by electron microscope, use both low- and high-magnification recordings of the target cell previously

obtained by the light microscope. Use the three-dimensional Z-stacks to assess the position of the centrioles in *XY* and *Z* planes. Note that serial sections, if collected by following the above procedure, will be flipped with respect to the images collected on the light microscope. Thus, we recommend printing the flipped versions of the original images in preparation for the search on the electron microscope. Using the recordings in both low and high magnification, and knowing at which region and cell depth the centrioles were at the time of fixation, one can easily find the target cell and the centrioles within. As long as serial sectioning was preformed parallel to the confocal plain, it will be relatively simple to translate the *Z* position of the centrioles into the sectioning depth.

2. Start imaging from the first to the last section of the first grid. Once the centrioles are identified, acquire images of them at low magnification (2.5–5K) and then at higher magnification (10–12K). Lower magnifications will assist during the alignment of the images from consecutive serial sections, as images of each section will be slightly rotated and shifted with respect to the previous one. Images in lower magnification contain more cellular landmarks which are often used to assist the alignment of the images obtained on higher magnification.
3. After images of the centrioles are captured from serial sections, the centrioles can be aligned (see Figure 5) using Photoshop, if no specialized image aligning software is available.

1.7 PREPARATION OF FORMVAR-COATED SLOT GRIDS

Slot grids need to be coated with a thin layer of formvar supporting film before they are used to pick up serial sections. The procedure for preparing the coated grids is illustrated in Figure 6.

1. Preclean the slot grids by putting them in a clean glass beaker filled with 10–20 mL of pure ethanol or acetone. Soak the grids for 15 min at room temperature, swirling several times. Pure out the ethanol or acetone and dry the grids well before use.
2. Clean both sides of a microscopy glass slide with ethanol, air dry, and dip into 0.5% formvar solution. Pull the slide steadily out of the solution, and air dry the slide for several minutes. The thin film of formvar will form on the glass slide (see Video 2).
3. Fill a big dish with dH_2O. To detach formvar from the glass slide, scrape off the formvar film from the edges of the slide with a razor blade (see Video 3).
4. Slowly dip the glass slide in the water. The formvar film will detach from the slide and stay afloat on the water. Gently pull the slide out of the water (see Video 4).
5. Place precleaned slot grids one by one on the top of the floating film with their shiny sides facing toward the film. The film will permanently stick to the cooper grid.

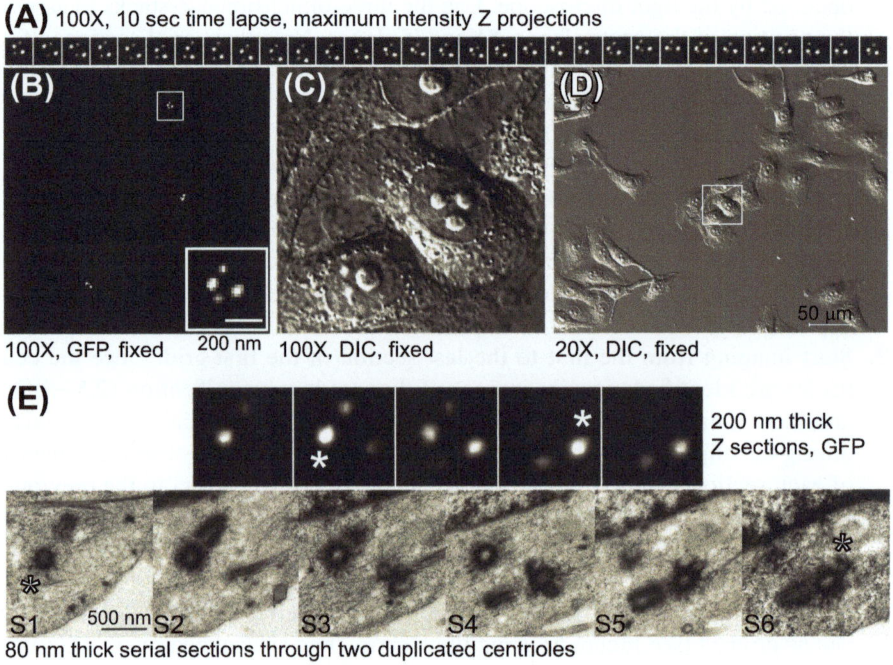

FIGURE 5 CLEM analysis of the centrioles from Centrin1-GFP expressing HeLa cell arrested in G2.

(A) A series of maximal density projections illustrating the movement of the centriole pairs. A Z stack spanning the entire centrosome content was collected using spinning disc confocal and 100× objective lens, every 10 s (B and C) The same cells imaged by time-lapse microscopy were fixed with 2.5% glutaraldehyde and returned to the microscope. The position of the centrioles within the cell was then recorded in fluorescence and DIC. (D) Low-magnification image of the target cell and its neighbor cells in DIC. (E) The centrioles recorded by time-lapse imaged on the electron microscope. Four consecutive immunofluorescence Z sections and six serial EM sections are presented (S1–S6). The asters in S1 and S6 correspond to the immunofluorescence Z section 2 and 4, with maximal intensity for centrin-GFP signal belonging to the mother centrioles (indicating the position of the distal part of the centriole).

6. To collect coated grids from the water, wrap another microscope slide with parafilm. Submerge one edge of the slide under the water, and establish the contact with the film carrying grids. Slowly push the slide deeper into the water until the film with the grids is attached to the parafilm. Then pull the slide with the grids attached to it from the water (see Video 5).
7. Dry coated grids in a clean glass dish at room temperature. Protect the grids from dust. Note that the film and the shiny side of the slot grids are now facing up.

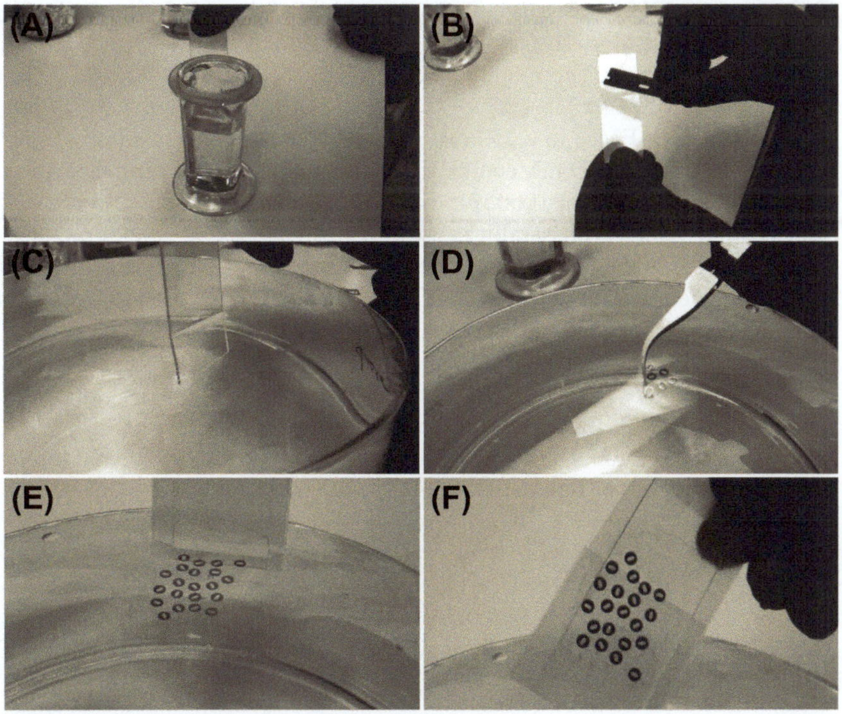

FIGURE 6 Preparation of the formvar-coated slot grids.

Selected frames from Videos 2–5 illustrating the key steps in the coating of the slot grids with formvar. (A) Dipping microscope slide in 0.5% formvar solution. (B) Scraping formvar film off the edges of the slide. (C) Detaching formvar film from the slide. (D) Putting the grids on the top of the floating formvar film. (E and F) Collecting coated grids from the water.

1.8 CHEMICALS, BUFFERS, AND MEDIA

- CO_2-independent medium (18045-088; Life Technologies).
- Formvar (15800; Electron Microscopy Sciences (EMS)): 0.5% (m/v) solution in chloroform; can be reused many times.
- Fixative: 2.5% Glutaraldehyde in 1X-PBS pH 7.2, or 2.5% glutaraldehyde in 0.1 M cacodylate buffer, pH 7.4. The fixative should be prepared fresh each time.
- Glutaraldehyde (G5882; Sigma-Aldrich); once opened, a bottle of glutaraldehyde should be stored at 4 °C and used within 2 months.
- Hydrofluoric acid, 48% (339261; Sigma-Aldrich)
- Lead citrate (178000; EMS): put 0.01 g (or 0.04 g) lead citrate powder into 9 mL dH_2O (boiled and cooled down to room temperature), add 1 mL of freshly made 1 N sodium hydroxide (21160; EMS), and mix for several minutes until lead citrate is completely dissolved. Filter the solution with a 0.22 μm filter before use.
- Osmium tetroxide (19100; EMS): 2% (m/v) solution in dH_2O.

- Poly-L-lysine (P1524; Sigma): Prepare a stock of 1 mg/mL in sterile dH_2O. Aliquot and store at −20 °C.
- Resin for embedding: Thoroughly mix 25 mL EMbed-812 (14900; EMS), 15 mL Araldite 502 (10900; EMS), 55 mL DDSA (13710; EMS), and 1.5−1.9 mL DMP-30 (13600; EMS) in a 250 mL glass flask on a magnetic stirrer. Transfer the resin into several 50 mL conical tubes, and get rid of the air bubbles in the resin by centrifugation at 1000 rpm for 5 min. The resin is ready for use or the tubes can be sealed with parafilm and stored at −20 °C for several months.
- Resin blocks: Fill the flat bottom embedding capsules (70021; EMS) with Embed 812 resin, and put them into the 60 °C oven to polymerize for 48 h. After that, take the resin blocks out of the capsules. Prepare these resin blocks before trimming and sectioning.
- Sodium cacodylate (12300; EMS): stock solution: 0.4 M in dH_2O, pH 7.4).
- Tannic acid (21700; EMS): 0.15% (m/v) solution in 1X-PBS.
- Uranyl acetate (22400; EMS): 1 or 2% (m/v) solution in dH_2O filtered through a 0.22 μm syringe filter unit (SLGS033GS; Millipore).

1.9 INSTRUMENTATION

Coverslips, 25-mm round, 0.17 mm thick (640715; Warner Instruments). Most high-end objectives are infinity-corrected for a glass coverslip of 0.17 mm thickness. This thickness corresponds to coverslip # 1.5.

- Glass slides (48312-003; VWR International)
- Grid staining matrix system kit (71179-01; EMS)
- Diamond knife with a boat (Ultra 45°, 3 mm, Diatome)
- Diamond knife cleaning kit (70600; EMS)
- Microscope Objective Marker (MBW10000; Nikon)
- Loctite liquid super glue
- Objective Diamond Scriber (Leica), a pen with a diamond tip (various vendors)
- Glass-grid-bottom Petri dishes, 35 mm (35G-2-14-C-grid; MatTek Corporation)
- Razor blades for manual trimming (71980; EMS)
- Rose chambers (homemade) of noncorrosive aluminum, with FDA approved silicone rubber spacer. Syringes (3 and 20 mL) and needles (305165; BD) for perfusion
- Slot grids (2330P-XA; SPI Supplies)
- Tweezers (72864-D; EMS)
- Ultramicrotome, Leica EM UC7 (Leica Microsystems)

ACKNOWLEDGMENTS

We thank Dr Anil Shukla for his enthusiasm and help in acquiring all videos and numerous pictures that we used during image preparation. We also thank Ms Veronica Farmer for reading of the manuscript. JL feels beholden to her former mentor Dr Alexey Khodjakov for sharing his knowledge of correlative light and electron microscopy, and to Dr Tatiana Vinogradova for a generous offer of a diamond scriber. Work in the lab of JL is supported by the Intramural Research Program of the NIH, National Cancer Institute, Center for Cancer Research.

SUPPLEMENTARY DATA

Supplementary data related to this article can be found online at http://dx.doi.org/10.1016/bs.mcb.2015.03.003.

REFERENCES

Alvey, P. L. (1985). An investigation of the centriole cycle using 3T3 and CHO cells. *Journal of Cell Science, 78*(1), 147–162.

Carvalho-Santos, Z., Azimzadeh, J., Pereira-Leal, J. B., & Bettencourt-Dias, M. (2011). Tracing the origins of centrioles, cilia, and flagella. *Journal of Cell Biology, 194*(2), 165–175. http://dx.doi.org/10.1083/jcb.201011152.

Keller, D., Orpinell, M., Olivier, N., Wachsmuth, M., Mahen, R., Wyss, R., et al. (2014). Mechanisms of HsSAS-6 assembly promoting centriole formation in human cells. *Journal of Cell Biology, 204*(5), 697–712. http://dx.doi.org/10.1083/jcb.201307049.

Kong, D., Farmer, V., Shukla, A., James, J., Gruskin, R., Kiriyama, S., et al. (2014). Centriole maturation requires regulated Plk1 activity during two consecutive cell cycles. *Journal of Cell Biology, 206*(7), 855–865. http://dx.doi.org/10.1083/jcb.201407087.

Lau, L., Lee, Y. L., Sahl, S. J., Stearns, T., & Moerner, W. E. (2012). STED microscopy with optimized labeling density reveals 9-fold arrangement of a centriole protein. *Biophysical Journal, 102*(12), 2926–2935. http://dx.doi.org/10.1016/j.bpj.2012.05.015.

Lawo, S., Hasegan, M., Gupta, G. D., & Pelletier, L. (2012). Subdiffraction imaging of centrosomes reveals higher-order organizational features of pericentriolar material. *Nature Cell Biology, 14*(11), 1148–1158. http://www.nature.com/ncb/journal/v14/n11/abs/ncb2591.html-supplementary-information.

Leung, B. O., & Chou, K. C. (2011). Review of super-resolution fluorescence microscopy for biology. *Applied Spectroscopy, 65*(9), 967–980. http://dx.doi.org/10.1366/11-06398.

Lončarek, J., Hergert, P., & Khodjakov, A. (2010). Centriole reduplication during prolonged interphase requires procentriole maturation governed by Plk1. *Current Biology, 20*(14), 1277–1282. http://dx.doi.org/10.1016/j.cub.2010.05.050.

Loncarek, J., Hergert, P., Magidson, V., & Khodjakov, A. (2008). Control of daughter centriole formation by the pericentriolar material. *Nature Cell Biology, 10*(3), 322–328.

Mennella, V., Agard, D. A., Huang, B., & Pelletier, L. (2014). Amorphous no more: subdiffraction view of the pericentriolar material architecture. *Trends in Cell Biology, 24*(3), 188–197.

Mennella, V., Keszthelyi, B., McDonald, K. L., Chhun, B., Kan, F., Rogers, G. C., et al. (2012). Subdiffraction-resolution fluorescence microscopy reveals a domain of the centrosome critical for pericentriolar material organization. *Nature Cell Biology, 14*(11), 1159–1168.

Pereira, A., Matos, I., Lince-Faria, M., & Maiato, H. (2009). Dissecting mitosis with laser microsurgery and RNAi in Drosophila cells. In A. D. McAinsh (Ed.), *Mitosis* (Vol. 545, pp. 145–164). Humana Press.

Rattner, J. B., & Phillips, S. G. (1973). Independence of centriole formation and DNA synthesis. *Journal of Cell Biology, 57*(2), 359–372. http://dx.doi.org/10.1083/jcb.57.2.359.

Rieder, C. L., & Bowser, S. S. (1985). Correlative immunofluorescence and electron microscopy on the same section of epon-embedded material. *Journal of Histochemistry & Cytochemistry, 33*(2), 165–171. http://dx.doi.org/10.1177/33.2.3881520.

Rizzo, R., Parashuraman, S., & Luini, A. (2014). Correlative video-light–electron microscopy: development, impact and perspectives. *Histochemistry and Cell Biology, 142*(2), 133–138. http://dx.doi.org/10.1007/s00418-014-1249-3.

Rose, G. G., Pomerat, C. M., Shindler, T. O., & Trunnell, J. B. (1958). A cellophane-strip technique for culturing tissue in multipurpose culture chambers. *Journal of Biophysical and Biochemical Cytology, 4*(6), 761–764. http://dx.doi.org/10.1083/jcb.4.6.761.

Sillibourne, J. E., Specht, C. G., Izeddin, I., Hurbain, I., Tran, P., Triller, A., et al. (2011). Assessing the localization of centrosomal proteins by PALM/STORM nanoscopy. *Cytoskeleton, 68*(11), 619–627. http://dx.doi.org/10.1002/cm.20536.

Sonnen, K. F., Schermelleh, L., Leonhardt, H., & Nigg, E. A. (2012). 3D-structured illumination microscopy provides novel insight into architecture of human centrosomes. *Biology Open, 1*(10), 965–976. http://dx.doi.org/10.1242/bio.20122337.

Spiegelhalter, C., Laporte, J., & Schwab, Y. (2014). Correlative light and electron microscopy: from live cell dynamic to 3D ultrastructure. In J. Kuo (Ed.), *Electron microscopy* (Vol. 1117, pp. 485–501). Humana Press.

Stout, J., Rizk, R., & Walczak, C. (2009). Protein inhibition by microinjection and RNA-mediated interference in tissue culture cells: complementary approaches to study protein function. In D. J. Carroll (Ed.), *Microinjection* (Vol. 518, pp. 77–97). Humana Press.

Tsou, M.-F. B., Wang, W.-J., George, K. A., Uryu, K., Stearns, T., & Jallepalli, P. V. (2009). Polo kinase and separase regulate the mitotic licensing of centriole duplication in human cells. *Developmental Cell, 17*(3), 344–354. http://dx.doi.org/10.1016/j.devcel.2009.07.015.

Vorobjev, I. A., & Chentsov, Y. (1982). Centrioles in the cell cycle. I. Epithelial cells. *Journal of Cell Biology, 93*(3), 938–949. http://dx.doi.org/10.1083/jcb.93.3.938.

Yamanaka, M., Smith, N. I., & Fujita, K. (2014). Introduction to super-resolution microscopy. *Microscopy, 63*(3), 177–192. http://dx.doi.org/10.1093/jmicro/dfu007.

CHAPTER

Generation of a conditional analog-sensitive kinase in human cells using CRISPR/Cas9-mediated genome engineering

2

Tyler C. Moyer, Andrew J. Holland[1]

Department of Molecular Biology and Genetics, Johns Hopkins University School of Medicine, Baltimore, MD, USA

[1]*Corresponding author: E-mail: aholland@jhmi.edu*

CHAPTER OUTLINE

Introduction **20**
1. CRISPR/Cas System **21**
2. Analog-Sensitive Kinases **21**
3. Methods **22**
3.1 Designing a Guide RNA for Sequence-Specific DNA Cleavage by SpCas9 ... 22
3.2 Cloning Oligonucleotides into the PX459 Vector 24
3.2.1 Vector preparation 25
3.2.2 Oligonucleotide annealing 26
3.2.3 Oligonucleotide phosphorylation 26
3.2.4 Ligation and transformation 26
3.3 Design of a Repair Template 27
3.3.1 Insertion of the point mutation 27
3.3.2 Mutation of the PAM site 28
3.3.3 Insertion of a restriction enzyme cleavage site 28
3.4 Transfection and Screening 28
3.4.1 Day 1 28
3.4.2 Day 2 29
3.4.3 Day 4 29
3.4.4 Limiting dilution 29
3.4.5 Genomic DNA extraction 30

Methods in Cell Biology, Volume 129, ISSN 0091-679X, http://dx.doi.org/10.1016/bs.mcb.2015.03.017

3.4.6 Screening 31
3.4.7 PCR amplification 32
3.4.8 Restriction enzyme digest 32
3.4.9 Sequencing clones 33
4. Functional Analysis **33**
Conclusion **34**
References **34**

Abstract

The ability to rapidly and specifically modify the genome of mammalian cells has been a long-term goal of biomedical researchers. Recently, the clustered, regularly interspaced, short palindromic repeats (CRISPR)/Cas9 system from bacteria has been exploited for genome engineering in human cells. The CRISPR system directs the RNA-guided Cas9 nuclease to a specific genomic locus to induce a DNA double-strand break that may be subsequently repaired by homology-directed repair using an exogenous DNA repair template. Here we describe a protocol using CRISPR/Cas9 to achieve bi-allelic insertion of a point mutation in human cells. Using this method, homozygous clonal cell lines can be constructed in 5–6 weeks. This method can also be adapted to insert larger DNA elements, such as fluorescent proteins and degrons, at defined genomic locations. CRISPR/Cas9 genome engineering offers exciting applications in both basic science and translational research.

INTRODUCTION

Genome engineering is a term used to describe the process of making specific, targeted alterations in the genome of a living organism. Genome engineering exploits the repair of a DNA double-strand break (DSB) through the endogenous pathway of homologous recombination (HR). By providing an exogenous DNA repair template that contains homology to the targeted site, it is possible to exploit the HR machinery to create defined alterations close to the site of a DSB. However, mammalian genomes comprise billions of base pairs and there is a low probability of a spontaneous DSB occurring close to the region to be targeted; as a consequence, desired recombination events occur extremely infrequently (Capecchi, 1989). A major breakthrough came with the demonstration that targeted DSBs greatly increase the frequency of homology-directed repair (HDR) at a specific locus (Choulika, Perrin, Dujon, & Nicolas, 1995; Plessis, Perrin, Haber, & Dujon, 1992; Rouet, Smih, & Jasin, 1994; Rudin, Sugarman, & Haber, 1989). This discovery has spurred the development of programmable endonucleases that can be exploited to promote site-specific cleavage of the genome.

Zinc finger nucleases (ZFNs) and transcription activator-like effector nucleases (TALENs) are artificial restriction enzymes produced by fusing customizable DNA binding domains to the sequence-independent nuclease domain of the restriction enzyme Fok1 (Boch et al., 2009; Christian et al., 2010; Miller et al., 2007, 2011;

Moscou & Bogdanove, 2009; Urnov et al., 2005). Fok1 requires dimerization for its activity, and thus a pair of ZFNs or TALENs is required to bind to opposite strands of DNA on either side of a target site to allow Fok1 dimerization and DNA cleavage. While ZFNs and TALENs have been shown to be capable of creating targeted DNA breaks and introducing genomic sequence changes through HDR, difficulties in protein design and synthesis proved to be a barrier to their widespread use (Hsu, Lander, & Zhang, 2014).

1. CRISPR/Cas SYSTEM

Recently, a new tool based on clustered, regularly interspaced, short palindromic repeats (CRISPR) systems from bacteria have been exploited for genome engineering in human cells and have generated considerable excitement (Hsu et al., 2014). CRISPR systems have the distinct advantage of using RNA-guided nuclease activity to target cleavage of DNA and thereby eliminate the need for protein engineering and optimization.

CRISPR/Cas modules were identified in bacteria as part of an adaptive immune system that enables hosts to recognize and cleave foreign invading DNA (Horvath & Barrangou, 2010; Marraffini & Sontheimer, 2010). CRISPR modules comprise arrays of short nucleotide repeats interspersed with unique spacers that share homology with foreign phage or plasmid DNA. Of the three CRISPR/Cas systems that have evolved in bacteria, the type II system is the simplest and involves only three components: a processed RNA that is complementary to the spacers, known as a CRISPR-RNA (crRNA), a trans-activating tracrRNA that hybridizes to the crRNA, and the Cas9 nuclease. The crRNA and the tracrRNA form an RNA double-strand structure that directs Cas9 to generate DSBs at a site complementary to the targeting region of the crRNA (Brouns et al., 2008; Deltcheva et al., 2011; Garneau et al., 2010). The RNA components of the CRISPR/Cas9 system (the crRNA and the tracrRNA) can be combined into a singular guide RNA (gRNA) (Jinek et al., 2012). The gRNA directs Cas9 to induce DSBs in the genome of cells at sites complementary to a $\sim$20 base pair targeting sequence in the gRNA. The simplicity of these RNA-guided nucleases has allowed scientists to repurpose the CRISPR/Cas9 system to create site-specific DNA breaks in a variety of eukaryotic cells (Cong et al., 2013; Mali et al., 2013).

2. ANALOG-SENSITIVE KINASES

Nearly one-third of the proteome is subject to phosphorylation by protein kinases. Adenosine triphosphate (ATP)-competitive small molecule inhibitors are powerful tools for probing the function of kinases in living cells. However, many kinases possess a similar catalytic core, and thus achieving specificity in inhibiting kinase activity in cells is a major challenge. One method to overcome this limitation is to exploit a chemical genetic strategy in which a kinase is engineered to accept ATP analogs that are not efficiently utilized by wild-type kinases. These engineered kinases are referred to as analog-sensitive (AS) kinases (Bishop et al., 2000). This is achieved through the mutation of a bulky hydrophobic 'gatekeeper' amino acid in the ATP binding pocket to a smaller amino acid (alanine or glycine) (to identify the gatekeeper residue for a kinase see http://sequoia.ucsf.edu/ksd) (Liu et al., 1999). An AS kinase can be specifically inhibited with generic nonhydrolyzable bulky ATP analogs, allowing rapid and reversible control of kinase activity in cells. Despite broad utility, the use of the AS kinase approach in mammalian cells has been hampered by the difficulty of functionally replacing an endogenous kinase with appropriate levels of an AS kinase. The development of CRISPR/Cas9 offers a facile method for introducing AS mutations into endogenous mammalian kinases.

Here, we outline a method for using CRISPR/Cas9 genome engineering to introduce an AS point mutation in a single step into both alleles of Polo-like kinase 4 (Plk4). This protocol can be used to introduce point mutations into any target gene of choice and could also be adapted to insert larger DNA elements, such as fluorescent proteins and degrons, at defined genomic locations.

3. METHODS

3.1 DESIGNING A GUIDE RNA FOR SEQUENCE-SPECIFIC DNA CLEAVAGE BY SpCas9

The CRISPR/Cas9 system is capable of generating targeted DSBs in the genome of mammalian cells. The *Streptococcus pyogenes* (Sp) CRISPR/Cas9 system is the most widely used system and will be the focus of this protocol. The specificity of SpCas9 targeting is determined by a 20-nucleotide (nt) targeting sequence within the gRNA that is complementary to the genomic target sequence. The genomic target sequence must precede an "NGG" sequence known as the protospacer adjacent motif (PAM), which is necessary for target cleavage, but is not encoded within the gRNA (Figure 1) (Mojica, Diez-Villasenor, Garcia-Martinez, & Almendros, 2009; Shah, Erdmann, Mojica, & Garrett, 2013). The PAM sequence has evolved to ensure that the CRISPR/Cas9 system does not self-target the CRISPR modules in the bacterial genome (Shah et al., 2013). SpCas9 usually cleaves the DNA 3-nt upstream of (i.e., 5′ to) the PAM to produce a blunt-ended DSB (Figure 1). Breaks can either be repaired by HDR or through error-prone nonhomologous end-joining (NHEJ) pathway, which usually introduces insertions and deletions (InDels) of bases at the cut site.

Several plasmid constructs are available for SpCas9/gRNA expression in mammalian cells. In our experiments we have used the PX459 vector (available from Addgene, vector #62988), which enables expression of a gRNA, SpCas9, and a puromycin resistance gene from a single vector. In this section we describe how to design a gRNA to direct cleavage at a specific genomic site.

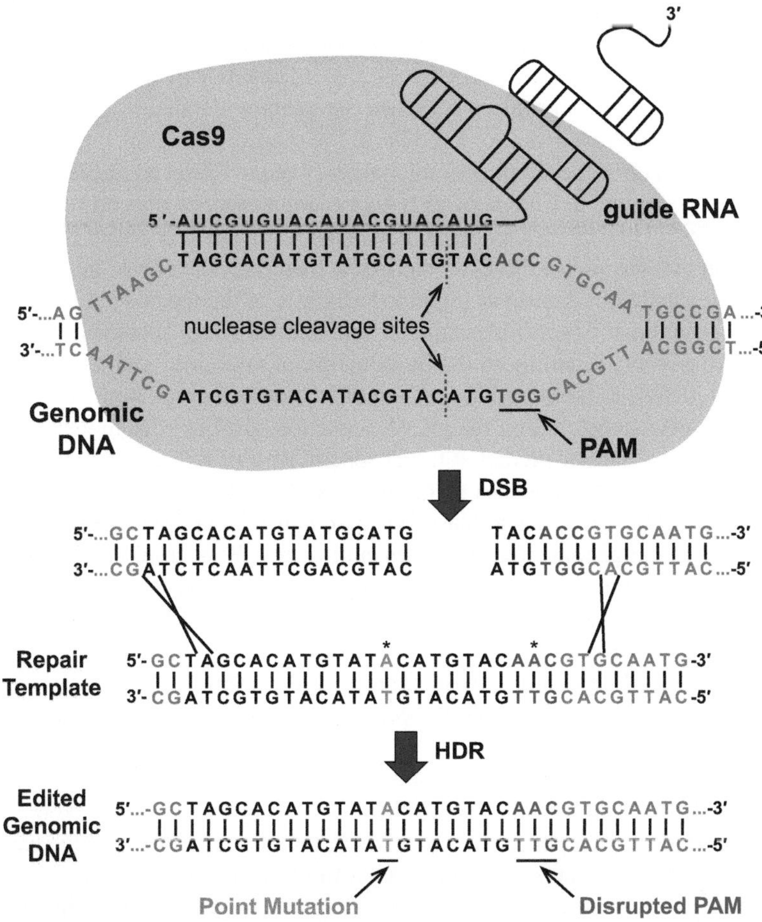

FIGURE 1 Genome Editing Using the CRISPR/Cas9 System.

SpCas9 nuclease is directed to a specific locus through base-pairing of the targeting sequence (underlined) of its associated guide RNA (gRNA) with a genomic target sequence. The genomic target sequence is followed by a protospacer-adjacent motif (PAM, red (light gray in print versions)) that is required for SpCas9 recognition and cleavage. Double-stranded cleavage usually occurs three base pairs upstream of the PAM. The double-strand break (DSB) may be repaired through homology directed repair (HDR) using an exogenously supplied repair template that contains a mutation of interest (green (dark gray in print versions)). Inserting a second mutation (blue (dark gray in print versions)) in the PAM prevents additional rounds of SpCas9/gRNA cutting at this locus. Asterisks represent locations of point mutations in the repair template.

1. Download the genomic sequence of the target gene from the National Center for Biotechnology Information (http://www.ncbi.nlm.nih.gov/gene).
2. Identify the codon for the amino acid that will be mutated and copy a sequence of 50 nucleotides on either side of the desired mutation.

3. Visit crispr.mit.edu and paste the copied sequence into the query box. Select "other region" as the sequence type and choose the correct target genome. Click "Submit Query" and then select "Guides & offtargets." Results may take some time to appear as the program searches the genome database.
4. The results will show a rank-ordered list of the potential genomic targets for gRNA recognition. All of the 20-nt genomic target sites are followed at the 3′ end by a PAM shown in green. Choice of gRNA targeting sequence depends on two main parameters: (1) HDR efficiency decreases as the distance between the desired mutation and the site of the DSB increases; therefore, the site of genomic target cleavage should be as close as possible to the site of the introduced mutation. (2) gRNAs can direct SpCas9 cleavage of nonidentical target sequences in the genome, possibly resulting in the introduction of undesired mutations. For each selected gRNA, a list of "off-target" binding sites is shown along with the position of mismatches within the gRNA sequence. gRNAs with higher "quality scores" have a greater predicted target specificity. For a detailed analysis of the effect of mismatches on gRNA recognition, see Hsu et al. (2013). We recommend choosing a gRNA that has at least four base pair mismatches to any other sequence in the genome and promotes cutting at <20-nt from the site of the intended mutation. For some genomic target sequences, it is not possible to achieve these parameters. In this case we select the highest scoring gRNA that directs cutting within 20-nt on either side of the desired mutation site.
5. We recommend selecting two to three gRNAs and testing each for cleavage efficiency using the SURVEYOR® Mutation Detection Kit (Transgenomic), as described in Ran et al. (2013).
6. Copy the 20-nt sequence of the genomic target sequence. The PX459 vector uses a U6 promoter to transcribe the gRNA and this requires that a G be the first nucleotide in the transcript. In cases where the genomic target sequence does not begin with a G, append an extra G at the 5′ end of the gRNA.
7. Generate the reverse complement of the genomic target sequence (including the 5′ G if it was added) (Figure 2(A)).
8. Add the overhang sequence 5′-CACC-3′ to the 5′ end of genomic target and the sequence 5′-AAAC-3′ to the 5′ end of the reverse complement (Figure 2(A)). These sequences will produce the correct overhangs for cloning into the PX459 vector.
9. Order single-stranded DNA oligonucleotides for the two sequences generated in Step 8.

3.2 CLONING OLIGONUCLEOTIDES INTO THE PX459 VECTOR

The PX459 vector contains two BbsI cleavage sites that allow for the insertion of annealed oligonucleotides containing the gRNA target sequence. BbsI cleaves DNA outside of its recognition site to produce overhangs complementary to those added in Step 8 above. Below we describe how to clone the gRNA into the PX459 expression vector.

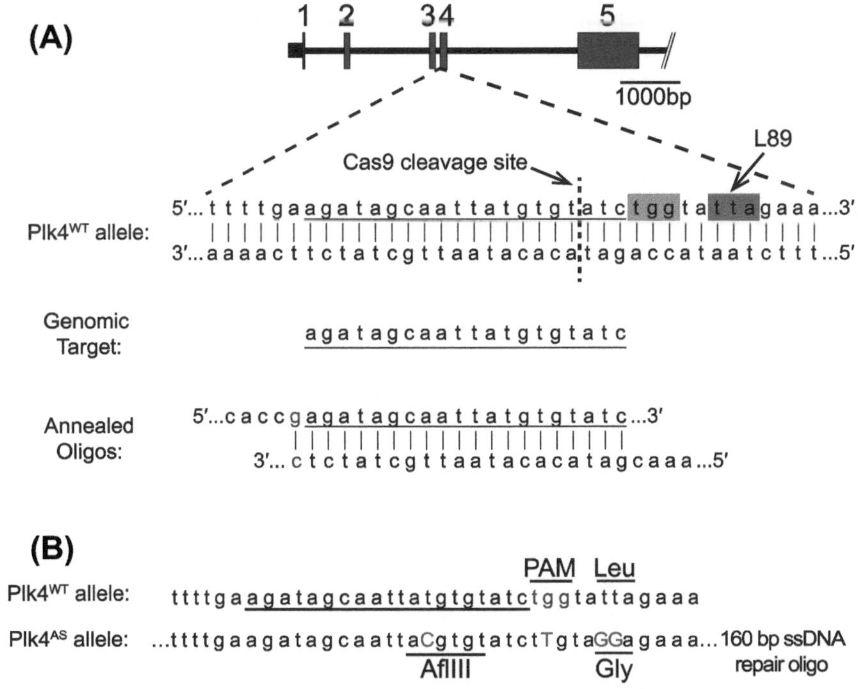

FIGURE 2 Guide RNA and Repair Template Design.

(A) A Plk4 genomic target sequence (underlined) is chosen due to the close proximity of the cut site to the desired mutation (gatekeeper residue L89G, green (gray in print versions)). The PAM site required for SpCas9/gRNA cleavage (red (light gray in print versions)) lies immediately downstream of the genomic target sequence. The 5′ end of the genomic target sequence must start in a G for efficient transcription from the U6 promoter on the PX459 vector. When the genomic target sequence does not begin with a G, this must be added to the gRNA targeting sequence (blue (dark gray in print versions)). To clone the gRNA targeting sequence into the BbsI-digested PX459 vector, the sequence 5′-CACC-3′ must be added onto the 5′ end of the genomic target sequence and the sequence 5′-AAAC-3′ must be added onto the reverse complement of the genomic target sequence. (B) The Plk4^{AS} repair template contains the gatekeeper residue mutation (L89G), a PAM site mutation, and introduces an AflIII restriction site to facilitate identifying clones that have undergone HDR (all mutations are shown in green (light gray in print versions)). Note that the PAM mutation and restriction enzyme recognition site insertion do not alter the coding sequence.

3.2.1 Vector preparation

1. Digest 1 μg of the PX459 vector with BbsI for 2 h at 37 °C.
2. To remove terminal phosphates, add 0.1 μL of calf intestinal phosphatase (CIP) to the reaction and incubate at 37 °C for 30 min. Since the two overhangs produced following BbsI digestion are not complementary this step is not required, but can reduce background.

3. Purify the cut vector using a standard polymerase chain reaction (PCR) cleanup kit. Note that BbsI digestion of PX459 produces a small 22-nt fragment that will pass through the column, leaving a 9153-nt, linear piece of vector DNA. The purified linear vector can be stored at −20 °C until ready for use.

3.2.2 Oligonucleotide annealing

4. Combine:
   a. 43 μL of H_2O of molecular biology grade.
   b. 1 μL of each oligonucleotide from a 100 μM stock.
   c. 5 μL of New England Biolabs Buffer 3 (www.neb.com/).
5. Anneal oligonucleotides in a thermocycler with the following protocol:
   a. 4 min at 95 °C.
   b. 10 min at 70 °C.
   c. Cool to 4°C at 1 °C/min.

   Annealed oligonucleotides can be stored at −20 °C until ready for use.

3.2.3 Oligonucleotide phosphorylation

We use T4 polynucleotide kinase (PNK) from New England Biolabs to add terminal phosphates to the annealed oligonucleotides.

6. Combine:
   a. 5 μL of H_2O of molecular biology grade
   b. 2 μL of the annealed oligonucleotides from above
   c. 1 μL of T4 PNK buffer (www.neb.com)
   d. 1 μL of ATP (10 mM stock)
   e. 1 μL of T4 PNK (www.neb.com)
7. Allow reaction to proceed in a thermocycler with the following protocol:
   a. 30 min at 37 °C
   b. 10 min at 70 °C (to inactivate PNK)
   c. Quick cool to 4 °C

   Phosphorylated oligonucleotides can be stored at −20 °C until ready for use.

3.2.4 Ligation and transformation

We use a 2X stock of Takara T4 DNA Ligase (DNA ligation kit, Version 2.1) and homemade TOP10 competent cells.

8. In a 0.5 mL tube, combine the following:
   a. 1 μL BbsI digested vector
   b. 4 μL phosphorylated annealed oligonucleotides
   c. 5 μL 2X Takara T4 ligase
9. As a control, set up the same reaction as above but substitute 4 μL of molecular biology grade H_2O for the oligonucleotides.
10. Allow ligation to proceed for 1 h at room temperature.
11. Hand-thaw a frozen aliquot of competent bacteria and keep on ice.

12. Add the entire 10 μL reaction mixture to the competent bacteria and incubate on ice for 20–30 min.
13. Heat shock the bacteria for 1 min at 42 °C and return to ice for at least 1 min.
14. Plate the bacteria on prewarmed ampicillin (or carbenicilin) agar plates and incubate at 37 °C for 16 h. A successful ligation should produce few (if any) colonies on a control (vector alone) plate and many-fold more colonies on an experimental (vector with insert) plate.
15. Select a colony from the experimental plate and prepare a 1–5 mL culture in lysogeny broth containing ampicillin. Shake at 37 °C for at least 16 h and perform a plasmid purification as normal.
16. To check for correct oligonucleotide insertion, sequence the plasmid using the U6 forward primer (5′-ACTATCATATGCTTACCGTAAC-3′).
17. PX459 plasmid DNA containing a correctly cloned gRNA can be stored at −20 °C until ready for use.

3.3 DESIGN OF A REPAIR TEMPLATE

HDR can be exploited to generate defined edits at the site of a DSB introduced by SpCas9. As a substrate for HDR, cells are provided with an exogenous repair template containing the desired alteration. Repair templates can be single-stranded (ssDNA) or double-stranded DNA (dsDNA) with homology arms flanking the cut site. ssDNA repair templates have higher efficiency of HDR than identical dsDNA templates (Lin, Staahl, Alla, & Doudna, 2014). Therefore, for small insertions or point mutations we use single-stranded oligonucleotides with approximately 80 nucleotides of homology on either side of the cut site (Figure 2(B)). Thc ssDNA repair template can be designed to be complementary to either the sense or antisense strand. Along with the desired modification, a repair template should also possess a mutation in the PAM site that will prevent re-cutting by SpCas9 after HDR. In addition, to simplify the downstream screening of cell line clones, we strongly recommend introducing a silent restriction site along with the desired mutation. In the example below, we outline the steps for designing the ssDNA oligonucleotide repair template for introducing a point mutation (Figure 2(B)).

1. Identify the specific site of cleavage by SpCas9 in the genomic target sequence. This will be 3-nts upstream of the PAM sequence (Figures 1 and 2(A)).
2. Select 80 nucleotides on either side of the cut site to act as the homology for the repair template. Make the following three modifications to the sequence:

3.3.1 Insertion of the point mutation

3. Identify the codon for the amino acid to be mutated and change the sequence to code for a new amino acid.

3.3.2 Mutation of the PAM site

4. Identify the PAM in the repair template and replace one or both of the G bases with a C or T to create a silent mutation. For SpCas9, the "NGG" PAM sequence can be mutated to anything other than "NAG" to prevent SpCas9 cleavage (Hsu et al., 2013; Jiang, Bikard, Cox, Zhang, & Marraffini, 2013). If it is not possible to make a silent point mutation in the PAM sequence then at least four silent point mutations should be introduced into the targeting sequence of the gRNA to prevent re-cutting. On the whole, SpCas9 is less tolerant of mismatches that lie close to the PAM (Hsu et al., 2013).

3.3.3 Insertion of a restriction enzyme cleavage site

To rapidly screen clonal cell lines (see below), it is desirable to introduce a silent restriction enzyme site into the repair template.

5. Use the online restriction enzyme site finder WatCut, (http://watcut.uwaterloo.ca/template.php) to identify a region of the repair template that will allow the creation of a restriction enzyme site that does not change the coding sequence. Introduce the mutation into the repair template. To form a reliable marker to track insertion of the point mutation, the restriction site should be positioned as close as possible to the mutation. In addition, we recommend choosing a restriction enzyme that is not blocked by CpG methylation and shows good activity in a range of restriction enzyme buffers.
6. Order the ssDNA repair oligonucleotide from IDT (http://www.idtdna.com/site). In our experience polyacrylamide gel electrophoresis purification is not necessary.

3.4 TRANSFECTION AND SCREENING

For genome editing experiments we recommend selecting a stably diploid cell line, as this simplifies the process of achieving homozygous gene targeting. While we have successfully targeted aneuploid cell lines using the CRISPR/SpCas9 system, determining the genotype of the resulting clones is more complex. Since puromycin is used to achieve rapid killing of cells that do not receive the PX459 expression vector, it is important that the chosen cell line is also puromycin sensitive. An alternative approach is to use the PX458 expression vector (available from Addgene, vector #48138) that co-expresses SpCas9, gRNA, and GFP, allowing fluorescent transfected cells to be directly sorted into individual wells of a 96-well plate.

There are a number of methods for delivering DNA to mammalian cells in culture. Here, we describe a method using Roche's X-tremeGENE 9 transfection reagent. Depending on the cell line to be used, other DNA delivery methods (e.g., nucleofection) may be required.

3.4.1 Day 1

1. Seed cells for transfection at 2×10^5 cells/well in 2 mL of media in a six-well plate. Seed at least three wells per transfection, with one of these wells serving

as a nontransfected control. Transfections are carried out in the presence of serum and, if desired, antibiotics.

3.4.2 Day 2

2. Change media on cells 30 min prior to transfection.
3. In a 1.5 mL tube prepare the following for each transfected well in the order written below:
   a. 100 μL serum-free media at room temperature
   b. 3 μL X-tremeGENE 9 transfection reagents
   c. 1 μg of total DNA (20:1 molar ratio of repair template:PX459 plasmid)
4. Flick tube gently 10 times to mix.
5. Incubate at room temperature for 15–20 min.
6. Add the transfection mixture drop-wise to cells.
7. Return cells to the incubator.

3.4.3 Day 4

The PX459 plasmid contains a puromycin resistance marker to select for transfected cells.

8. Change media on all cells (including controls) with fresh media containing puromycin (1–5 μg/mL is usually sufficient).
9. Return cells to the incubator.

Cells in the untransfected control wells should all die within 1–2 days in puromycin. When all control cells are dead proceed with limiting dilution. Cells should not remain in puromycin more than 2–3 days as the PX459 plasmid does not integrate into the genome and resistance to puromycin is lost over time.

3.4.4 Limiting dilution

After puromycin selection, single-cell clones should be isolated and screened to identify those that have undergone HDR. Either single-cell sorting or dilution cloning may be used to isolate single cells. Below we outline how to obtain single clones using dilution cloning.

Cells are diluted into 96-well plates to obtain wells containing a single cell. Since the clonogenic survival of cell lines varies greatly, we recommend seeding multiple 96-well plates with varying cell densities (1, 5, 10, and 50 cells/well). A 96-well plate that has growth in 10% of the wells will have a ~90% probability of a given well having a single colony.

1. Add 15 mL of puromycin-free media to a sterile reagent reservoir.
2. Add the desired number of cells to the media. For example, to achieve 10 cells/well, 1000 cells would be added to the media.
3. Mix the cells in the media by pipetting up and down five times with a 10 mL pipette.

4. Using a multichannel pipette, add 150 μL of media to each well of the 96-well plate.
5. Repeat Steps 1–4 for each cell density.
6. Wrap the 96-well plates in plastic film and return to the incubator.

It will take 2–4 weeks for single cell to grow into colonies large enough to isolate sufficient quantities of genomic DNA.

3.4.5 Genomic DNA extraction

To extract genomic DNA from a small number of clones, we use the Sigma GenElute Mammalian Genomic DNA Miniprep Kit (G1N350). For extracting genomic DNA from a large number of clones, we use a protocol adapted to use with 96-well plates and outlined below.

1. Using a tissue culture microscope, identify wells containing a single colony. Transfer 96 individual clones into 24-well plates. Return cells to the incubator.
2. When the clones are confluent, trypsinize cells in 200 μL of 0.05% trypsin.
3. Remove 160 μL of the trypsinized cell suspension and place into a single well of a 96-well plate with U-bottom wells.
4. Add 1 mL of media to the cells remaining in the 24-well plates and return plates to the incubator.
5. Spin the 96-well plate at 2000 RPM in a swinging-bucket rotor for 10 min to pellet cells.
6. To remove supernatant, quickly invert the plate to remove media and remove excess liquid by blotting on paper towels.
7. Resuspend cells in each well with 150 μL phosphate buffered saline (PBS) and spin at 2000 RPM for 10 min. Remove PBS as in Step 6.
8. To lyse cells, add 50 μL of lysis buffer (10 mM Tris–HCl, pH 7.5, 10 mM of ethylene diamine tetraacetic acid, 0.5% of sodium dodecyl sulfate, 10 mM of NaCl, 1 mg/mL of proteinase K) and seal the plate with parafilm. Put plate into a humidified chamber at 60 °C overnight. A humidified chamber can be created by placing a few inches of water in a small plastic container with a sealable lid. The 96-well plate is placed in the sealed container on top of a test tube rack so that it rests above the water level.
9. The next day, remove the plate from the humidified chamber and cool to room temperature.
10. Add 100 μL of ice-cold EtOH/NaCl mix (75 mM of NaCl in ~100% EtOH; it forms a cloudy solution) to precipitate DNA and mix well.
11. Incubate at room temperature for 30 min.
12. Spin at 4000 RPM in a swinging-bucket rotor for 20 min to pellet precipitated DNA.
13. Decant liquid as in Step 6.
14. Rinse pellet with 150 μL cold 70% EtOH and spin for 10 min at 4000 RPM.
15. Decant liquid as in Step 6.
16. Repeat washing step (#14–15) and air-dry DNA for 10 min.

17. Add 50 μL TE, pH 8.0 (10 mM of Tris−HCl, pH 8.0, 1 mM EDTA) to genomic DNA pellet.
18. Cover plate with parafilm and incubate at 50 °C for 2 h.

Genomic DNA is ready for further screening or may be stored at 4 °C until ready for use.

3.4.6 Screening

HDR can be tracked by PCR amplification of genomic DNA followed by restriction enzyme digest. Forward and reverse primers are designed to bind to regions of the genome outside of the homology arms of the repair template to amplify a region of 250−1000-nt (Figure 3(A)). Primers are designed using the Primer3 program

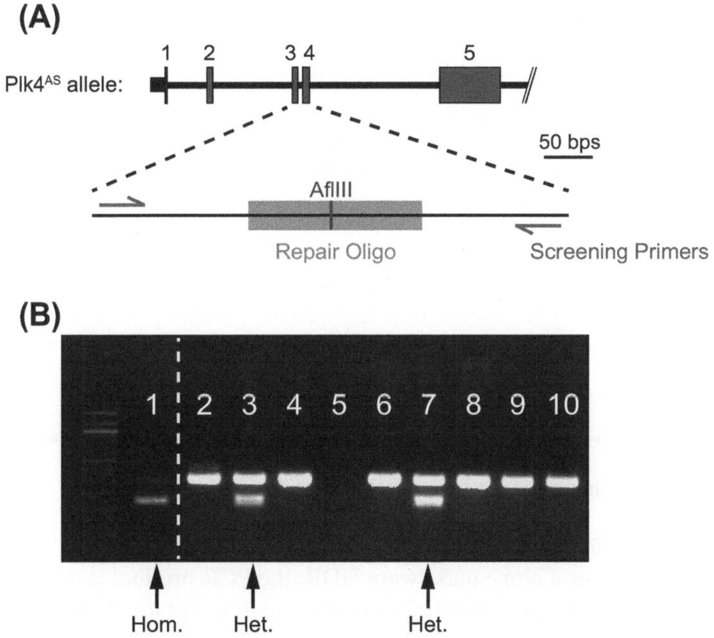

FIGURE 3 Screening of the Plk4^{AS} Allele.

(A) Forward and reverse screening primers (green (light gray in print versions)) should be designed to bind outside of the sequence of the repair template (gray). PCR-amplified Plk4^{AS} alleles will be digested with the restriction enzyme AflII whereas Plk4^{WT} alleles will not. (B) Genomic DNA extracted from individual clones is PCR-amplified and subject to AflII digestion. Clone 1 shows near complete digestion of the PCR product and is homozygous for the Plk4^{AS} allele. Clones 3 and 7 show partial digestion of the PCR products and are heterozygous for the Plk4^{AS} allele. Remaining clones were either not cut by SpCas9 or repaired through non-homologous end joining (NHEJ). The genomic DNA extraction or PCR failed for clone 5.

(biotools.umassmed.edu). We recommend screening at least three different forward and reverse primers in all possible combinations to identify a primer pair that yields a strong and specific PCR product from low quantities of DNA (Figure 3(B)). In most cases, the PCR protocol outlined below produces good results. However, PCR optimization may be required.

3.4.7 PCR amplification

1. For each clone, add 2 μL of genomic DNA to a single well of a 96-well plate.
2. Prepare a master mix (100 reaction per plate) as follows:

a. Molecular biology grade H_2O	1120 μL
b. 5X GC buffer (www.neb.com)	400 μL
c. Forward primer (10 μM stock)	100 μL
d. Reverse primer (10 μM stock)	100 μL
e. dNTPs (stock with each dNTP at 10 mM)	40 μL
f. Phusion polymerase (www.neb.com/)	40 μL

3. Add 18 μL of the master mix to each well, cover the plate with a piece of sealing film and perform a PCR reaction using the following parameters:

a. 98 °C	30 s
b. 98 °C	10 s
c. 57 °C	15 s
d. 72 °C	2 kb/min
e. Repeat steps b to d	34X
f. 72 °C	5 min
g. 12 °C	hold

The PCR product is ready for restriction enzyme digest (below).

3.4.8 Restriction enzyme digest

To determine whether a clone underwent HDR, the PCR product is digested with the restriction enzyme that was introduced into the repair template. If the genome underwent HDR with the repair template, digestion products smaller than the length of the amplicon will be observed (Figure 3(B)). Prepare a digestion for each clone's PCR product as below:

- 8.5 μL genomic PCR product
- 1 μL 10X restriction enzyme buffer (www.neb.com)
- 0.5 μL restriction enzyme

Note that for some restriction enzymes it is possible to perform the digestion directly in PCR buffer.

4. Incubate at 37 °C for 2 h.

5. Add 10 μL of a DNA loading dye to each restriction enzyme digest and remaining undigested PCR products and load the entire samples into wells of a 2% agarose DNA gel.
6. Run gel and image with a gel documentation system.

There are three possible results for each reaction (Figure 3(B)).

1. **Only an undigested band**: There are two possibilities here (Figure 3(B), Lanes 2, 4, 6, 8, 9, and 10): (1) Both alleles are wild type. In this case SpCas9 may have failed to cut, or the DSB was repaired by HDR using the sister chromatid, rather than the repair template. (2) One or both alleles are mutated. The DSB was repaired through (NHEJ) resulting in the creation of InDels that prevent further SpCas9 cutting.
2. **Both undigested and digested bands**: There are two possibilities here that are not mutually exclusive (Figure 3(B), Lanes 3 and 7). (1) The clone is heterozygous. In this case, the SpCas9/gRNA complex cleaved the genomic DNA and repaired one allele by HDR using the repair template. The other allele is either wild type or has been repaired by NHEJ. In our experience, many cells repair one allele by HDR using the repair template and the second allele undergoes NHEJ producing InDels. InDels frequently result in frameshift mutations that generate a null allele. (2) The cells are polyclonal. This may occur when two or more parental cells grow in a single well.
3. **Only (or mostly) digested bands**: This clone is homozygous for insertion of the point mutation (Figure 3(B), Lane 1).

3.4.9 Sequencing clones

To verify insertion of the desired mutation, the undigested PCR product can be cloned into a vector for sequencing. We recommend using the ZeroBlunt® TOPO® Cloning Kit from Invitrogen, which allows for easy cloning of blunt-end PCR products into the pCR-Blunt II-TOPO vector. We usually sequence ~10 clones to ensure coverage of both alleles.

Clones that are homozygous for the insertion of the point mutation can be taken forward for further analysis. However, heterozygous clones that contain a point mutation in one allele and a frameshift mutation in the second allele may also prove useful if the frameshift mutation creates a null allele.

4. FUNCTIONAL ANALYSIS

Depending on the point mutation introduced and the predicted phenotype, a variety of assays can be conducted. In the case of an AS kinase, treatment with bulky ATP-analogs (1NM-PP1, 3 MB-PP1) should inhibit kinase activity. In this case, cells can be treated with varying concentrations (0.1–20 μM) of a bulky ATP-analog and assayed for kinase activity.

CONCLUSION

The ease of use and high efficiency of the CRISPR/Cas9 system has propelled its use in genome engineering in a wide array of organisms and cultured cells (Hsu et al., 2014). However, there is competition between NHEJ and HDR for DSB repair (Heyer, Ehmsen, & Liu, 2010). While NHEJ is well suited for disrupting gene expression, introducing precise alterations in the genome requires efficiently engaging the HDR machinery. Increasing the frequency with which DSBs are repaired by HDR will greatly facilitate genome editing applications. An additional consideration for using the CRISPR/SpCas9 system is the potential of off-target cutting and mutagenesis. This concern is elevated in cell lines where there is no opportunity to breed out unintended mutations. Evaluating and improving the specificity of CRISPR/SpCas9 genome editing remains an important future goal.

REFERENCES

Bishop, A. C., Ubersax, J. A., Petsch, D. T., Matheos, D. P., Gray, N. S., Blethrow, J., et al. (2000). A chemical switch for inhibitor-sensitive alleles of any protein kinase. *Nature, 407*(6802), 395–401. http://dx.doi.org/10.1038/35030148.

Boch, J., Scholze, H., Schornack, S., Landgraf, A., Hahn, S., Kay, S., et al. (2009). Breaking the code of DNA binding specificity of TAL-type III effectors. *Science, 326*(5959), 1509–1512. http://dx.doi.org/10.1126/science.1178811.

Brouns, S. J., Jore, M. M., Lundgren, M., Westra, E. R., Slijkhuis, R. J., Snijders, A. P., et al. (2008). Small CRISPR RNAs guide antiviral defense in prokaryotes. *Science, 321*(5891), 960–964. http://dx.doi.org/10.1126/science.1159689.

Capecchi, M. R. (1989). Altering the genome by homologous recombination. *Science, 244*(4910), 1288–1292.

Choulika, A., Perrin, A., Dujon, B., & Nicolas, J. F. (1995). Induction of homologous recombination in mammalian chromosomes by using the I-SceI system of Saccharomyces cerevisiae. *Molecular and Cellular Biology, 15*(4), 1968–1973.

Christian, M., Cermak, T., Doyle, E. L., Schmidt, C., Zhang, F., Hummel, A., et al. (2010). Targeting DNA double-strand breaks with TAL effector nucleases. *Genetics, 186*(2), 757–761. http://dx.doi.org/10.1534/genetics.110.120717.

Cong, L., Ran, F. A., Cox, D., Lin, S., Barretto, R., Habib, N., et al. (2013). Multiplex genome engineering using CRISPR/Cas systems. *Science, 339*(6121), 819–823. http://dx.doi.org/10.1126/science.1231143.

Deltcheva, E., Chylinski, K., Sharma, C. M., Gonzales, K., Chao, Y., Pirzada, Z. A., et al. (2011). CRISPR RNA maturation by trans-encoded small RNA and host factor RNase III. *Nature, 471*(7340), 602–607. http://dx.doi.org/10.1038/nature09886.

Garneau, J. E., Dupuis, M. E., Villion, M., Romero, D. A., Barrangou, R., Boyaval, P., et al. (2010). The CRISPR/Cas bacterial immune system cleaves bacteriophage and plasmid DNA. *Nature, 468*(7320), 67–71. http://dx.doi.org/10.1038/nature09523.

Heyer, W. D., Ehmsen, K. T., & Liu, J. (2010). Regulation of homologous recombination in eukaryotes. *Annual Review of Genetics, 44*, 113–139. http://dx.doi.org/10.1146/annurev-genet-051710-150955.

Horvath, P., & Barrangou, R. (2010). CRISPR/Cas, the immune system of bacteria and archaea. *Science, 327*(5962), 167–170. http://dx.doi.org/10.1126/science.1179555.

Hsu, P. D., Lander, E. S., & Zhang, F. (2014). Development and applications of CRISPR-Cas9 for genome engineering. *Cell, 157*(6), 1262–1278. http://dx.doi.org/10.1016/j.cell.2014.05.010.

Hsu, P. D., Scott, D. A., Weinstein, J. A., Ran, F. A., Konermann, S., Agarwala, V., et al. (2013). DNA targeting specificity of RNA-guided Cas9 nucleases. *Nature Biotechnology, 31*(9), 827–832. http://dx.doi.org/10.1038/nbt.2647.

Jiang, W., Bikard, D., Cox, D., Zhang, F., & Marraffini, L. A. (2013). RNA-guided editing of bacterial genomes using CRISPR-Cas systems. *Nature Biotechnology, 31*(3), 233–239. http://dx.doi.org/10.1038/nbt.2508.

Jinek, M., Chylinski, K., Fonfara, I., Hauer, M., Doudna, J. A., & Charpentier, E. (2012). A programmable dual-RNA-guided DNA endonuclease in adaptive bacterial immunity. *Science, 337*(6096), 816–821. http://dx.doi.org/10.1126/science.1225829.

Lin, S., Staahl, B. T., Alla, R. K., & Doudna, J. A. (2014). Enhanced homology-directed human genome engineering by controlled timing of CRISPR/Cas9 delivery. *eLife, 4*. http://dx.doi.org/10.7554/eLife.04766.

Liu, Y., Bishop, A., Witucki, L., Kraybill, B., Shimizu, E., Tsien, J., et al. (1999). Structural basis for selective inhibition of Src family kinases by PP1. *Chemistry & Biology, 6*(9), 671–678.

Mali, P., Yang, L., Esvelt, K. M., Aach, J., Guell, M., DiCarlo, J. E., et al. (2013). RNA-guided human genome engineering via Cas9. *Science, 339*(6121), 823–826. http://dx.doi.org/10.1126/science.1232033.

Marraffini, L. A., & Sontheimer, E. J. (2010). CRISPR interference: RNA-directed adaptive immunity in bacteria and archaea. *Nature Reviews Genetics, 11*(3), 181–190. http://dx.doi.org/10.1038/nrg2749.

Miller, J. C., Holmes, M. C., Wang, J., Guschin, D. Y., Lee, Y. L., Rupniewski, I., et al. (2007). An improved zinc-finger nuclease architecture for highly specific genome editing. *Nature Biotechnology, 25*(7), 778–785. http://dx.doi.org/10.1038/nbt1319.

Miller, J. C., Tan, S., Qiao, G., Barlow, K. A., Wang, J., Xia, D. F., et al. (2011). A TALE nuclease architecture for efficient genome editing. *Nature Biotechnology, 29*(2), 143–148. http://dx.doi.org/10.1038/nbt.1755.

Mojica, F. J., Diez-Villasenor, C., Garcia-Martinez, J., & Almendros, C. (2009). Short motif sequences determine the targets of the prokaryotic CRISPR defence system. *Microbiology, 155*(Pt 3), 733–740. http://dx.doi.org/10.1099/mic.0.023960-0.

Moscou, M. J., & Bogdanove, A. J. (2009). A simple cipher governs DNA recognition by TAL effectors. *Science, 326*(5959), 1501. http://dx.doi.org/10.1126/science.1178817.

Plessis, A., Perrin, A., Haber, J. E., & Dujon, B. (1992). Site-specific recombination determined by I-SceI, a mitochondrial group I intron-encoded endonuclease expressed in the yeast nucleus. *Genetics, 130*(3), 451–460.

Ran, F. A., Hsu, P. D., Wright, J., Agarwala, V., Scott, D. A., & Zhang, F. (2013). Genome engineering using the CRISPR-Cas9 system. *Nature Protocols, 8*(11), 2281–2308. http://dx.doi.org/10.1038/nprot.2013.143.

Rouet, P., Smih, F., & Jasin, M. (1994). Introduction of double-strand breaks into the genome of mouse cells by expression of a rare-cutting endonuclease. *Molecular and Cellular Biology, 14*(12), 8096–8106.

Rudin, N., Sugarman, E., & Haber, J. E. (1989). Genetic and physical analysis of double-strand break repair and recombination in *Saccharomyces cerevisiae*. *Genetics, 122*(3), 519–534.

Shah, S. A., Erdmann, S., Mojica, F. J., & Garrett, R. A. (2013). Protospacer recognition motifs: mixed identities and functional diversity. *RNA Biology, 10*(5), 891–899. http://dx.doi.org/10.4161/rna.23764.

Urnov, F. D., Miller, J. C., Lee, Y. L., Beausejour, C. M., Rock, J. M., Augustus, S., et al. (2005). Highly efficient endogenous human gene correction using designed zinc-finger nucleases. *Nature, 435*(7042), 646–651. http://dx.doi.org/10.1038/nature03556.

CHAPTER

Studying centrosome function using three-dimensional cell cultures

3

Teresa Arnandis, Susana A. Godinho[1]

Barts Cancer Institute, Queen Mary University of London, London, UK

[1]Corresponding author: E-mail: s.godinho@qmul.ac.uk

CHAPTER OUTLINE

Introduction ... **38**
1. Methods ... **41**
1.1 Propagation of MCF-10A in Monolayer Cultures ... 41
1.1.1 Important notes ... 41
1.2 Preparation of the Overlay 3D Culture ... 42
1.2.1 Important notes ... 43
1.3 Plating the Cells in 3D Cultures ... 45
1.3.1 Important notes ... 45
1.4 Indirect Immunofluorescence Staining of MCF-10A Acini Cultured in Matrigel ... 45
1.4.1 Buffer recipes ... 47
1.5 Live-Cell Imaging ... 48
1.6 Guidelines for Image Acquisition ... 48
Conclusion ... **49**
Acknowledgments ... **49**
References ... **49**

Abstract

Three-dimensional (3D) cell cultures have long been recognized as a tool for the study of tissue architecture, polarity, and invasion. However, only recently these systems have been used to study centrosome and cilia functions. Studying these organelles in 3D cultures has elucidated new functions that otherwise would have been overlooked, demonstrating the value of these experimental systems to the field. Here we describe a culture method to study mammary epithelial cells in a 3D environment.

Methods in Cell Biology, Volume 129, ISSN 0091-679X, http://dx.doi.org/10.1016/bs.mcb.2015.03.010

INTRODUCTION

Two-dimensional (2D) culture techniques have been extensively used to characterize and manipulate cells since their development in 1907 (Harrison, 1907). However, while traditional 2D cultures have been extremely important to dissect molecular mechanisms regarding many aspects of cellular function, limitations of their use have also become clear. In particular, when asking questions regarding how cells organize themselves into well-controlled organs and how disruption of the tissue architecture impacts diseases, such as cancer, 2D cultures are far from being the ideal system. To overcome the limitations imposed by monolayer cell culture, a wide range of 3D culture assays have been developed in order to recapitulate the 3D organization of cells and extracellular matrix within tissues and organs (Shamir & Ewald, 2014). Three-dimensional cultures constitute an excellent model to address complex problems using simplified systems that recapitulate several aspects of tissue organization *in vivo*. One of such 3D models is the culture of glandular epithelial cells. Indeed, many histological features of the glandular epithelium (e.g., mammary gland) can be mimicked *in vitro*, such as polarized morphology (apicobasal polarity axis), attachment to a basement membrane, and the formation of a hollow lumen (Gudjonsson et al., 2002; Petersen, Ronnov-Jessen, Howlett, & Bissell, 1992; Weaver et al., 2002). Thus, this system constitutes a *bona fide* model to study epithelium glandular architecture (Figure 1) (Debnath & Brugge, 2005).

The growth of cells in 3D cultures requires the use of hydrogels (viscoelastic matrices) containing exogenous extracellular matrix proteins. One commonly used exogenous matrix is that derived from the Engelbreth–Holm–Swarm (EHS) murine tumor, commercially available as Matrigel™ (Inman & Bissell, 2010). Many breast cancer cell lines can grow in Matrigel-derived 3D cultures (Kenny et al., 2007); however, it is challenging to recapitulate the formation of acinar structures. There are, however, few nontransformed epithelial cell lines that can form acinar structures in 3D cultures, such as the MCF-10A and HMT-3522-S1 cell lines featured in the initial publication that described the 3D culture assay for human mammary gland (Petersen et al., 1992). Our laboratory uses MCF-10A cells that were isolated from fibrocystic breast tissue (Soule et al., 1990) and are one of the most common cell lines used to study the pathways involved in mammary morphogenesis and breast neoplasia (Debnath, Muthuswamy, & Brugge, 2003; Shaw, Wrobel, & Brugge, 2004). Shortly after plating in 3D cultures, MCF-10A cells form structures that develop an axis of apicobasal polarity, illustrated by the basal secretion of matrix components (collagen-IV and laminin-V), the apical orientation of the Golgi (and centrosome, see Figure 2), and the appropriate localization of junctional complexes (Debnath et al., 2002). The structures subsequently become unresponsive to proliferative signals, and a *bona fide* lumen is formed by cavitation, involving the removal of centrally localized cells via multiple cell death processes (Mailleux, Overholtzer, & Brugge, 2008). The resulting acini resemble those of normal breast tissue, which are clustered in lobules that connect to intralobular ductules that, in turn, connect to interlobular ducts (Figure 1).

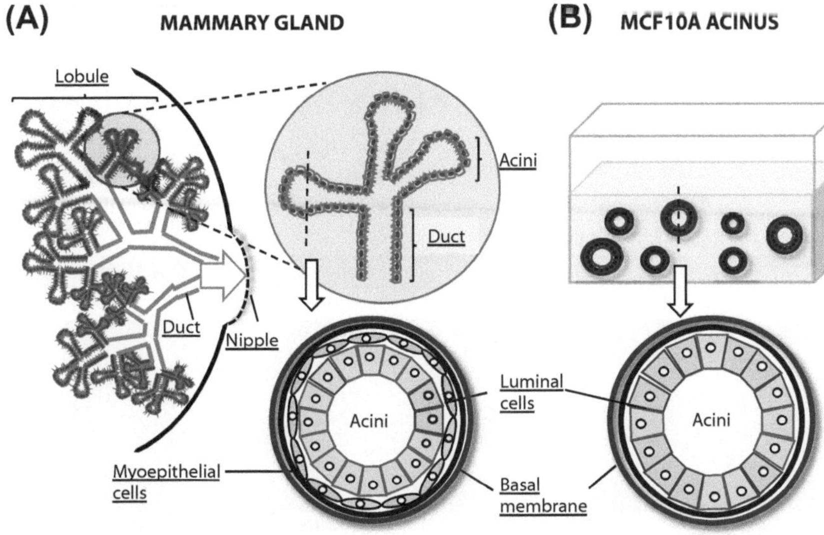

FIGURE 1 Comparative structure of the mammary gland and the acini formed by MCF-10A in 3D cultures.

(A) Schematic representation of the anatomy of the mammary gland. The mammary gland is a compound tubulo-alveolar gland embedded within a connective tissue. It consists of a series of lobules (15–25), each of which is drained by a collecting duct that ends in the nipple. The collecting duct has several branches, which are subdivided into small segments of terminal ducts and a cluster of globular structures named acini (acinus is the secretory unit that contains milk producing cells). A cross-section of the bilayered epithelia of acini is shown in magnification. Luminal epithelial cells display apicobasal polarity and are surrounded by the myoepithelial cells that directly interact with the basement membrane. (B) Structure of a cross-section from 3D cultures of MCF-10A cells. Cells form a single-layered acinus-like structure exhibiting apicobasal polarity. In this 3D monoculture system lacking myoepithelial cells, luminal cells interact directly with the basement membrane.

There are clear advantages of using 3D cultures that address some of the limitations of classical 2D assays, mouse models, or human tissues (Debnath & Brugge, 2005). First, 3D cultures provide an invaluable tool since they are easy to manipulate and are amenable to microscopy and biochemical analyses (Debnath & Brugge, 2005). Secondly, 3D cultures can reproduce *in vivo* phenotypes, such as maintaining the glandular architecture in the case of MCF-10A cells, while reducing the complexity of *in vivo* systems (Mailleux et al., 2008). Thirdly, several laboratories have used the 3D culture of nontransformed mammary epithelial cells to assess the biological effects of the activation of specific oncogenes, individually or in defined combinations (Shaw et al., 2004). In fact, pioneering work from Bissell and others has demonstrated that normal and tumorigenic mammary tissue can be readily distinguished when grown in a laminin-rich environment (Petersen et al.,

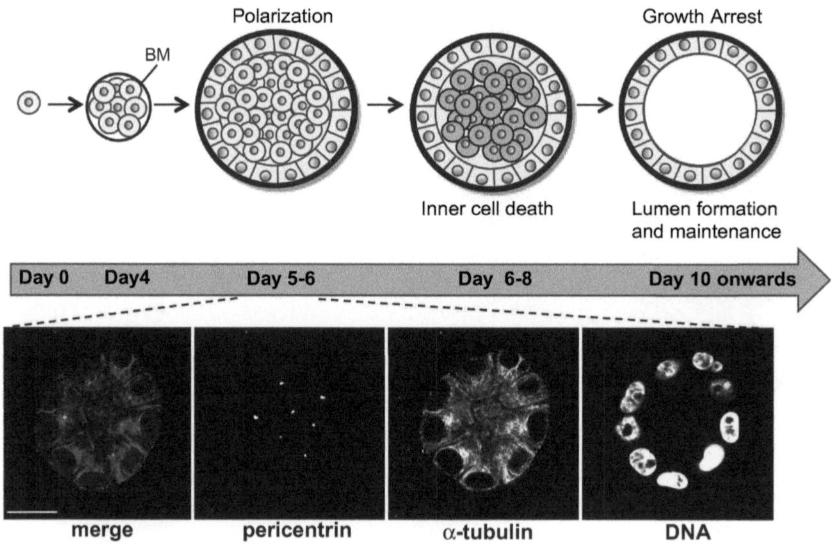

FIGURE 2 Events in MCF-10A acinar morphogenesis.

During the development of MCF-10A mammary epithelial acini in 3D cultures, a sequence of biological events is observed. In the early stages of 3D culture, apicobasal polarization becomes evident within cell clusters (day 5–6), before the appearance of bona fide lumen formation. This can be observed in the depicted IF images where the centrosome is localized apically. At day 6–8, two distinct populations of cells differentiate within each acini: an outer layer of cells in direct contact with the matrix and an inner subset of cells lacking matrix contact. Throughout morphogenesis, this outer cell layer remains polarized. From day 8 onwards, the centrally located cells undergo cell death, which contributes to the formation of a hollow lumen. The acini remain hollow thereafter. Cells were stained for microtubules (α-tubulin; green), centrosomes (pericentrin; red), and DNA (Hoechst; blue). Scale bar: 20 μm. (See color plate)

1992). Moreover, induction of HER2 in MCF-10A cells leads to a multiacinar phenotype that resembles human ductal carcinoma *in situ* with amplified *HER2* gene (Debnath et al., 2002; Muthuswamy, Li, Lelievre, Bissell, & Brugge, 2001).

Despite the unquestionable value of 3D cultures, only recently they have been applied to study centrosome function. The centrosome is the main microtubule-organizing center in animal cells and has important roles in the regulation of cell polarity, cell shape, and cilia formation (Ishikawa, Thompson, Yates, & Marshall, 2012). Thus, a 3D environment is crucial to fully understand centrosome function. When plated in 3D cultures, inner medullary collecting duct cells, IMCD-3, containing extra centrosomes display disrupted epithelial organization. This phenotype was attributed to defects in ciliary signaling as a result of the formation of extra cilia in cells with extra centrosomes (Mahjoub & Stearns, 2012). In addition, we showed that centrosome amplification in MCF-10A cells induces the formation of invasive structures by a mechanism involving RAC1 activation (Godinho et al., 2014). The invasive phenotype is reminiscent of what has been observed *in vivo* in mouse models and in fragments of human tumors grown in 3D cultures (Cheung, Gabrielson, Werb, & Ewald, 2013; Friedl, Locker, Sahai, & Segall, 2012), further reinforcing the value of this approach to study centrosome function.

Here we describe a detailed protocol to perform 3D cultures of mammary epithelial cells.

1. METHODS

1.1 PROPAGATION OF MCF-10A IN MONOLAYER CULTURES

As mentioned above, MCF-10A is a spontaneously immortalized, but nontransformed human mammary epithelial cell line, which exhibits numerous features of normal breast epithelium, including lack of tumorigenicity in nude mice, lack of anchorage-independent growth, and dependence on growth factors and hormones for proliferation and survival (Soule et al., 1990). When plated in 2D monolayers, confluent MCF-10A cells adopt cobblestone morphology, typical of mammary epithelial cells. Any deviation from this morphology should be cause for concern, since they can spontaneously undergo epithelial to mesenchymal transition; if the cells appear fibroblastic or spindle cell-like, it is unlikely that they will properly form acinar structures in 3D culture. A protocol for passaging cells grown on 75 cm^2 dishes is described below:

1. Aspirate the growth medium (Table 1), and rinse with phosphate-buffered saline (PBS).
2. Aspirate the PBS and add 2 mL of trypsin solution (0.25% Trypsin–ethylenediaminetetraacetic acid). It is not recommended to use high-concentration trypsin solutions when passaging MCF-10A cells.
3. Incubate in a 5% CO_2 humidified incubator at 37 °C for 10–15 min. Check the extent of trypsinization after 10 min, by gently tapping the plate to dislodge the cells. Continue every few minutes. The cells should be completely dissociated from the plate to avoid clonal selection of nonadherent cells.
4. Once cells are dislodged, add 5 mL of resuspension medium (Table 1) and pipette to break up cell clumps.
5. Spin down the cells at 150 g in a tissue culture centrifuge for 3–5 min.
6. Aspirate the medium and resuspend the cells in 5 mL of MCF-10A growth medium. Plate 1 mL of cells per 75 cm^2 flask in a total of 15 mL of MCF-10A growth medium (1:5 dilution from the original flask). MCF-10A cells passage at 1:5 ratio to reach 70–80% confluence in 2 days.

1.1.1 Important notes

- Never let cells grow very confluent as this can induce morphological changes: 70–80% confluence is recommended.
- Always keep cells in fresh medium (change every 2–3 days even if cells have not been trypsinized).

Table 1 Medium Recipes for MCF-10A Cells

Component	Growth Medium	Resuspension Medium	Assay Medium
DMEM/F12 Sigma Aldrich (#D8437)	500 mL	500 mL	500 mL
Horse serum Sigma Aldrich (#H1138)	25 mL (5% final)	50 mL (10% final)	10 mL (2% final)
EGF (25 μg/mL) Sigma Aldrich (#E4127)	400 μL (20 ng/mL final)		400 μL (20 ng/mL final)
Hydrocortisone (1 mg/mL) Sigma Aldrich (#H0888)	250 μL (0.5 μg/mL final)		250 μL (0.5 μg/mL final)
Cholera toxin (1 mg/mL) Sigma Aldrich (#C8052)	50 μL (100 ng/mL final)		50 μL (100 ng/mL final)
Insulin (4 mg/mL) Invitrogen (#12585)	1250 μL (10 μg/mL final)		1250 μL (10 μg/mL final)
Pen/Strep (100X) Sigma Aldrich (#P4333)	5 mL	5 mL	5 mL

Once prepared, all types of medium are stored at 4 °C. EGF, hydrocortisone, and insulin stocks are kept at −20 °C. Cholera toxin is kept at 4 °C.

1.2 PREPARATION OF THE OVERLAY 3D CULTURE

There are two well-described methods to plate the cells in 3D cultures using reconstituted basement membrane: the totally embedded assay and the overlay method (Figure 2). In 3D embedded cultures, cells are completely surrounded by a thick extracellular matrix gel, which recapitulates a broader and more complex range of tissue architectures than is typically observed in the overlay method where cells are only partially embedded in the matrix (Shamir & Ewald, 2014). However, we decided to perform the overlay method for several beneficial aspects: the acini can easily be stained *in situ*, eliminating the need for cryosectioning; proteins are easier to extract for immunoblotting analysis; and finally, smaller amounts of Matrigel, an expensive reagent, are used compared to those used in total embedment protocols. To assess invasion it is recommended to add collagen-I to the Matrigel to increase matrix stiffness (protocol described below), but the same procedure could be used with Matrigel alone (Godinho et al., 2014).

1. Thaw the Matrigel (Table 2) on ice overnight in a cold room (4 °C). It is important to keep the lot number of the Matrigel used for each experiment, since

Table 2 Reagents to Prepare the Overlay 3D Culture

Component (Stock Concentration)	Amount Required to Prepare One 8-Well Slide (500 μL)	Source
Matrigel	300 μl from the original concentration	BD Biosciences
Collagen-I (4.04 mg/mL)	200 μL (1.6 mg/mL) of the neutralized collagen-I	Corning

Before using, Matrigel should be resuspended thoroughly on ice and aliquoted to avoid repeated thaw and freezing. Aliquots are kept at −20 °C.

there is a lot-to-lot variability in terms of protein concentration and composition (see important notes below).

2. Prepare the collagen-I (Table 2), taking into account that the final concentration in the mixture will be 1.6 mg/mL. Since the original stock of collagen-I is dissolved in an acidic medium in order to obtain its highest solubility (Zhang, Duan, Huang, Song, & Regenstein, 2014), it is essential to add the appropriate concentration of PBS and NaOH in order to reach a physiological pH (between 7 and 7.5). Stock solutions of PBS 10X and 0.1 M of NaOH, both filtered and sterile, should be added to collagen-I prior to the plating: 62.5 μL of PBS 10X and 62.5 μL of 0.1 M NaOH should be added to 500 μL of collagen-I. The pH of the collagen solution can be assessed using pH strips prior to preparing the Matrigel/collagen-I mixture. Neutralization of the pH is important to avoid loss of cell proliferation.
3. Prepare the mixture of Matrigel and collagen-I by adding the pH-neutralized collagen-I to the Matrigel. Pipette gently until a homogeneous mixture is formed, avoiding bubbles. Since our standard procedure involves plating 42 μL per well, we prepare 500 μL of the mixture for a slide with eight wells/chambers (BD Falcon culture slides, #354108) (see Table 2). Add 42 μL of this mixture in the center of each well of the slide. With a 10 μL pipette tip spread the mixture until the blue edges of the well without overspread; this will form a high meniscus on the border (Figure 3). To polymerize the matrix, put the slides into the incubator for a maximum of 1 h. If more time is needed, add on the top of the polymerized gelled beds 200 μL of Dulbecco's modified Eagle's medium (DMEM)/F12 to prevent the matrix from drying.

1.2.1 Important notes

- It is important to keep in mind that EHS-derived matrices, such as the growth factor reduced Matrigel (catalog #356230) that we use in the laboratory, are biological products purified from mouse tumors. Therefore, the exact composition is unknown and can vary between lots. We test individual lots, prior to purchasing large quantities for experiments. In determining the appropriateness of each lot for 3D assays, we prefer those with protein concentrations ranging from 10 to 12 mg/mL and an endotoxin level less than 2 units/mL.
- When preforming the casting of the Matrigel/collagen-I gelled beds at room temperature it is important to do it rapidly to prevent uneven polymerization.

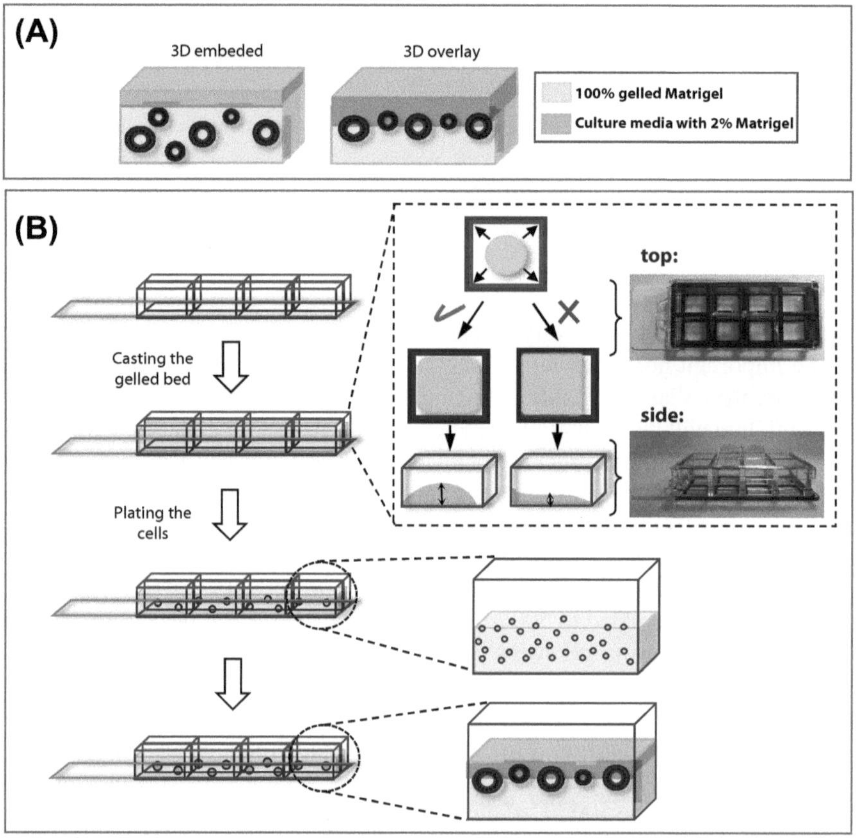

FIGURE 3 Procedure to perform the in vitro MCF-10A 3D culture assay.

(A) Two methods can be used to grow MCF-10A cells in 3D cultures: total embedded (left) and overlay/on top (right). (B) Scheme of the overlay method for 3D culture of MCF-10A cells. Each well of the chamber slide is initially coated with a Matrigel (or Matrigel + collagen-I; see Section 1.2 for details) and allowed to solidify at 37 °C inside the cell culture incubator for 45–60 min. Precautions should be taken when performing the casting of the bed, since overspreading of the Matrigel could induce the formation of a meniscus in the border of the well that could impair acini formation. Ideally, the spreading of the Matrigel should be done until it touches the blue rim inside the well but not the walls. MCF-10A cells are seeded onto this bed as a single-cell suspension in assay medium containing 2% Matrigel. The assay medium is replaced every 4 days.

Alternatively, if more time is needed, the casting can be performed on top of ice to prevent polymerization.

- The stiffness of the extracellular matrix is important to regulate normal cellular function (Paszek et al., 2005; Roskelley, Desprez, & Bissell, 1994). Normally, Matrigel gelled beds have similar elastic modules of that of the mouse mammary gland; however, mammary tumors are usually stiffer (higher elastic modulus) than normal tissue (Weigelt, Ghajar, & Bissell, 2014). Addition of low concentrations of collagen-I (maximum 2 mg/mL) to the Matrigel increases the stiffness of the gelled beds, important when studying invasion, without disrupting the normal acinar morphology. However, higher concentrations of collagen-I (>3 mg/mL) prevent the formation of normal acini and induce an invasive behavior in normal cells (Paszek et al., 2005).

1.3 PLATING THE CELLS IN 3D CULTURES

The protocol continues as follows:

4. While the Matrigel mixture is solidifying, trypsinize a plate as described in Section 1.1, resuspending the cell pellet in assay medium. Count the cells at this stage using an hemocytometer or any specific device for this purpose.
5. Prepare a suspension of cells with assay medium to achieve a final concentration of 10,000 cells/mL (in each chamber plate 400 μL corresponds to 4000 cells). Make sure to resuspend thoroughly the cells to make a single-cell suspension before plating.
6. Add 2% Matrigel to the media containing the cells, just before plating. Plate the cells on the top of each chamber by adding carefully 400 μL of this mixture.
7. Allow the cells to grow in a 5% CO_2 humidified incubator at 37 °C for 24 h. The day that the assay is set up corresponds to Day 0.
8. After 24 h of plating (Day 1), the cells should be refed with assay medium containing 2% Matrigel. Any drug or pharmacological inhibitor used for the experiment should be added at this stage. Before adding the Matrigel to the fresh media, take the medium out of the wells. Add 400 μL of the new medium carefully on top of the wells.
9. Depending on the duration of the experiment, the media should be replaced every 4 days, thus feed on days 4, 8, 12, and 16. The end point of the experiment will differ on the biological parameter under study. In our laboratory we analyze cells on Day 4 to evaluate invasion.

1.3.1 Important notes

- Make sure to resuspend thoroughly the cells to make a single-cell suspension using a p1000 pipette before plating
- Remember to leave an aliquot of Matrigel in a cold room (4 °C) on ice after plating cells on Day 0 to re-feed the cells with new assay media containing 2% Matrigel on Day 1.

1.4 INDIRECT IMMUNOFLUORESCENCE STAINING OF MCF-10A ACINI CULTURED IN MATRIGEL

A basic indirect immunofluorescence (IF) technique for 3D cultures is provided in this section; take into account that some particular antibodies might require optimization. In most cases, but not all, antibodies and staining conditions that work for indirect IF of MCF-10A cells grown as monolayers or on glass coverslips, also work in 3D cultures (Table 3).

1. **Fixation**: Aspirate media from each well of the chamber slide, wash with 1X PBS and immediately fix the structures with 5% Formalin (original stock 10% then diluted 1:1 in PBS) for 20 min at 37 °C. If fixation with methanol or 1:1 methanol/acetone is required, then fix the samples for 10–20 min at −20 °C. For eight well/chamber slides, 400–500 μL volumes are appropriate for the fixation, permeabilization, and all subsequent washing steps in this protocol.
2. **Glycine rinse**: Rinse wells three times, 10 min each, at room temperature (RT) with glycine buffer (composition detailed below). It is advisable to keep the glycine solution at 4 °C to avoid contamination.
3. **Permeabilization**: 0.5% Triton X-100 in PBS at RT for 10 min. Depending on the antibody used for immunostaining, the detergent concentration or duration of permeabilization may require optimization.
4. **Rinse** once for 10 min with IF wash buffer (composition detailed below) at RT. When kept at 4 °C the IF buffer can form crystals; thus make sure the solution is completely dissolved and homogenous prior to use.
5. **Blocking**: Incubate with 200 μL of 10% goat serum in IF wash (Blocking Solution) for 1 h at RT.
6. **Primary antibody**: Remove the blocking solution with a pipette and incubate the primary antibody diluted in 150 μL of Blocking Solution overnight at 4 °C. To prevent evaporation, seal the slide wells with parafilm and incubate

Table 3 Useful Antibodies and Dyes for the Analysis of MCF-10A in 3D Cultures

Antibody	Company	Dilution	Species	Fixation
Active (Cleaved) Caspase 3	Cell signaling	1:100	Rabbit	Formalin or PFA
E-Cadherin	BD Biosciences	1:100	Mouse	M/A
Laminin-V Alexa 488 conjugated	Millipore	1:100	Mouse (D4B5)	Formalin or PFA
β-Catenin	BD Biosciences	1:200	Mouse (14)	Formalin or M/A
α-tubulin	Sigma Aldrich	1:1000	Mouse	Formalin or PFA
Polyglutamylated tubulin (GT335)	Adipogen	1:100	Mouse	Formalin or PFA
Pericentrin	Abcam	1:500	Rabbit	Formalin or PFA
Dyes	**Company**	**Dilution**		**Fixation**
Phalloidin Alexa 568	Molecular Probes	1:500		PFA only
Hoechst 33342	Molecular Probes	1:2500		Formalin and M/A

PFA: 4% Paraformaldehyde. M/A: 1:1 Methanol and acetone solution.

slides inside a humid chamber. For better results use conjugated primary antibodies.

7. **Rinse** three times, 20 min each with IF wash at RT.
8. **Secondary antibody**: Incubate the secondary antibodies diluted in 150 μL of Blocking Solution for 50–60 min at RT. Almost all of our laboratory secondary antibodies are Alexa-conjugated from Invitrogen/Molecular Probes and are used at 1:500 dilution. From this point onwards the experiment proceeds in dark (wrapped in aluminum foil).
9. **Rinse** three times, 20 min each, with 1X IF wash at RT.
10. **Nuclear stain**: Incubate with 1:2500 Hoechst 33342 in 1X PBS for 5–10 min at RT.
11. **Rinse** once for 10 min, with PBS.
12. Mount slides with ProLong® Antifade (P7481) reagent and allow to dry overnight at RT. Once dried, the slides can be stored at 4 °C for a couple of months. For long-term storage, −20 °C is recommended, although the slides can dehydrate. To image acinar 3D structure, we preferentially use a confocal microscope (see Section 1.6). As shown in Figure 4, confocal imaging of 3D acinar structures can be used to visualize centrosomes and microtubules during interphase and mitosis.

1.4.1 Buffer recipes

Glycine buffer: 130 mM NaCl, 13 mM Na_2HPO_4, 3.5 mM NaH_2PO_4, 100 mM glycine at pH 7.4.

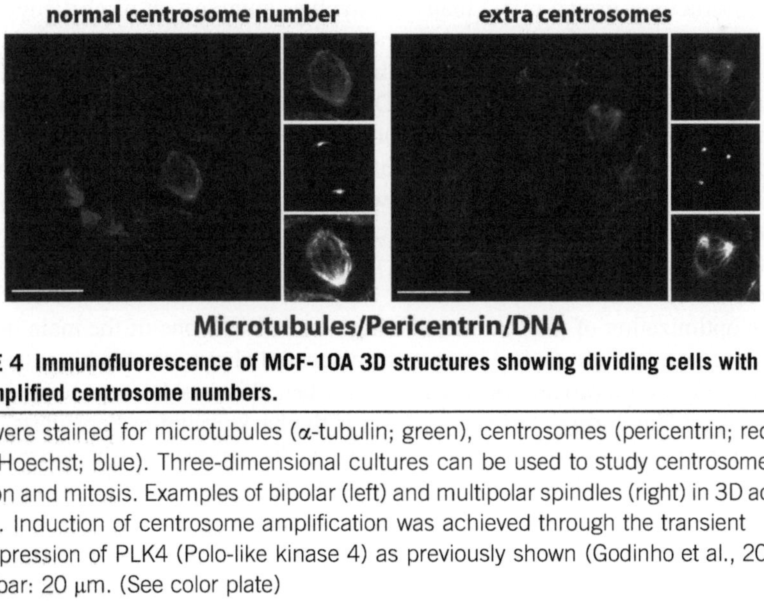

FIGURE 4 Immunofluorescence of MCF-10A 3D structures showing dividing cells with normal and amplified centrosome numbers.

Cells were stained for microtubules (α-tubulin; green), centrosomes (pericentrin; red), and DNA (Hoechst; blue). Three-dimensional cultures can be used to study centrosome function and mitosis. Examples of bipolar (left) and multipolar spindles (right) in 3D acini are shown. Induction of centrosome amplification was achieved through the transient overexpression of PLK4 (Polo-like kinase 4) as previously shown (Godinho et al., 2014). Scale bar: 20 μm. (See color plate)

IF wash buffer 1X: 130 mM NaCl, 13 mM Na_2HPO_4, 3.5 mM NaH_2PO_4, 7.7 mM NaN_3, 0.1% bovine serum albumin, 0.2% Triton X-100, 0.05% Tween-20 at pH 7.4.

1.4.1.1 Important notes

- All aspirations should be performed using a pipette and all solutions should be added carefully to avoid disruption of the Matrigel and acini detachment.
- Incubation of the primary antibodies at 4 °C could liquefy the Matrigel leading to loss of the acini during the IF protocol. Addition of collagen-I to the Matrigel prevents this from happening.
- IF in 3D cultures can have high levels of background, especially associated with the Matrigel. This can be prevented by using primary antibodies already conjugated with a fluorophore.

1.5 LIVE-CELL IMAGING

We use the same protocol described in Section 1.2 to prepare the gelled beds for live cell imaging. It is important, however, to use special chamber slides with bottom coverglass instead of the classical eight-chamber slides (Lab-Tek chamber slide cover #154534). Imaging can then be carried out using an inverted microscope, where the temperature, CO_2, humidity, and evaporation can be controlled. It is also important to use microscope lens with sufficient working distance to allow imaging the acinar 3D structures throughout the culture.

1.6 GUIDELINES FOR IMAGE ACQUISITION

Imaging structures grown in 3D matrices involves a unique set of challenges when compared to imaging of cells grown as monolayers on glass coverslips. One important difference is the thickness of the sample that can range from 50 to 100 μm. Therefore, to account for the whole 3D structure, confocal microscopy will be required. In normal structures, the equatorial cross-section is easily found and could provide the most representative pattern of the polarized hollow glandular architecture (Debnath et al., 2003). However, whenever a different feature needs to be analyzed, such as invasive protrusions or centrosome staining, the complex architecture is best appreciated by taking a series of optical cross-sections (Z-stack) through the structure (Godinho et al., 2014).

The optimization of the immunostaining conditions is one of the main determinants in acquiring high-quality images; as discussed above, establishing these conditions is highly empirical. However, even when this has been accomplished, a common problem during confocal analysis is the background or "noisiness" of the image, due to antibody retaining on the Matrigel (better results using primary conjugated antibodies) and because of the out-of-focus light in these thick specimens. To improve image acquisition, we often increase the acquisition time and also line-averaging while acquiring the image. By reducing the pinhole size, we can decrease the background signal; however, this is usually accompanied by a decrease in the overall fluorescence signal, which might not be suitable for some stainings (Debnath & Brugge, 2005).

CONCLUSION

We presented here a simple method to study cells in a 3D environment. Three-dimensional cultures of MCF-10A cells provide a physiological system to study normal morphogenesis and also the impact of oncogenes on glandular architecture. This simplified system already proved to be useful to understand how centrosome amplification can impact epithelial architecture and invasion (Godinho et al., 2014; Mahjoub & Stearns, 2012). However, the value of this method to the field is certainly not limited to studying the role of centrosome abnormalities and should also be used as a valuable tool for the understanding of normal centrosome function, both in interphase and mitosis.

ACKNOWLEDGMENTS

We thank Alex Pool and Dinis Calado for carefully reading this manuscript. This work is supported by the Higher Education Funding Council for England (T.A. and S.A.G.) and Medical Research Council (S.A.G.).

REFERENCES

Cheung, K. J., Gabrielson, E., Werb, Z., & Ewald, A. J. (2013). Collective invasion in breast cancer requires a conserved basal epithelial program. *Cell, 155*, 1639–1651.

Debnath, J., & Brugge, J. S. (2005). Modelling glandular epithelial cancers in three-dimensional cultures. *Nature Reviews Cancer, 5*, 675–688.

Debnath, J., Mills, K. R., Collins, N. L., Reginato, M. J., Muthuswamy, S. K., & Brugge, J. S. (2002). The role of apoptosis in creating and maintaining luminal space within normal and oncogene-expressing mammary acini. *Cell, 111*, 29–40.

Debnath, J., Muthuswamy, S. K., & Brugge, J. S. (2003). Morphogenesis and oncogenesis of MCF-10A mammary epithelial acini grown in three-dimensional basement membrane cultures. *Methods, 30*, 256–268.

Friedl, P., Locker, J., Sahai, E., & Segall, J. E. (2012). Classifying collective cancer cell invasion. *Nature Cell Biology, 14*, 777–783.

Godinho, S. A., Picone, R., Burute, M., Dagher, R., Su, Y., Leung, C. T., et al. (2014). Oncogene-like induction of cellular invasion from centrosome amplification. *Nature, 510*, 167–171.

Gudjonsson, T., Ronnov-Jessen, L., Villadsen, R., Rank, F., Bissell, M. J., & Petersen, O. W. (2002). Normal and tumor-derived myoepithelial cells differ in their ability to interact with luminal breast epithelial cells for polarity and basement membrane deposition. *Journal of Cell Science, 115*, 39–50.

Harrison, R. G. (1907). Observations on the living developing nerve fiber. *Proceedings of the Society for Experimental Biology and Medicine New York, 4*, 140–143.

Inman, J. L., & Bissell, M. J. (2010). Apical polarity in three-dimensional culture systems: where to now? *Journal of Biology, 9*, 2.

Ishikawa, H., Thompson, J., Yates, J. R., III, & Marshall, W. F. (2012). Proteomic analysis of mammalian primary cilia. *Current Biology: CB, 22*, 414–419.

Kenny, P. A., Lee, G. Y., Myers, C. A., Neve, R. M., Semeiks, J. R., Spellman, P. T., et al. (2007). The morphologies of breast cancer cell lines in three-dimensional assays correlate with their profiles of gene expression. *Molecular Oncology, 1*, 84–96.

Mahjoub, M. R., & Stearns, T. (2012). Supernumerary centrosomes nucleate extra cilia and compromise primary cilium signaling. *Current Biology: CB, 22*, 1628–1634.

Mailleux, A. A., Overholtzer, M., & Brugge, J. S. (2008). Lumen formation during mammary epithelial morphogenesis: insights from in vitro and in vivo models. *Cell Cycle, 7*, 57–62.

Muthuswamy, S. K., Li, D., Lelievre, S., Bissell, M. J., & Brugge, J. S. (2001). ErbB2, but not ErbB1, reinitiates proliferation and induces luminal repopulation in epithelial acini. *Nature Cell Biology, 3*, 785–792.

Paszek, M. J., Zahir, N., Johnson, K. R., Lakins, J. N., Rozenberg, G. I., Gefen, A., et al. (2005). Tensional homeostasis and the malignant phenotype. *Cancer Cell, 8*, 241–254.

Petersen, O. W., Ronnov-Jessen, L., Howlett, A. R., & Bissell, M. J. (1992). Interaction with basement membrane serves to rapidly distinguish growth and differentiation pattern of normal and malignant human breast epithelial cells. *Proceedings of the National Academy of Sciences of the United States of America, 89*, 9064–9068.

Roskelley, C. D., Desprez, P. Y., & Bissell, M. J. (1994). Extracellular matrix-dependent tissue-specific gene expression in mammary epithelial cells requires both physical and biochemical signal transduction. *Proceedings of the National Academy of Sciences of the United States of America, 91*, 12378–12382.

Shamir, E. R., & Ewald, A. J. (2014). Three-dimensional organotypic culture: experimental models of mammalian biology and disease. *Nature Reviews Molecular Cell Biology, 15*, 647–664.

Shaw, K. R., Wrobel, C. N., & Brugge, J. S. (2004). Use of three-dimensional basement membrane cultures to model oncogene-induced changes in mammary epithelial morphogenesis. *Journal of Mammary Gland Biology and Neoplasia, 9*, 297–310.

Soule, H. D., Maloney, T. M., Wolman, S. R., Peterson, W. D., Jr., Brenz, R., McGrath, C. M., et al. (1990). Isolation and characterization of a spontaneously immortalized human breast epithelial cell line, MCF-10. *Cancer Research, 50*, 6075–6086.

Weaver, V. M., Lelievre, S., Lakins, J. N., Chrenek, M. A., Jones, J. C., Giancotti, F., et al. (2002). Beta4 integrin-dependent formation of polarized three-dimensional architecture confers resistance to apoptosis in normal and malignant mammary epithelium. *Cancer Cell, 2*, 205–216.

Weigelt, B., Ghajar, C. M., & Bissell, M. J. (2014). The need for complex 3D culture models to unravel novel pathways and identify accurate biomarkers in breast cancer. *Advanced Drug Delivery Reviews, 69–70*, 42–51.

Zhang, J., Duan, R., Huang, L., Song, Y., & Regenstein, J. M. (2014). Characterisation of acid-soluble and pepsin-solubilised collagen from jellyfish (*Cyanea nozakii* Kishinouye). *Food Chemistry, 150*, 22–26.

CHAPTER

Analysis of centrosomes in human cancer

4

Stefan Duensing

Division of Molecular Urooncology, Department of Urology, University of Heidelberg School of Medicine, Heidelberg, Germany

E-mail: stefan.duensing@med.uni-heidelberg.de

CHAPTER OUTLINE

Introduction 52
Centrosomes and Cancer 52
Centrosome Aberrations and Mitotic Outcomes 53
Not all Centrosome Aberrations are Created Equal 53
1. Methods 54
1.1 Evaluation of Centrosome Aberrations in Human Cancer Tissue—Difficulties and Challenges 54
1.1.1 Tissue 54
1.1.2 Counting 54
1.1.3 Evaluation 55
1.1.4 Immunostaining 55
1.2 Analysis 56
Conclusions 58
Acknowledgments 58
References 58

Abstract

Centrosomes are small cytoplasmic organelles that function as major microtubule-organizing centres during interphase and mitosis. In cancer cells, centrosomes are frequently abnormal in number, size, and structure. Numerous studies have reported centrosome aberrations in human tumors where they frequently increase with malignant progression and advanced disease stage. However, there are a number of caveats when analyzing centrosomes in human tissue. Besides the actual immunodetection and quantification of centrosomes, which can be difficult and cumbersome, centrosome

Methods in Cell Biology, Volume 129, ISSN 0091-679X, http://dx.doi.org/10.1016/bs.mcb.2015.03.002

aberrations require a careful evaluation in the cellular context in which they occur. This chapter highlights the importance of careful interpretation of centrosome aberrations in human tumors.

INTRODUCTION

CENTROSOMES AND CANCER

Centrosomes are small cytoplasmic organelles that have critical functions in a number of important cellular processes. During interphase, the single centrosome of a cell functions as a major microtubule-organizing center. In quiescent cells, the centrosome participates in the formation of the primary cilium. Prior to cell division, the single centrosome duplicates precisely once in order to organize a bipolar mitotic spindle. Normal centrosome duplication involves the disengagement of the two preexisting centrioles and the formation of a single procentriole (daughter centriole) in close proximity to the maternal or mother centriole. In the past decade, it has become clear that centriole duplication is a highly organized process in which protein kinases, phosphatases, components of the ubiquitin–proteasome machinery, and structural proteins cooperate to ensure that one and only one daughter centriole is produced at each preexisting maternal centriole per cell division cycle (Arquint, Gabryjonczyk, & Nigg, 2014; Frat-Karalar & Stearns, 2014).

Centrosomes are not only important for physiologic cellular processes, they have also been implicated in a number of human diseases including ciliopathies and cancer (Bettencourt-Dias, Hildebrandt, Pellman, Woods, & Godinho, 2011; Chavali, Pütz, & Gergely, 2014; Lingle et al., 2002; Pihan et al., 1998). Cancer cells frequently harbor an abnormal number of centrosomes and centrioles (also referred to as centrosome/centriole amplification), which are also frequently altered on the structural level. Numerous malignant tumors across a large variety of histological subtypes have been found to contain malignant cells with centrosome aberrations (Duensing et al., 2009). At the same time, many tumors contain aneuploid populations, in particularly at later stages of malignant progression. Given the stunning link between centrosome aberrations, aneuploidy and malignant progression, centrosome abnormalities have been suggested to function as a driving force for numerical chromosomal imbalances and disease progression. This notion, however, is not new and dates back more than a century ago to Theodor Boveri and David von Hansemann and has been notoriously difficult to prove. Nevertheless, there are a number of recent findings in human tumors and model organisms that underscore the link between centrosome aberrations and cancer, which may be more than merely coincidental (Basto et al., 2008; Duensing, Duensing, Crum, & Munger, 2001; Godinho et al., 2014).

In conclusion, the detection of centrosome aberrations in human cancer tissue may confer prognostic information and centrosome aberrations could potentially be used as surrogate marker for aneuploidy, which is commonly associated with poor patient outcome.

CENTROSOME ABERRATIONS AND MITOTIC OUTCOMES

When considering centrosome aberrations as driver for aneuploidization and malignant progression, it is useful to take a closer look at the steps that follow centrosome amplification i.e., entry and progression through mitosis. This notion is underscored by the fact that the link between centrosome aberrations and aneuploidy is not perfect (Duensing, 2005). Chromosomally unstable tumors have been reported to show no or very few centrosome abnormalities (Martín-Subero et al., 2003). Conversely, there are examples where extremely altered centrosomes were detected in a genomically stable tumor (Duensing, Lee, Dal Cin, & Münger, 2003). It is therefore important not only to analyze the frequency of interphase cells with centrosome aberrations, but also of dividing cell with abnormal spindle poles. It is likewise critical to assess whether multipolarity persists at later stages of cell division i.e., anaphase. Surprisingly, there is evidence that highly multipolar mitosis may be very unlikely, in fact may never, proceed to anaphase (Duensing, Chin, Wang, Kuan, & Duensing, 2008). In contrast to sometimes dramatic and impressive multipolar mitotic figures, tripolar metaphases as well as tripolar anaphases are more frequent findings in human tumors suggesting that at least a proportion of tripolar metaphases are able to proceed to anaphase (Duensing et al., 2008). This does still not prove that daughter cells originating from a tripolar anaphase are viable and can produce daughter cells themselves. The latter is an important prerequisite for the establishment of tumor subclone with an enhanced risk for aneuploidy and more aggressive behavior. Lastly, particular attention should be given to so-called pseudo-bipolar mitoses. This form of mitotic anomaly is characterized by a basically bipolar mitosis despite the presence of more than two spindle poles and can involve centrosome clustering (Krämer, Maier, & Bartek, 2011; Quintyne, Reing, Hoffelder, Gollin, & Saunders, 2005). Not only may daughter cells originating from a pseudo-bipolar mitosis acquire gains and or losses of chromosomes that are more tolerable than changes occurring during a tripolar mitosis, recent results highlight that this event can cause chromosome segregation errors due to faulty microtubule–kinetochore attachments (Ganem, Godinho, & Pellman, 2009). A long-standing question is whether a tetraploid intermediate stage is necessary for cells to become aneuploid or whether they arise from diploid cells. There is evidence to support the notion that both possibilities exist (Duensing et al., 2001; Meraldi, Honda, & Nigg, 2002).

In conclusion, centrosome aberrations in human cancers should not be analyzed in an isolated manner but a careful evaluation of mitotic defects is required to ensure that centrosome aberrations are in fact functional.

NOT ALL CENTROSOME ABERRATIONS ARE CREATED EQUAL

As illustrated above, a key question when assessing centrosome aberrations in tumor cells is whether the cell is viable, able to enter mitosis, progress through mitosis, and finally give rise to a viable and replication-competent daughter cell. In all other cases, centrosomes are unlikely to drive aneuploidy and malignant progression. Obviously, this notion is impossible to prove in fixed, archival tissue specimens discussed here. However, there are a number of hints that can be drawn from the cellular context in which centrosome aberrations occur.

There are two major mechanisms behind centrosome amplification that need to be taken into consideration: centrosome accumulation and centrosome overduplication (Duensing, 2005).

Centrosome accumulation may arise as a secondary event due to cellular insults unrelated to the centrosome duplication cycle. These can include abortive mitosis or cytokinesis defects after which cells with more than the normal centrosomal content arise. Obviously, such cells have experienced a proliferative block and the key question is whether such cells can enter and complete mitosis in a subsequent attempt to divide (Duensing et al., 2000, 2001). A genuine centrosome overduplication, however, entails the overproduction of daughter centrioles. In this case, a true disruption of the centrosome duplication cycle is involved and the cells may hence be less damaged and more likely to give rise to viable daughter cells (Duensing et al., 2001). Supernumerary daughter centrioles can form simultaneously at the same maternal centriole as a number of previous studies have shown (Duensing et al., 2007; Habedanck, Stierhof, Wilkinson, & Nigg, 2005). Alternatively, daughter centrioles may serve prematurely as mothers for new daughter centrioles while skipping the normal maturation process. It is also possible that maternal centrioles give rise to multiple rounds of daughter centrioles without intermittent cell division. In each of these scenarios, there will be an excessive number of daughter centrioles in comparison to maternal centrioles. During centrosome accumulation, however, the ratio should in principal be balanced. A distinction between centrosome accumulation and centrosome overduplication can hence be made when the ratio between mature maternal centrioles and immature daughter centrioles is assessed. One problem here is that centrioles are even more difficult to detect in tissue specimens than centrosomes.

1. METHODS

1.1 EVALUATION OF CENTROSOME ABERRATIONS IN HUMAN CANCER TISSUE—DIFFICULTIES AND CHALLENGES

1.1.1 Tissue

One of the key difficulties with all tissue-based studies is preservation and storage time. Both under- or overfixation as well as prolonged storage can lead to deterioration of epitope reactivity that may, in extreme cases, make any immunostaining impossible.

1.1.2 Counting

Centrosomes and spindle poles are small structures and subject to loss during tissue sectioning (Figure 1). It is hence normal and not unexpected that not all cells in a

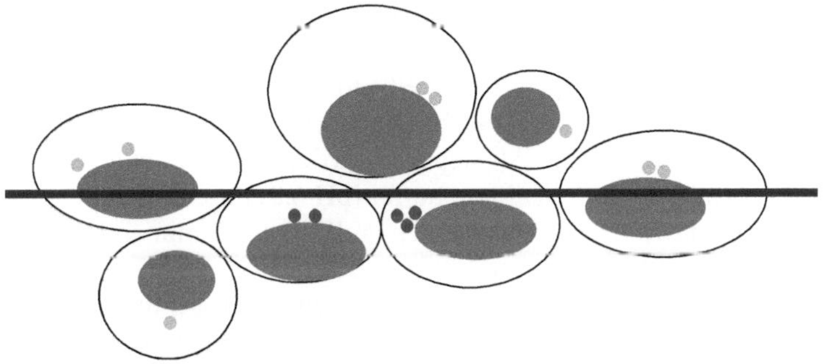

FIGURE 1 Tissue sectioning affects centrosome analysis.

Due to their small size, tissue sectioning (blue line (black in print versions)) will affect the ability to visualize centrosomes in a section leading to a situation where only a proportion of centrosomes are visible (red (dark gray in print versions)) whereas others are lost or not visible (light red (light gray in print versions)).

tissue specimen show a visible centrosome or spindle pole staining. In addition, supernumerary centrosomes sometimes localize very close to each other and it can be difficult to count the precise number. This problem can be aggravated by the fact that centrosomal signals sometimes vary significantly in size.

1.1.3 Evaluation

Centrosomes need to be evaluated in the cellular context in which they occur. It can therefore be useful to count centrosome overduplication versus centrosome accumulation by using a marker for mature, maternal centrioles such as CEP170 or ninein (see below) (Guarguaglini et al., 2005; Mogensen, Malik, Piel, Bouckson-Castaing, & Bornens, 2000). As a first approximation, it can be helpful to simply inspect the nuclear size and shape. Big, irregular, and multiple nuclei point to an accumulation whereas cells containing a single, smaller, and less abnormal nucleus may show rather an overduplication of centrosomes than an accumulation. This can further be corroborated by combining FISH with immunostaining for centrosomes in cultured cells (Duensing et al., 2001) but this approach is not suitable for tissue sections.

1.1.4 Immunostaining

Centrosomes can be visualized in formalin-fixed, paraffin-embedded (FFPE) tissue either by immunohistochemistry or immunofluorescence. In our hands, immunofluorescence allows a more accurate quantification of centrosome numbers (Korzeniewski, Wheeler, Chatterjee, Duensing, & Duensing, 2010). However, this advantage comes at the cost of losing tissue information provided by H&E staining. The following protocol describes a double-immunofluorescence staining for a PCM marker, γ-tubulin, and a marker for mature, maternal centrioles, CEP170 (Guarguaglini et al., 2005). A single staining for γ-tubulin for a simple assessment of centrosome numbers is likewise possible and may be used as a first assessment of centrosome aberrations in a given tumor.

1. Bake FFPE tissue sections overnight at 65 °C.
2. Deparaffinize sections thrice in xylene for 10 min each.
3. Dehydrate sections twice in 100% ethanol for 10 min each. Rehydrate in a graded ethanol series (90%, 70%, 50%) for 5 min each followed by washing in dH_2O twice for 2 min each. All steps are carried out at room temperature (RT).
4. Microwave slides in 0.01 M citrate buffer (pH 6.0) thrice for 10 min each. Allow slides to cool down to RT. Wash slides in dH_2O for 1 min and phosphate-buffered saline (PBS) twice for 1 min each.
5. Block sections with 10% normal donkey serum (Jackson ImmunoResearch) for 30 min at RT, wash in PBS and incubated with an anti-γ-tubulin monoclonal antibody (GTU-88; Sigma) at a 1:500 dilution for 2 days at 4 °C.
6. Incubate slides at 37 °C for at least 1 h before proceeding to the secondary antibody. Wash slides in PBS for 5 min at RT, then incubate with an FITC-conjugated donkey-anti-mouse secondary antibody at a 1:100 dilution (Jackson ImmunoResearch) for at least 2 h at 37 °C.
7. Wash slides twice in PBS for 15 min followed by incubation with a rabbit anti-CEP170 antibody (Invitrogen) (Guarguaglini et al., 2005) at a 1:1000 dilution overnight at 4 °C. Warm up slides to 37 °C for at least 1 h followed by incubation with a Rhodamine Red anti-rabbit secondary antibody (Jackson ImmunoResearch) at a 1:1000 dilution for 2 h at 37 °C.
8. Wash slides twice in PBS for 15 min, stain with DAPI (Vector Laboratories) and store at 4 °C in the dark until analysis (Duensing et al., 2008).

1.2 ANALYSIS

Because of the small size of centrosomes, it is necessary to use a 100×/immersion oil objective for the assessment of centrosome numbers. First, use a low power view for general orientation and identification of cancerous areas. Then, switch to 100×/oil and focus up and down until centrosomal signals appear. Adjacent normal tissue can be used to find centrosomes since they are typically more easily to detect in more organized tissue such as epithelia (Figure 2). In tissue, as in cultured cells, centrosomes typically are present as a single centrosome or, more commonly, a centrosome doublet (Figure 2). Centrosomes are usually found in close proximity to the nucleus but this does not always have to be the case, in particular, in noncancerous ciliated/secretory epithelia (Figure 2). Nonetheless, in tumor cells, the centrosomes are typically in a juxtanuclear position (Figure 2). It is now possible to count the number of centrosomes in those cells, in which centrosomal signals appear. Note that many cells do not show any centrosomal signal due to tissue sectioning (Figure 1) and are hence noninformative. One possibility of data analysis is to calculate the proportion of cells with an abnormal centrosome number (>2 centrosomes) as compared with a normal centrosomal content (1 or 2) from a certain number of informative cells counted (more than 100 cells is recommended). Most cells with an abnormal centrosomal content will show three or four centrosomes, high numbers are detectable albeit less frequently.

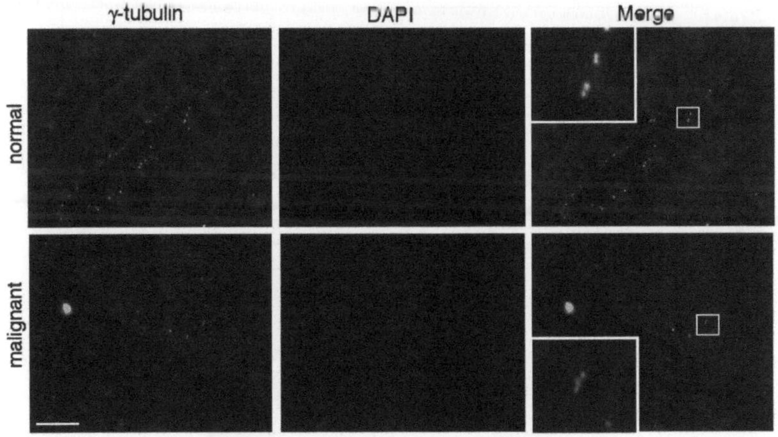

FIGURE 2 Centrosomes in benign and malignant human tissue.

Immunofluorescence microscopic analysis of centrosomes stained for γ-tubulin in benign colorectal mucosa (top) and a squamous cell carcinoma (bottom). Note apical location of centrosomes in the normal tissue and the juxtanuclear position in the malignant tissue. Nuclei stained with DAPI. Scale bar = 25 μm.

If a double-immunofluorescence analysis was performed as described above, it is possible to switch to a different filter and to determine the number of γ-tubulin signals that colocalize with a marker for mature, maternal centrioles, for example, CEP170 (Figure 3). The most frequent finding under normal conditions will be two centrosomes with one centrosome containing an older, fully mature mother centriole until late G2 when the second mother becomes fully mature (Figure 3). In case of centrosome overduplication, there will be more CEP170-negative centrosomes than CEP170-positive centrosomes (typically only one) whereas during centrosome accumulation, the number of CEP170-positive centrosomes exceeds those that are negative (Figure 3) (Duensing et al., 2008).

As pointed out above, mitotic cells should likewise undergo a detailed quantification. However, the problem is that due to tissue sectioning and a tilted mitotic plane, not all spindle poles may be visible. This is particularly problematic when cancer cells with increased chromosomal content divide since it will increase the spindle dimensions in comparison with normal diploid cells. Hence, a quantitative assessment of the proportion of normal and abnormal mitotic cells will likely lead to an underestimate of the actual number of cells undergoing multipolar mitosis. Nevertheless, a careful evaluation of the proportion of cells with abnormal centrosomes in interphase with or without genuine overduplication and abnormal spindle poles in metaphase and later mitotic stages, respectively, and a correlation between these phenotypes appears to be the most accurate approach for the assessment of centrosomes in cancer.

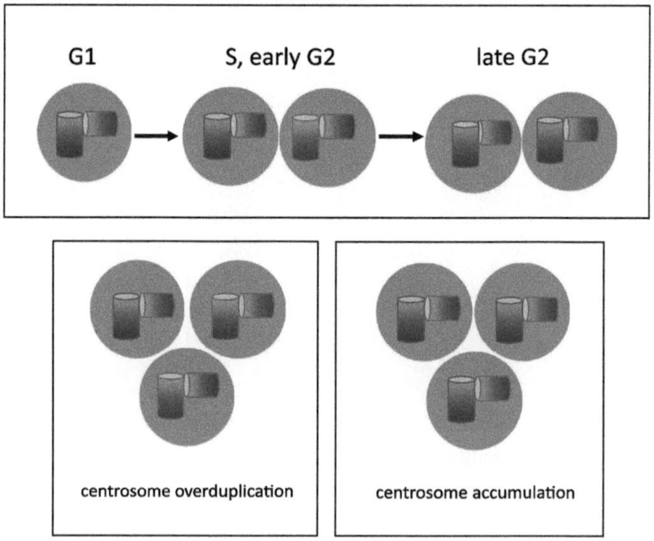

FIGURE 3 Detection of centrosome overduplication in human tissue.

Graphical display of centrosome overduplication and centrosome accumulation using a marker for centrosomes/pericentriolar material (γ-tubulin; green (very light gray in print versions)) in combination with a marker for older, mature mother centrioles (e.g., CEP170; red (gray in print versions)). Immature centrioles (blue (light gray in print versions)) are not visualized but displayed graphically for clarification purposes. Centrosomes containing older mature centrioles will appear yellow in color due to overlap of the red and green fluorescence.

Adapted from Duensing et al. (2008).

CONCLUSIONS

Centrosome aberrations are frequent findings in cancer cells and correlate with aneuploidy and malignant progression. Therefore, centrosome aberrations may be useful prognostic biomarkers but a careful interpretation of centrosomal staining results in the cellular context in which they occur is mandatory.

ACKNOWLEDGMENTS

Work on centrosomes and cancer in the author's laboratory is supported by the Medical Faculty Heidelberg and the German Cancer Aid (Grant No. 111247).

REFERENCES

Arquint, C., Gabryjonczyk, A.-M., & Nigg, E. A. (September 5, 2014). Centrosomes as signalling centres. *Philosophical Transactions of the Royal Society of London. Series B: Biological Sciences, 369*(1650).

Basto, R., Brunk, K., Vinadogrova, T., Peel, N., Franz, A., Khodjakov, A., et al. (June 13, 2008). Centrosome amplification can initiate tumorigenesis in flies. *Cell, 133*(6), 1032–1042.

Bettencourt-Dias, M., Hildebrandt, F., Pellman, D., Woods, G., & Godinho, S. A. (August 2011). Centrosomes and cilia in human disease. *Trends in Genetics, 27*(8), 307–315.

Chavali, P. L., Pütz, M., & Gergely, F. (September 5, 2014). Small organelle, big responsibility: the role of centrosomes in development and disease. *Philosophical Transactions of the Royal Society of London. Series B: Biological Sciences, 369*(1650).

Duensing, S. (May 2005). A tentative classification of centrosome abnormalities in cancer. *Cell Biology International, 29*(5), 352–359.

Duensing, A., Chin, A., Wang, L., Kuan, S.-F., & Duensing, S. (March 1, 2008). Analysis of centrosome overduplication in correlation to cell division errors in high-risk human papillomavirus (HPV)-associated anal neoplasms. *Virology, 372*(1), 157–164.

Duensing, S., Duensing, A., Crum, C. P., & Munger, K. (March 15, 2001). Human papillomavirus type 16 E7 oncoprotein-induced abnormal centrosome synthesis is an early event in the evolving malignant phenotype. *Cancer Research, 61*(6), 2356–2360.

Duensing, S., Lee, B. H., Dal Cin, P., & Münger, K. (September 8, 2003). Excessive centrosome abnormalities without ongoing numerical chromosome instability in a Burkitt's lymphoma. *Molecular Cancer, 2*, 30.

Duensing, S., Lee, L. Y., Duensing, A., Basile, J., Piboonniyom, S., Gonzalez, S., et al. (August 29, 2000). The human papillomavirus type 16 E6 and E7 oncoproteins cooperate to induce mitotic defects and genomic instability by uncoupling centrosome duplication from the cell division cycle. *Proceedings of the National Academy of Sciences of the USA, 97*(18), 10002–10007.

Duensing, A., Liu, Y., Perdreau, S. A., Kleylein-Sohn, J., Nigg, E. A., & Duensing, S. (September 20, 2007). Centriole overduplication through the concurrent formation of multiple daughter centrioles at single maternal templates. *Oncogene, 26*(43), 6280–6288.

Duensing, A., Spardy, N., Chatterjee, P., Zheng, L., Parry, J., Cuevas, R., et al. (October 2009). Centrosome overduplication, chromosomal instability, and human papillomavirus oncoproteins. *Environmental and Molecular Mutagenesis, 50*(8), 741–747.

Frat-Karalar, E. N., & Stearns, T. (September 5, 2014). The centriole duplication cycle. *Philosophical Transactions of the Royal Society of London. Series B: Biological Sciences, 369*(1650).

Ganem, N. J., Godinho, S. A., & Pellman, D. (July 9, 2009). A mechanism linking extra centrosomes to chromosomal instability. *Nature, 460*(7252), 278–282.

Godinho, S. A., Picone, R., Burute, M., Dagher, R., Su, Y., Leung, C. T., et al. (June 5, 2014). Oncogene-like induction of cellular invasion from centrosome amplification. *Nature, 510*(7503), 167–171.

Guarguaglini, G., Duncan, P. I., Stierhof, Y. D., Holmström, T., Duensing, S., & Nigg, E. A. (March 2005). The forkhead-associated domain protein Cep170 interacts with Polo-like kinase 1 and serves as a marker for mature centrioles. *Molecular Biology of the Cell, 16*(3), 1095–1107.

Habedanck, R., Stierhof, Y.-D., Wilkinson, C. J., & Nigg, E. A. (November 2005). The Polo kinase Plk4 functions in centriole duplication. *Nature Cell Biology, 7*(11), 1140–1146.

Korzeniewski, N., Wheeler, S., Chatterjee, P., Duensing, A., & Duensing, S. (2010). A novel role of the aryl hydrocarbon receptor (AhR) in centrosome amplification – implications for chemoprevention. *Molecular Cancer, 9*, 153.

Krämer, A., Maier, B., & Bartek, J. (August 2011). Centrosome clustering and chromosomal (in)stability: a matter of life and death. *Molecular Oncology, 5*(4), 324–335.

Lingle, W. L., Barrett, S. L., Negron, V. C., D'Assoro, A. B., Boeneman, K., Liu, W., et al. (February 19, 2002). Centrosome amplification drives chromosomal instability in breast tumor development. *Proceedings of the National Academy of Sciences of the USA, 99*(4), 1978–1983.

Martín-Subero, J. I., Knippschild, U., Harder, L., Barth, T. F. E., Riemke, J., Grohmann, S., et al. (November 2003). Segmental chromosomal aberrations and centrosome amplifications: pathogenetic mechanisms in Hodgkin and Reed-Sternberg cells of classical Hodgkin's lymphoma? *Leukemia, 17*(11), 2214–2219.

Meraldi, P., Honda, R., & Nigg, E. A. (February 15, 2002). Aurora-A overexpression reveals tetraploidization as a major route to centrosome amplification in p53−/− cells. *EMBO Journal, 21*(4), 483–492.

Mogensen, M. M., Malik, A., Piel, M., Bouckson-Castaing, V., & Bornens, M. (September 2000). Microtubule minus-end anchorage at centrosomal and non-centrosomal sites: the role of ninein. *Journal of Cell Science, 113*(Pt 17), 3013–3023.

Pihan, G. A., Purohit, A., Wallace, J., Knecht, H., Woda, B., Quesenberry, P., et al. (September 1, 1998). Centrosome defects and genetic instability in malignant tumors. *Cancer Research, 58*(17), 3974–3985.

Quintyne, N. J., Reing, J. E., Hoffelder, D. R., Gollin, S. M., & Saunders, W. S. (January 7, 2005). Spindle multipolarity is prevented by centrosomal clustering. *Science, 307*(5706), 127–129.

CHAPTER

5

In vivo examination of the cortical cytoskeleton in multiciliated cells using electron tomography

Daniel K. Clare*, Maud Dumoux*, Delphine Delacour[§,1]

*Institute of Structural and Molecular Biology, Birkbeck College and University College of London, London, UK

[§]Cell Adhesion and Mechanics Group, Jacques Monod Institute, CNRS-UMR7592, Paris Diderot University, Paris Cedex, France

[1]Corresponding author: E-mail: delacour.delphine@ijm.univ-paris-diderot.fr

CHAPTER OUTLINE

Introduction **62**

1. Rationale **63**

2. Methods **65**

2.1 Mouse Dissection and Tissue Preparation 65

2.2 TEM Tissue Sample Preparation 67

2.3 Sectioning and Preparation for Electron Tomography 68

2.4 Electron Tomography 69

2.5 Data Processing 71

2.5.1 Alignment and reconstruction of tilt series 71

2.5.2 Filtering of the tomographic volume 75

2.5.3 Modeling/segmentation of features from tomograms 76

3. Materials **78**

3.1 Mouse Dissection 78

3.2 TEM Sample Preparation 78

3.3 Sample Sectioning and Preparation for Electron Tomography 79

3.4 Electron Tomography 79

3.5 Data Processing 79

Conclusion **79**

List of Abbreviations **80**

Acknowledgments **80**

References **80**

Methods in Cell Biology, Volume 129, ISSN 0091-679X, http://dx.doi.org/10.1016/bs.mcb.2015.03.001

Abstract

Multiciliated cells are characterized by coordinated arrays of motile cilia. In the respiratory tract, the maintenance of this array is essential to ensure proper ciliary and mucus clearance. The establishment and the maintenance of the ciliary set are mediated by the correct positioning of basal bodies at the cell cortex. While microtubule and actin cytoskeletons have been reported to regulate basal body lattices, an understanding of their detailed organization was missing until recently. Here, we describe how electron tomography can highlight the arrangement of the cytoskeletal networks and their interplay with basal bodies in ciliated cells in their tissular environment. Thanks to this approach, information in fine detail on large parts of the cell, dense in organelles, is provided. In combination with other approaches, such as transgenic animal models, electron tomography constitutes a powerful technique giving an overview of tissues and cells concomitantly with acquisition of three-dimensional detail.

INTRODUCTION

Cilia are organelles composed of a centriole, called the basal body (BB), anchored at the plasma membrane, from where emerges an axoneme made of the plasma membrane surrounding an organized core of microtubule (MT) doublets. Cilia are qualified as primary or motile (Ishikawa & Marshall, 2011). Primary cilia are nonmotile cilia with a classical 9 + 0 MT doublet arrangement. Most cell types develop a single primary cilium, which acts as a mechanosensor of environmental stimuli and is required for cell signal transduction from external cues (Singla & Reiter, 2006; Veland, Awan, Pedersen, Yoder, & Christensen, 2009). Primary cilium defects have been robustly correlated with embryonic disorders and pathologies such as *situs inversus* or polycystic kidney disease (Hildebrandt, Benzing, & Katsanis, 2011; Nigg & Raff, 2009).

In contrast to primary cilia, motile cilia are composed of a 9 + 2 MT doublet assembly. Ciliary beating is attributed to the additional central MT doublet and dynein arms present in the axonemal core. Motile cilia are found on a restricted number of cell types, such as ependymal cells, oviduct cells, or respiratory tract cells. Up to 300 motile cilia can be found per cell, which are consequently called multiciliated cells. Compared to primary cilia, their function is rather mechanical. Because of their motile nature, they generate a flow above cell layers facing the organ lumen, and thus ensure cerebrospinal fluid turnover or mucus clearance in the airway tract, for instance. Aberrant motile cilium motion or organization leads to impaired fluid flow and development of hydrocephaly or respiratory disorders (Brooks & Wallingford, 2014; Knowles & Boucher, 2002; Lee, 2013; Randell & Boucher, 2006). However, their restricted mechanical function has been recently challenged, as several studies have attributed a sensory role to motile cilia. Welsh and colleagues demonstrated the association of bitter taste receptors with axonemes and first proposed a chemosensory function for airway motile cilia (Shah, Ben-Shahar, Moninger, Kline, & Welsh, 2009). Additionally, the serum response transcription factor was localized at motile cilia and was proposed to modulate cellular responses to tissue injury (Nordgren et al., 2014).

The flow generated by motile cilia in organ lumens is orientated within the tissue plane. For instance, the flow generated in respiratory tracts is directed along a lung-to-throat axis for particle and mucus removal. This cilium coordination between cells within the tissue plane is named planar polarity. In addition, a second level of cilium coordination within individual cells occurs and is named rotational polarity (Wallingford, 2010). Molecular and cellular mechanisms ordering the polarized arrangement of motile cilia have been extensively studied in the last decade. Planar cell polarity proteins such as Dishevelled (Dvl) and Vangl-1 and -2 have been reported to be required for planar cell polarity (Kishimoto & Sawamoto, 2012; Marshall & Kintner, 2008; Park, Mitchell, Abitua, Kintner, & Wallingford, 2008; Vladar & Axelrod, 2008; Vladar, Bayly, Sangoram, Scott, & Axelrod, 2012). Recent work suggests the existence of a feedback loop of regulation, exerted by the fluid flow applied on ciliated cells, allowing refinement in the arrangement of BBs (Guirao et al., 2010; Mitchell, Jacobs, Li, Chien, & Kintner, 2007). Cytoskeletal networks have been widely described as key players in cellular organization and they are in close contact with BBs. Therefore, cytoskeletal networks constitute good candidates for the establishment and maintenance of the coordinated positioning and motion of cilia at the apical cell cortex (Clare et al., 2014; Kunimoto et al., 2012; Vladar & Axelrod, 2008; Vladar et al., 2012; Werner et al., 2011; Wright, Brown, McCandlish, Thompson, & Cornwell, 1983). It appears that sequential participation of actin and MT networks drives BB cortical positioning. First, targeting, docking, and spacing of BBs are driven by actin. Second, the cortical MT network is crucial for the rotational polarity of BBs, based on drug-based experiments or transgenic mouse models.

1. RATIONALE

Studying the organization of the cytoskeleton at the BB level on mouse tissue is a challenge as it necessitates the following:

- Easy identification of multiciliated cell in the tissue,
- Localization of the area of interest in the cell, and
- Resolution of fine details of cytoskeleton arrangement at the cell cortex where numerous BBs develop together with MTs, actin, and other cellular components in a restricted volume.

For that, the approach should combine the following:

- A large range of magnification, which enables the concomitant analysis of tissue organization and the arrangement of actin filaments,
- A staining procedure that allows identification of cells as well as intracellular components, and

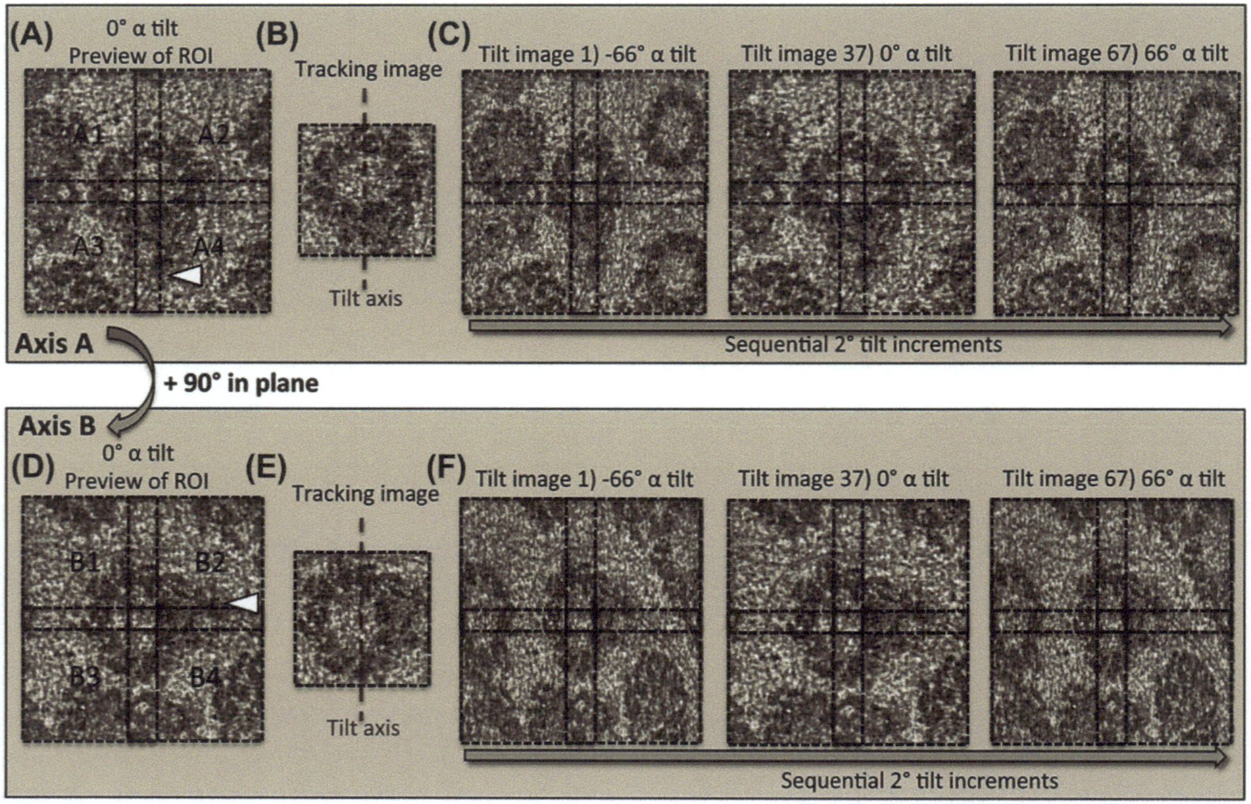

FIGURE 1 Schematic for the collection of montaged dual-axis tilt series.

The initial ROI (region of interest) is selected at 0° α tilt and then a preview using the *Montage panel* can be collected. This example shows a two-by-two montage in which each tilt image will contain four images (A1–4). SerialEM uses image shift to acquire the four images per angle of the tilt series, with the four images aligned via cross-correlation using their overlapping regions. (B) Then using the *Walkup* feature, a tracking image is used to make sure the ROI remains centered when the microscope holder is tilted to the highest negative tilt. (C) Then at each tilt image through the series (−66° to +66° in 2° steps) four images are acquired and this series is saved as axis A. Then the grid is manually rotated by an in-plane rotation of 90° using the sample holder, (D) the ROI is recentered and then tracked (E) while tilting to highest negative tilt using the *Walkup* feature. (F) The B axis is acquired using the same parameters as axis A (the tilt range does not have to be the same as axis A). White stars point out the basal foot cap.

- Preservation of the sample over the time needed to survey the tissue and cell prior to acquisition of high resolution data (4–8 nm).

Immunofluorescence techniques offer the advantage of being versatile and widely accessible (Vladar & Axelrod, 2008; Vladar et al., 2012; Werner et al., 2011). However, for this study, the need for multiple parallel labeling on tissue samples, low magnification range, and photo bleaching compromise this approach.

Ultrastructural experiments using conventional electron microscopy (EM) coupled to electron tomography are suitable choices in this field of research. Single axis tomography is based on the acquisition of a series of images from resin sections up to 400 nm thick, tilted over a range of angles (usually $-60°$ to $+60°$). The images collected at different tilt angles represent two-dimensional projections of the section volume, which can be aligned and back-projected to generate a 3D map of the section. However, as it is not possible to tilt to 90°, there is a "missing wedge" of information. Dual-axis tomography was developed to reduce this information loss and consists of collecting two tilt series of the same region of interest (ROI), with the second axis collected after a 90° rotation of the ROI in the xy plane (Figure 1). This variant in tomographic acquisition reduces the amount of missing information from a wedge to a pyramid (Mastronarde, 1997). This is particularly important when studying fibrillar structures such as the cytoskeleton. In addition, montaged tomograms allow the observation of large areas of a cell while preserving the image resolution.

Over the last decade, electron tomography has contributed to the observation of cellular and organelle organization in cultured cells (Donohoe, Mogelsvang, & Staehelin, 2006; Dumoux, Clare, Saibil, & Hayward, 2012; Noske, Costin, Morgan, & Marsh, 2008). Nevertheless, despite the very exciting possibilities of electron tomography, applications for direct observations on tissue remain sporadic; probably due to the difficulty in preparing tissue samples for EM and the fact that electron tomography still remains a marginal technique for most cell biologists.

Here, we provide a detailed protocol on how to observe the organization of the cell cortex in mouse ciliated cells at high resolution (4–8 nm), with a particular emphasis on the arrangements of MT and actin at BBs. We will explain how to dissect an intact mouse trachea and how to stain and embed it in resin prior to sectioning, followed by a section detailing the acquisition of montaged dual-axis tomograms. Finally, we will explain how to align and segment the data to highlight the organization of the MT and actin networks at the apical cortex of multiciliated cells.

2. METHODS

2.1 MOUSE DISSECTION AND TISSUE PREPARATION

We describe here a dissection method for mouse trachea. Mice are handled in accordance with the French regulations for animal care.

1. Sacrifice the mice by cervical dislocation.
2. After opening the rib cage (RC), remove the submandibular salivary glands (SG) and the infrahyoid musculature (IM) (Figure 2(A)). The trachea is then visible (Figure 2(B)).
3. The trachea is carefully cut at the border of the bronchial bifurcation (Figure 2(C)).
4. Handle the previous cut border with fine tweezers and cut the lining esophagus with microscissors (Figure 2(D)–(E)).
5. The trachea is finally removed by cutting the border of the larynx (Figure 2(E)–(F)).

Note associated to point (NAP) 1: Be gentle during cervical dislocation, in order to not break the trachea.

NAP2: Before the dissection, splash mouse fur with 70% ethanol to avoid contamination of the tissues. During the dissection of the RC, SG, and IM, take

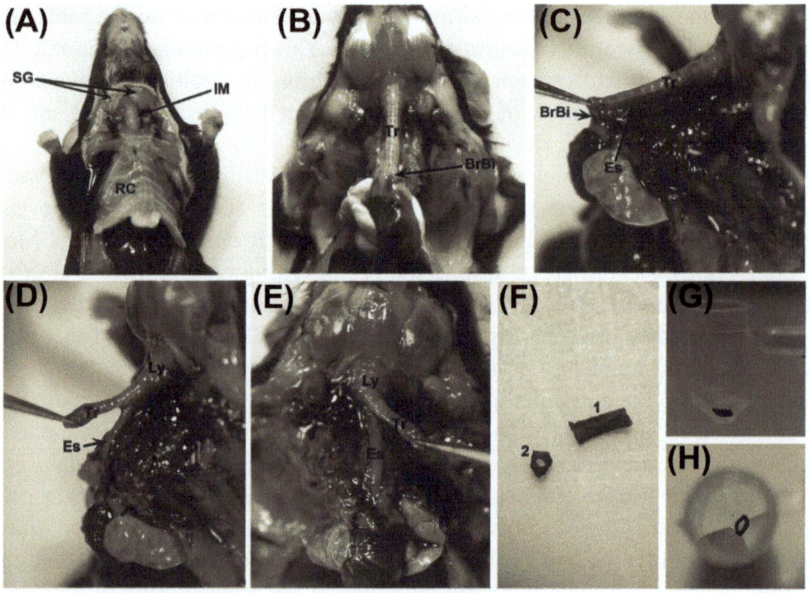

FIGURE 2 Mouse trachea dissection for TEM preparation.

(A) After sacrifice, the adult mouse is laid on its back and the ventral skin is cut, revealing the submandibular salivary glands (SG) and infrahyoid musculature (IM). (B) Once the rib cage, SG, and IM are removed, trachea and bronchial bifurcation (BrBi) are visible. (C) The esophagus is affixed behind the trachea. (D–E) The trachea is cut at the level of the BrBi and progressively isolated from the esophagus, up to the larynx (Ly). (F) Once totally isolated, the trachea (1) is placed on a Teflon plate and tracheal rings (2) are generated. (G–H) Tracheal rings are placed at the bottom of BEEM capsules for subsequent resin embedding (G side view, H top view).

care to keep your scissors as parallel as possible to the mouse body axis, in order to not damage the trachea.

2.2 TEM TISSUE SAMPLE PREPARATION

We provide here a detailed protocol for EM trachea preparation. Solution dilutions and incubation time may vary according to the tissue.

In the following section, buffers and solutions are 0.2 μm filtered prior to use. Most of them are freshly prepared. Exceptions to this are the Millonig buffer and the epoxy resin solution which can be prepared in advance and stored at 4 or −20 °C, respectively.

1. Once isolated, the trachea is placed on a Teflon plate for manipulation (Figure 2(F)).
2. One-mm-thick tracheal rings are cut with a single-edged razor blade (Figure 2(F)) and washed three times in PBS-containing microtubes in order to remove blood cells and debris.
3. For fixation, immerse tracheal rings in 1.6% glutaraldehyde in Millonig buffer overnight at 4 °C.
4. Rinse the samples in Millonig buffer three times for 15 min each at 4 °C.
5. Incubate the samples in 2% osmium tetroxide in Millonig buffer for 1 h at 4 °C.
6. Rinse the samples three times for 5 min each in Millonig buffer.
7. Dehydrate the samples by sequential 10-min incubations in 30%, 50%, 70%, and 80% ethanol in water.
8. Incubate the samples in 95% ethanol in water overnight at 4 °C.
9. Finalize the dehydration by incubating three times for 20 min in 100% ethanol at room temperature.
10. Incubate the samples in 100% propylene oxide for 10 min.
11. Start the sample embedding by incubating 30 min in propylene oxide–epoxy resin progressively at ratios 1:4, 1:3, and finally 1:2 (v:v).
12. Incubate the samples in 100% epoxy resin overnight at room temperature with the lid open under the fume hood in order to evaporate residual propylene oxide.
13. Place the samples at the bottom of BEEM embedding capsules in a perpendicular plane to the capsule walls (Figure 2(G)–(H)). Fill the capsules with epoxy resin and leave them for 30 min on the bench.
14. After 30 min, correct the positioning of the samples in the embedding capsules if necessary.
15. Incubate the samples for 3 days in a wet incubator at 60 °C to allow epoxy resin polymerization.
16. Three days later, remove the plastic BEEM embedding capsules with a scalpel.

NAP1 and 2: Always handle the trachea at their termini. Since the terminal parts will be damaged, trim 1 mm from each end to prepare the tracheal ring.

NAP2: Do not compress the tissue and perform one clear cut when you generate the tracheal rings to avoid sample damage.

NAP3: Millonig buffer is prepared as follows: 0.1 M Na_2HPO_4, 0.1 M NaH_2PO_4, pH 7.3.

NAP11: For 100 mL epoxy resin solution, mix 23.2 mL Epon 812, 49.6 mL DDSA, 18.8 mL epoxy resin Araldite CY212, and 3 mL DMP30. Centrifuge at 25,000 g to remove air bubbles before −20 °C storage.

2.3 SECTIONING AND PREPARATION FOR ELECTRON TOMOGRAPHY

Once the specimen is embedded in epoxy resin, sections are cut using an ultramicrotome. The thickness of the ultrathin sections depends on the analysis and the voltage of the transmission electron microscope to be used. For immunogold labeling, 50 nm sections are sufficient. For tomography, the section thickness ranges from 150 to 400 nm, depending on the microscope voltage. Typically, for a 200 keV microscope, the section should not be thicker than 200–250 nm. The procedure for cutting ultrathin sections is described in the following.

1. Mount the epoxy resin block securely in the microtome chuck.
2. Trim the four sides of the resin block and the front face to generate a small block for sectioning. Each side is usually trimmed to a depth of 30–50 μm using a cutting speed of 100 mm/s and a feed of 500 nm.
3. Retract the trimming knife from the prepared block and mount the diamond sectioning knife.
4. Align the sectioning knife with the block face. You should see a clear reflection of the knife blade on the block face. Then fill the water bath of the sectioning knife with 0.2 μm filtered water.
5. Once the knife is positioned, set the desired section thickness and the speed to 1 mm/s and start sectioning.
6. Break the ribbons in the water bath into grid-sized pieces (usually 3–4 sections) and arrange them for placement on the grid.
7. Then pick the sections up with the Perfect Loop and deposit them on formvar or carbon-coated finder grids.
8. Apply fiducial markers to the section side of the grid (2 μL of 10 nm protein-A-gold), then wash 3 times in 30 μL of 0.2 μm filtered water and blot dry. Repeat this procedure for the opposite side of the grid.

NAP2: When using a diamond trimming knife, the two corners of the blade can be used to generate two parallel sides of the block, and the front of the blade to create the front face on the block. Then the block is rotated by 90°, using the microtome chuck, to remove the remaining two sides on the block. It is crucial that the block is well trimmed (square, clean sides), otherwise sections will not come off as continuous ribbons.

NAP4: Position the knife first, prior to filling the water bath, to avoid getting water on the block face and the back of the knife (specifically for hydrophilic resins).

You should make sure that the water covers the knife blade but has a concave meniscus.

NAP5: If the sections are not flat when they come off the knife, absorb chloroform on a piece of filter paper and then place it over the top of the water bath (do not touch the water bath with the filter paper) and the chloroform vapor will flatten the sections.

NAP6: To break the ribbons, you will need a fine hair attached to a cocktail stick.

NAP7: When picking up the sections, the Perfect Loop should be parallel to the surface of the water bath. You must maintain the contact with the water surface for a few seconds before moving the loop away. Then place this on the surface of an EM grid with a piece of filter paper placed under the grid. Formvar provides a more robust support layer than carbon, which makes the application of fiducial markers more reproducible. Finder grids are also essential if you are planning to collect dual-axis tilt series as they will help you to locate the ROI after rotating the grid by 90° for the second axis.

NAP8: Coating both sides of the sections with gold beads will allow for correction of distortions in the section during reconstruction. The water washes are important, to remove salts and additional components present in the fiducial marker buffer.

2.4 ELECTRON TOMOGRAPHY

We use SerialEM software for tomographic data collection (Mastronarde, 2005). Both video and written tutorials are available for the operation of SerialEM (http://bio3d.colorado.edu/SerialEM/). SerialEM automates a number of steps during tilt series collection including eucentric height determination, setting the image focus, and tracking of the object. To get the best tomograms, it is advisable to collect two tilt series (dual-axis) of the ROI with the second series collected following a 90° in-plane rotation of the grid using a dual-axis tilt holder (Figure 1). This will reduce the amount of missing information in the z direction thus improving the reconstructed volume. This is particularly important if you want to study fibrillar specimens (Figure 3). It is also possible to acquire montaged tomograms, where image shifts are used to acquire multiple areas for each tilt image. The benefit of taking montaged tomograms is that you can capture a larger area but still maintain a higher magnification (better sampling of your ROI). The next section will describe the collection of dual-axis tomograms in a two-by-two montage.

1. Grids are mounted in a Fischione dual-axis high tilt holder (Model 2040).
2. Insert the holder into the microscope (in this case a Tecnai F20) and align the microscope.
3. Select the magnification you will use for collecting the tilt series and perform a gain reference of the digital detector to remove inherent detector features (needs to be done on an empty grid square). In these experiments, we used a Gatan US4000 4K CCD camera.
4. Start SerialEM and set up imaging conditions in the *Camera and macro control* panel on the left hand side of the SerialEM display.

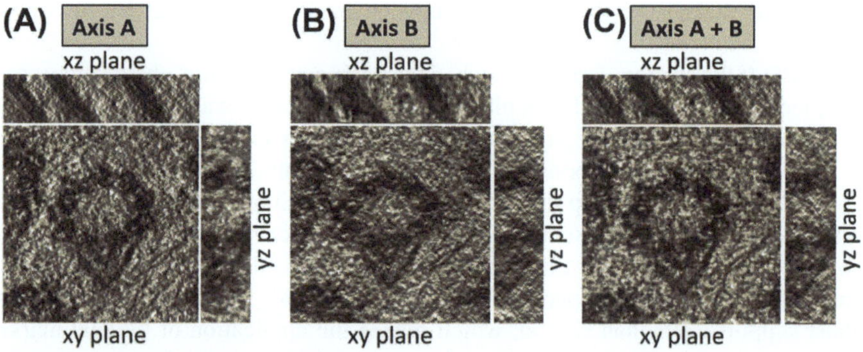

FIGURE 3 Single axis compared to dual axis.

10 section averages from axis A (A), axis B (B), and both axes combined (C) from a filtered volume shown in the xy, xz, and yz directions. The volumes were calculated from a tilt series collected from −65° to 63° in the A axis and −61° to 63° in the B axis. The B axis volume has been rotated by 90° around z so that it matches the orientation of axis A and the combined volume (for display purposes). The xz plane in axis A and the yz plane in axis B look inferior as they show the major effects of the missing wedge. In the combined volume, the xz and yz slices look similar as the missing wedge is now a missing pyramid.

5. Lower the magnification and find an ROI.
6. Set the ROI to eucentric height using the eucentric rough and fine routines under the *Tasks* menu.
7. Assess the tilt range that you can achieve for the ROI (before grid bars obscure it). We routinely use −66° to +66° with 2° angle steps.
8. Lower the magnification and leave the beam on the ROI until section shrinkage is complete and then return to the magnification used for data collection.
9. Under the *File* menu, select *Montage setup* and in the panel that appears, set the number of panels in x and y to 2. Click *Ok* and give the montage a filename, for example, *Tiltseries1a.st*, and click *Ok*.
10. Center your ROI at 0° tilt using a trial image (click *Trial* on the *Camera and macro control* panel and shift the image with right mouse button). Then use the *Walkup feature*, under the *Task* menu, go to the highest possible negative tilt angle (−66°).
11. Set up the tilt series under the *Tilt series* menu, making sure that the tilt range, tilt step, desired defocus, and tracking parameters are set up.
12. After the tilt series is finished, close the tilt series file using the *Tilt series* menu option *Terminate*.
13. After collecting the tilt series for axis A, the grid is rotated by 90° in the xy plane, using the dual-axis tilt holder, in order to collect the second, B axis, tilt series.
14. Repeat operations (5)–(12), except that the area should not need to be pre-irradiated and the montaged series is now given the filename *Tiltseries1b.st.*

NAP 1: The advantage of a Fischione dual-axis holder is that the position of the grid bars with respect to the tilt axis can be optimized by freely rotating the grid. Once this is done, you can lock the holder such that it will only rotate in 90° steps from that position.

NAP2: Follow your local instructions when aligning the microscope.

NAP3: This is very important for both CCD and direct electron detectors. It is also important when collecting montaged tomograms as any gradient across the image from a poorly gain referenced detector will be clearly visible.

NAP4: SerialEM needs to be correctly calibrated for automatic functions to work. Setting up the imaging conditions is done with the *Image setup* panel. For record mode, we routinely binned the CCD pixels by 2 and took a 1s exposure. For trial and focus modes, it is best to bin the CCD pixels by 4 and use shorter exposure times. The image shift calibration should be done at the magnification used to collect the tilt series (particularly important for montaged tilt series).

NAP5: Ciliated cells present either a ciliary set or an array of BBs depending on the orientation of the section. Transversal sections are useful for observing the cytoskeleton organization between BBs but the depth of view is limited by the section thickness. Longitudinal sections allow observation of the axonemes, the entire BBs and the cytoskeleton organization in the apical cortex.

NAP6: If the ROI is not at eucentric height, the area will move laterally out of view when the stage starts to tilt.

NAP7: If a grid bar appears before the end of the tilt range, you can change the rotation angle of the grid using the Fischione holder, or reduce the tilt range of the title series by a few degrees. Alternatively you can find another ROI.

NAP8: For montaged tomograms, increase the size of the pre-irradiated area to cover the entire ROI.

NAP9: Once the montage is set up, use the *Montage* control panel, on the left hand side of the SerialEM display, to collect a preview of the ROI. This can also be used to center your ROI.

NAP11: The defocus value used for resin sections is generally −0.5, which is enough to make sure that the sample will not be overfocused at high tilts. It is also a good idea to autofocus every 4°−6° up to −/+50° and every 2° up to −/+66°. The tracking was routinely done before and after autofocusing.

NAP13: In order to make sure that the B axis is collected using the same ROI as the A axis, read in the 0° tilt image from axis A using the *File* menu option *Read* (select 0° tilt image, usually around number 33) and rotate it by 90° using the *Process* menu option *Rotate left* or *Rotate right.*

2.5 DATA PROCESSING

2.5.1 Alignment and reconstruction of tilt series

IMOD is a well referenced and widely used software for the alignment and reconstruction of tomograms from either single or dual-axis tilt series (including montaged tilt series) (Kremer, Mastronarde, & McIntosh, 1996; Mastronarde,

1997). There is an excellent tutorial with a large troubleshooting section present on the software Web page (http://bio3d.colorado.edu/imod/). The next section will describe the alignment procedure for a dual-axis montaged tomogram using the sequential processing steps organized within the IMOD software.

1. In your terminal window, start *etomo*, IMOD's tomogram processing interface, and press the *New tomogram* button. Then select the tilt series for axis A in the *Data set name* box and press the *Scan the header* button. Then fill in the box with the size of the fiducial markers used, and make sure that both the dual axis and montage options are checked. Then press the *Create com script* button.
2. The *Pre-processing step* is used to remove any extreme pixels from the tilt series. Press the *Create fixed stack* button followed by the *Use fixed stack* button.
3. The *Coarse alignment* step aligns the tilt images in the tilt series to each other using cross-correlation. The *Calculate cross-correlation* button cross-correlates the tilt images, and then the aligned tilt series is generated using the *Generate coarse aligned stack* button. To view the aligned tilt series in 3dmod (IMOD's viewer), press the *View aligned stack* button.
4. The *Fiducial model generation* step is used to pick and track gold beads through the tilt series. The initial picking of the gold beads is done on the 0° tilt image after pressing the *Seed fiducial model* button. After picking the gold beads save the model in 3dmod, using the 3dmod *file* menu and the option *Save model.* Finally, use the *Track seed model* button to track them through the rest of the tilt series.
5. The *Fine alignment* step uses the tracked gold beads to find the optimum alignment between the tilt images. First, under the *General* tab, generate the global alignment and check that there is an even distribution of gold beads on both surfaces of the section using the *Compute alignment* button and *View 3D model* button. Next, if the beads are well distributed on both surfaces, tick the *Enable local alignments* box and go to the *Local variable* tab and under the *Local distortion solution type* select *Full solution* and then press the *Compute alignment* button. Finally, manually refine the position of the beads with the highest deviation from their predicted position, in order to minimize the error of the alignment using the *View/edit fiducial model* button and the *Bead fixer* menu (see **NAP5**).
6. The *Tomogram positioning* step corrects for any x-axis tilt, and/or tilts around the tilt axis and sets the top and bottom of the volume in the z direction. Press the *Create sample tomograms* button followed by the *Create boundary model* button. Then in 3dmod track the top and bottom of the section in each of the three volumes using two model points (open contour). Once this is done, save the model (as before). Then press the *Compute z shift and pitch angle* button and finally the *Create final alignment* button.
7. The *Final aligned stack* step uses the fine alignment to generate the final aligned tilt series. Press the *Create full aligned stack* button.

8. The *Tomogram generation* step uses back projection to generate the tomogram. Press the *Generate tomogram* button.
9. In the next series of steps, the operations from (2) to (8) are repeated with the second, B axis. To go to the B axis press the *Axis B* button.
10. The *Tomogram combination* step aligns and then combines the A and B axis volumes. Set the z axis minimum and maximum (see *NAP10*), then press the *Create combine scripts* followed by the *Start combine* buttons.
11. In the *Postprocessing* module, the combined volume is cropped, rotated around its x axis, and converted to 8-bit gray scale, to reduce the file size of the volume on disk. Initially press the *3dmod full volume* button, then using the *Rubber band tool* in the 3dmod window select the region of the tomogram (press the *Lo* and *Hi* buttons to exclude unwanted z sections and draw a box to crop in xy) and press the *Get xyz volume range from 3dmod* button. Then select the region for scaling the gray levels and press the *Get xyz subarea from 3dmod* button followed by the *Trim volume* button.

NAP1: You should create a separate directory for each tomogram, containing the tilt series from both axes. You will not need to fill in a box for axis B as IMOD will detect it directly if it is named properly. You should examine both axes of the tilt series using the *View raw image stack button*, so that any bad tilt images can be ignored during the alignment and reconstruction procedure.

NAP2: Extreme pixel values need to be truncated, as they can create lines radiating throughout the volume.

NAP3: If the cross-correlation procedure does not work you can try changing the low-pass filter radius in order to exclude more high frequencies. If the alignment is still poor you will need to use *Midas* to manually fix the misalignments. If you have problems with the alignment of the individual images in the montaged tilt image you can try to manually fix them using the *Fix edge with midas* button.

NAP4: It is advisable to pick gold beads evenly over the tilt image and to make sure to pick gold beads on both surfaces of the section. 3dmod has a nice display to help with selecting gold beads from both surfaces, as it displays the gold beads on the 0° tilt image in pink and the gold beads from the 8° image in green. Pink beads from one surface will have their equivalent green bead to the right, whereas beads from the opposite surface will have their equivalent green beads on the left. To activate this display, turn on the overlay function in the *Bead fix* menu. Make sure that the *Auto-centre* option is also selected, and that a new contour is started automatically for each picked gold bead. If there are any gaps, where a bead was not tracked on a tilt image, they can be filled in by pressing the *Use tracked beads as a starting model* button. If this does not fill all the gaps, then press the *Fix fiducial model* button, which will allow you to manually fill in the gaps. If it appears that a gold bead is lost from some of the tilt images, you can remove the contour for that bead (ctrl-D).

NAP5: When local alignments are used (particularly for montaged tilt series), the tilt series is divided up into user-defined patches and the best alignment solution for that patch determined. If the patches do not contain at least eight gold beads, their size will be increased. To solve for local specimen distortion, while using local alignments, at least three of the eight beads should be present on one side of the section. The *Distortion correction* adjusts for x-axis stretch and for rotation differences between x and y axis. If there are not enough beads or poorly distributed, do not use the distortion correction. The mean error reported by IMOD is the average distance in pixels between the measured position of the fiducial markers and their predicted position in the alignment solution. When trying to minimize this error, beads with an error above a user set threshold need to be checked in the *3dmod using the bead fixer* menu. For local alignments, you need to press the *Go to first local set* button followed by the *Go to next big residual* button in the bead fixer menu. This will take you to the first large residual in the first local area. Then if the suggested adjustment to the position of the gold bead is correct, press the *Move point by residual* button. If it is not then press *Undo move* button followed by the *Go to next big residual* button. Repeat this for all the residuals in all the local areas and then press the *Save and run tilt align* button. This procedure is iterative, so you may have to go through the residuals list multiple times.

NAP6: By default, three small volumes (display the different volumes using the *4th* dimension button in 3dmod) are calculated from the top, middle, and bottom of the tilt series. The top and bottom of the section need to be set for each volume reconstructed at the top, middle, and bottom of the tilt series. The default is 200 z-sections but this may not incorporate the whole section so you may need to be increased.

NAP8: Multiple central processing units (CPUs) should be used when generating the 3D volume, to speed up the process.

NAP9: During step 3, the gold beads will be initially transferred from axis A to axis B by pressing the *Transfer fiducials from other axis* button. At least 8–10 beads should be common between axis A and B (IMOD will write in the log window how many beads were transferred from axis A to axis B). To view the transferred gold beads, press the *View seed model* button. At this point, you can add more beads if not enough were transferred. The gold beads are tracked through the tilt series as before.

NAP10: It is important to set the z-limit to the number of z slices that contain clear features in the axis A volume. When tilt series are reconstructed, it is routine to reconstruct them in a larger volume, and therefore the first few and the last few z slices will normally be empty. You may also need to increase or decrease the size of patches that IMOD uses to match the volumes: the more feature-rich the volume, the smaller the patches that can be used. The program initially uses the gold beads to get a rough alignment, and then uses 3D cross-correlation between patches from the two volumes to optimize the match between them. If you have trouble in matching using the whole volume, you can also select a smaller region over which to match the two volumes.

NAP11: When selecting an area for scaling the output densities, make sure that it does not contain any fiducial markers as this will cause a compression of the gray levels used to represent your stained specimen in the 8-bit gray scale. Open the final volume and make sure that the densities are well spread over the 8-bit gray scale.

2.5.2 Filtering of the tomographic volume

For filtering the tomographic volumes we used nonlinear anisotropic diffusion (NAD), as it increases the contrast between the edges of stained membrane/protein and the background (Frangakis & Hegerl, 2001). The filter applies a user determined *K* value, which controls whether the gray values of the voxels will diffuse evenly in all directions or in the direction of the smallest density variation along edges such as membranes or proteins. The *K* value is applied in repeated iterations which act to smooth the filtered volume (Figure 4). The disadvantage to NAD filtering is that it is very slow and computer intensive. The following section will describe the NAD filtering procedure.

1. In your terminal window, start *Etomo* and press the *NAD* button. Then select the tilt series you want to filter.
2. Press the *View full volume* button, and then using 3dmod, select a small region of the tomogram using the *Rubber band* tool. Then press the *Get test volume range from 3dmod* button followed by the *Extract test volume* button.
3. Enter a range of *K* values (for example, 0.2, 2, 4, 8, 16, 32, 64, or see http://bio3d.colorado.edu/imod/doc/man/nad_eed_3d.html#TOP) into the box and press the *Run with different K values* button. To view the results in 3dmod, press the *View different K values test results* button.
4. Select the *K* value which gives the best contrast and perform a finer search around that value.

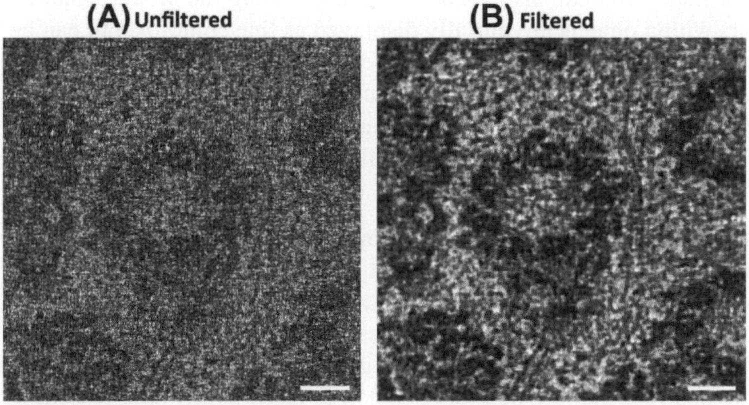

FIGURE 4 Filtering the tomographic volumes.

5 section averages shown in the xy plane of a volume reconstructed from a dual-axis tilt series, unfiltered (A) and filtered (B). The filtering was done using NAD (nonlinear anisotropic diffusion) using a *K* value of 32 with 15 iterations. Scale bar: 100 nm.

5. Once the *K* value is fixed, enter a number of different iterations (for example, 10, 15, 20, and 30) into the box and press the *Run with different iterations* button. To view the results in 3dmod, press the *View different iteration test results* button.
6. Enter the *K* value and number of iterations into the final box and press the *Filter full volume* button. This will take some time. Once done, you can view the result in 3dmod (*View filtered volume*).

NAP2: As the *K* value and the number of iterations used need to be empirically determined, a small volume (roughly 100 pixels cube) needs to be extracted from the tomogram. It is best to have a feature of interest in the subvolume.

NAP5: The more iteration you run, the smoother the resulting tomogram will be.

NAP6: This procedure breaks down the tomogram into smaller subvolumes, filters them, and then recombines them. You need to select the *overlap chunks* option when using NAD, otherwise you may see grid artifacts caused by the breaking up/stitching back together of the subvolumes. It is recommended to use multiple CPUs when filtering the whole volume.

2.5.3 Modeling/segmentation of features from tomograms

We used 3dmod to segment the volumes. 3dmod has a number of accessory viewers designed to help with segmenting tomograms such as the *Model viewer*, *Slicer*, and *x-y-z display*. *Slicer* is particularly useful, as it allows the user to average multiple slices and to look at oblique slices through the volume, which can be convenient when trying to orientate the ROI. 3dmod has three contour types that can be assigned to an object: *Open contours*, *Closed contours*, and *Scattered points*. The contours are displayed on the z section of the tomographic volume as you draw them. You can also view the contours using the *Model window* (crtl-V). Contours within objects are initially displayed as line contours, but can be meshed to produce smooth-rendered surfaces. We created one object with multiple closed contours per BB and basal foot cap (Figure 5). The MTs and actin filaments were two separate objects, each containing multiple open contours representing single MT or actin filaments (Figure 5). The following section will describe the segmentation procedure.

1. Display the tomogram in 3dmod. Then go to the z section in which you want to start segmenting your first component of interest.
2. Open up a new object (*3dmod menu > Edit > Object > New*), name the object and select the contour type you will use. We drew contours every third z-slice for BBs and basal foot caps. For straight MTs and actin filaments, we marked either end (use the middle mouse button to mark points on the section). For curved ones, we manually marked them along their entire length.
3. Once you have segmented the first component, you need to open a new object and select the contour type you want to use for the next component of interest.

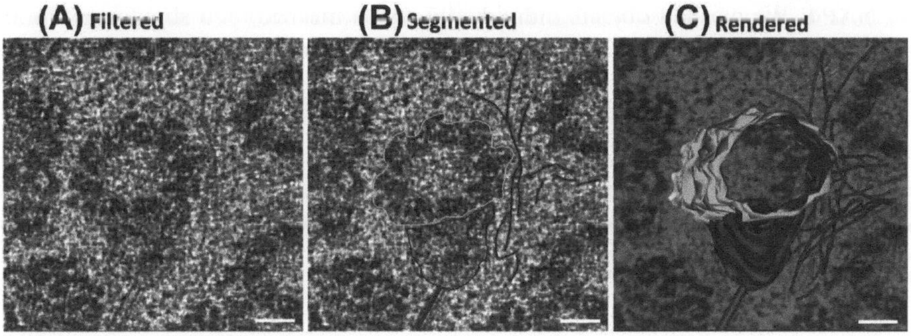

FIGURE 5 Segmenting tomograms.

Z section shown in the xy plane of a volume reconstructed from a dual-axis tilt series after nonlinear anisotropic diffusion filtering (A), then segmented (B), and rendered (C). The microtubules (MTs) are shown in green, the basal body (BB) in cyan, the basal foot in red, and the actin in gold. The MT and actin filaments were modeled with open contours, while the BB and basal foot cap were modeled with closed contours. The MT and actin filaments were rendered as tubes (20 pixel and 6 pixels, respectively), the BB and the basal foot were rendered as surfaces. Scale bar: 100 nm. (See color plate)

4. When all the objects are segmented, the objects are then meshed using the menu in the *Model window* (*Model menu > Edit > Object select the meshing tab*). The MT or actin filaments were meshed as tubes with a diameter of 20 and 6 pixels, respectively (pixel size was 1.1 nm) (Figure 5). The BBs and basal foot caps were meshed as a surface at every third z section.
5. In the *Model window*, using the *Edit > Image* menu, z-slices from the tomogram can be displayed with the rendered model. You should select an opacity that allows you to observe the model below the z-slices.

NAP1: If your component is difficult to track in the z sections, you can try using *Slicer* (*3dmod > Image menu > Option slicer*) to reorient the component, using different x, y, and z angles, so that it is easier to track. This was particularly useful when segmenting the MT and actin networks. Make sure the *Model* option is selected in the 3dmod control window.

NAP2: When generating models be careful to select the correct contour type for the particular component you are segmenting. 3dmod regularly autosaves your model file (**.mod* is the model file, and **.mod~* is the autosave).

NAP3: For *Open contours*, you can select the start and the end of a component in the volume, and then use the fill in z under the *3dmod menu > Object > Point*, which will plot the points between the start and end point of the component. This is particularly useful for segmenting MTs. When modeling multiple components using *Open contours*, such as multiple MTs or actin filaments, remember to initiate a new *Open contour* for each filament. When using *Closed contours*, as for the BBs, a new contour will be automatically opened on each new z section.

NAP4: We meshed objects individually. Once meshed, you should see the objects change in the *Model window*. The color of the object can be changed using the *Colour menu*.

3. MATERIALS

3.1 MOUSE DISSECTION

Materials:	Fine tweezers, Electron Microscopy Sciences (EMS): Cat. #72702-D
	Fine iris scissors, fine science tools (FST): Cat.#14094-11
	Vannas Spring Scissors, FST: Cat.#15000-03
	Teflon plate, home made
Reagents:	70% ethanol, VWR: Technisolve Cat. #CONFTBC570DE

3.2 TEM SAMPLE PREPARATION

Instrumentation:	Incubator at 60 °C, Memmert: Cat. #IF30
	Centrifuge at 25,000 g, Eppendorf: Cat. # Centrifgue 5417R
Materials:	Teflon plate, home made
	Single-edge razor blades, EMS: Cat. #71962
	Microtubes, Eppendorf: Cat. #0030 121.023
	BEEM embedding capsules, EMS: Cat. #69911-05
	Scalpels, Paramount: Cat. #SN-SM0501
Reagents:	25% glutaraldehyde, EMS: Cat. #16400
	4% osmium tetroxide, EMS: Cat. #19190
	30% ethanol, home made (with 0.2 μm filtered water and 95% ethanol)
	50% ethanol, home made (with 0.2 μm filtered water and 95% ethanol)
	70% ethanol, VWR: Technisolve Cat. #CONFTBC570DE
	80% ethanol, home made (with 0.2 μm filtered water and 95% ethanol)
	95% ethanol, VWR: Normapur Cat. #20823.362
	100% ethanol, Sigma: Cat. #51976
	Propylene oxide, EMS, Cat. #20412-20414
	Epon-815 resin, EMS: Cat. #14910
	DDSA, EMS: Cat. #13710
	Araldite CY212, TAAB: Cat. #E006
	DMP30, EMS: Cat. #13600

3.3 SAMPLE SECTIONING AND PREPARATION FOR ELECTRON TOMOGRAPHY

Instrumentation:	Ultramicrotome Leica EM UC7, Leica
Materials:	Diamond knife sectioning 45°, Leica: Diatome Cat. #16DIADU4530
	Diamond trimming knife, Leica: Diatome Cat. #16DIADTB45
	Perfect loop, Agar Scientific: Cat. #AGT5112
	Scalpels, VWR: Cat. #233-0034
	Formvar-coated copper finder grids, Agar Scientific: Cat. #S138-F1
	Hair on a cocktail stick, home made
	Whatman grade 4 filter paper, Fisher Scientific: Cat. #11372845
Reagents:	Protein-A-gold (10 nm), EMS: Aurion Cat. #25284
	0.2 μm filtered water

3.4 ELECTRON TOMOGRAPHY

Instrumentation:	FEI Tecnai F20 Electron microscope
	Gatan US4000 4K CCD camera
	Fischione dual-axis high tilt holder (Model 2040)
Software:	SerialEM (http://bio3d.colorado.edu/SerialEM/)
Materials:	Fine tweezers, Agar Scientific: Cat. #AGT5288

3.5 DATA PROCESSING

Instrumentation:	Computer
Software:	IMOD (http://bio3d.colorado.edu/imod/)

CONCLUSION

Analyzing the structure of the cell cortex using dual-axis electron tomography, we were able to extend previous results obtained by classical TEM and immunofluorescence analyses, and refine the organization of cytoskeletal networks at the bases of mouse motile cilia (Sandoz et al., 1988; Werner et al., 2011). It would be interesting to investigate whether this cytoskeleton arrangement also occurs at primary cilium and kinocilium bases (Ezan et al., 2013). Electron tomography has been used for different aspects of cilium biology, such as intraflagellar transport particles (Pigino et al., 2013, 2009). Recently Dixon and colleagues analyzed the axonemal MT translocation in cases of primary cilium dyskinesia (Burgoyne et al., 2014). However, the method we provide here is not restricted to ciliated cell analyses and could be easily adapted and used for other organelle or cytoskeleton investigations on mouse tissues.

LIST OF ABBREVIATIONS

BBs Basal bodies
BrBi Bronchial bifurcation
CPU Central processing unit
Dvl Dishevelled
EM Electron microscopy
Es Esophagus
IM Infrahyoid musculature
Ly Larynx
MT Microtubule
NAD Nonlinear anisotropic diffusion
PCP Planar cell polarity
RC Rib cage
ROI Region of interest
SG Submandibular salivary glands
SRF Serum response factor
Tr Trachea

ACKNOWLEDGMENTS

We are grateful to Helen Saibil (Birkbeck College, University College of London) for support (Welcome Trust, Cryoelectron tomography equipment grant (WT086018 and WT079605) awarded to H.S.) and comments on the manuscript. This work was supported by grants from GEFLUC, from the Fondation pour la Recherche Médicale (FRM), from the Association pour la Recherche contre le Cancer and from La Ligue, Comité de Paris (to D.D.).

REFERENCES

Brooks, E. R., & Wallingford, J. B. (2014). Multiciliated cells. *Current Biology, 24*, R973–R982.

Burgoyne, T., Lewis, A., Dewar, A., Luther, P., Hogg, C., Shoemark, A., et al. (2014). Characterizing the ultrastructure of primary ciliary dyskinesia transposition defect using electron tomography. *Cytoskeleton, 71*, 294–301.

Clare, D. K., Magescas, J., Piolot, T., Dumoux, M., Vesque, C., Pichard, E., et al. (2014). Basal foot MTOC organizes pillar MTs required for coordination of beating cilia. *Nature Communications, 5*, 4888.

Donohoe, B. S., Mogelsvang, S., & Staehelin, L. A. (2006). Electron tomography of ER, Golgi and related membrane systems. *Methods (San Diego, California), 39*, 154–162.

Dumoux, M., Clare, D. K., Saibil, H. R., & Hayward, R. D. (2012). Chlamydiae assemble a pathogen synapse to hijack the host endoplasmic reticulum. *Traffic (Copenhagen, Denmark), 13*, 1612–1627.

Ezan, J., Lasvaux, L., Gezer, A., Novakovic, A., May-Simera, H., Belotti, E., et al. (2013). Primary cilium migration depends on G-protein signalling control of subapical cytoskeleton. *Nature Cell Biology, 15*, 1107–1115.

Frangakis, A. S., & Hegerl, R. (2001). Noise reduction in electron tomographic reconstructions using nonlinear anisotropic diffusion. *Journal of Structural Biology, 135*, 239–250.

Guirao, B., Meunier, A., Mortaud, S., Aguilar, A., Corsi, J. M., Strehl, L., et al. (2010). Coupling between hydrodynamic forces and planar cell polarity orients mammalian motile cilia. *Nature Cell Biology, 12*, 341–350.

Hildebrandt, F., Benzing, T., & Katsanis, N. (2011). Ciliopathies. *The New England Journal of Medicine, 364*, 1533–1543.

Ishikawa, H., & Marshall, W. F. (2011). Ciliogenesis: building the cell's antenna. *Nature Reviews, 12*, 222–234.

Kishimoto, N., & Sawamoto, K. (2012). Planar polarity of ependymal cilia. *Differentiation, 83*(2), S86–90.

Knowles, M. R., & Boucher, R. C. (2002). Mucus clearance as a primary innate defense mechanism for mammalian airways. *The Journal of Clinical Investigation, 109*, 571–577.

Kremer, J. R., Mastronarde, D. N., & McIntosh, J. R. (1996). Computer visualization of three-dimensional image data using IMOD. *Journal of Structural Biology, 116*, 71–76.

Kunimoto, K., Yamazaki, Y., Nishida, T., Shinohara, K., Ishikawa, H., Hasegawa, T., et al. (2012). Coordinated ciliary beating requires Odf2-mediated polarization of basal bodies via basal feet. *Cell, 148*, 189–200.

Lee, L. (2013). Riding the wave of ependymal cilia: genetic susceptibility to hydrocephalus in primary ciliary dyskinesia. *Journal of Neuroscience Research, 91*, 1117–1132.

Marshall, W. F., & Kintner, C. (2008). Cilia orientation and the fluid mechanics of development. *Current Opinion in Cell Biology, 20*, 48–52.

Mastronarde, D. N. (1997). Dual-axis tomography: an approach with alignment methods that preserve resolution. *Journal of Structural Biology, 120*, 343–352.

Mastronarde, D. N. (2005). Automated electron microscope tomography using robust prediction of specimen movements. *Journal of Structural Biology, 152*, 36–51.

Mitchell, B., Jacobs, R., Li, J., Chien, S., & Kintner, C. (2007). A positive feedback mechanism governs the polarity and motion of motile cilia. *Nature, 447*, 97–101.

Nigg, E. A., & Raff, J. W. (2009). Centrioles, centrosomes, and cilia in health and disease. *Cell, 139*, 663–678.

Nordgren, T. M., Wyatt, T. A., Sweeter, J., Bailey, K. L., Poole, J. A., Heires, A. J., et al. (2014). Motile cilia harbor serum response factor as a mechanism of environment sensing and injury response in the airway. *American Journal of Physiology Lung Cellular and Molecular Physiology, 306*, L829–L839.

Noske, A. B., Costin, A. J., Morgan, G. P., & Marsh, B. J. (2008). Expedited approaches to whole cell electron tomography and organelle mark-up in situ in high-pressure frozen pancreatic islets. *Journal of Structural Biology, 161*, 298–313.

Park, T. J., Mitchell, B. J., Abitua, P. B., Kintner, C., & Wallingford, J. B. (2008). Dishevelled controls apical docking and planar polarization of basal bodies in ciliated epithelial cells. *Nature Genetics, 40*, 871–879.

Pigino, G., Cantele, F., Vannuccini, E., Lanzavecchia, S., Paccagnini, E., & Lupetti, P. (2013). Electron tomography of IFT particles. *Methods in Enzymology, 524*, 325–342.

Pigino, G., Geimer, S., Lanzavecchia, S., Paccagnini, E., Cantele, F., Diener, D. R., et al. (2009). Electron-tomographic analysis of intraflagellar transport particle trains in situ. *The Journal of Cell Biology, 187*, 135–148.

Randell, S. H., & Boucher, R. C. (2006). Effective mucus clearance is essential for respiratory health. *American Journal of Respiratory Cell and Molecular Biology, 35*, 20–28.

Sandoz, D., Chailley, B., Boisvieux-Ulrich, E., Lemullois, M., Laine, M. C., & Bautista-Harris, G. (1988). Organization and functions of cytoskeleton in metazoan ciliated cells. *Biology of the Cell/Under the Auspices of the European Cell Biology Organization, 63*, 183–193.

Shah, A. S., Ben-Shahar, Y., Moninger, T. O., Kline, J. N., & Welsh, M. J. (2009). Motile cilia of human airway epithelia are chemosensory. *Science (New York, NY), 325*, 1131–1134.

Singla, V., & Reiter, J. F. (2006). The primary cilium as the cell's antenna: signaling at a sensory organelle. *Science (New York, NY), 313*, 629–633.

Veland, I. R., Awan, A., Pedersen, L. B., Yoder, B. K., & Christensen, S. T. (2009). Primary cilia and signaling pathways in mammalian development, health and disease. *Nephron, 111*, p39–53.

Vladar, E. K., & Axelrod, J. D. (2008). Dishevelled links basal body docking and orientation in ciliated epithelial cells. *Trends in Cell Biology, 18*, 517–520.

Vladar, E. K., Bayly, R. D., Sangoram, A. M., Scott, M. P., & Axelrod, J. D. (2012). Microtubules enable the planar cell polarity of airway cilia. *Current Biology, 22*, 2203–2212.

Wallingford, J. B. (2010). Planar cell polarity signaling, cilia and polarized ciliary beating. *Current Opinion in Cell Biology, 22*, 597–604.

Werner, M. E., Hwang, P., Huisman, F., Taborek, P., Yu, C. C., & Mitchell, B. J. (2011). Actin and microtubules drive differential aspects of planar cell polarity in multiciliated cells. *The Journal of Cell Biology, 195*, 19–26.

Wright, N. G., Brown, R. M., McCandlish, I. A., Thompson, H., & Cornwell, H. J. (1983). Patterns of cilia formation in the lower respiratory tract of the dog: a scanning electron microscopic study. *Research in Veterinary Science, 34*, 340–346.

CHAPTER

Tackling centrosome biology through gene targeting in chicken B cells

6

Pavithra L. Chavali, Fanni Gergely[1]

Li Ka Shing Centre, Cancer Research UK Cambridge Research Institute, Cambridge, UK

[1]*Corresponding author: E-mail: fanni.gergely@cruk.cam.ac.uk*

CHAPTER OUTLINE

Introduction **84**
1. Methods **86**
1.1 Genetic Manipulation of DT40 Cells 86
1.2 Design and Preparation of Targeting Constructs 88
1.2.1 Notes 88
1.2.2 Materials 88
1.3 Generation of Heterozygous Mutant Cells 89
1.3.1 Notes 90
1.3.2 Materials 90
1.4 Generation of Homozygous Mutant Cells 91
1.4.1 Notes 91
1.5 Excision of Antibiotic Resistance Genes by Cre Recombinase 91
1.5.1 Notes 92
1.5.2 Materials 92
1.6 In Situ Tagging of One or Both Alleles 92
1.6.1 Notes 93
2. Analysis of DT40 Cells **93**
2.1 Immunofluorescence 95
2.1.1 Notes 97
2.1.2 Materials 97
2.2 Time-lapse Imaging 97
2.2.1 Notes 98
2.2.2 Materials 98

Methods in Cell Biology, Volume 129, ISSN 0091-679X, http://dx.doi.org/10.1016/bs.mcb.2015.03.008

2.3 Centrosome Purification from DT40 Cells 98
2.3.1 Notes 99
2.3.2 Materials 99
Conclusion **100**
Acknowledgments **100**
References **100**

Abstract

The centrosome proteome comprises hundreds of proteins whose function at the organelle and in the cellular context is unknown. Loss-of-function studies present a powerful tool to probe the roles of these individual constituents and hence improve our insight into key questions of centrosome biology such as how centrosomes are built, how they duplicate, and which cellular processes they partake in. In cultured cells ribonucleic acid (RNA) interference remains the most widely used method to achieve protein depletion, but due to the remarkable stability of many centrosome components depletion is often incomplete. In such instances genome editing provides a viable alternative. The exceptionally high homologous recombination rate of chicken DT40 cells makes this lymphocytic cell line ideal for genetic manipulation. Here we describe methods for the design and generation of knockouts and in situ tagging of genes in these cells. Furthermore, we report an optimized technique that allows isolation of centrosomes from DT40 cells for use in in vitro functional assays and proteomic analysis. Gene editing by CRISPR-Cas9 technology is fast replacing RNA interference as a method of choice for loss-of-function studies, but the combination of the fast cell cycle, the robustness in culture and ease of gene targeting, will continue to make DT40 cells a useful model system for studies of vertebrate protein function.

INTRODUCTION

Loss-of-function techniques are extensively used to probe gene function in cells and organisms. Ribonucleic acid interference (RNAi) is the most popular method to achieve protein depletion, but despite its universal applicability, it has some major drawbacks. Of these, incomplete depletion and off-target effects are particularly serious shortcomings. Furthermore, RNAi is not well suited for elucidating the roles of protein domains and single amino acids, since in such cases RNAi needs to be combined with expression of transgenes carrying the desired deletions or mutations, a setup that can generate significant experimental noise. The ability to manipulate the endogenous locus of a gene offers solutions to many of these drawbacks, but it requires the use of genetically tractable model systems. Homologous recombination is an efficient tool of genome engineering, and thus models that provide a high targeted to random integration ratio have proven extremely useful for reverse genetic studies, a prime example being yeast (Orr-Weaver, Szostak, & Rothstein, 1981). The majority of vertebrate cells however exhibit the exact opposite profile, namely high random integration and very low targeted integration rates, with the notable exceptions of mouse embryonic stem (ES) cells and the chicken DT40 B cell line (Bollag, Waldman, & Liskay, 1989; Buerstedde & Takeda, 1991; te Riele, Maandag, & Berns, 1992). Although ES cells have been invaluable in generating transgenic animals, spontaneous differentiation and nonstraightforward culture conditions render these cells less well suited for in vitro studies. By contrast, DT40 cells grow rapidly and have both a stable karyotype and phenotype in culture, thereby providing almost limitless amount of isogenic cell material (Molnar et al., 2014). Another advantage of DT40 cells is that the size of targeting constructs do not need to exceed ~10 kb, which makes for easy propagation of the plasmid in bacteria. This is because homology arms of 1–3 kb are usually sufficient for homologous recombination to occur. Introns are also shorter in the chicken genome than in mammals, and this facilitates the targeting of several exons within a single recombination event (Burt, 2005). Finally, DT40 cells lack functional p53, which makes it possible to generate genotypes that predispose cells to genome instability (Yamazoe, Sonoda, Hochegger, & Takeda, 2004). Indeed, studies in DT40 have been critical to our understanding of deoxyribonucleic acid (DNA) repair mechanisms, chromatin organization and centromeric gene function (Mocciaro et al., 2010; Nakayama & Takami, 2001; Nishihashi et al., 2002; Regnier et al., 2005; Shang et al., 2013; Sonoda et al., 2001; Sonoda et al., 1999; Sudo et al., 2001; Takami, Takeda, & Nakayama, 1997; Takata et al., 1998; Vagnarelli et al., 2006).

DT40 cells are derived from bursal lymphomas induced by Avian Leucosis Virus in *Gallus gallus domesticus* (Baba, Giroir, & Humphries, 1985). The potential use of this cell line for genome engineering came to the fore when studies uncovered a high ratio of targeted versus random integration, orders of magnitudes greater than in mammalian cells (Buerstedde & Takeda, 1991). DT40 cells are ~10 μm in diameter with a large nucleus to cytoplasm ratio. They grow in suspension, exhibiting a generation time of 8–10 h when kept at 40 °C, with a small increase in doubling time at 37 °C. DT40 cells have a relatively stable karyotype; they contain $2n = 80$ chromosomes that include two sex chromosomes: ZW. Of these chromosomes 1–4 and Z are macrochromosomes, whereas the others are minichromosomes, characteristic of birds. DT40 cells are predominantly diploid with the exception of the trisomic chromosome 2 (Molnar et al., 2014). Targeting construct design in DT40 has been aided by the availability of the chicken genome sequence along with cDNA sequences in public databases such as BBSRC ChickEST and U.D. Chick EST. The high homologous recombination rate together with the ease of propagation make DT40 ideal for characterizing the biochemical roles of protein domains and residues. Indeed, large molecular complexes, protein interactomes and even whole organelle proteomes have been successfully identified from wild type and various mutant DT40 cell lines (Hall, Hester, Griffin, Lilley, & Jackson, 2009; Mosedale et al., 2005; Ohta et al., 2010). In the context of centrosome biology, the use of DT40 cells have proved to be equally valuable, helped by the fact that centrosome morphology and centrosomal genes are highly conserved between vertebrates and mammals (Barr, Kilmartin, & Gergely, 2010; Dantas, Wang, Lalor, Dockery, & Morrison, 2011; Inanc et al., 2013; Shang et al., 2013; Sir et al., 2011, 2013; Wang, Dantas,

Lalor, Dockery, & Morrison, 2013). Moreover, by generating multiple independent acentriolar DT40 cell lines we have recently shown that centrosomes are not essential in DT40 cells, and thus the cell line represents an ideal tool for studying centrosomal genes without the need for inducible knockout approaches (Sir et al., 2013).

Below we describe detailed methodology for generating deletions and in situ tagging of endogenous genes in DT40 cells and outline a protocol optimized for centrosome isolation from these chicken cells.

1. METHODS

1.1 GENETIC MANIPULATION OF DT40 CELLS

Successful gene targeting requires careful design of a targeting strategy. Although deletion of the entire coding sequence is the best way to create protein null mutants, this is not always possible for genes that span large genomic distances. In such instances, a knockout can be achieved by frameshifting the transcript or by introducing a premature STOP codon into the sequence. Targeting constructs normally comprise an antibiotic resistance gene flanked by a left and right homology arms (termed LA and RA, respectively)(Figure 1). Regions of up to 7 kilobases (kb) have been successfully deleted and replaced by antibiotic resistance genes in our laboratory. Typically the greater the region to be deleted the less efficient homologous recombination is, although this can be improved by the use of longer homology arms, the trade-off being the preparation of a larger targeting construct. Ideally both homology arms should be greater than 1.5 kb: we usually pair arms of ~1.5–2 kb with arms of 3–4 kb. Once the region to be deleted is identified, a targeting construct is designed. The two arms are amplified from DT40 genomic DNA using polymerase chain reaction (PCR) and cloned in a sequential manner into pBluescript vector followed by antibiotic resistance cassette of choice. To target both alleles, constructs with identical homology arms, but different antibiotic resistance cassettes must be generated. Antibiotic resistance genes flanked by mutant loxP sites are available, which allow removal of the resistance genes through transient expression of cre recombinase (Arakawa, Lodygin, & Buerstedde, 2001). For generating conditional knockouts, DT40 cells carrying inducible cre recombinase can be used, a strategy particularly valuable when a target gene is essential for cell survival and proliferation (Arakawa et al., 2001). Although DT40 cells exhibit a high homologous targeting rate, it is not nearly high enough to obtain homozygously targeted cells in a single step. Thus, cells need to be sequentially targeted; first cells become heterozygous and then retargeting these heterozygous cells lead to homozygous colonies. Heterozygous cells are often helpful in studies and these are more straightforward to obtain by homologous recombination in DT40 cells than through CrispR-Cas9 technology. We normally transfect homozygously targeted cells with cre recombinase to excise antibiotic resistance cassettes (flanked by loxP sites), but if desired this can be carried out in heterozygous cells. For example, if only a

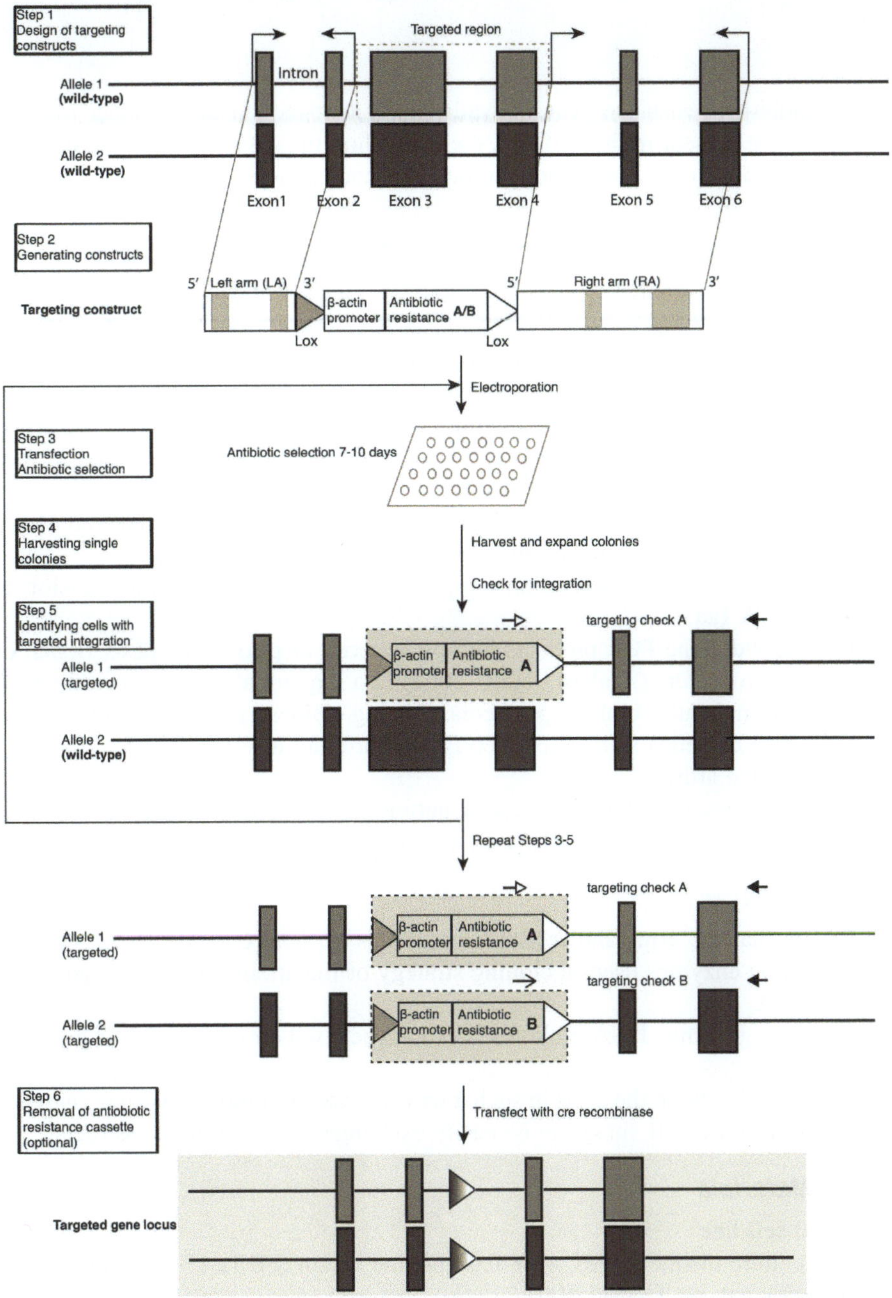

FIGURE 1 Outline of gene targeting in DT40 cells.

When modifying both alleles of a gene, different antibiotic resistance genes must be used for targeting, which is depicted here as antibiotic resistance genes A and B. For checking integration to genomic locus, PCR primers specific to antibiotic resistance cassettes A and B need to be used. Since β-actin promoter is difficult to amplify, design primers to the 3′ of the cassette and pair these with primers recognizing sequences outside of RA or LA depending on orientation of cassettes in the construct. In homozygous cells, successful gene targeting can be confirmed using a primer pair against the deleted region.

single antibiotic resistance gene is available for transfection due to other markers already being present in the cell line, then recycling the cassette from heterozygous cells could be the only viable mode to obtain a homozygous cell line.

1.2 DESIGN AND PREPARATION OF TARGETING CONSTRUCTS

1. Design and synthesize primers for the left and the right arms (RA and LA, respectively). Gene targeting strategy can be devised using genome browsers such as http://www.ensembl.org/Gallus_gallus/Info/Index. Primers must include restriction enzyme sites for subcloning of arms. Care must be taken to identify a unique restriction site within the vector that falls outside of the arms and antibiotic resistance genes, as this will be required for linearization of the final targeting construct.
2. Extract genomic DNA from DT40 cell line using Puregene cell and tissue isolation kit as per manufacturer's instructions.
3. PCR amplify LA and RA using a high-fidelity polymerase such as Phusion or Takara LA Taq.
4. Extract and clone PCR products into pBluescript or pLox Neo, pLox Blasti, or pLox Puro vectors (Arakawa et al., 2001) in a sequential manner according to standard molecular biology protocols. If using pBluescript, excise antibiotic resistance cassettes from pLox Neo/Blasti/Puro and insert these into construct between the arms.
5. Confirm presence of LA, RA, and antibiotic resistance cassette in the final targeting construct using restriction digest, PCR, and sequencing.

1.2.1 Notes

- Ensure that the arms and resistance cassettes do not contain internal sites for restriction enzymes used in cloning strategy or linearizing the final targeting construct.
- In-house genomic DNA isolation kits can be used to extract DT40 genomic DNA.
- We normally clone the arms in such a way that the antibiotic resistance cassettes (flanked by BamHI sites) can be easily exchanged between the constructs.

1.2.2 Materials

DT40 cell line
Cell culture media (RPMI, 10% fetal bovine serum (FBS), 1% chicken serum (Biosera, cat. no. CH515-500)
T75, T25 flasks and 24-well and 96-well flat bottom plates (tissue culture treated)
Cell culture incubator (40 °C)
Puregene cell kit for genomic DNA extraction (Qiagen, cat. no.1042601)
Primers to amplify left and right arms
PCR tubes
pBluescript or similar vector (~3–4 kb) appropriate for restriction sites

pLoxNeo/Puro/Blasti vectors
Cloning reagents (T4 ligase, restriction enzymes, antibiotics, *Escherichia coli* strains for transformation)
DNA and protein electrophoresis systems and buffers

1.3 GENERATION OF HETEROZYGOUS MUTANT CELLS

1. After thawing DT40 cells, culture them for at least 4–7 days prior to transfection in RPMI supplemented with 10% FBS and 1% chicken serum, 4 μL β-mercaptoethanol.
2. Linearize at least 60 μg of the targeting construct using the appropriate restriction enzyme for 3–4 h or overnight.
3. Purify the linearized vector by ethanol precipitation and resuspend in 300 μL sterile 1X phosphate buffered saline (PBS) at 55 °C for 1 h.
4. Centrifuge 2×10^7 cells at 1000 rpm and wash with ice cold PBS. Resuspend cell pellet in 500 μL cold PBS.
5. Add 300 μL of linearized vector to the cell suspension and leave on ice for 10 min.
6. Transfer to a prechilled 4-mm electroporation cuvette and pulse using Biorad gene pulser at 550 V, 25 μF.
7. Transfer the cuvette to ice and leave for 10 min.
8. Meanwhile prepare 70-mL warm DT40 culture media.
9. Transfer the contents of the cuvette gently into the media and mix gently.
10. Aliquot 100 μL of this suspension per well into appropriate number of 96-well plates and incubate overnight at 40 °C.
11. Twenty to twenty-four hours later, add 100 μL of the appropriate antibiotic (made up as a 2X solution) and incubate the plates for 7–10 days or until the colonies become visible.
12. When a well with a single colony is identified, after gentle pipetting its entire content is transferred into 300 μL fresh medium. We normally use 24-well plates for this initial expansion of the clones. An additional 300 μL fresh medium is added after ~1–2 days to the wells.
13. After 3–4 days or when cells reach appropriate density, cells will be split between two 24-well plates. About 50–100 μL of cell suspension will be placed on the first plate to be continuously cultured. The second plate is to provide a frozen stock. This is prepared by transferring 300 μL of cell suspension per well onto a prechilled plate where each well contains 300 μL media and 70 μL DMSO. After ~30 min on ice, freeze plate at −80 °C for storage of up to 2 months. The remaining cell suspension can be used for extraction of genomic DNA on the same day or if required placed on a third plate to be expanded for a few more days in culture.
14. Genomic DNA is isolated from each clone and analyzed by PCR. Correct integration is checked for using diagnostic primers. These are designed so that one recognizes a sequence within the resistance cassette, whereas the other

recognizes the genomic region just outside an arm. Since the antibiotic resistance cassettes can be cloned in forward or reverse orientation in the targeting construct, this dictates the choice of primer pairs. Note that the β-actin promoter contains G-C rich sequences that are difficult to amplify by PCR.

15. Clones with correct integration of antibiotic resistance genes are expanded for second round of gene targeting.

1.3.1 Notes

- Digestion is performed for at least 3 h or overnight using high-fidelity restriction enzymes in a 500 μL reaction mixture. Run 5 μL of the digestion mixture on an agarose gel to confirm complete linearization before ethanol precipitation.
- We use 50 μL of 3 M sodium acetate (pH 5.2) and 2.5 volumes of 100% ethanol, incubate at −20 °C for a minimum of 1 h before precipitating the DNA by centrifugation. This DNA is washed again in 70% ethanol to remove excess salt, then dried and solubilized in sterile PBS.
- DT40 cells should be passaged the day before transfection to ensure a viability of over 90% at the time of transfection.
- Performing a kill curve for each antibiotic is highly recommended even when different batches are to be used from the same supplier. As for guidance we use the following final antibiotic concentrations for selection: Neomycin (G418, Life Technologies, cat. no. 10131035) 1.5 mg/mL; Puromycin (Acros Organics, cat. no. 227420100) 0.5 μg/mL; Blasticidin (Acros Organics, cat. no. BP-2467-25) 30 μg/mL.
- It is not recommended to disturb the plates for the first 7 days of incubation as this can result in cells moving around in the well giving the appearance of multiple clones being present per well.
- It is crucial to follow the guidelines for cell numbers and plating volume, as the plating of too many cells cause formation of more than one colony per well.
- When transferring colonies into a 24-well plate, use fresh pipette tips for each well to avoid mixing of cell populations.
- It is crucial to establish the orientation of the antibiotic resistance gene in the targeting vector before diagnostic PCRs on genomic DNA.
- The β-actin promoter contains G-C rich sequences that are difficult to amplify by PCR, and this needs to be taken into account when designing diagnostic primer pairs.

1.3.2 Materials

Cell counter (hemocytometer/Vi-Cell)
Primers to check integration
Sterile PBS
Biorad Gene pulser and cuvettes (4 mm thick)

Selection antibiotics: G418 (Life Technologies, cat. no. 10131035), Puromycin (Acros Organics, cat. no. 227420100), Blasticidin S HCl (Acros Organics, cat. no. BP-2467-25)

1.4 GENERATION OF HOMOZYGOUS MUTANT CELLS

1. The targeting of the second allele is carried out as described under the previous section "Generation of heterozygous mutant cells." This targeting is performed in the heterozygous cell line and thus requires a construct with a different antibiotic resistance gene to the one already present in the first allele.
2. For selection, we recommend adding this new antibiotic together with the antibiotic used to select heterozygous cells. This ensures that all positive colonies will retain the original antibiotic resistance gene in their first targeted allele. If only the second antibiotic is used, half the colonies showing correct integration will have their first allele retargeted with the new antibiotic resistance gene and thus remain heterozygously targeted.
3. The plates are closely monitored for colony appearance after 7–10 days. If the gene to be knocked out is considered important for cell growth, colonies need to be monitored for up to 14 days, since homozygous clones may be smaller and take longer to become visible.
4. Single colonies are then transferred to 24-well plates for expansion as before and screened for correct integration of the antibiotic resistance genes in both alleles.

1.4.1 Notes

- If no positive clones are obtained by day 10–14 after transfection, check that the correct antibiotic resistance genes are present in the constructs and that antibiotics are used at appropriate concentrations.
- If random integration occurs at similar frequency during first and second allele targeting, yet the second transfection fails to result in correct targeting event, this could indicate that the gene of interest is essential for cell viability. In such cases, inducible knockouts can be generated.
- Testing for homozygous integration after second targeting can also be carried out using PCR primers against the deleted region, confirming absence of this genomic sequence. Regardless of screening strategy, it is good practice to do this PCR reaction to exclude rare cases of polyploid cells being targeted.
- If antibodies are available to the protein of interest, targeting can be confirmed by western blotting or immunofluorescence.

1.5 EXCISION OF ANTIBIOTIC RESISTANCE GENES BY CRE RECOMBINASE

Once correct targeting of one or both alleles is confirmed, the antibiotic resistance cassette(s) can be excised. Antibiotic resistance genes can interfere with transcription, alter splicing pattern, and frameshift the transcript, so in general it is preferable to remove them. This is particularly important when the aim is to delete specific domains of proteins rather than to create protein null cell lines. The method below describes how to excise antibiotic resistance genes flanked by loxP sites (Figure 1).

1. Wash 2×10^6 DT40 cells in PBS and resuspend in 100 μL of Ingenio electroporation solution (Mirus Bio LLC, cat. no. MIR50114).
2. Add 15 μg of plasmid encoding cre recombinase under appropriate promoter for chicken cells.
3. Transfect cells by placing in Amaxa nucleofector, program B-009.
4. Immediately resuspend cells in 5 mL prewarmed media in a 6-well plate and incubate for 24 h.
5. Serially dilute cells in DT40 medium such that the density of 1 cell per 100 μL is achieved. Plate 100 μL of this mix into each well of 96-well plate.
6. Incubate plates for a week at standard growth conditions without disturbing them.
7. Select wells with single colonies and transfer these one by one into a 24-well plate. We normally select 24 colonies.
8. Expand cells for 3–4 days and replica plate them into two 24-well plates. Keep 1 plate for expanding cells and use 2 plates for screening colonies.
9. Since cre recombinase excises antibiotic resistance genes, cells that transiently expressed the recombinase should no longer be resistant to original antibiotics. Thus, the easiest way to screen colonies is to treat one plate with the first antibiotic used for selection of heterozygous cells and the other plate with the second antibiotic used for generation of homozygous cells and identify cell clones that lost resistance to both antibiotics.
10. Expand these clones and confirm loss of antibiotic resistance cassettes by PCR. It is also worth sequencing the region around the loxP sites to ensure that recombination took place as expected.

1.5.1 Notes

- Instead of Ingenio electroporation solution, Amaxa solution T can also be used.
- Warming the electroporation solution before use yields better transfection efficiency.

1.5.2 Materials

Amaxa nucleofector
Ingenio electroporation solution (Mirus Bio) or Amaxa Solution T

1.6 IN SITU TAGGING OF ONE OR BOTH ALLELES

Tagging a gene in situ allows us to visualize protein localization in fixed or live cells or to perform affinity purification of endogenous macromolecular complexes (Mosedale et al., 2005). In situ tagging is particularly useful in chicken cells, where species-specific antibodies are difficult to obtain commercially. Tags such as Flag, HA, or GFP have all been used successfully in our laboratory along with tandem affinity purification tags (Sir et al., 2011).

The methodology to in situ tag genes is largely the same as generating knockouts, except for the design of the constructs. In the case of gene deletions, the LA and RA are normally chosen so that the sequence between the arms is deleted upon recombination of the antibiotic resistance cassette into this region. When in situ tagging genes at the N-terminus, the tag must be present between the antibiotic resistance cassette and the RA in the construct, whereas inserting a tag at the C terminus requires the tag to be located between the LA and the antibiotic resistance gene. It is paramount that the design takes reading frames into consideration and when inserting a tag to the C terminus the STOP codon must be removed from the endogenous gene and be present in the tag.

Figure 2 depicts the strategy to generate C-terminal epitope tagging in DT40. Briefly, LA and RA are designed such that they flank the STOP codon of the gene of interest. Although it requires an extra cloning step, the 3′ untranslated region (UTR) of the gene can be included in the LA, ensuring that any potential regulatory sequences encoded in this region are contained in the tagged transcript. LA and RA are sequentially cloned into pBluescript vector with the appropriate tag inserted in between the arms followed by antibiotic resistance cassette of choice. The rules for designing LA and RA are the same as before; the arms have to be long enough to allow targeted homologous recombination.

Gene targeting with these constructs is performed as in previous sections; heterozygous cells can be obtained after one round of transfection, and if desired, these can be targeted again to get homozygous cells. As before, constructs containing different antibiotic resistance genes must be used for the two targeting events.

1.6.1 Notes

- You may choose to in situ tag genes in various mutant DT40 cell lines, so care must be taken that each construct to be transfected uses a distinct antibiotic resistance gene from the ones already present or that antibiotic resistance genes are recycled prior to a new transfection.
- Instead of screening by PCR, positive clones can be expanded and analyzed for expression of tagged proteins using western blots or immunofluorescence.
- We find that in most cases C-terminal tagging works well; however, N terminal tagging might sometimes be preferable. In such cases care should be taken not to disrupt the gene promoter and the 5′ UTR, which can impede transcription and translation.

2. ANALYSIS OF DT40 CELLS

If gene targeting is successful, one can proceed to establish the cellular phenotype of the targeted DT40 cells. In our experience, chicken cells are amenable to all standard

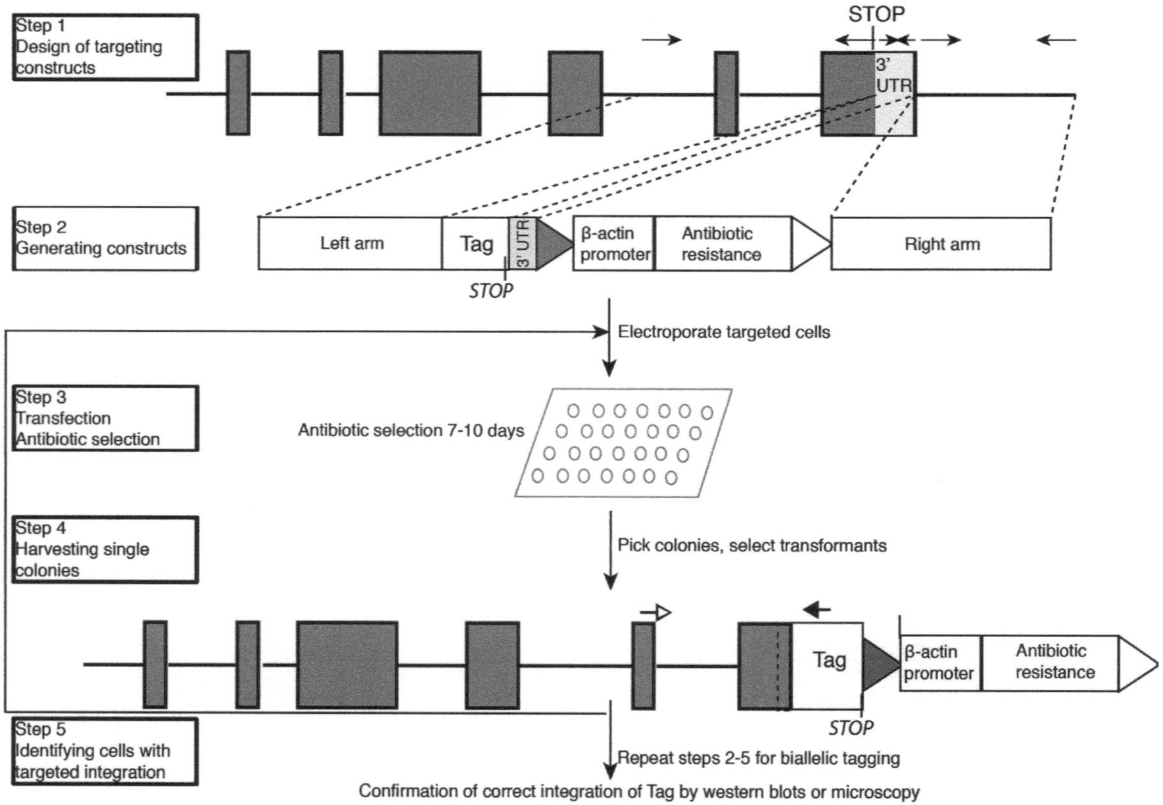

FIGURE 2 Outline of in situ gene tagging in DT40 cells.

When aiming for a C-terminal tag, the STOP codon of the gene needs to be removed. To this end, the left arm should terminate just before the STOP codon so that the last exon is in frame with the tag. 3′ UTR of the gene can be inserted after the tag in construct as depicted.

biochemical, molecular, and cell biology techniques. The lack of commercially available antibodies against chicken proteins may be a limitation, but this can be overcome by in situ gene tagging as described in Section 1.6. Moreover, the nonadherent nature and small size of DT40 cells can make visualization challenging. Hardly twice the size of budding yeast, chicken B cells will never be the ideal tool for high-resolution time-lapse microscopy, nonetheless the approaches described here will allow imaging both fixed and live DT40 cells.

The lymphoblastic cell line KE-37 has been used for uncovering the human centrosome proteome, but gene manipulation in this cell line is not trivial (Andersen et al., 2003; Bornens & Moudjou, 1999). By contrast, gene targeting in DT40 cells is straightforward and therefore it is feasible to combine proteomic approaches with gene editing in these cells. To this end, we have optimized existing centrosome isolation protocols for DT40 cells and provide the detailed methodology below (Figure 3). In vitro functional assays can also be performed on purified centrosomes. Collectively, these techniques enable comparisons between protein composition, phosphorylation patterns, and microtubule nucleation capacity of centrosomes isolated from wild-type and mutant DT40 cells.

2.1 IMMUNOFLUORESCENCE

1. Coat coverslips with 2 mg/mL poly-L-lysine by placing a large drop onto coverslip for 30 min at 37 °C.
2. Wash coverslips extensively with Milli-Q water and air-dry.
3. Take 1–2 × 10^6 cells and spin them at 900 g for 5 min. Remove supernatant and gently resuspend cells in 150 μL of fresh warm media.
4. Transfer ~30–40 μL of cell suspension onto each dry coverslip in a 4- or 24-well dish and leave cells to settle for 10 min at 40 °C.
5. Remove excess media and nonadherent cells, wash gently once with room temperature PBS.
6. Use fixative appropriate for antigen. For instance, fix centrosomal proteins with 100% ice-cold methanol for 5 min or microtubules and kinetochore proteins with 4% formaldehyde for 10 min at 37 °C.
7. Permeabilize cells with PBS containing 0.1% Tween (PBST) or 0.5% Triton X-100 for 5–10 min.
8. Nonspecific binding of antibodies can be minimized by incubating with the blocking solution for 10 min at room temperature.
9. Add appropriate concentration of primary antibodies in blocking solution and incubate for 2 h at room temperature or 37 °C. This step can also be performed overnight at 4 °C.
10. Remove primary antibody, wash 3X with PBST for 5 min each at room temperature.
11. Proceed with secondary antibodies conjugated to fluorophores of choice as with primary antibodies. Incubate in dark for 1 h at room temperature or 37 °C.

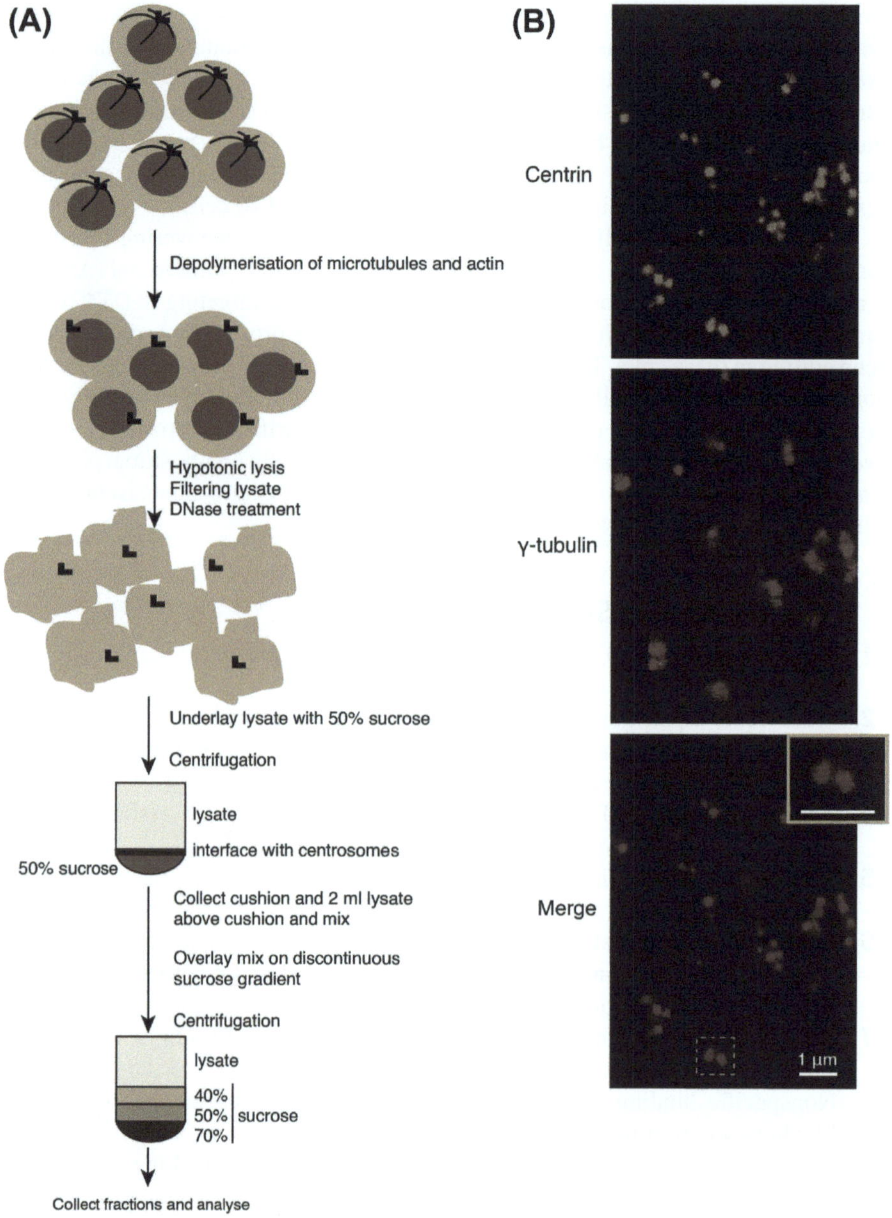

FIGURE 3

(A) Protocol for centrosome isolation from DT40 cells. (B) A field of centrosomes is shown after sedimentation onto coverslips. Centrioles are labeled by anticentrin antibodies (green in merge), whereas the pericentriolar matrix (PCM) is labeled with anti-γ-tubulin antibodies (red in merge). Scale bar = 1 μm. (See color plate)

12. Wash 3X with PBST for 5 min each at room temperature. If required add 5 μg/mL Hoescht in PBS to coverslips after first wash and incubate for 5 min at room temperature.
13. Perform final rinse with water.
14. Mount coverslip onto glass slide. A 4 μL mounting medium should be spotted onto glass slide, and coverslips should be flipped over so that cells face downward. Any standard mounting media preparations can be used (i.e., propyl gallate, Prolong Antifade reagent, mowiol).

2.1.1 Notes

- If antibody shows high background, blocking can be performed overnight. Blocking solutions can be prepared with horse or bovine sera.
- Stainings of centriolar proteins can be improved by including 0.05% sodium dodecyl sulfate (SDS) in permeabilizing solution.

2.1.2 Materials

Glass coverslips 12 mm
Poly l lysine (Sigma Aldrich, cat. no. P9155-5 MG)
Fixatives (ice cold methanol, formaldehyde)
Blocking solution: 5% bovine serum albumin in PBS
Primary and secondary antibodies
0.1% Tween 20 in PBS
Hoescht DNA stain
Mounting medium

2.2 TIME-LAPSE IMAGING

1. Transfect 1×10^6 cells with 5 μg of GFP-tubulin using the Amaxa nucleofector with program B-009.
2. Transfer immediately into 2.5 mL prewarmed medium in a 6-well plate and incubate for 24 h. Transfection efficiency should be at least 10–30%.
3. Spin 1.5 mL of cells at 900 g for 5 min at room temperature.
4. Since DT40 are suspension cells, they need to be immobilized for imaging. There are multiple ways to achieve this; cells can be seeded onto Ibidi dishes (cat. no. 81156) or concanavalin A–coated Lab-tek chamber slides (Mat-tek Corporation, cat. no P35G-1.5-14C). Resuspend the pellet in 200 μL warm media and place it as a liquid droplet onto dish or coverslip of choice, then allow cells to settle for 10 min at 40 °C.
5. Gently remove media with floating cells and fill dish with appropriate volume of prewarmed media. Adding FBS and chicken serum are optional when filming for less than 3 h. Leibovitz's medium (Life Technologies, cat. No. 11415-064) is recommended under CO_2-free conditions, but otherwise use standard growth media.

6. Set temperature-controlled chamber to 40 °C for imaging, especially if the DT40 cells have been maintained at 40 °C.

2.2.1 Notes

- To increase number of mitotic cells, synchronize cells with 9 μM RO-3306 (Sigma Aldrich, cat. no. 872573-93-8), a cyclin-dependent kinase 1 inhibitor, for 3–6 h. Release block by washing cells 3 × 2 min with warm medium; cells normally enter mitosis 10–15 min later.
- For time lapse imaging, cotransfecting cells with a DNA marker like histone H2B can help identification of prophase cells, which is otherwise difficult for DT40 cells round throughout the cell cycle.
- During live cell imaging keep laser exposure to minimum as this can have a profound effect on mitotic duration and outcome.

2.2.2 Materials

Amaxa nucleofector
Ingenio electroporation solution (Mirus Bio) or Amaxa Solution T
Leibovitz's medium
35 mm Ibidi dish (or Labtek chambers) or concanavalin A (0.3 mg/mL)-coated dish
Spinning disk confocal microscope with temperature-controlled chamber
Plasmids with fluorescently tagged tubulin or histone H2B

2.3 CENTROSOME PURIFICATION FROM DT40 CELLS

1. Grow 500 mL of DT40 cells at a density of 2.5–3 × 10^6 cells/mL. Feed cells the day before centrosome preparation.
2. Add nocodazole to 1 μg/mL and cytochalasin D to 1 μg/mL for 1 h (see Figure 3(A) for summary of protocol).
3. Spin cells at 2000 rpm for 5 min at 4 °C. Aspirate to remove supernatant.
4. Resuspend cells in 50 mL cold 1X TBS, transfer to 50 mL tube, spin at 1500 rpm for 5 min at 4 °C.
5. Resuspend pellet in 25 mL of cold 8% sucrose in 0.1X TBS, spin at 1000 rpm, for 5 min at 4 °C. Resuspend pellet gently but fast.
6. Squirt 10 mL lysis buffer with protease inhibitor cocktail onto the pellets in each tube. Lysis needs to be instant for successful centrosome preparation.
7. Invert suspension 3–5× and gently pipette it 2–3× to ensure complete lysis. Avoid bubbles at this step.
8. Leave on ice for 5 min, spin at 3000 rpm for 10 min at 4 °C.
9. Filter the lysate through a nylon mesh (pH at this point is around 5.8).
10. Adjust pH of lysate to 6.8 by adding 1M K-PIPES pH7.2, 1 mM EDTA.
11. Add 300 μL DNase and leave on ice for 15 min.
12. Underlay the lysate with 2 mL of 50% sucrose solution.
13. Spin at 11,000 rpm in an SW28 rotor or equivalent for 30 min at 4 °C to sediment centrosomes onto sucrose cushion.

14. Meanwhile prepare discontinuous sucrose gradients in tubes that fit SW40 rotor or equivalent: 2 mL of 70% sucrose, 1.5 mL of 50% sucrose, and 1.5 mL of 40% sucrose.
15. Following the spin from step 13, aspirate away supernatant leaving 2 mL lysate above the cushion. Take this 2 mL lysate +2 mL of the underlying cushion, gently vortex.
16. Transfer this lysate/cushion mix onto the discontinuous sucrose gradient and spin for 2 h at 25,000 rpm in an SW40 rotor at 4 °C.
17. Remove supernatant above the 40% sucrose solution.
18. Puncture the bottom of the centrifuge tube with a 26G needle. Collect fractions of 400 μL and analyze them with refractometer. Normally centrosomes are present at 45–50% sucrose.

2.3.1 Notes

- If mitotic centrosomes are needed, pretreat cells with 200 ng/mL nocodazole for 10 h or 100 μM monastrol for 6 h.
- Fractions can be stored at −80 °C for up to a month.
- Centrosomes can be sedimented onto coverslips by centrifuging at 22,000 g for 10 min at 4 °C and fixed in ice cold methanol (Figure 3(B)).
- To concentrate centrosomes for western blotting or mass-spectrometry, the fractions are mixed with 1 mL of 10 mM PIPES pH 6.8 and centrifuged for 20 min at 35,000 rpm in MLA130 rotor or equivalent. The centrosomes appear as glassy precipitate, which can be mixed with SDS loading buffer.

2.3.2 Materials

Cellbind roller bottles (Corning) and apparatus (at 2 rpm)
Nocodazole (5 mg/mL stock)
Cytochalasin D (5 mg/mL stock)
Tris-Buffered Saline (TBS)
Ultra pure sucrose
Cold methanol
Sucrose dissolving buffer: 10 mM K-PIPES pH 7.2, 1 mM EDTA, 0.1% β-mercapto-ethanol, 0.1% Triton X-100
Lysis buffer: 1 mM Tris pH 8.0, 0.1% β-mercaptoethanol (freshly added before use), 0.5% NP-40, 0.5 mM $MgCl_2$, 1mM PMSF, 10 μg/mL Aprotinin, 1 μg/mL Leupeptin, 1 μg/mL pepstatin
Nylon mesh (FACS strainer 70 μm)
DNase (10 mg/mL)
1M K-PIPES
Beckman or equivalent ultracentrifuge with rotors and tubes (e.g., SW40, SW28, MLA130)
Refractometer

CONCLUSION

We have described experimental protocols for homologous gene targeting in DT40 cells. The methods outlined above enable genome editing in vertebrate cells; isogenic cell lines can be generated at ease that carry deletions in one to several centrosomal genes or disruptions of specific domains and residues. Moreover, genes, both wild type and mutants, can be tagged in situ, which aids purification of endogenous protein complexes and allows visualization of their behavior in vivo. Lastly, we have provided a detailed method to isolate functionally active centrosomes from DT40 cells. The combination of genome editing with in vitro functional analysis such as microtubule nucleation assays is a powerful tool to delineate the role of individual centrosome components both in the organelle and cellular contexts.

ACKNOWLEDGMENTS

We thank John Kilmartin for his advice regarding centrosome sedimentation protocols. Our laboratory acknowledges the support of Cancer Research UK, the University of Cambridge and Hutchison Whampoa Ltd.

REFERENCES

Andersen, J. S., Wilkinson, C. J., Mayor, T., Mortensen, P., Nigg, E. A., & Mann, M. (2003). Proteomic characterization of the human centrosome by protein correlation profiling. *Nature, 426*, 570–574.

Arakawa, H., Lodygin, D., & Buerstedde, J. M. (2001). Mutant loxP vectors for selectable marker recycle and conditional knock-outs. *BMC Biotechnology, 1*, 7.

Baba, T. W., Giroir, B. P., & Humphries, E. H. (1985). Cell lines derived from avian lymphomas exhibit two distinct phenotypes. *Virology, 144*, 139–151.

Barr, A. R., Kilmartin, J. V., & Gergely, F. (2010). CDK5RAP2 functions in centrosome to spindle pole attachment and DNA damage response. *Journal of Cell Biology, 189*, 23–39.

Bollag, R. J., Waldman, A. S., & Liskay, R. M. (1989). Homologous recombination in mammalian cells. *Annual Review of Genetics, 23*, 199–225.

Bornens, M., & Moudjou, M. (1999). Studying the composition and function of centrosomes in vertebrates. *Methods in Cell Biology, 61*, 13–34.

Buerstedde, J. M., & Takeda, S. (1991). Increased ratio of targeted to random integration after transfection of chicken B cell lines. *Cell, 67*, 179–188.

Burt, D. W. (2005). Chicken genome: current status and future opportunities. *Genome Research, 15*, 1692–1698.

Dantas, T. J., Wang, Y., Lalor, P., Dockery, P., & Morrison, C. G. (2011). Defective nucleotide excision repair with normal centrosome structures and functions in the absence of all vertebrate centrins. *Journal of Cell Biology, 193*, 307–318.

Hall, S. L., Hester, S., Griffin, J. L., Lilley, K. S., & Jackson, A. P. (2009). The organelle proteome of the DT40 lymphocyte cell line. *Molecular & Cellular Proteomics, 8*, 1295–1305.

Inanc, B., Putz, M., Lalor, P., Dockery, P., Kuriyama, R., Gergely, F., et al. (2013). Abnormal centrosomal structure and duplication in Cep135-deficient vertebrate cells. *Molecular Biology of the Cell, 24*, 2645–2654.

Mocciaro, A., Berdougo, E., Zeng, K., Black, E., Vagnarelli, P., Earnshaw, W., et al. (2010). Vertebrate cells genetically deficient for Cdc14A or Cdc14B retain DNA damage checkpoint proficiency but are impaired in DNA repair. *Journal of Cell Biology, 189*, 631–639.

Molnar, J., Poti, A., Pipek, O., Krzystanek, M., Kanu, N., Swanton, C., et al. (2014). The genome of the chicken DT40 bursal lymphoma cell line. *G3 (Bethesda), 4*, 2231–2240.

Mosedale, G., Niedzwiedz, W., Alpi, A., Perrina, F., Pereira-Leal, J. B., Johnson, M., et al. (2005). The vertebrate Hef ortholog is a component of the Fanconi anemia tumor-suppressor pathway. *Nature Structural & Molecular Biology, 12*, 763–771.

Nakayama, T., & Takami, Y. (2001). Participation of histones and histone-modifying enzymes in cell functions through alterations in chromatin structure. *Journal of Biochemistry, 129*, 491–499.

Nishihashi, A., Haraguchi, T., Hiraoka, Y., Ikemura, T., Regnier, V., Dodson, H., et al. (2002). CENP-I is essential for centromere function in vertebrate cells. *Developmental Cell, 2*, 463–476.

Ohta, S., Bukowski-Wills, J. C., Sanchez-Pulido, L., Alves Fde, L., Wood, L., Chen, Z. A., et al. (2010). The protein composition of mitotic chromosomes determined using multiclassifier combinatorial proteomics. *Cell, 142*, 810–821.

Orr-Weaver, T. L., Szostak, J. W., & Rothstein, R. J. (1981). Yeast transformation: a model system for the study of recombination. *Proceedings of the National Academy of Sciences of the United States of America, 78*, 6354–6358.

Regnier, V., Vagnarelli, P., Fukagawa, T., Zerjal, T., Burns, E., Trouche, D., et al. (2005). CENP-A is required for accurate chromosome segregation and sustained kinetochore association of BubR1. *Molecular and Cellular Biology, 25*, 3967–3981.

te Riele, H., Maandag, E. R., & Berns, A. (1992). Highly efficient gene targeting in embryonic stem cells through homologous recombination with isogenic DNA constructs. *Proceedings of the National Academy of Sciences of the United States of America, 89*, 5128–5132.

Shang, W. H., Hori, T., Martins, N. M., Toyoda, A., Misu, S., Monma, N., et al. (2013). Chromosome engineering allows the efficient isolation of vertebrate neocentromeres. *Developmental Cell, 24*, 635–648.

Sir, J. H., Barr, A. R., Nicholas, A. K., Carvalho, O. P., Khurshid, M., Sossick, A., et al. (2011). A primary microcephaly protein complex forms a ring around parental centrioles. *Nature Genetics, 43*, 1147–1153.

Sir, J. H., Putz, M., Daly, O., Morrison, C. G., Dunning, M., Kilmartin, J. V., et al. (2013). Loss of centrioles causes chromosomal instability in vertebrate somatic cells. *Journal of Cell Biology, 203*, 747–756.

Sonoda, E., Matsusaka, T., Morrison, C., Vagnarelli, P., Hoshi, O., Ushiki, T., et al. (2001). Scc1/Rad21/Mcd1 is required for sister chromatid cohesion and kinetochore function in vertebrate cells. *Developmental Cell, 1*, 759–770.

Sonoda, E., Sasaki, M. S., Morrison, C., Yamaguchi-Iwai, Y., Takata, M., & Takeda, S. (1999). Sister chromatid exchanges are mediated by homologous recombination in vertebrate cells. *Molecular and Cellular Biology, 19*, 5166–5169.

Sudo, T., Ota, Y., Kotani, S., Nakao, M., Takami, Y., Takeda, S., et al. (2001). Activation of Cdh1-dependent APC is required for G1 cell cycle arrest and DNA damage-induced G2 checkpoint in vertebrate cells. *EMBO Journal, 20*, 6499–6508.

Takami, Y., Takeda, S., & Nakayama, T. (1997). An approximately half set of histone genes is enough for cell proliferation and a lack of several histone variants causes protein pattern changes in the DT40 chicken B cell line. *Journal of Molecular Biology, 265*, 394–408.

Takata, M., Sasaki, M. S., Sonoda, E., Morrison, C., Hashimoto, M., Utsumi, H., et al. (1998). Homologous recombination and non-homologous end-joining pathways of DNA double-strand break repair have overlapping roles in the maintenance of chromosomal integrity in vertebrate cells. *EMBO Journal, 17*, 5497–5508.

Vagnarelli, P., Hudson, D. F., Ribeiro, S. A., Trinkle-Mulcahy, L., Spence, J. M., Lai, F., et al. (2006). Condensin and Repo-Man-PP1 co-operate in the regulation of chromosome architecture during mitosis. *Nature Cell Biology, 8*, 1133–1142.

Wang, Y., Dantas, T. J., Lalor, P., Dockery, P., & Morrison, C. G. (2013). Promoter hijack reveals pericentrin functions in mitosis and the DNA damage response. *Cell Cycle, 12*, 635–646.

Yamazoe, M., Sonoda, E., Hochegger, H., & Takeda, S. (2004). Reverse genetic studies of the DNA damage response in the chicken B lymphocyte line DT40. *DNA Repair (Amst), 3*, 1175–1185.

CHAPTER

Centriole biogenesis and function in multiciliated cells

7

Siwei Zhang, Brian J. Mitchell[1]

Department of Cell and Molecular Biology, Feinberg School of Medicine, Northwestern University, Chicago, IL, USA

[1]*Corresponding author: E-mail: brian-mitchell@northwestern.edu*

CHAPTER OUTLINE

Introduction **104**

1. Visualization of Centrioles and Related Structures in *Xenopus* MCCs **106**

1.1 Visualization of Centrioles via Expression of Fluorescently Labeled Centriole-Related Protein from Exogenous Messenger RNA 106

1.2 Visualization of Centrioles in Fixed Embryos 108

1.3 Visualization of Centrioles in Live Embryos 108

1.4 Visualization of Centrioles via Antibody Staining and Immunofluorescence 112

2. Driving Specific Expression in *Xenopus* MCCs **112**

3. Generating Mosaic *Xenopus* Embryos to Circumvent Embryo Lethality **113**

4. Ciliated Epithelial Spheroid Culture **114**

5. Manipulating MCC Formation **115**

6. Fully-Automatic, Program-Assisted Counting of Centrioles in Confocal Images Using ImageJ **118**

Conclusion **123**

References **124**

Abstract

The use of *Xenopus* embryonic skin as a model system for the development of ciliated epithelia is well established. This tissue is comprised of numerous cell types, most notably the multiciliated cells (MCCs) that each contain approximately 150 motile cilia. At the base of each cilium lies the centriole-based structure called the basal body. Centriole biogenesis is typically restricted to two new centrioles per cell cycle, each templating from an existing "mother" centriole. In contrast, MCCs are post-mitotic cells in which the majority of centrioles arise "de novo" without templating from a mother centriole, instead, these centrioles nucleate from an electron-dense structure termed the

Methods in Cell Biology, Volume 129, ISSN 0091-679X, http://dx.doi.org/10.1016/bs.mcb.2015.03.015

deuterostome. How centriole number is regulated in these cells and the mechanism by which the deuterosome templates nascent centrioles is still poorly understood. Here, we describe methods for regulating MCC cell fate as well as for visualizing and manipulating centriole biogenesis.

INTRODUCTION

The ability of cells to duplicate their centrioles each cell cycle is a critical step in proper chromosome segregation and cell division. During interphase, centrioles act as important microtubule organizing centers (MTOCs) creating inward-outward directionality to microtubule-based transports. Moreover, in many differentiated cells the older of the two centrioles, termed the "mother" centriole, acts as the basal body nucleating a primary cilium. When this process goes awry, supernumerary centrioles can form, which has been observed in numerous types of cancer and is considered to correlate with cancer progression (Vitre & Cleveland, 2012). Additionally, genetic disruption of numerous centriolar proteins results in a wide range of phenotypes, most notably microcephaly and dwarfism (Nigg, Cajanek, & Arquint, 2014). For the reasons above, the study of centriole biogenesis has been the focus of many labs and there are a number of elegant systems with which to study centriole formation. The focus of this chapter will be to highlight a variation on the main theme of centriole biogenesis. Specifically, in post-mitotic MCCs there is a massive centriole amplification process that results in the production of more than 100 centrioles that nucleate the numerous motile cilia that project from the apical surface of these cells (Figure 1(A) and (B)).

During cell cycle-regulated centriole duplication, the two centrioles that comprise the centrosome separate, and a new daughter centriole nucleates orthogonally from the side of each mother centriole (Figure 2(A)). In cells in which key

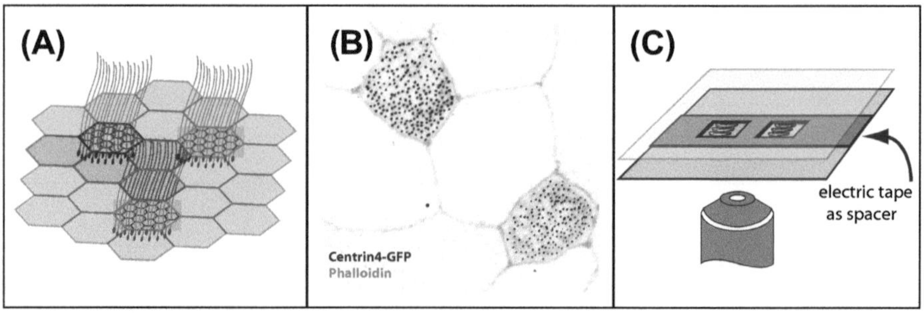

FIGURE 1 ***Xenopus*** **MCCs.**

(A) Illustration of the ciliated epithelia that covers the skin of *Xenopus* embryos. (B) Confocal image of multiciliated cells with centrin4-GFP marking centrioles/basal bodies (green) and rhodamine-phalloidin marking actin in the cell periphery (red). (C) Illustration of the mounting of *Xenopus* embryos using electrical tape as a spacer. (See color plate)

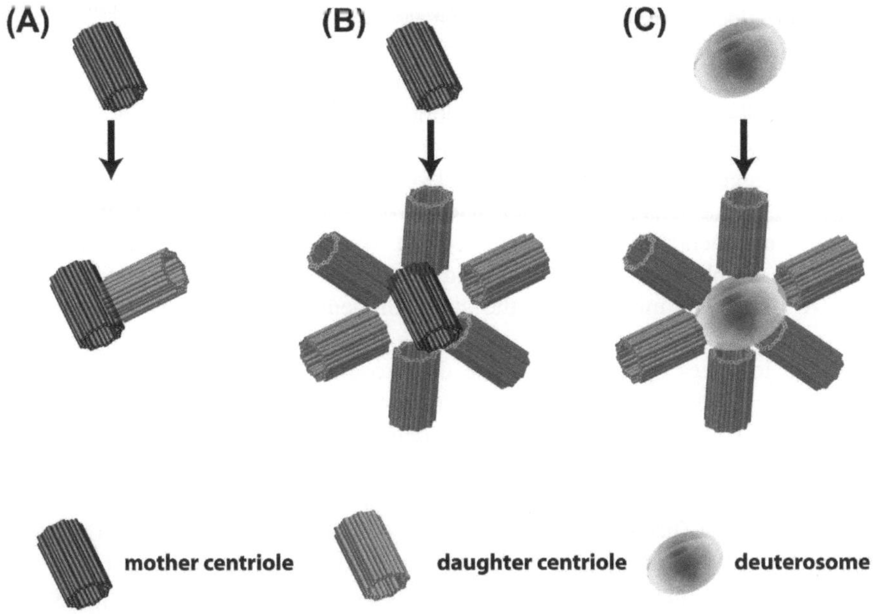

FIGURE 2 Centriole biogenesis.

(A) Standard centriole duplication with a single daughter (green (light gray in print versions)) nucleating from a mother centriole (pink (dark gray in print versions)). (B) Mother centriole-derived centriole amplification with numerous daughters (green (light gray in print versions)) simultaneously nucleating from a mother centriole (magenta (dark gray in print versions)). (C) Deuterosome-mediated centriole amplification with numerous daughters (green (light gray in print versions)) simultaneously nucleating from the deuterosome (orange (gray in print versions)).

regulators of centriole duplication (e.g., Plk4 or Cep152) are disregulated, it has been observed that numerous centrioles can simultaneously nucleate off of a single mother centriole (Figure 2(B)) (Brownlee & Rogers, 2013). Interestingly, in MCCs it has been observed that numerous centrioles simultaneously nucleate from the mother centriole, suggesting that the critical regulators might be differentially regulated. However, it has been suggested from the Electron-Microscopy literature that this form of centriole biogenesis accounts for only a small portion of the centrioles that are generated. In contrast, the vast majority of centrioles in MCCs arise "de novo" without a mother centriole to template from. Instead, these centrioles nucleate from a nondescript electron dense structure termed the deuterosome (Figure 2(C)) (Kalnins & Porter, 1969; Sorokin, 1968; Steinman, 1968). Recent papers have described several molecular components of the deuterosome; however, the functional regulation of this structure is still poorly understood (Klos Dehring et al., 2013; Zhao et al., 2013). The ciliated epithelium that lines the skin of *Xenopus* embryos represents an ideal system for characterizing this structure due to the facile nature of molecular manipulations and the ease of microscopic analysis (Figure 1(A) and (B)). Here, we present a detailed summary of the tools and techniques available to study centriole biogenesis in the MCCs of the *Xenopus* skin.

The well-characterized centriole duplication regulatory proteins Plk4 and Cep152, as well as the core structural protein Sas6, have all been observed at the deuterosome suggesting that key elements between centriole-centriole and deuterosome-centriole biogenesis are maintained (Klos Dehring et al., 2013). However, there are obvious fundamental differences that account for the abilities to (1) generate centrioles in the G0 phase of the cell cycle and (2) generate greater than a hundred centrioles. A significant milestone in the study of deuterosomes in *Xenopus* was the observation that a green fluorescent protein (GFP)-tagged version of the protein CCDC78 localizes specifically to the deuterosome (Klos Dehring et al., 2013). This allowed the first detailed characterization of the structure. Importantly, the depletion of CCDC78 leads to a loss of deuterosome localization of Cep152 and subsequently a severe reduction in centriole amplification. Another deuterosome-specific protein CCDC67, which has been named Deup1, also localizes to the deuterosome in mouse (Zhao et al., 2013). Deup1 is thought to provide a counterbalance to the mother centriole-driven centriole biogenesis from its paralogue Cep63 (Zhao et al., 2013). This again suggests that there are both similarities and unique features that distinguish centriole-centriole and deuterosome-centriole biogenesis. Finally, it has been recently proposed from work in the MCCs of mouse ependymal cells that deuterosomes nucleate exclusively from the daughter centrioles (Al Jord et al., 2014). This suggests that centriole biogenesis is not entirely "de novo" but, instead, deuterosomes act as amplifiers that acquire some signal from existing centrioles to increase the centriole nucleating capacity. While these recent studies have begun to provide a better picture on this structure, clearly there is still much to be learned. Many of the techniques used to study centriole biology in *Xenopus* rely on standard approaches that have been described in detail elsewhere (Mimoto & Christian, 2011; Sive, Grainger, & Harland, 1998, 2007c, 2007e, 2010a; Werner & Mitchell, 2013). Here, we aim to highlight specific tools, reagents, and experimental conditions that will facilitate the study of centriole amplification in MCCs.

1. VISUALIZATION OF CENTRIOLES AND RELATED STRUCTURES IN *XENOPUS* MCCs

1.1 VISUALIZATION OF CENTRIOLES VIA EXPRESSION OF FLUORESCENTLY LABELED CENTRIOLE-RELATED PROTEIN FROM EXOGENOUS MESSENGER RNA

One significant advantage of using *Xenopus* as a model organism is the ease of inducing localized protein expression via direct injection of in vitro synthesized messenger RNA (mRNA) without the requirement of genome-level manipulations (Mimoto & Christian, 2011; Sive et al., 2007e, 2010a; Sive, Grainger, & Harland,

2010b). The detailed protocols for manipulating early *Xenopus* embryos using mRNA injection method has been extensively described and profiled. Here, we will briefly describe these methods and focus on the slight modifications that we have found useful. These methods can be used to introduce proteins that will modulate developmental processes (e.g., transcription factors or signaling molecules) or simply to introduce fluorescently tagged proteins for visualization. In MCCs, the ability to fluorescently tag centrioles and deuterosomes allows the initiation of analysis of the process of centriole amplification.

Prior to mRNA synthesis, it is recommended to use column-based DNA purification of linearized plasmid DNA to prevent carry-over contamination of residual phenol-chloroform that may severely inhibit mRNA synthesis (e.g., Qiagen PCR Purification kit; Cat. No. 28104). The quality of mRNA for injections is critical for reproducibility. Techniques for generating mRNA are well established and there are numerous commercial kits available to facilitate optimal production (e.g., Life Technologies mMessage mMachine kits Cat. No. #1340/1341/1342). Additionally, column-based mRNA clean-up kits such as Life Technologies MegaClear kit (Cat. No. #AM1908) or Qiagen minElute clean-up kit (Cat. No. 74204) are preferable over the traditional phenol-chloroform method to minimize loss and prevent carry-over contamination. Column-based purification is essential if mRNA is to be combined with morpholino oligos (MOs) for injection. While these purification methods are not always essential and they do slightly increase the cost of mRNA synthesis, it is our experience that quality, yield, and reproducibility are all significantly enhanced with these techniques.

Eggs collected from mature adult females are in vitro fertilized with isolated testes (Sive et al., 1998, 2007e; Sive, Grainger, & Harland, 2007b, 2007d). Typically, injections are performed at the two or four cell stage. This is done so that the quality of the eggs can be evaluated by the quality of the first and second cell divisions. It is not recommended to inject embryos with poor cleavage planes as these rarely develop properly. Since the ectoderm gives rise to the ciliated epithelium, injection sites should be located on the animal hemisphere of the embryo, away from the midline and towards the center of the embryo within each blastomere. While the proper amount of mRNA to inject is empirical and requires titration, a reasonable starting point is a 10–20 pL injection of 50–200 pg mRNA. (Helpful hint: if using fluorescent proteins within the blue spectrum [BFP/CFP] it is preferable to use albino frogs since pigment granules exhibit strong autofluorescence in that range.)

Injected embryos are allowed to develop until the desired stage. The developmental rate of embryos can be artificially controlled by incubating embryos under different temperatures (Khokha et al., 2002). Generally, the earliest expression of MCC-specific genes initiates at stage 11.5, with high levels of expression observed at stage 14. Radial intercalation, the process by which MCCs interdigitate from an inner layer to the superficial layer, starts at stage 14 and can be optimally visualized at stage 17–18 (Stubbs, Davidson, Keller, & Kintner, 2006; Werner et al., 2014). At this stage, the formation and dynamics of centrioles, as well as the acentriolar structure, the deuterosome, can be observed. Centriole amplification typically occurs between stage 15 and 18 and ciliogenesis occurs after cells have completed radial intercalation (stage 19–20). While cilia are formed around stage 20, the process of properly polarizing these cilia occurs over time based on a combination of planar polarity signaling and a flow-mediated positive feedback loop, both of which impinge on the regulation of cytoskeletal dynamics (Mitchell, Jacobs, Li, Chien, & Kintner, 2007; Mitchell et al., 2009; Park, Mitchell, Abitua, Kintner, & Wallingford, 2008; Werner et al., 2011). Generally, the structure and formation of centrioles in *Xenopus* MCCs can be conveniently observed using fluorescently tagged centriolar proteins.

A list of molecular markers for centrioles and related structures as well as fixation conditions and optimal concentrations has been provided in Table 1.

1.2 VISUALIZATION OF CENTRIOLES IN FIXED EMBRYOS

Several techniques have been described for fluorescent imaging in whole *Xenopus* embryos (Kieserman, Lee, Gray, Park, & Wallingford, 2010; Lee, Kieserman, Gray, Park, & Wallingford, 2008). A simple method of mounting the relative-flat embryos for imaging can be achieved by sandwiching the fixed embryos between two pieces of coverglass, while using several layers of electrical tape as a spacer (Figure 1(C)) (Werner & Mitchell, 2013). It should be noted that for optimal imaging of samples fixed in MeOH or Dent's, sufficient rehydration time (approximately 4 h) should be allowed for fluorescent proteins to return to their native conformations.

For successful observation of centrioles in mounted embryos, appropriate microscope configuration for confocal microscopy is highly recommended. Traditional epifluorescent microscopy may be challenging due to excessive fluorescence emitted from the underlying tissue. As a general guideline, a 60–100x oil objective (1.4 n. a.) with sufficient working distance is recommended for taking images on centrioles and subcortical structures. In order to obtain clear images on centrioles without interference from background and cytosolic particles, it is recommended to use Z-stacks (10–15 images) with a step size of 0.2–0.5 μm. Collecting too few images may result in decreased Z-axis resolution on centrioles and associated subcortical structures, while collecting too many stacks may risk excessive photobleaching to fluorescent proteins, especially red fluorescent protein (RFP). This balance is particularly challenging while visualizing MCCs amplifying their centrioles as this occurs below the surface of the embryo (see below).

1.3 VISUALIZATION OF CENTRIOLES IN LIVE EMBRYOS

Batches of injected embryos should be prescreened under a fluorescent dissecting microscope to identify embryos with sufficient fluorescence intensities. After which, embryos can be mounted for live imaging. Mounting techniques and methods in *Xenopus* embryos vary between stages of the embryos, as well as the desired timespan in which the embryos are to be imaged. Generally, a detailed and comprehensive review of over mounting *Xenopus* embryos has been published (Kieserman et al.,

Table 1 Useful Molecular Markers for the Study of Multiciliated Cells

Marker Name	Description	Labeled Structure	Fixatives	Suggested Concentration
Centrioles and basal bodies				
γ-tubulin	Antibody sigma #T-6557	Label basal bodies in fixed tissue (Park et al., 2008)	100% MeOH; Dent's Variable results: 3% PFA/EtOH	1:300–1:500
Centrin-RFP	Tagged centrin4 (Xl. 50473)	Strongly labels basal bodies (Mitchell et al., 2009; Park et al., 2008)	3% PFA or 3% PFA/EtOH Variable results: 100% MeOH; Dent's	40 pg mRNA
GFP-HYLS	Tagged HYLS (Xl. 13199)	Strongly labels basal bodies (Dammermann et al., 2009)	Dent's 3% PFA/EtOH 3% PFA	250 pg mRNA; 20 pg DNA
Sas6-RFP	Tagged SAS6 (Xl. 33005)	Labels active deuterosomes and centrioles (Klos Dehring et al., 2013)	Dent's 3% PFA/EtOH 3% PFA	250 pg mRNA
RFP-Cep152	Tagged Cep152 (Xl. 13956)	Labels deuterosomes and nascent centrioles (Klos Dehring et al., 2013)	Dent's 3% PFA/EtOH 3% PFA	250 pg mRNA
GFP-Plk4	Tagged Plk4 (Xl. 5660)	Labels deuterosomes and centrioles (Klos Dehring et al., 2013)	Dent's 3% PFA/EtOH 3% PFA	250 pg mRNA
GFP-Ccdc78	N-terminal GFP tagged ccdc78 (Xl. 4890)	Specifically labels deuterosomes (Klos Dehring et al., 2013)	Dent's 3% PFA/EtOH 3% PFA	200 pg mRNA
Ccdc78-GFP	C-terminal GFP tagged ccdc78 (Xl. 4890)	Labels both deuterosomes and centrioles (Klos Dehring et al., 2013)	Dent's 3% PFA/EtOH 3% PFA	200 pg mRNA
GFP-Deup1	N-terminal GFP tagged Deup1 (CCDC67 cloned from Xenopus laevis)	Strongly labels deuterosomes and weakly labels centrioles (unpublished observation)	3% PFA	200pg mRNA

Continued

Table 1 Useful Molecular Markers for the Study of Multiciliated Cells—cont'd

Marker Name	Description	Labeled Structure	Fixatives	Suggested Concentration
GFP/RFP CLAMP (Spef1)	Tagged CLAMP (Xl. 26316)	Labels striated rootlets. Can be used to score cilia orientation (Mitchell et al., 2009; Park et al., 2008)	3% PFA or 3% PFA/EtOH Variable results: 100% MeOH; Dent's	250 pm mRNA; 20 pg DNA
GFP-mid1ip1 (Mig12)	Tagged mig12 (Xl. 47645)	Labels rootlet, striated. Can be used to score cilia orientation (Park et al., 2008)	3% PFA or 3% PFA/EtOH Variable results: 100% MeOH; Dent's	250 pm mRNA; 30 pg DNA
Vinculin	GFP-tagged vinculin (Xl. 21637)	Labels rootlets in ciliated cells (Antoniades, Stylianou, & Skourides, 2014)	Live imaging is recommended 3% PFA if fixation is required	250 pg mRNA
FAK	GFP-tagged FAK (Xl. 6819)	Labels rootlets in ciliated cells (Park, Shen, Chien, & Guan, 2009)	Live imaging is recommended 3% PFA if fixation is required	250 pg mRNA
Paxilin	GFP-tagged Paxilin (Xl. 724)	Labels rootlets in ciliated cells (Antoniades et al., 2014)	Live imaging is recommended 3% PFA if fixation is required	250 pg mRNA
Dvl2	GFP-tagged Dvl2, (Xl. 7670)	Labels adjacent to the basal body (Park et al., 2008)	3% PFA	250 pg mRNA
Bbof1	GFP-tagged Bbof1 (Xl. 66678)	Transiently associated with basal bodies (Chien et al., 2013)	3.7% Formaldehyde + 0.25% glutaraldehyde	250 pg mRNA

2010). Essentially, mounted embryos are pressed against a flat coverglass to maintain a flat field of view for live imaging, while maintaining sufficient level of access to the culture media to ensure embryo survival and growth, especially for extended imaging during time course analysis. Alternatively, instead of whole embryos, explanted tissue such as animal cap explants can also be used for imaging, especially if the observed tissue is expected to undergo growth, extension, or movement within one or more dimensions (Figure 3(A)). For a detailed protocol on customized chamber making and embryo mounting, please refer to Ref. Werner and Mitchell (2013). The "electrical tape" method described in the previous section can also be used in live embryos as well as more traditional half-rose chambers or glass bottom Petri dishes (Wadsworth, 2007; Werner & Mitchell, 2013).

For imaging centriole biogenesis in MCCs, several special considerations should be made. As mentioned above, centriole biogenesis initiates around stage 16, prior to the cells undergoing radial intercalation and joining the outer epithelium. This means that the MCCs are approximately 10 μm below the surface when centrioles are amplifying. While this is within the range of confocal microscopy, it is challenging to get high-resolution images. Molecular solutions to this issue will be discussed below, but from an imaging perspective there are several helpful hints.

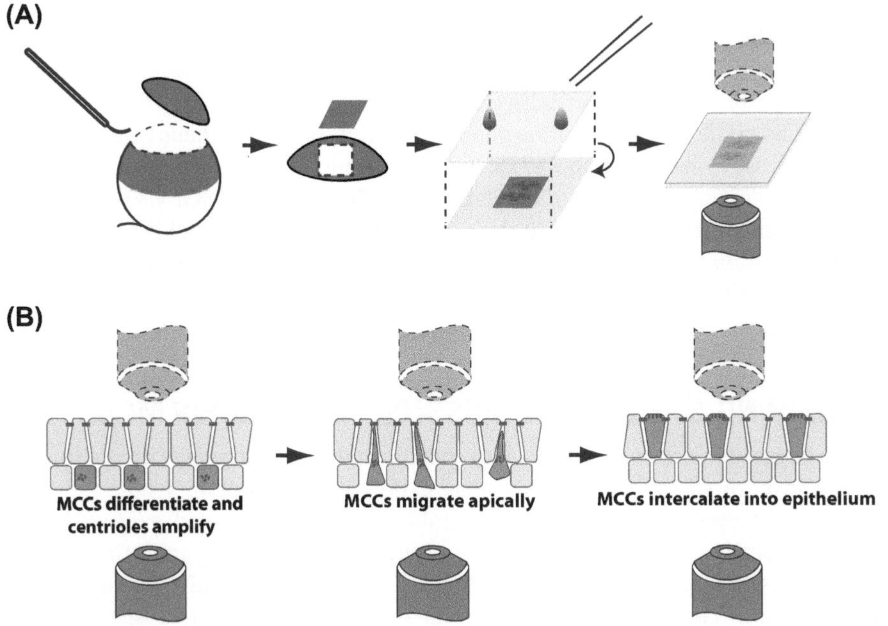

FIGURE 3 Ciliated epithelial explants.

(A) Illustration of dissecting off the animal cap of the embryo to generate an epithelial explant. (B) Process of visualizing multiciliated radial intercalation in which the cells migrate into the outer epithelium.

The use of skin explants provides two advantages: First, the sample is much flatter than whole embryos making deeper images more facile (Figure 3(A)). Second, the explants can be flipped upside down and imaged from underneath, which in some situations allows higher-resolution imaging (Figure 3(B)) (Stubbs et al., 2006). While other imaging techniques such as two-photon or light sheet illumination would also be beneficial, the detailed description of these methods is beyond the scope of this chapter. Additionally, other groups have successfully imaged MCCs from mouse using super-resolution techniques, and this will obviously be beneficial to translate into *Xenopus* (Al Jord et al., 2014; Zhao et al., 2013).

1.4 VISUALIZATION OF CENTRIOLES VIA ANTIBODY STAINING AND IMMUNOFLUORESCENCE

In addition to the fluorescent-tagged protein-based methods of visualizing centrioles as described above, it is also possible and practical to visualize centrioles and associated structures using immunofluorescence. For a detailed protocol in performing antibody staining and immunofluorescence in *Xenopus* epithelium, please refer to Ref. Werner and Mitchell (2013). In addition to the general protocol, several particular points should be noted when performing immunofluorescence in *Xenopus* embryos, as described below.

Fixatives need to be optimized for each individual antigen. Generally, it is advisable to try the following fixatives as a starting point: 3% paraformaldehyde (PFA) in phosphate-buffered saline (PBS), and MEMFA (one volume of 37% formaldehyde, one volume of MEM salt including 1 M MOPS, 20 mM EGTA, 10 mM $MgSO_4$, pH 7.4, and eight volumes of ultra-pure H_2O). High-purity, methanol-free formaldehyde, such as Sigma–Aldrich F8775, is recommended since it will not interfere with phalloidin staining, or Dent's fixative (80% MeOH, 20% DMSO, v/v).

Different antibodies may have their own preferences over blocking buffers. Five percent BSA in PBST (PBS +0.1% Triton X-100) or 10% heat-inactivated goat serum in PBST are two good starting points for testing antibodies.

2. DRIVING SPECIFIC EXPRESSION IN *XENOPUS* MCCs

While targeted injection of mRNA into specific embryo blastomeres does allow for a certain level of spatial control over gene expression and protein synthesis in early *Xenopus* embryos, it is sometimes desirable to achieve targeted gene expression both temporally and spatially in MCCs alone. To achieve this goal, the α-tubulin promoter is placed upstream of the desired gene in a plasmid construct (Stubbs et al., 2006). The α-tubulin promoter is transcriptionally active specifically in MCCs starting at stage 13 and remains active throughout MCC development. Generally, injection of 10–50 pg of plasmid DNA containing the α-tubulin promoter is sufficient to drive the expression of the desired gene in MCCs. However, it should be taken into account that a certain level of mosaicism may arise from the injection as part of the inherited problems of DNA injection into early *Xenopus* embryos. It is possible to circumvent this problem by using transgenic methods to generate stable transgenic lines; however, significantly more effort and extended time are expected (minimum 18 months for each *Xenopus laevis* generation and 6 months for each *Xenopus tropicalis* generation) (Amaya & Kroll, 1999; Loeber, Pan, & Pieler, 2009). Alternatively, F0 transgenic animals can also be generated and analyzed. While a certain degree of mosaicism still presents under these conditions, one can typically identify samples that contain consistently strong expression. As will be discussed below, a certain level of mosaicism is often preferable, as it provides an internal control for cell biological analysis.

3. GENERATING MOSAIC *XENOPUS* EMBRYOS TO CIRCUMVENT EMBRYO LETHALITY

Either overexpression or knockdown of certain genes or proteins at the whole embryo level may result in developmental defects and/or poor embryo survival, since the desired genes may be crucial for proper cellular functioning. These risks are especially prone to happen during the investigation of centriolar proteins and present a challenge for analysis of MCCs, as these cells only emerge later in development. Hence, it is often desirable to circumvent such developmental defects by introducing a degree of mosaicism into the embryos to be investigated in order to partially compensate for the lethal effects. Generally, the injection of exogenous mRNAs or MOs can be restricted to one or more cells at the 4, 8, 16, or 32-cell stage, therefore restraining the overexpression or knockdown effects within a subset of cells in the developed embryos. A more targeted approach, which may even allow population-specific expressions within certain tissues or regions can be executed at the 16–32 cell stage with the reference from a fate map of individual blastomeres (Dale & Slack, 1987; Moody, 1987a, 1987b). In all cases, it is recommended to label the cells that have been injected using molecular tracers such as fluorescein isothiocyanate (FITC)-labeled dextran (Sigma–Aldrich 46945) or rhodamine-labeled dextran (micro-ruby, Life Technologies D-7162). Strong fluorescence can be observed immediately after injection, and the fluorescence persists after fixation and remains strong during immunofluorescence. Injection of mRNAs encoding fluorescent proteins such as eGFP or mCherry as tracers can also be adopted in experiments, although such fluorescent proteins usually result in lower signal intensities compared to labeled dextrans and may not perform well after fixation and immunofluorescence. If MOs are to be used in the mosaic knockdowns, MOs can be prelabeled by either FITC or lissamine-rhodamine upon purchase for tracing the MO presence within embryos, and the fluorescence is maintained after fixation. Finally, it is often desirable to have centriolar markers (e.g., Centrin-RFP) in all cells and a specific MO in a subset of cells. To accomplish this, one set of injections can be performed at the 2-cell stage and a second set of injections can be performed at the 4, 8, 16, or 32-cell stage. This allows for a direct comparison of "wild-type" and mutant cells in the same embryos and can be a particularly powerful approach.

4. CILIATED EPITHELIAL SPHEROID CULTURE

During embryonic development, MCCs of *Xenopus* embryos undergo radial intercalation into the outer epidermal layer, and the numerous centrioles within are subsequently distributed to the apical surface (Figure 3(B)). This event provides a unique opportunity to study radial intercalation and the essential role that centrioles play during this process (Werner et al., 2014). However, performing time-lapse confocal imaging on intercalating cells presents a challenge, since intercalation generally occurs perpendicular to the coverglass surface in mounted embryos, which results in a severe decrease in resolution due to the absorption of both excitation lasers and emission light from the tissue itself. In addition, to obtain sufficient resolution of cell movements during the intercalation process, deep Z-stacks are required, which further increases the potential threat of photobleaching. Hence, it may be desired to mount the intercalating epidermis in a way that the plane of intercalation is parallel to the coverglass and, therefore, parallel to the XY axis. To achieve this, *Xenopus* epithelial spheres derived from ectodermal explants can be used, which are much smaller in size compared to the whole embryo and, more importantly, enable confocal imaging of intercalating cells from the side of the sphere, thus greatly reducing scan time and hence the risk of photobleaching (Werner & Mitchell, 2013).

1. Use agarose-coated dishes in all experiments. The surfaces of certain high-quality dishes for cell culture are specifically precoated with polymers and thus become extremely adhesive to any exposed tissues. In such cases, isolated animal cap explants will "cling" to the bottom of the dish and are difficult to manipulate and transfer. To make agarose-coated dishes, pour hot 1% agar dissolved in 0.1x MMR into the dishes to be used, let stand for 30 s, and aspirate out the remaining agar solution. A very thin layer of agar coating will remain on the inner surface of the dish and prevent any adhesion to explants.
2. Isolate animal cap explants around stage 10 (Figure 3(A)) (Sive, Grainger, & Harland, 2007a). Remove any residual white remnants that derive from the yolk-rich inner embryo mass from the inside of semispherical ectodermal animal cap explants. This is best done using an eyebrow knife to scrape off the white mass.
3. Cut the ectodermal explant into small pieces of approximately 20–30 μm in diameter using an eyebrow knife. This can be done in 0.1x MMR to circumvent high salt concentration-induced quick involution of the explants, especially when the operator is relatively inexperienced.
4. Transfer the small tissue explants obtained into another dish with 0.5x MMR plus gentamycin. Use a wide-bore glass pipette and avoid getting explants into

contact with air or the surface tension will rupture the explants. Under normal conditions, explants will naturally form spheres in 30 min–1 h.

5. Incubate explants until control (nonexplanted) embryos reach stage 16–20.
6. Prepare agarose pads for mounting and imaging. Melt 2% agarose in 0.1x MMR to 90 °C in a heat block or microwave oven. Apply a drop of agarose to a glass microscopic slide that has been positioned between two other slides. The other two slides must be fixed to the bench surface using two layers of electrical tape. Place another glass slide on top of the agarose and perpendicular to all other three slides. A thin layer of agarose will form inside the buffer zone created by two layers of electrical tape between the two slides. For a graphical illustration of the procedure, see Ref. Werner and Mitchell (2013).
7. Carefully remove the top slide from the agarose pad by sliding it along the other ones. In most cases, some air bubbles will be trapped inside the buffer zone and leaving small holes on the agarose pad, which can be immediately used for mounting the epithelial spheres inside. Small holes can also be cut using fine forceps or punching the pad with pipette tips to sit the spheres within.
8. Carefully transfer the epithelial spheres into the holes on the agarose pad using a wide-bore glass pipette. Gently place a coverslip on the agarose pad and seal the rim with melted petroleum jelly or silicon grease. The spheroid cultures are now ready to be used for live imaging.

5. MANIPULATING MCC FORMATION

The MCCs that form on the surface of *Xenopus* embryos exhibit a stereotypical spacing pattern that reflects the requirement for cell vertices during the process of radial intercalation (Figure 4). For example, one can greatly increase the number of MCC precursors, but only a fraction of these will intercalate due to the limitations of the number of proper vertices (Stubbs et al., 2006). The transcriptional program that leads to MCC development has been the focus of numerous studies and several factors are known to modulate MCC formation. Here, we will describe several of these factors and how they can be exploited to study centriole formation and function.

Notch signaling appears to provide one of the first levels of regulation for MCCs at around stage 12. Activation of the Notch pathway using mRNA injections to overexpress the intracellular domain of Notch (pCS2-NICD) leads to a severe decrease in the number of MCCs, while conversely the inhibition of the Notch pathway via overexpression of a dominant negative Mastermind (pCS2-dnMM) construct leads to an overproduction of MCCs (Figure 4 and Table 2) (Deblandre, Wettstein, Koyano-Nakagawa, & Kintner, 1999). However, the effect of each of these constructs is restricted to the initial cell lineage in that ectopic expression within outer epithelial cells fails to convert these cells to the MCC's lineage. In stark contrast, the loss of multicilin (MCIDAS), which is a member of the gemini family of proteins, results in a loss of MCCs. More importantly, ectopic expression of MCIDAS using a

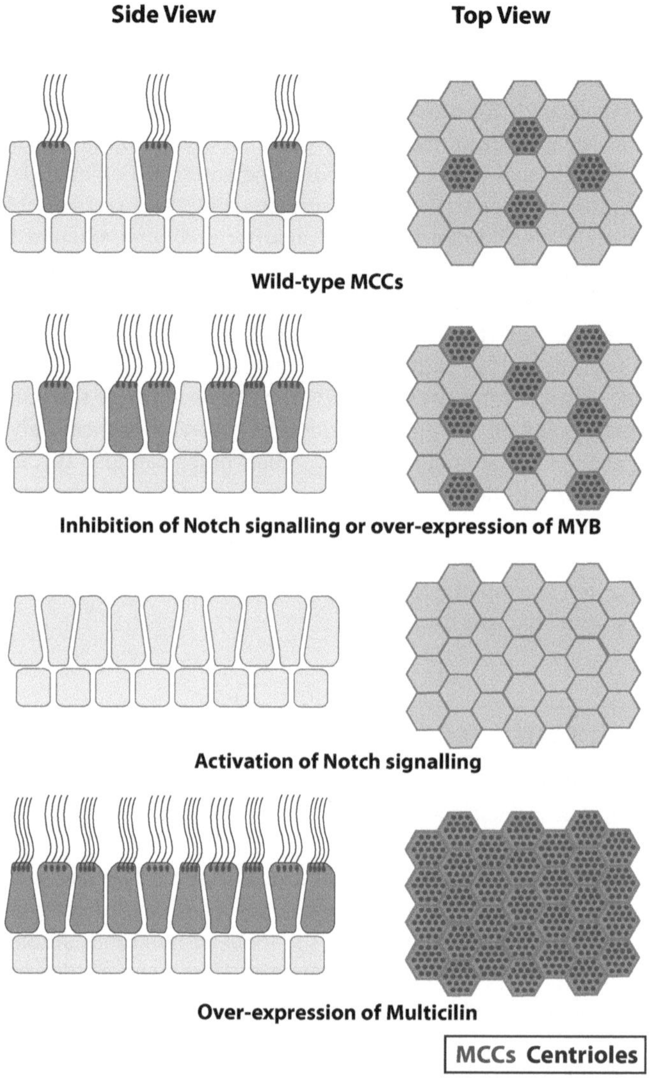

FIGURE 4 Modulation of multiciliated cell fate.

Illustration of multiciliated cell phenotypes obtained with the manipulation of various transcriptional regulators (as labeled).

dexamethasone inducible construct (pCS2-hGR-MCIN) leads to the conversion of outer epithelial cells into MCCs, such that the entire epithelium is now covered with MCCs (Figure 4 and Table 2) (Stubbs, Vladar, Axelrod, & Kintner, 2012). Two downstream targets of MCIDAS are the transcription factors MYB and

FoxJ1. MYB controls centriole duplication and FoxJ1 controls ciliogenesis. Ectopic expression of MYB leads to more MCCs but fails to convert outer cells into MCCs (similar to dnMM) (Tan et al., 2013). In contrast, ectopic expression of FoxJ1 does not create more MCCs due to a lack of centriole amplification but rather lead to the formation of ectopic cilia such that outer cells will use their existing centrioles to nucleate one to two cilia per cell (Stubbs, Oishi, Izpisua Belmonte, & Kintner, 2008).

While each of these factors can provide useful tools for the study of centrioles, MCIDAS is particularly useful. The formation of ectopic MCCs (with temporal control using human glucocorticoid receptor-fused version of MCIDAS, hGR-MCI together with dexamethasone induction) eliminates the issues of imaging MCCs undergoing centriole amplification deep in the tissue (see above). When outer cells are converted to MCCs, one can easily visualize the formation of nascent centrioles budding off of the deuterosome. Additionally, it has been observed that deuterosomes are formed in MCCs ectopically expressing both MCIDAS and MYB, suggesting that these MCCs undergo typical deuterosome-mediated centriole amplification. Table 2 outlines the optimal dose and experimental use of these constructs as well as the expected outcomes.

Table 2 Tools for Manipulating Multiciliated Cell Formation

Regulator	Dose	Effects
Notch-ICD (intracellular domain of Notch) (Deblandre et al., 1999)	175 pg mRNA	Overexpression in early embryos activates Notch signaling, thus reducing or preventing MCC formation
dnHMM (dominant-negative version of mastermind) (Fryer, Lamar, Turbachova, Kintner, & Jones, 2002)	1 ng mRNA	Overexpression in early embryos inhibits Notch signaling, thus promoting MCC formation
FoxJ1 (Stubbs et al., 2008)	1–5 ng mRNA	Overexpression induces nodelike ciliated cells;
	75 ng Morpholino oligo	knockdown inhibits formation of both node and epidermal ciliated cells
MYB (Tan et al., 2013)	500 pg mRNA	Overexpression promotes MCC formation in epidermis
hGR-MCI (inducible version of multicilin) (Stubbs et al., 2012)	40 pg mRNA; 2 μM of dexamethasone for induction (Ryan et al., 2004);	Overexpression will convert all epidermal cells into MCCs;
	50 ng MO	knockdown prevents MCC formation in epidermis

6. FULLY-AUTOMATIC, PROGRAM-ASSISTED COUNTING OF CENTRIOLES IN CONFOCAL IMAGES USING ImageJ

When investigating the molecular mzechanisms related to centriole proliferation and/or development in ciliated cells, it is often necessary to make statistical analysis on the number of centrioles within each ciliated cell in control versus experimental group(s). In most cases, manual counting is applied to collect centriole counts from samples, which is both time-consuming and labor-intensive. Several commercially available software packages, such as Nikon NIS-Documentation™, have built-in functions that allow automatic cell border tracking, image segmentation, and particle counting within each of the region of interests (ROI) that greatly reduces both experiment time and human labor. However, one drawback is that commercial software is often instrument-bound and/or expensive to buy, which restricts their widespread application. Here, we introduce a method using the built-in macro function of the GNU Freeware ImageJ for both fully automatic and program-assisted centriole counting.

Essentially, two critical steps define the core conception of automatic centriole counting. First, maximum-projected Z-stack confocal images need to be segmented into single cells that contribute to multiple ROIs. Secondly, centrioles within each ROI (i.e., cell) need to be counted separately without interference from neighboring cells.

1. Prepare embryos or animal cap explants for confocal microscopy. Generally, at least two channels are required in samples that will be imaged. The first channel (C0) labels centrioles using centriole-specific markers such as GFP-Centrin4, while the second channel (C1) labels cell boundaries using membrane-specific markers such as memRFP, LifeAct-RFP, or fluorescently labeled phalloidin (recommended).
2. Prepare confocal images for analysis. An image size of at least 512 × 512 pixels (1024 × 1024 or higher is ideal) is required for successful image analysis. Instrument and scanning parameters: 40–60× oil lens, 2 μm thick Z-stack in total, 10 optical sections with 0.2 μm distance in between. Final output images should be multichannel (not RGB) images that have been superimposed using maximum projection along the Z-axis with the centriole channel in the C0 position.
3. Program requirements for imaging analysis: all the codes used within the section have been tested using ImageJ 1.49h 64-bit version with adaptiveThr plugin installed (downloadable at https://sites.google.com/site/qingzongtseng/adaptivethreshold).
4. If using machine-assisted counting (which is more useful when the signals from cell boundaries channel is not very strong), cell boundaries are manually drawn using the built-in ROI function of ImageJ and added into the ROI manager on the cell boundaries channel. Images need to be presplit into two independent images before the ROIs are drawn (Image > Color > Split Channels). After all the ROIs have been set up, transfer all ROIs to the centrioles channel (Edit > Selection > Restore Selection or Ctrl + Shift + E).

5. Load the macro scripts below (Box 1; this only needs to be done once and the code will remain open for all subsequent analyses) using Plugins > Macros > Edit. Make sure the centriole window has the focus and press Ctrl + R to run the script. The script will attempt to count the number of centrioles within each ROI in an exhaustive pattern until all ROIs have been counted. After counting, the centriole counts within each cell will be displayed in the summary window and can be subsequently exported (column 2). If desired, the area size of each ROI (i.e., cell) can be calculated by dividing the Total Area of the centrioles (column 3) over the %Area (column 5). The threshold level in line 2 and centriole size in line 9 need to be determined empirically.
6. The fully automatic script has the ability to batch-process whole folders containing unlimited number of images, make analysis, and save the results into designated folders as tab-separated text files. Several parameters, as displayed in the program initiation page, need to be empirically determined and may differ on images obtained from separate sets of experiments. Hence, it is suggested to use images from the same set of experiments during each batch run. The code itself is extensively annotated and minor adjustments can be performed by referring to the in-line comments when editing the macro.

BOX 1 CODE FOR AUTOMATED QUANTIFICATION OF CENTRIOLE NUMBER

```
//---------------------Set global variables
var minimum_cell_size;
var maximum_cell_size;
var adaptiveThr_pixel_block_size_cell;
var adaptiveThr_subtract_value_cell;
var gaussian_radius;

var minimum_centriole_size;
var maximum_centriole_size;
var adaptiveThr_pixel_block_size_centriole;
var adaptiveThr_subtract_value_centriole;

var edge_count;
var min_count_threshold;
var output_2_file;
var watershed_count;

get_all_parameters(); //set all parameters for counting

dir_Input = getDirectory("Choose Input Directory "); // Get Input
directory
dir_Output = getDirectory("Choose Output Directory "); // Get Output
directory
list=getFileList(dir_Input); //Get file name
```

Continued

BOX 1 CODE FOR AUTOMATED QUANTIFICATION OF CENTRIOLE NUMBER—cont'd

```
output_2_file = File.open(dir_Output + "output.txt"); //set output
summary file

for (i=0; i<list.length; i++)
{
   // load image for processing
   showProgress(i+1, list.length);
   open(dir_Input + list[i]);
   Current_File_title = list[i];

   //-------------------------------------Beginning of image processing

   // Note: the 1st Channel is Centrin while the 2nd Channel is Phalloidin.

   selectWindow(Current_File_title); //Get image

   run("8-bit"); //adaptiveThr only works on 8-bit images
   run("Split Channels"); // separate Phalloidn and centrin channels

   selectWindow("C2-" + Current_File_title); // process Phalloidin to
identify cell boundaries
   run("adaptiveThr                    ",                    "using=Mean
from=adaptiveThr_pixel_block_size_cell
then=adaptiveThr_subtract_value_cell"); // Threshold

   selectWindow("C2-" + Current_File_title); // Reconfirm window
   run("Gaussian Blur...", "sigma=gaussian_radius"); // Apply Gaussian
Blur

   wait(1000); //System wait time for medium-slow computers, necessary,
otherwise the current window may lose focus.

   run("Make Binary"); // Transform to binary
   run("Invert LUT"); //Need BLACK cells over WHITE cell boundaries

   //Determine whether edge counts should be included
   if (edge_count == true)
   {
   run("Analyze Particles...", "size=minimum_cell_size-maximum_cell_size
circularity=0.20-1.00 clear add"); //Identify single cells, at least 500
in size
   }
```

BOX 1 CODE FOR AUTOMATED QUANTIFICATION OF CENTRIOLE NUMBER—cont'd

```
    else
    {
        run("Analyze Particles...", "size=minimum_cell_size-maximum_cell_
size circularity=0.20-1.00 clear exclude add"); //Identify single cells,
at least 500 in size
    }

    selectWindow("C1-" + Current_File_title); //Switch to basal bodies
window
        run("adaptiveThr ", "using=[Weighted mean] from
=adaptiveThr_pixel_block_size_centriole then
=adaptiveThr_subtract_value_centriole "); // Apply threshold
    run("Invert LUT"); // Invert LUT, need to be BLACK centrin signals over
WHITE background
    run("Make Binary"); // Transform to binary, need to be BLACK cells over
WHITE cell boundaries
    if (watershed_count == true)
        run("Watershed"); // Try to separate touching particles

    selectWindow("C2-" + Current_File_title);
    run("Select All"); //Get Selection from Phalloidin channel

    selectWindow("C1-" + Current_File_title);
    run("Restore Selection"); //Transfer selection to Centrin channel

    n = roiManager("count"); // Count centrin signals (i.e.: basal bodies)

    for (j=0; j<n; j++)
    {
        roiManager("select", j);
        run("Analyze Particles...", "size=min_centriole_size-
maximum_centriole_size clear summarize");
    }

    // Save Summary window
    selectWindow("Summary");
    saveAs("text", dir_Output + Current_File_title + ".txt");

    lines = split(getInfo(), "\n"); //save output to one large summary
file
    print(output_2_file, lines[0] + ' \t' + "Mother Cell Area"); //Write
form titles
```

Continued

BOX 1 CODE FOR AUTOMATED QUANTIFICATION OF CENTRIOLE NUMBER—cont'd

```
    for (m=1; m<lengthOf(lines);m++)
    {
        values = split(lines[m], "\t"); //False counts filter
        centriole_count = values[1];
        if (centriole_count > min_count_threshold)
        {
            centriole_area = parseFloat(values[2]);
            centriole_area_percentage = parseFloat(values[4]);
            cell_area = centriole_area / centriole_area_percentage * 100;
//Count the total area of mother cell
            print(output_2_file, lines[m] + ' \t' + cell_area);//Write
mother cell area
        }
    }

    //--------------------------------------End of Image processing
    // Cleanup
    close(); //Close Channel 1d
    close(); //Close Channel 2
    // close(dir_Input+list[i]);
    //setBatchMode(true);
   }

   //------------------File selection function

   function get_all_parameters()
   {
       Dialog.create("Please set program parameters");
       Dialog.addMessage("Parameters for detecting cell boundaries");
       Dialog.addSlider("Minimum Cell Size", 2500, 10000, 5000);
       Dialog.addSlider("Maximum Cell Size", 3000, 100000, 50000);
       Dialog.addSlider("Pixel Block Size for Cell Boundary", 3, 1000, 437);
       Dialog.addSlider("Subtraction Value for Cell Boundary", -100,
100, -12);
       Dialog.addSlider("Gaussian Blur Radius", 0, 30, 3);

       Dialog.addMessage("Parameters for detecting basal bodies")
       Dialog.addSlider("Minimum Basal Body Size", 10, 100, 10);
       Dialog.addSlider("Maximum Basal Body Size", 10, 100, 40);
       Dialog.addSlider("Pixel Block Size for Basal Bodies", 3, 100, 60);
       Dialog.addSlider("Subtraction Value for Basal Bodies", -100,
100, -11);
```

BOX 1 CODE FOR AUTOMATED QUANTIFICATION OF CENTRIOLE NUMBER—cont'd

```
        Dialog.addMessage("Misc");
        Dialog.addCheckbox("Including Edges?", false);
        Dialog.addCheckbox("Watershed?", false);
        Dialog.addSlider("Minimum Basal Bodies Count Threshold Per Cell?",
0, 50, 10);

        Dialog.addMessage("Basal Body Counter v0.1 by Siwei Zhang, 16 Oct
2014");

        Dialog.show();

        minimum_cell_size = Dialog.getNumber();
        maximum_cell_size = Dialog.getNumber();
        adaptiveThr_pixel_block_size_cell = Dialog.getNumber();
        adaptiveThr_subtract_value_cell = Dialog.getNumber();
        gaussian_radius = Dialog.getNumber();

        minimum_centriole_size = Dialog.getNumber();
        maximum_centriole_size = Dialog.getNumber();
        adaptiveThr_pixel_block_size_centriole = Dialog.getNumber();
        adaptiveThr_subtract_value_centriole = Dialog.getNumber();
        edge_count = Dialog.getCheckbox();
        watershed_count = Dialog.getCheckbox();
        min_count_threshold = Dialog.getNumber();

        return;
    }
```

CONCLUSION

Centriole biogenesis has been the focus of research in many labs. The vast majority of studies have been performed in cell culture or the early *Caenorhabditis elegans* embryo. This is due to the fact that many deletions of proteins involved in centriole formation are embryonic lethal. In addition to the role that centrioles play in organizing microtubules during interphase (MTOC) and mitosis (spindle pole), they also have the capacity to act as basal bodies nucleating the microtubule-based cilium. While the regulation of cell cycle-dependent centriole duplication has been widely studied, the massive amplification of centrioles that occurs within MCCs is only starting to be addressed at a molecular level. MCCs are found in a variety of tissues including the mammalian lung, oviduct, and ependyma. These tissues, as well as primary cultures, have proven to be extremely useful for the study of MCC formation and function. Here, we have described the use of the skin of *Xenopus* embryos as a source for addressing the molecular regulation of centriole amplification. *Xenopus* has the advantage that the ciliated tissue develops on the external surface of the animal, thus eliminating the need for tissue dissection or long-term tissue culture. The ease of molecular manipulations and the speed of development have also made *Xenopus* an appealing system with which to address the biology of MCCs and, specifically, the fascinating question of how these cells uncouple centriole biogenesis from the cell cycle and how they utilize the poorly characterized structure the deuterosome to nucleate centrioles de novo. *Xenopus* has been instrumental in recent advances in understanding both the transcriptional program leading to MCCS as well as the initial molecular characterization of the deuterosome. However, clearly there is much to be learned about these unique cells and we hope that the techniques provided here will be a spring board for future work in this area.

REFERENCES

Al Jord, A., Lemaitre, A. I., Delgehyr, N., Faucourt, M., Spassky, N., & Meunier, A. (2014). Centriole amplification by mother and daughter centrioles differs in multiciliated cells. *Nature, 516*(7529), 104–107. http://dx.doi.org/10.1038/nature13770.

Amaya, E., & Kroll, K. L. (1999). A method for generating transgenic frog embryos. *Methods in Molecular Biology, 97*, 393–414. http://dx.doi.org/10.1385/1-59259-270-8:393.

Antoniades, I., Stylianou, P., & Skourides, P. A. (2014). Making the connection: ciliary adhesion complexes anchor basal bodies to the actin cytoskeleton. *Developmental Cell, 28*(1), 70–80. http://dx.doi.org/10.1016/j.devcel.2013.12.003.

Brownlee, C. W., & Rogers, G. C. (2013). Show me your license, please: deregulation of centriole duplication mechanisms that promote amplification. *Cellular and Molecular Life Sciences, 70*(6), 1021–1034. http://dx.doi.org/10.1007/s00018-012-1102-6.

Chien, Y. H., Werner, M. E., Stubbs, J., Joens, M. S., Li, J., Chien, S., et al. (2013). Bbof1 is required to maintain cilia orientation. *Development, 140*(16), 3468–3477. http://dx.doi.org/10.1242/dev.096727.

Dale, L., & Slack, J. M. (1987). Fate map for the 32-cell stage of *Xenopus laevis*. *Development, 99*(4), 527–551.

Dammermann, A., Pemble, H., Mitchell, B. J., McLeod, I., Yates, J. R., 3rd, Kintner, C., et al. (2009). The hydrolethalus syndrome protein HYLS-1 links core centriole structure to cilia formation. *Genes & Development, 23*(17), 2046–2059. http://dx.doi.org/10.1101/gad.1810409.

Deblandre, G. A., Wettstein, D. A., Koyano-Nakagawa, N., & Kintner, C. (1999). A two-step mechanism generates the spacing pattern of the ciliated cells in the skin of *Xenopus* embryos. *Development, 126*(21), 4715–4728.

Fryer, C. J., Lamar, E., Turbachova, I., Kintner, C., & Jones, K. A. (2002). Mastermind mediates chromatin-specific transcription and turnover of the Notch enhancer complex. *Genes & Development, 16*(11), 1397–1411. http://dx.doi.org/10.1101/gad.991602.

Kalnins, V. I., & Porter, K. R. (1969). Centriole replication during ciliogenesis in the chick tracheal epithelium. *Zeitschrift fur Zellforschung und mikroskopische Anatomie, 100*(1), 1–30.

Khokha, M. K., Chung, C., Bustamante, E. L., Gaw, L. W., Trott, K. A., Yeh, J., et al. (2002). Techniques and probes for the study of *Xenopus tropicalis* development. *Developmental Dynamics, 225*(4), 499–510. http://dx.doi.org/10.1002/dvdy.10184.

Kieserman, E. K., Lee, C., Gray, R. S., Park, T. J., & Wallingford, J. B. (2010). High-magnification in vivo imaging of *Xenopus* embryos for cell and developmental biology. *Cold Spring Harbor Protocols, 2010*(5). http://dx.doi.org/10.1101/pdb.prot5427. pdb prot5427.

Klos Dehring, D. A., Vladar, E. K., Werner, M. E., Mitchell, J. W., Hwang, P., & Mitchell, B. J. (2013). Deuterosome-mediated centriole biogenesis. *Developmental Cell, 27*(1), 103–112. http://dx.doi.org/10.1016/j.devcel.2013.08.021.

Lee, C., Kieserman, E., Gray, R. S., Park, T. J., & Wallingford, J. (2008). Whole-mount fluorescence immunocytochemistry on *Xenopus* embryos. *Cold Spring Harbor Protocols, 2008*. http://dx.doi.org/10.1101/pdb.prot4957. pdb prot4957.

Loeber, J., Pan, F. C., & Pieler, T. (2009). Generation of transgenic frogs. *Methods in Molecular Biology, 561*, 65–72. http://dx.doi.org/10.1007/978-1-60327-019-9_4.

Mimoto, M. S., & Christian, J. L. (2011). Manipulation of gene function in *Xenopus laevis*. *Methods in Molecular Biology, 770*, 55–75. http://dx.doi.org/10.1007/978-1-61779-210-6_3.

Mitchell, B., Jacobs, R., Li, J., Chien, S., & Kintner, C. (2007). A positive feedback mechanism governs the polarity and motion of motile cilia. *Nature, 447*(7140), 97–101. http://dx.doi.org/10.1038/nature05771.

Mitchell, B., Stubbs, J. L., Huisman, F., Taborek, P., Yu, C., & Kintner, C. (2009). The PCP pathway instructs the planar orientation of ciliated cells in the *Xenopus* larval skin. *Current Biology, 19*(11), 924–929. http://dx.doi.org/10.1016/j.cub.2009.04.018.

Moody, S. A. (1987a). Fates of the blastomeres of the 16-cell stage *Xenopus* embryo. *Developmental Biology, 119*(2), 560–578.

Moody, S. A. (1987b). Fates of the blastomeres of the 32-cell-stage *Xenopus* embryo. *Developmental Biology, 122*(2), 300–319.

Nigg, E. A., Cajanek, L., & Arquint, C. (2014). The centrosome duplication cycle in health and disease. *FEBS Letters, 588*(15), 2366–2372. http://dx.doi.org/10.1016/j.febslet.2014.06.030.

Park, T. J., Mitchell, B. J., Abitua, P. B., Kintner, C., & Wallingford, J. B. (2008). Dishevelled controls apical docking and planar polarization of basal bodies in ciliated epithelial cells. *Nature Genetics, 40*(7), 871–879. http://dx.doi.org/10.1038/ng.104.

Park, A. Y., Shen, T. L., Chien, S., & Guan, J. L. (2009). Role of focal adhesion kinase Ser-732 phosphorylation in centrosome function during mitosis. *Journal of Biological Chemistry, 284*(14), 9418–9425. http://dx.doi.org/10.1074/jbc.M809040200.

Ryan, K., Russ, A. P., Levy, R. J., Wehr, D. J., You, J., & Easterday, M. C. (2004). Modulation of eomes activity alters the size of the developing heart: implications for in utero cardiac gene therapy. *Human Gene Therapy, 15*(9), 842–855. http://dx.doi.org/10.1089/hum.2004.15.842.

Sive, H. L., Grainger, R. M., & Harland, R. M. (1998). *The early development of* Xenopus laevis*: A laboratory manual*. Plainview, New York: Cold Spring Harbor Laboratories.

Sive, H. L., Grainger, R. M., & Harland, R. M. (2007a). Animal cap isolation from *Xenopus laevis*. *Cold Spring Harbor Protocols, 2007*. http://dx.doi.org/10.1101/pdb.prot4744. pdb prot4744.

Sive, H. L., Grainger, R. M., & Harland, R. M. (2007b). Isolating *Xenopus laevis* testes. *Cold Spring Harbor Protocols, 2007*. http://dx.doi.org/10.1101/pdb.prot4735. pdb prot4735.

Sive, H. L., Grainger, R. M., & Harland, R. M. (2007c). Synthesis and purification of digoxigenin-labeled RNA probes for in situ hybridization. *Cold Spring Harbor Protocols, 2007*. http://dx.doi.org/10.1101/pdb.prot4778. pdb prot4778.

Sive, H. L., Grainger, R. M., & Harland, R. M. (2007d). *Xenopus laevis* egg collection. *Cold Spring Harbor Protocols, 2007*. http://dx.doi.org/10.1101/pdb.prot4736. pdb prot4736.

Sive, H. L., Grainger, R. M., & Harland, R. M. (2007e). *Xenopus laevis* in vitro fertilization and natural mating methods. *Cold Spring Harbor Protocols, 2007*. http://dx.doi.org/10.1101/pdb.prot4737. pdb prot4737.

Sive, H. L., Grainger, R. M., & Harland, R. M. (2010a). Microinjection of RNA and preparation of secreted proteins from *Xenopus* oocytes. *Cold Spring Harbor Protocols, 2010*(12). http://dx.doi.org/10.1101/pdb.prot5538. pdb prot5538.

Sive, H. L., Grainger, R. M., & Harland, R. M. (2010b). Microinjection of *Xenopus* embryos. *Cold Spring Harbor Protocols, 2010*(12). http://dx.doi.org/10.1101/pdb.ip81. pdb ip81.

Sorokin, S. P. (1968). Reconstructions of centriole formation and ciliogenesis in mammalian lungs. *Journal of Cell Science, 3*(2), 207–230.

Steinman, R. M. (1968). An electron microscopic study of ciliogenesis in developing epidermis and trachea in the embryo of *Xenopus laevis*. *American Journal of Anatomy, 122*(1), 19–55. http://dx.doi.org/10.1002/aja.1001220103.

Stubbs, J. L., Davidson, L., Keller, R., & Kintner, C. (2006). Radial intercalation of ciliated cells during *Xenopus* skin development. *Development, 133*(13), 2507–2515. http://dx.doi.org/10.1242/dev.02417.

Stubbs, J. L., Oishi, I., Izpisua Belmonte, J. C., & Kintner, C. (2008). The forkhead protein Foxj1 specifies node-like cilia in *Xenopus* and zebrafish embryos. *Nature Genetics, 40*(12), 1454–1460. http://dx.doi.org/10.1038/ng.267.

Stubbs, J. L., Vladar, E. K., Axelrod, J. D., & Kintner, C. (2012). Multicilin promotes centriole assembly and ciliogenesis during multiciliate cell differentiation. *Nature Cell Biology, 14*(2), 140–147. http://dx.doi.org/10.1038/ncb2406.

Tan, F. E., Vladar, E. K., Ma, L., Fuentealba, L. C., Hoh, R., Espinoza, F. H., et al. (2013). Myb promotes centriole amplification and later steps of the multiciliogenesis program. *Development, 140*(20), 4277–4286. http://dx.doi.org/10.1242/dev.094102.

Vitre, B. D., & Cleveland, D. W. (2012). Centrosomes, chromosome instability (CIN) and aneuploidy. *Current Opinion in Cell Biology, 24*(6), 809–815. http://dx.doi.org/10.1016/j.ceb.2012.10.006.

Wadsworth, P. (2007). Studying mitosis in cultured mammalian cells. *Cold Spring Harbor Protocols, 2007*. http://dx.doi.org/10.1101/pdb.prot4674. pdb prot4674.

Werner, M. E., Hwang, P., Huisman, F., Taborek, P., Yu, C. C., & Mitchell, B. J. (2011). Actin and microtubules drive differential aspects of planar cell polarity in multiciliated cells. *Journal of Cell Biology, 195*(1), 19–26. http://dx.doi.org/10.1083/jcb.201106110.

Werner, M. E., & Mitchell, B. J. (2013). Using *Xenopus* skin to study cilia development and function. In W. F. Marshall (Ed.), *Methods in Enzymology, volume 525: Cilia, Part B* (Vol. 525).

Werner, M. E., Mitchell, J. W., Putzbach, W., Bacon, E., Kim, S. K., & Mitchell, B. J. (2014). Radial intercalation is regulated by the Par complex and the microtubule-stabilizing protein CLAMP/Spef1. *Journal of Cell Biology, 206*(3), 367–376. http://dx.doi.org/10.1083/jcb.201312045.

Zhao, H., Zhu, L., Zhu, Y., Cao, J., Li, S., Huang, Q., et al. (2013). The Cep63 paralogue Deup1 enables massive de novo centriole biogenesis for vertebrate multiciliogenesis. *Nature Cell Biology, 15*(12), 1434–1444. http://dx.doi.org/10.1038/ncb2880.

CHAPTER

8

Subdiffraction resolution microscopy methods for analyzing centrosomes organization

Vito Mennella*,§,¶,1, Rachel Hanna*,§, Moshe Kim*,§

**Department of Biochemistry, University of Toronto, Toronto, ON, Canada*

§Cell Biology Program, The Hospital for Sick Children, Toronto, ON, Canada

¶Peter Gilgan Centre for Research and Learning, Toronto, ON, Canada

[1]Corresponding author: E-mails: vito.mennella@sickkids.ca; vito.mennella@utoronto.ca

CHAPTER OUTLINE

Introduction **130**

1. Methods **133**

1.1 Sample Choice and Preparation 133

1.1.1 Cells, embryos, isolated centrosomes 133

1.1.2 Fixation 135

1.1.3 Fluorescence labeling method 135

1.1.4 Primary antibody selection 136

1.1.5 Reference markers 136

1.1.6 Secondary antibodies and multicolor imaging 136

1.1.7 Sample mounting 138

1.2 Image Acquisition and Data Analysis 140

1.2.1 3DSIM microscope image acquisition 140

1.2.2 STORM microscopy image acquisition 140

1.3 Data Analysis 141

1.3.1 3D volume alignment and averaging 141

1.3.2 Analysis of intensity profiles 143

1.4 Reagents, Protocols 144

1.4.1 Sample preparation 144

1.4.2 Fixation 144

1.4.3 Fluorescent antibody staining 145

1.4.4 Imaging buffers 145

Methods in Cell Biology, Volume 129, ISSN 0091-679X, http://dx.doi.org/10.1016/bs.mcb.2015.03.009

Summary and Outlook 146
Acknowledgments 146
References 146

Abstract

In this chapter, we describe the current methods of examining the structure of centrosomes by fluorescence subdiffraction microscopy. By using recently developed microscopy techniques, centrosomal proteins can now be examined in cells with a resolution of only a few nanometers, a level of molecular detail beyond the reach of traditional cell biology methods as confocal and widefield microscopy. We emphasize imaging by three-dimensional structured illumination microscopy, stochastic optical reconstruction microscopy, and quantitative approaches to image data analysis.

INTRODUCTION

In recent years, proteomic analyses (Andersen, Wilkinson, & Mayor, 2003; Keller, Romijn, Zamora, Yates, & Marshall, 2005; Müller et al., 2010), comparative genomics (Avidor-reiss et al., 2004; Azimzadeh, 2014; Li et al., 2004), and loss of function studies (Alves-Cruzeiro, Nogales-Cadenas, & Pascual-Montano, 2014) have identified a large inventory of proteins involved in different facets of centrosome function. With this new wealth of information, a complex and more realistic picture of the centrosome's role in cells is emerging. The centrosome is no longer viewed just as a microtubule-organizing center but also as a hub at the crossroads of signal transduction pathways (Arquint, Gabryjonczyk, & Nigg, 2014; Bornens & Gönczy, 2014). As new centrosome functions are revealed, it is critical to determine the mechanisms through which the centrosome's structural plasticity is determined (Brito, Gouveia, & Bettencourt-Dias, 2012; Gönczy, 2012); how the centrosome, either by acquiring or exchanging structural and functional modules (Brito et al., 2012), coordinates many functions such as the organization of the microtubule cytoskeleton (Bornens, 2012; Nigg & Raff, 2009), ciliogenesis (Ishikawa & Marshall, 2011; Kobayashi & Dynlacht, 2011), cell polarity and migration (Bornens, 2012; Tang & Marshall, 2012), formation of the immunological synapse (Stinchcombe & Griffiths, 2014), and DNA damage control (Arquint et al., 2014).

It is an open and fascinating question as to how hundreds of protein components that make up the centrosome are first assembled in discrete dynamic protein complexes, and then integrated into higher-order supramolecular assemblies (Gopalakrishnan et al., 2011; Habermann & Lange, 2012; Mennella, Agard, Huang, & Pelletier, 2013; Woodruff, Wueseke, & Hyman, 2014; Wueseke et al., 2014). Addressing this issue will allow us to develop a comprehensive, mechanistic view of how centrosomes function in cells. Toward this goal, it is important to build a multiresolution scale structural and functional map (Alber, Förster, Korkin, Topf, & Sali, 2008; Ryan et al., 2013), which integrates cell biology studies, the angstrom resolution maps derived from structural methods, and nanometer-resolution information, which has been largely missing from studies in centrosome biology (Habermann & Lange, 2012; Mennella et al., 2013).

Subdiffraction resolution fluorescence microscopies are powerful methods for obtaining information on the cellular organization at the nanometer-resolution scale. The most common implementations—three-dimensional structured illumination microscopy (3DSIM) (Gustafsson et al., 2008), stochastic optical reconstruction microscopy (STORM) (Rust, Bates, & Zhuang, 2006), photoactivated localization microscopy (PALM) (Betzig et al., 2006), and stimulated emission depletion (STED) (Klar, Jakobs, Dyba, Egner, & Hell, 2000)—circumvent Abbe's diffraction limit (Ball, Parton, Hamilton, & Davis, 2012). This was previously considered a hard physical limit of fluorescence optical microscopy, which set the resolution to approximately 250 nm in the plane x/y direction and approximately 500 nm in the axial (z) direction (when imaging with a high numerical aperture (1.4 NA) objective and fluorescence emission ~520 nm). In this review, we highlight 3DSIM and STORM applications for the subdiffraction imaging analysis of centrosomes. These methods break the diffraction barrier by employing different optical principles and have different strengths in terms of resolution and multicolor imaging capabilities, thereby providing an excellent dual combination approach.

The 3DSIM is in essence a modified wide-field microscope in which the sample is illuminated with a striped pattern—"spatially structured light illumination" (Gustafsson et al., 2008; Schermelleh et al., 2008). This pattern is used to create 3D interference fringes on the sample, which in turn shifts high-frequency sample information not normally observable by the microscope to the observable range of frequencies (Gustafsson, 2000; Gustafsson et al., 2008; Schermelleh et al., 2008). 3DSIM offers the advantage of straightforward sample preparation and implementation of multicolor imaging but provide an increase in resolution of only a factor of 2 (125 nm in the x/y plane and 250 nm in z plane). Conversely, single molecule localization techniques, such as STORM and PALM, break the diffraction barrier by detecting stochastically activated diffraction-limited single molecules, which have to be first separated in space and time, and then localized with high resolution by Gaussian fitting (Huang, Bates, & Zhuang, 2009). These methods can reach a higher resolution (routinely ~25-nm x/y plane and 50-nm z plane), and they require a simple experimental setup, easily built in the laboratory. However, they necessitate a careful control of fluorophore photoactivation and density, and require extra effort in terms of sample preparation (Dempsey, 2013; Huang et al., 2009). We refer to recent reviews for a detailed discussion of the biophysical principles of these technologies, which are commonly referred to as super resolution microscopies (Hell, 2007; Huang et al., 2009; Schermelleh, Heintzmann, & Leonhardt, 2010).

Many properties of the centrosome make it an ideal organelle to investigate by subdiffraction resolution imaging. Centrosomes are symmetrical organelles with a centriole core whose dimensions - 250 nm in diameter and 150 nm to microns in length depending on the species and tissue- are close to the diffraction limit (Winey & O'Toole, 2014). Individual protein components are often arranged in well-ordered structures with high-molecular density, a property that facilitates subdiffraction fluorescence imaging (because of the high signal to noise for 3DSIM and the large number of localizations available in STORM/PALM). Centrosomes also include an unusual number of large coiled-coil proteins (Nido, Méndez, Pascual-García, Abia, & Bastolla, 2012), elongated alpha helical bundles that are frequently assembled in multimeric structures and are part of high-order assemblies, and therefore more easily amenable to domain position analysis (Lawo, Hasegan, Gupta, & Pelletier, 2012; Mennella et al., 2012).

Importantly, subdiffraction resolution methods investigate the structure and dynamics of centrosomes inside cells. There is no need for purification or a priori knowledge of protein assembly structures, information that is largely required in methods such as Förster resonance energy transfer (Juette et al., 2014). This is of particular importance in centrosome biology, because large coiled-coil proteins of the centrosome are quite challenging to purify, difficult to reconstitute as larger assemblies due to their size and they tend to aggregate likely due to an unusual amount of predicted disordered sequences (Nido et al., 2012).

For many years, the study of structural organization of centrosomes has been driven by electron microscopy (EM) experiments. EM imaging of intact centrosomes has revealed many of its structural elements, such as an insoluble protein matrix running throughout the pericentriolar material (PCM) made of 12- to 15-nm elements and other unknown structures (Moritz, Zheng, Alberts, & Oegema, 1998; Paintrand et al., 1992; Schnackenberg, Hull, Balczon, & Palazzo, 2000). More recently, EM has offered insights into the ultrastructural organization of the centriole cartwheel (van Breugel et al., 2011; Kitagawa et al., 2011), procentrioles (Guichard, Chrétien, Marco, & Tassin, 2010; Pelletier, O'Toole, Schwager, Hyman, & Müller-Reichert, 2006), or the gamma tubulin ring complex γTuRC (Kollman, Merdes, Mourey, & Agard, 2011). However, under the electron microscope, the rather homogeneous electron density, along with the antigen disruption caused by sectioning and antibody penetration issues, can have a negative effect on labeling and identification of specific proteins by immuno-EM. These limitations can be overcome in some instances by using smaller antibody probes for immuno-EM (Kilmartin, 2014) or by using protein tagging strategies in Cryo-EM (Choy, Kollman, Zelter, Davis, & Agard, 2009; Song et al., 2015). When compared to EM methods, subdiffraction fluorescence imaging has the overall advantage of a better labeling specificity and sensitivity, and it is relatively easy to implement, but it has the disadvantage of a lower resolution power. Future efforts should be aimed at integrating the strengths of both "subdiffraction imaging and EM modalities" in correlative approaches to investigate centrosome structure on a systematic scale (Jord et al., 2014; Klos Dehring et al., 2013; Shtengel et al., 2014).

Here, we present an overview of how to use subdiffraction imaging methods to study centrosome high-order structure. These methods have just begun to change the view of how centrosomes are organized by revealing the organization of the PCM (Fu & Glover, 2012; Lawo et al., 2012; Mennella et al., 2012; Sonnen, Schermelleh, Leonhardt, & Nigg, 2012), mechanistic details of centriole duplication (Comartin et al., 2013;

Fong, Kim, Yang, Liao, & Tsou, 2014; Klos Dehring et al., 2013; Sir et al., 2011) and amplification (Jord et al., 2014; Zhao et al., 2013), and the organization of ciliary structures (Lau, Lee, Sahl, Stearns, & Moerner, 2012). More discoveries are awaited.

1. METHODS

1.1 SAMPLE CHOICE AND PREPARATION

In particular when using subdiffraction imaging methods, the focal depth of the structure to image is a major issue to consider: the closer the centrosome is to the coverslip, the easier it is to image. In proximity to the objective, more photons can be collected from a single fluorophore, and less scattering and fewer optical aberrations are introduced. We have used a *Drosophila melanogaster* cell line of embryonic origin (Schneider S2). These cells have a relatively small volume and can be flattened either by placing an agarose pad on top of the cells in a glass bottom dish (Kner, Chhun, Griffis, Winoto, & Gustafsson, 2009) or by spreading the cells on a coverslip covered with a lectin such as concanavalin A (Rogers & Rogers, 2008). However, any tissue culture cell line (e.g., HeLa, retinal pigment epithelium (RPE), U2OS; Lawo et al., 2012; Mennella et al., 2012) or tissue (i.e., *Drosophila* syncytial blastoderm embryos; Conduit et al., 2014) can be imaged if it is close to the coverslip and has a relatively homogenous refractive index (RI). In practice, 3DSIM imaging is used routinely up to 10−15 μm from the coverslip, while single molecule imaging methods are more sensitive to depth (Juette et al., 2008; Kamiyama & Huang, 2012). New subdiffraction imaging strategies based on structured illumination patterns have been developed, which are capable of imaging to a depth of >15 μM (Gao et al., 2012; York et al., 2013).

1.1.1 Cells, embryos, isolated centrosomes

To minimize background fluorescence, it is recommended to preclean the high-precision glass coverslip #1.5 (0.17 mm) before sample preparation. A plasma cleaner apparatus is the best choice when available (H_2O_2 gas treatment for 2 min is sufficient). Alternatively, glass coverslips can be treated with 1 M KOH for 30′ in a sonicator bath followed by extensive washes with Milli-Q pure water. S2 cells are plated during the exponential growth phase for a minimum of 4 h to overnight on glass coverslips, 35-mm glass bottom dishes, or eight-well Lab-Tek glass bottom chambers, coated with concanavalin A to promote cell spreading. Human cell lines, such as HeLa and RPE cells, are directly cultured on the same type of supports before fixation.

Centrosomes can be easily isolated from cells or embryo extracts. We had good results by using success using a protocol to purify *Drosophila* centrosomes developed by Michelle Moritz at the University of California, San Francisco (Figure 1(A) and (B)) (Moritz et al., 1995). Excellent protocols for mammalian centrosome purification are also available (Jakobsen, Schrøder, Larsen, Lundberg, & Andersen, 2013; Mitchison & Kirschner, 1986). Once the centrosomes are

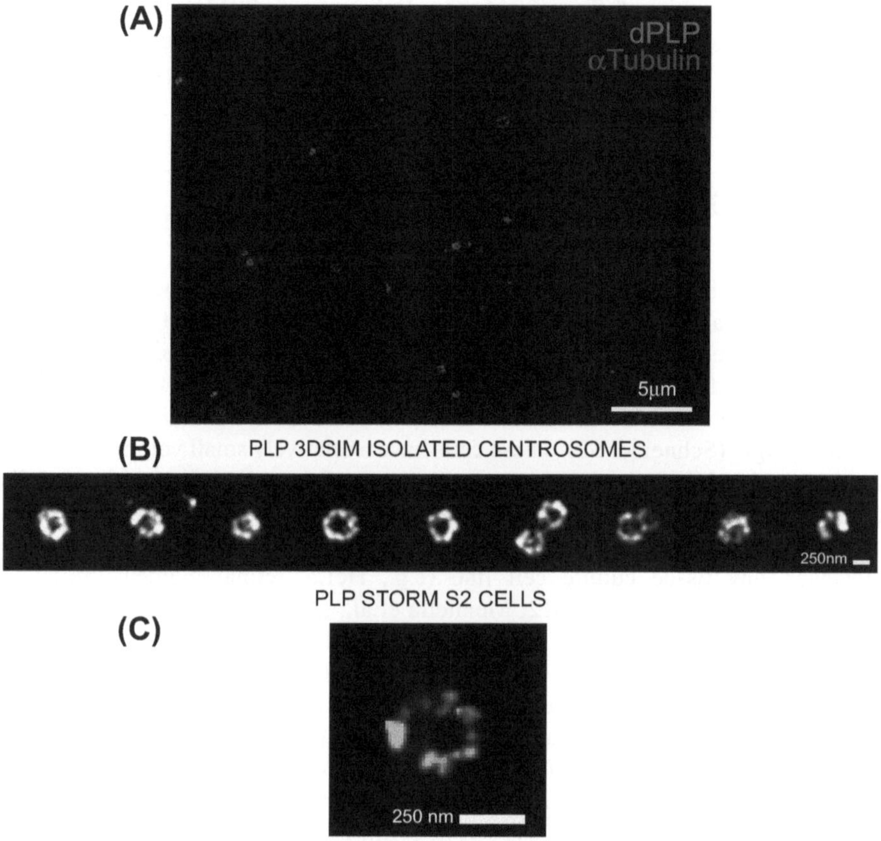

FIGURE 1

(A) 2D maximum intensity projections of SIM images of centrosomes isolated from *Drosophila* syncytial blastoderm embryos. After purification, centrosomes are allowed to regrow microtubules with rhodamine–tubulin and are labeled with rabbit antipericentrin like protein (Plp) antibodies recognized with antirabbit Alexa Fluor 488. Scale bars, 500 nm. (B) Examples of biochemically purified centrosomes labeled with rabbit anti-Plp recognized with antirabbit Alexa Fluor 488. Notice the variability in centrosome orientation and cell cycle stage. (C) STORM image of a centrosome in an S2 cell labeled with rabbit anti-Plp antibodies recognized with antirabbit Alexa Fluor 647. The secondary antibodies are labeled with a ratio of 0.3647 molecules per antibody. (See color plate)

Adapted from Mennella et al. (2012).

biochemically isolated, the purified material is made to adhere to the glass coverslip by centrifugation through a 30% glycerol cushion or by gravity ($\sim 30'$ at room temperature (RT) is generally sufficient for *Drosophila* centrosomes), on coverslips that have been pretreated with poly-L-lysine (Moritz et al., 1995).

1.1.2 Fixation

Imaging centrosomes in a live sample by subdiffraction microscopy is the ideal scenario because of the structural preservation of the sample. In this case, the major challenge is to acquire an entire image stack fast enough so that the centrosome does not move a distance larger than the limit of resolution of the particular imaging method we are using. Unfortunately, even with the introduction of fast frame-rate cameras, such as complementary metal–oxide–semiconductor, imaging centrosomes live by SIM for multiple time points remains challenging (Conduit et al., 2014). In most cells, centrosome movement is too fast to allow the acquisition of multiple images per stack of a 3DSIM experiment (~0.5- to 1-μm stack/sec for commercial SIM microscopes). Conversely, live PALM experiments are possible due to the combined use of very fast camera frame rates and single molecule tracking software (Huang et al., 2013; Jones, Shim, He, & Zhuang, 2011; Puchner, Walter, Kasper, Huang, & Lim, 2013; Sengupta, Van Engelenburg, & Lippincott-Schwartz, 2012; Shroff, Galbraith, Galbraith, & Betzig, 2008). However, they require a substantial investment in terms of time to characterize the blinking properties of the fluorescent proteins in the specific cell line under investigation and need careful control of imaging conditions and/or high laser power, which can be deleterious to the sample. Therefore, at this stage, fixed material remains the first choice of sample preparation for centrosome subdiffraction imaging.

The protocols for fixation and permeabilization vary depending on the antigen that needs to be recognized, and care should be taken in finding the best conditions to minimize the background and preserve centrosome ultrastructure. When available, EM images by high-pressure freezing and freeze-substituted material (Mcdonald & Morphew, 1993) or images from ice-frozen samples from Cryo-EM—the gold standards for cell architecture preservation—should be used as comparison to evaluate subdiffraction images (Guichard et al., 2013; Li, Fernandez, Marshall, & Agard, 2012). Microtubule-based structures are frequently well preserved with fixation in methanol; crosslinking reagents such as formaldehyde and glutaraldehyde should also be employed in parallel experiments because they can provide better preservation of the 3D structure (see below for protocols).

1.1.3 Fluorescence labeling method

For 3DSIM, the decision to employ antibodies or fluorescent proteins to analyze centrosomes is mostly a matter of reagent availability. In 3DSIM, all the fluorophores are excited simultaneously during acquisition; therefore, very often, we have enough signal to noise to obtain a successful image reconstruction. In single molecule imaging techniques, however, it is generally preferable to use an antibody labeling strategy. In fact, Alexa Fluor 647 is the organic dye of choice because of its photon efficiency, unless the goal is to perform live PALM experiments (Alexa Fluor 647 is not cell permeable) (Dempsey, Vaughan, Chen, Bates, & Zhuang, 2011). When antibodies are not available, a possible alternative is the use of an SNAP-tag strategy to label genetically encoded proteins with cyanine dyes (Jones et al., 2011). However, this approach can suffer from limitations of inefficient fusion protein incorporation and labeling. We refer to recent reviews in the literature comparing different fluorescent proteins when they are the sample of choice (Shcherbakova, Sengupta, Lippincott-Schwartz, & Verkhusha, 2014; Wang, Moffitt, Dempsey, Xie, & Zhuang, 2014).

1.1.4 Primary antibody selection

As a general rule, it is important to reduce the absolute distance of the antigen from the fluorescent dye that we are measuring by subdiffraction imaging. The use of smaller probes increases both the accuracy and the precision of measurements of fluorescence intensity. The minimization of the distance of the antigen from the fluorescent dye is particularly important when the objective of the experiment is to determine the position of protein domains (Lawo et al., 2012; Mennella et al., 2012). In the most common scenario of a primary secondary antibody complex recognizing an antigen, the distance between antigen and fluorophore can range from 15 to 20 nm (Figure 2). In addition, a variable number of secondary antibodies can bind a single primary antibody (Lau et al., 2012), further affecting the size of the antibody–antigen complex. The variability can also increase because the fluorescent dye can bind to the antibody in different regions of its structure. Therefore, when the amount of antibody is not limiting, preferable strategies are to label the fluorophore of interest directly to the primary antibody, or alternatively to antigen-binding fragments (Fabs), or single chain variable fragments (scFv) or nanobodies (Figure 2) (Szymborska et al., 2013). Phage and yeast display libraries allow for the fast and relatively inexpensive generation of such antibody reagents. Although protein stability and antigen affinity typically decrease along with the size of the affinity reagent, all of the above-mentioned formats have been successfully used to generate specific, low nanomolar affinity antibodies from naive libraries (Boder, Raeeszadeh-Sarmazdeh, & Price, 2012; De Meyer, Muyldermans, & Depicker, 2014; Holliger & Hudson, 2005; Miersch & Sidhu, 2012).

1.1.5 Reference markers

When building a map of several proteins, it is crucial to identify suitable reference markers, which allow for the determination of centrosome orientation (end-on, tilted, or side views) and aids in identifying a molecular ruler for measuring the distance of unknown components (Figure 3). For example, we have used antibodies that recognize the central hub of the centriole (Figure 3) or centriolar microtubules (Mennella et al., 2012) to map the position of different PCM components (Lawo et al., 2012; Mennella et al., 2012) (see below for a list of suggested markers).

1.1.6 Secondary antibodies and multicolor imaging

When the use of secondary antibodies is required to amplify the signal, such as in the case of labeling an antigen consisting of only a few molecules, most commercially available secondary antibodies can be used for 3DSIM imaging. However, in STORM experiments, it might be necessary to generate ad hoc antibodies with variable ratios of reporter dyes, which need to be tested for antigens with different molecule densities.

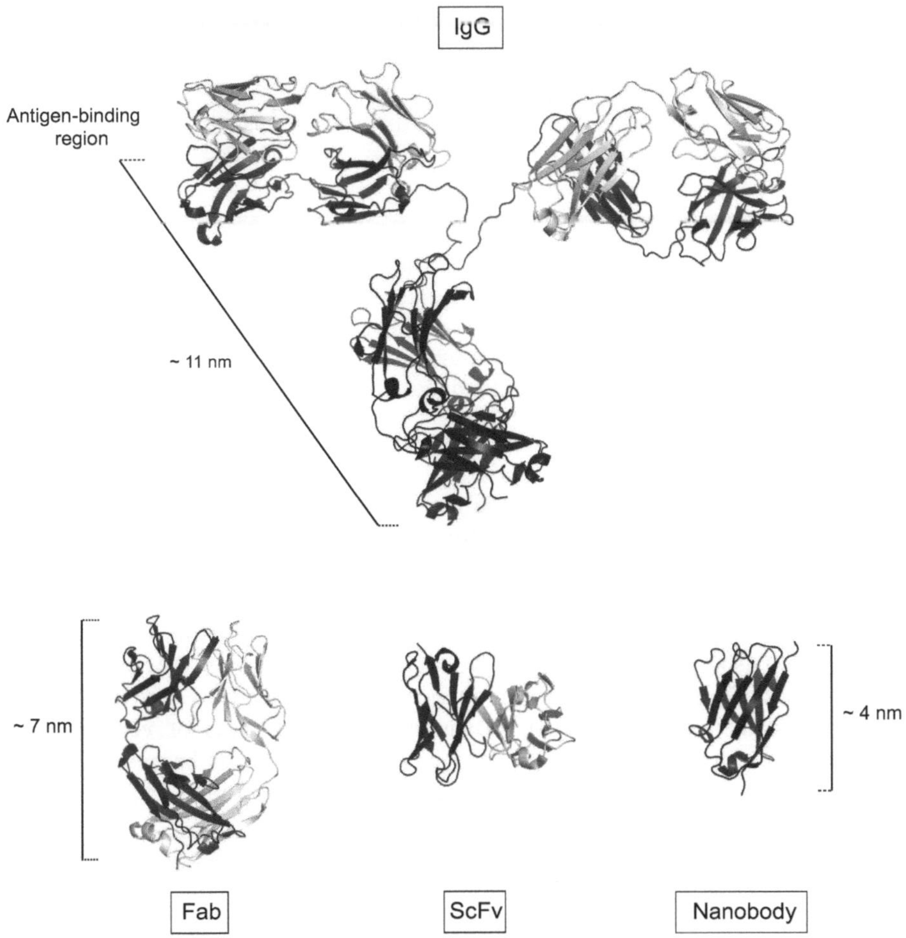

FIGURE 2

Representative crystal structures of antibody fragments. Images were created with the program PyMOL. IgG #1igt, Fab pdb #1tzi, scFv #1p4i, nanobody pdb #3ogo. Note that the antigen-binding region for Fab, scFv, and nanobodies is facing up toward the IgG structure. For all structures, the heavy chain is colored in dark gray, the light chain is colored in light gray.

Even though new powerful localization algorithms capable of separating overlapping fluorophores are available (Holden, Uphoff, & Kapanidis, 2011; Mukamel, Babcock, & Zhuang, 2012; Parthasarathy, 2012; Zhu, Zhang, Elnatan, & Huang, 2012), antibody labeling density remains a critical parameter for controlling resolution and for proper spatial and temporal separation of localization events (Dempsey, 2013; Durisic, Cuervo, & Lakadamyali, 2014; Huang et al., 2009). It is important to underscore that, when using a low ratio of fluorescent dye labeled over total antibodies, an

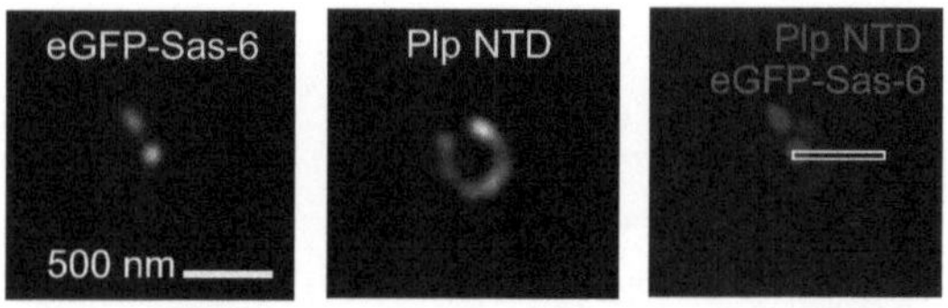

FIGURE 3

Example of centrosome reference marker (Sas-6) to determine centrosome orientation. Sas-6 is a known component of the central hub of centrioles. Note the end-on view and the side view of the Sas-6 fluorescence on the mother and daughter centriole, respectively. The end-on view is used as one end of a molecular ruler for measuring the distance of unknown components such as Plp. See the section on reagents for a list of other suggested markers. Fluorescence intensity profiles from the centre of the centriole image outward (reference marker Sas-6). Scale bar, 250 nm.

Adapted from Mennella et al. (2012).

appropriate molecule density has to be maintained to satisfy the Nyquist criterion (Dempsey, 2013; Huang et al., 2009; Shroff et al., 2008).

For multicolor image acquisition by STORM, we have successfully used photoswitchable dye pairs such as A405−A647 and A555−A647 (Bates, Huang, Dempsey, & Zhuang, 2007; Dempsey, 2013) (Figure 4). This approach has the major advantage of using the same reporter dye, but it is inefficient when the molecule density is very different between the two antigens; this thus leads to substantial molecules crosstalk (Dempsey, 2013). When using reporter dyes of very different wavelengths, results might be complicated by alignment issues because of optical aberration and different resolution in two different channels due to different photon emission properties (Dempsey, 2013).

1.1.7 Sample mounting

If the centrosome to image is located several microns away from the coverslip—such as in a tissue sample—there can be significant fluorescent light scattering that results in a reduction of the image signal to noise. The light scattering can be further accentuated if a mismatch of the RI of the sample and the objective immersion oil (or water) is present. This can have catastrophic results on the 3DSIM reconstruction process because it reduces the contrast of the SIM illumination pattern (Ball et al., 2012; Gao, Shao, Chen, & Betzig, 2014). To minimize this negative effect, we mount samples in a glycerol-based n-propyl gallate mounting medium (Science, 2015), which has an RI of 1.451, which is close to the RI of oil commonly used on objective lenses (1.518). To perfectly match the mounting media RI, different oils are available for purchase (Cargill). If an RI mismatch is present, it can be detected by the observation of an asymmetrical point spread function by looking at a bead sample in x/z or y/z sections (Ball et al., 2012; Wallace, Schaefer, & Swedlow, 2001). It is important to note that some common commercial mounting media can give suboptimal results because of their low RI or because of changes in their optical properties over time during the curing process.

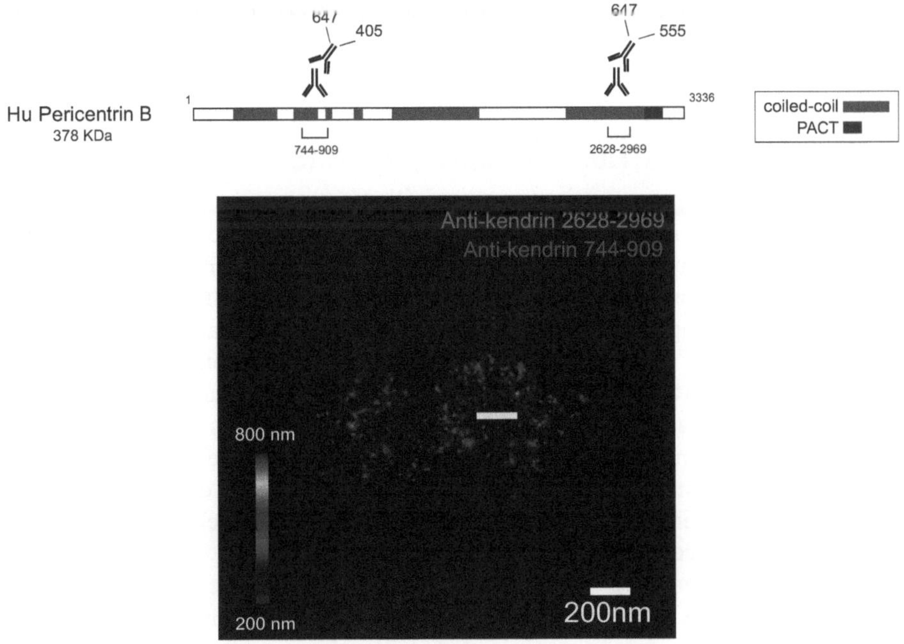

FIGURE 4

The 3D STORM image reconstruction obtained using two activators (Alexa Fluor 405 and 555) and a single reporter dye (Alexa Fluor 647). Primary antibodies recognize two different domains of Human Pericentrin B.

In STORM microscopy, a commonly used imaging buffer is composed of three parts: (1) a buffering solution such as phosphate saline buffer (PBS) pH 7.4, (2) an enzymatic oxygen scavenging system present to reduce photobleaching of the dye due to oxidation, and (3) a reducing agent that facilitates the recovery of dye molecules trapped in the dark state. In single molecule imaging, it is key to select the right imaging buffer, because it modulates the photoswitching properties of the fluorescent dyes, which in turn affects both the resolution and the capacity to resolve the structure. In particular, different reducing agents (and their concentrations) determine different photoswitching properties. This is especially important for centrosomal proteins, which can be tightly packed in high-density structures. In these circumstances, a buffer containing the reducing agent monoethanolamine (MEA) can be selected as a good starting point (see the recipe below). In this type of buffer, Alexa Fluor 647 has a lower on-off duty cycle, that is, it spends less time in a fluorescent state (and also emits lower photons). Other alternatives are 2-mercaptoethanol or 1,3,5,7-cyclooctatetraene, which can improve the number of photons extracted from a single 647 (Dempsey, 2013; Olivier, Keller, Gönczy, & Manley, 2013), but have higher on-off duty cycles. If imaging on glass bottom dishes, we suggest replacing the STORM imaging buffer frequently ($\sim$1 h) to maintain relatively constant imaging conditions—the oxygen-scavenging system lasts only a few hours when continuously in contact with air. Another option is to seal the coverslip to reduce the exposure to oxygen.

1.2 IMAGE ACQUISITION AND DATA ANALYSIS

1.2.1 3DSIM microscope image acquisition

Commercially available 3DSIM microscopes perform automated image acquisition steps and reconstruction, which makes the life of the end-user considerably easy. Here, we suggest some basic steps to ensure a successful 3DSIM image stack. Primarily, it is very important to have a sample with a good signal to noise. We have obtained successful reconstructions with as little as 1000 photon counts over the background, but >10,000 counts in an electron multiplying charge coupled device camera is a comfortable range. If higher counts are acquired, it is important to ensure that the camera is in the linear response range by adjusting the exposure time and laser power accordingly. The range of linear response can be different depending on the type of camera and acquisition mode (conventional vs. electron multiplying gain mode) and needs to be measured experimentally.

In a 3DSIM microscope acquisition, for each z position, we need to acquire 15 images (five images at different phases for three different grating angle rotations). Excellent 3DSIM raw data have minimal photobleaching and drift between the three different angle rotations. The raw images should be examined for the presence of phase shifts, intensity differences between angles, and overall image modulation contrast, which can be visually inspected by looking at the acquisition file. In the reconstructed image, the Fourier transform of the 3D volume should be analyzed for symmetry and homogeneity, which is lost due to incorrect sideband assignment (Gustafsson et al., 2008). A detailed discussion of the possible artifacts in SIM imaging is beyond the scope of this review, but a good reference guide and a user-friendly tool for performing image reconstruction quality control is SIM check, an image J plug-in that is publicly available (https://github.com/MicronOxford/SIMcheck).

In 3DSIM, it is necessary to apply translation, rotation, and both isotropic and anisotropic scaling for image alignment purposes. This procedure corrects for chromatic aberration and camera alignment, if more than one camera is used. The alignment parameters can be obtained from measurements of images of subdiffraction 100-nm multi-wavelength fluorescent beads. The alignment can be performed with Python scripts or software provided with commercially available 3DSIM microscopes. It is not necessary to obtain a calibration file after every acquisition, but it is good practice to do it every other week.

1.2.2 STORM microscopy image acquisition

General protocols for STORM imaging have been extensively covered in a recent issue of Methods in Cell Biology (Dempsey, 2013). Here, we focus on important details to take into account when specifically imaging centrosomes.

Sample drift is a critical parameter to control for optimal imaging. This is especially true in a PALM or STORM dataset acquisition, because of the tighter requirements due to a higher resolution and long image acquisition times. To correct for drift in the z plane, a perfect focus system is generally used, which is a part of the microscope hardware body, or, in a more sophisticated setup, a feedback loop system can be used with a dedicated camera (Pertsinidis, Zhang, & Chu, 2010). Drift in the x/y plane is corrected by fluorescence image based crosscorrelation methods by using structures in the image or fiducial markers (Dempsey, 2013; McGorty, Kamiyama, & Huang, 2013). Since centrosomes are relatively rare structures in a cell, drift correction might not be effective due to the limited number of fluorescent points/structures to correlate. This problem can be overcome by either using gold particles, which are added to the sample before imaging or gold fiducials that are embedded into the glass coverslip. However, this approach requires that the gold beads are in the same focal plane of the structure of interest. Alternatively, the acquisition of a bright field image can be added to the imaging sequence, which is subsequently used for crosscorrelation analysis (McGorty et al., 2013; Mennella et al., 2012).

1.3 DATA ANALYSIS

A considerable advantage of subdiffraction fluorescence microscopy is that it not only provides a crisper image of the centrosome compared to conventional fluorescence microscopy but it also allows for quantitative positional information of proteins in situ. The relative position of proteins in the context of larger assemblies as well as the position of protein domains can be determined with precision in the low nanometer scale (Dani, Huang, Bergan, Dulac, & Zhuang, 2010; Kanchanawong et al., 2010; Lawo et al., 2012; Mennella et al., 2012).

1.3.1 3D volume alignment and averaging

To obtain reliable information, two major issues need to be taken into account: (1) centrosomes are positioned at different orientations on the coverslip even when biochemically purified (Figures 1(A and B) and 5); (2) centrosomes can have heterogeneous structures, due to a multiplicity of factors (see below). One possible strategy to solve the orientation problem is to focus the analysis on a subset of centrosomes of similar orientations that are classified by using a marker indicating the centriole axis (Figures 3 and 5, see Section 1.1.5). This in part solves the problem of optical anisotropic resolution (when the x/y resolution is different from z), which affects all subdiffraction imaging techniques with the exception of iPALM (unfortunately, iPALM requires a rather complex optical setup). In the past, others and we have focused on end-on views of centrosomes when looking at the PCM structure, which has a centrosome with a diameter of approximately 200 nm (Klos Dehring et al., 2013; Mennella et al., 2012; Nido et al., 2012; Sonnen et al., 2012). Alternatively, centrosome side views can be selected to look at the distribution of proteins along the length of the centriole in species with a considerably longer length, such as

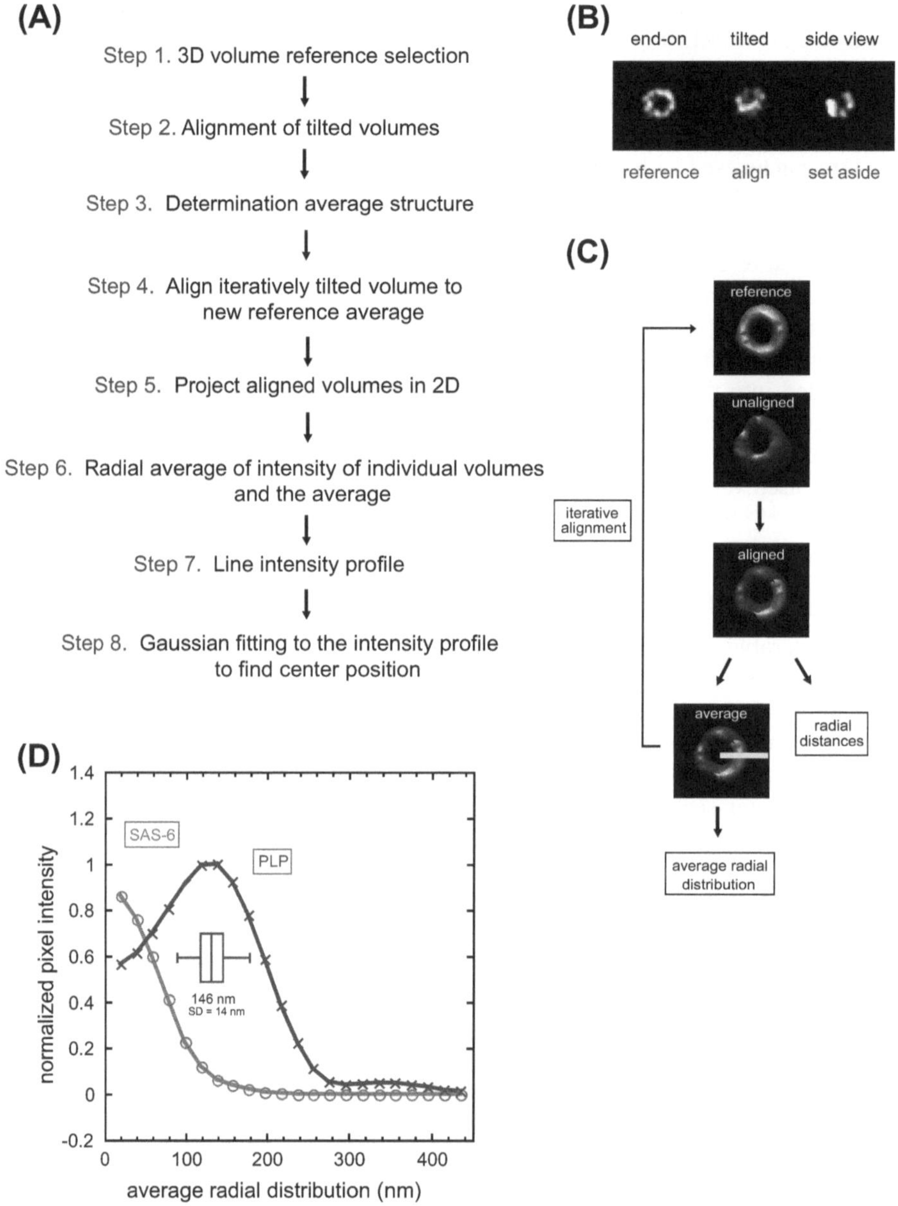

FIGURE 5

(A) Stepwise workflow of the alignment and averaging strategy of 3D image volumes of centrosomal proteins. (B) 3D volume reference selection from different centrosome orientations. (C) Volume alignment by crosscorrelation in the software Chimera. (D) Example of a line intensity profile distribution of a centrosome reference marker (sas-6) and a PCM marker (Plp).

Adapted from Mennella et al. (2012).

mammalian cells (Klos Dehring et al., 2013, Mennella et al., 2012). However, even when volumes roughly perpendicular to the imaging axis are selected, small differences in orientation can increase measurement errors due to anisotropic resolution.

To provide precise quantitative metrics of the spatial distribution of centrosomal components, it is useful to apply analysis tools borrowed from cryotomographic analyses (Frank, 2009) to align 3D subvolumes obtained from subdiffraction imaging. In this approach (Figure 5), an initial reference is selected from a large experimental dataset comprising many centrosomes, chosen on the basis of the optimal orientation of interest (Mennella et al., 2012, 2013). Subsequently, centrosome image volumes are translated, tilted, and in plane rotated to match the volume intensities of the reference by crosscorrelation of voxel intensities. A comparison of multiple volume references is useful to check for initial alignment bias. In addition, an average of the aligned volume can be used as a reference for further rounds of refinement by iterative alignment. This iterative process is quite fast compared to the one used with Cryo-EM images, likely because of the high signal to noise of the subdiffraction imaging data. This procedure can be done in image real space by the program Chimera (http://www.cgl.ucsf.edu/chimera) with the crosscorrelation function. In this program, unaligned 3D volumes can be translated and rotated in space to facilitate the alignment performed by the crosscorrelation function. To open the image volumes in Chimera, it is necessary to transform the image files into mrc format by using the software Priism (Chen, Hughes, Chan, Sedat, & Agard, 1996) (http://msg.ucsf.edu/IVE). More sophisticated approaches to image alignment and averaging rely on Fourier space filtering prior to crosscorrelation. In our experiments, we have not observed significant differences between these two alignment approaches when analyzing end-on views with small angle variation (0° to ~30° angle). However, a more detailed description of Fourier-based analysis methods at high tilt angles will be presented elsewhere.

The process of centrosome averaging can help with another important issue. Centrosomes can have heterogeneous structures due to cellular differences (i.e., cell cycle stage, dynamic structures, cell-to-cell variation), staining differences (antibody labeling, fluorescent protein levels), or deformation due to fixation and mechanical damage during sample preparation (Figure 1(B)). By deriving an average structure from many centrosomes, we eliminate small variable artifactual differences due to mechanical damage and fixation, because they are generally highly heterogeneous. However, information on significantly physiologically different structures is not lost, because it can be identified and classified as a distinct, averaged, well-populated subclass (Figure 5).

1.3.2 Analysis of intensity profiles

Once the 3D volumes are aligned in Chimera to the reference volume, radial averages about the center of mass can be calculated with image analysis programs such as Priism or Matlab from 2D algebraic projections of the average. Once the pixel-by-pixel intensity profiles are calculated, the center position of the intensity profile is determined by fitting a Gaussian function $y = y_0 + A/(\sigma \times \text{sqrt}(\pi/2)) \times (\exp(-2 \times ((x - x_c)/\sigma)^2))$ with A (height of the peak), y_0 (offset value), σ (width of the peak), and x_c (the position of the center of the peak) as parameters to fit (Mennella et al., 2012) (Figure 3(B)). Programs such as Origin or Kaleidagraph can be used for this purpose. When analyzing proteins that are present in different populations of the centrosome—centriole vs. PCM, for example, or centrosome versus centriolar satellite— it can be advantageous to restrict the analysis of fluorescence intensity distribution by using masks, which can easily be generated by an Image J plug-in (Fu & Glover, 2012) or ad hoc Matlab scripts (Lawo et al., 2012).

1.4 REAGENTS, PROTOCOLS

1.4.1 Sample preparation

Schneider S2 (*Drosophila* Genomics Resource Center, Invitrogen)
Human cell lines (ATCC)
High-precision glass coverslips, 0.17± mm (Schott # 1.5H)
Glass bottom dishes #1.5, 0.17 mm (Mattek, Bioptechs)
Lab-TEK eight-well chambers (Nunc)
Concanavalin A (MP Biomedicals, # 195283)
Poly-l-lysin solution (0.1% w/v in H_2O) (Sigma, #P8920)

1.4.2 Fixation

Protocol 1. MeOH fixation (Mennella et al., 2012)

- Wash in PBS, pH 7.4
- Fix in −20° C cold Methanol, 5′−15′
- Wash 1× in PBS, then 2× in PBST (PBS 0.05% Tween20)
- Block in 5% fetal bovine serum (FBS) in PBS, 20′

Protocol 2. PFA fixation (Garcia-Gonzalo et al., 2011)

- Wash in PBS, pH 7.4
- Fix in 1% PFA (Electron Microscopy Sciences, # 15713) in PBS, 20′
- Inactivate and permeabilize (25 mM glycine, 25 mM ammonium chloride, 0.1% Triton X-100 in PBS), 5′
- Block in 5% FBS in PBS, 20′

Protocol 3. PFA, then MeOH (Moisoi, Erent, Whyte, Martin, & Bayley, 2002)

- Wash in PBS, pH 7.4
- Fix in 1% PFA in PBS, 20′
- Inactivate PFA (25 mM glycine, 25 mM ammonium chloride)
- Wash 1× PBS
- Fix in ice-cold Methanol, 5′
- Wash 1× in PBS, then 2× in PBST
- Block in 5% FBS in PBS, 20′

Protocol 4. Triton X-100 preextraction and PFA fixation[108]

- Wash in PHEM (60-mM Pipes, 25 mM HEPES, 10 mM ethylene glycol tetra-acetic acid, 2 mM $MgCl_2$), pH 6.9
- Preextract and fix, 20′ (2% PFA, 0.1% TX-100 in PHEM buffer)
- Inactivate PFA in PHEM (25 mM glycine and ammonium chloride)
- Wash 1× in PHEM, then 2× in PBST
- Block in 5% FBS in PBS, 20′

Protocol 5. PFA, glutaraldehyde (Dani et al., 2010)

- Wash in PBS, pH 7.4
- Fix in 3% PFA and 0.05–0.1% glutaraldehyde (Electron Microscopy Sciences) in PBS, 10′
- Reduce with 0.1% sodium borohydride in PBS, 7′
- Wash 3× in PBS
- Block with bovine serum albumin (BSA) 3% and PBS Triton X-100 0.5%, 10′

1.4.3 Fluorescent antibody staining

Below is a short list of some commercially available antibodies recognizing centrosomal proteins that can be used as reference markers, antibodies directed against fluorescent proteins as well as secondary antibodies. Primary antibodies can be applied to the sample at 1–5 μg/mL final at RT in PBST 3% BSA, 1 h. Secondary antibodies can be used at the same concentration for 30′.

Antiacetylated Tubulin (6-11B-1, Sigma).
Antipolyglutamylated Tubulin (GT335, Enzo Biosciences)
Mouse and rabbit anti-GFP (A11121, A11122, Invitrogen)
Anti-GFP, anti-mCherry nanobody (GBA-488, GBA-594, Chromotek)
Antipericentrin (sc-68927, Santacruz)
Anti-cep152 (A302-480A, Bethyl Antibodies)
Anti-immunoglobulin (IgG) conjugated to Alexa 488,555 (Invitrogen)
Anti-IgG conjugated to Dylight 405 (Jackson)
4′,6-Diamidino-2-phenylindole 300 nm in PBST for 10 min (D1306, Invitrogen)

1.4.4 Imaging buffers

STORM Buffer with MEA

800 μL PBS, pH 7.4
100 μL of 50% (w/v) glucose solution (Sigma)
100 μL of 1 M MEA (Sigma #C30070)
50 μg/mL glucose oxidase (Sigma #G2133)
10 μg/mL catalase (Sigma #C100)

SIM mounting media

PBS, pH 7.4, 10% v/v

Ultrapure glycerol, 90% v/v (Invitrogen)
n-Propyl Gallate for microscopy, 0.05% w/v (Sigma)

Alignment tools

100-nm Multicolor beads (Tetraspeck, Invitrogen # T-7279)
Gold-embedded fiducial coverslips (Hestzig LLC # 600-30 AuF)
Gold beads (BB International, Microspheres—Nanospheres)

SUMMARY AND OUTLOOK

The future challenge is to integrate compositional, functional, and structural data into a comprehensive view of centrosome assembly. Subdiffraction methods can accelerate the reach of this goal by providing positional information on protein domains and their structural changes during the cell cycle or differentiation, clearer analysis of phenotypes, and faster data output compared to that in EM methods. Crucial to this task is to generate probes targeting various regions of all centrosomal proteins by protein tagging or antibody generation. Parallel efforts to reconstitute protein complexes from purified components or extracts along with other hybrid methods (Alber et al., 2008) are required to define the entire map of the structural organization of centrosomes.

ACKNOWLEDGMENTS

I would like to thank Lukman Winoto, Bryant Chhun, Sam Li, and members of the Agard, Sedat and Huang Laboratory for productive discussions over the years, which helped develop these imaging protocols. I also thank Andrew Sydor and Katherine Model for editorial help.

REFERENCES

Alber, F., Förster, F., Korkin, D., Topf, M., & Sali, A. (2008). Integrating diverse data for structure determination of macromolecular assemblies. *Annual Review of Biochemistry, 77*, 443–477.

Alves-Cruzeiro, J. M. D. C., Nogales-Cadenas, R., & Pascual-Montano, A. D. (2014). CentrosomeDB: a new generation of the centrosomal proteins database for human and *Drosophila melanogaster. Nucleic Acids Research, 42*, D430–D436.

Andersen, J. S., Wilkinson, C. J., & Mayor, T. (Dec 4, 2003). Proteomic characterization of the human centrosome by protein correlation profiling. *Nature, 426*(6966), 570–574.

Arquint, C., Gabryjonczyk, A.-M., & Nigg, E. A. (2014). Centrosomes as signalling centres. *Philosophical Transactions of the Royal Society of London B Biological Sciences, 369.*

Avidor-Reiss, T., Maer, A. M., Koundakjian, E., Polyanovsky, A., Keil, T., Subramaniam, S., et al. (May 14, 2004). Decoding cilia function: defining specialized genes required for compartmentalized cilia biogenesis. University of California at San Diego. *Cell, 117*(4), 527–539.

Azimzadeh, J. (2014). Exploring the evolutionary history of centrosomes. *Philosophical Transactions of the Royal Society of London B Biological Sciences, 369*.

Ball, G., Parton, R. M., Hamilton, R. S., & Davis, I. (2012). A cell biologist's guide to high resolution imaging. *Methods in Enzymology, 504*, 29–55. Elsevier Inc.

Bates, M., Huang, B., Dempsey, G. T., & Zhuang, X. (2007). Multicolor super-resolution imaging with photo-switchable fluorescent probes. *Science, 317*, 1749–1754.

Betzig, E., Patterson, G. H., Sougrat, R., Lindwasser, O. W., Olenych, S., Bonifacino, J. S., et al. (2006). Imaging intracellular fluorescent proteins at nanometer resolution. *Science, 313*, 1642–1645.

Boder, E. T., Raeeszadeh-Sarmazdeh, M., & Price, J. V. (2012). Engineering antibodies by yeast display. *Archives of Biochemistry and Biophysics, 526*, 99–106.

Bornens, M. (2012). The centrosome in cells and organisms. *Science, 335*, 422–426.

Bornens, M., & Gönczy, P. (Sep 5, 2014). Centrosomes back in the limelight. *Philosophical Transactions of the Royal Society of London B Biological Sciences, 369*(1650). http://dx.doi.org/10.1098/rstb.2013.0452. pii: 20130452.

van Breugel, M., Hirono, M., Andreeva, A., Yanagisawa, H. A., Yamaguchi, S., Nakazawa, Y., et al. (2011). Structures of SAS-6 suggest its organization in centrioles. *Science, 331*, 1196–1199.

Brito, D. A., Gouveia, S. M., & Bettencourt-Dias, M. (2012). Deconstructing the centriole: structure and number control. *Current Opinion in Cell Biology, 24*, 4–13.

Chen, H., Hughes, D. D., Chan, T. A., Sedat, J. W., & Agard, D. A. (Jan-Feb, 1996). IVE (Image Visualization Environment): a software platform for all three-dimensional microscopy applications. *Journal of Structural Biology, 116*(1), 56–60.

Choy, R. M., Kollman, J. M., Zelter, A., Davis, T. N., & Agard, D. A. (2009). Localization and orientation of the gamma-tubulin small complex components using protein tags as labels for single particle EM. *Journal of Structural Biology, 168*, 571–574.

Comartin, D., Gupta, G. D., Fussner, E., Coyaud, É., Hasegan, M., Archinti, M., et al. (2013). CEP120 and SPICE1 cooperate with CPAP in centriole elongation. *Current Biology, 23*, 1360–1366.

Conduit, P. T., Feng, Z., Richens, J. H., Baumbach, J., Wainman, A., Bakshi, S. D., et al. (2014). The centrosome-specific phosphorylation of Cnn by Polo/Plk1 drives Cnn scaffold assembly and centrosome maturation. *Developmental Cell, 28*, 659–669.

Dani, A., Huang, B., Bergan, J., Dulac, C., & Zhuang, X. (2010). Superresolution imaging of chemical synapses in the brain. *Neuron, 68*, 843–856.

De Meyer, T., Muyldermans, S., & Depicker, A. (2014). Nanobody-based products as research and diagnostic tools. *Trends in Biotechnology, 32*, 263–270.

Dempsey, G. T. (2013). A user's guide to localization-based super-resolution fluorescence imaging. *Methods in Cell Biology, 114*, 561–592. Elsevier Inc.

Dempsey, G. T., Vaughan, J. C., Chen, K. H., Bates, M., & Zhuang, X. (2011). Evaluation of fluorophores for optimal performance in localization-based super-resolution imaging. *Nature Methods, 8*, 1027–1036.

Durisic, N., Cuervo, L. L., & Lakadamyali, M. (2014). Quantitative super-resolution microscopy: pitfalls and strategies for image analysis. *Current Opinion in Chemical Biology, 20C*, 22–28.

Fluorescence microscopy: reduced photobleaching of rhodamine and fluorescein protein conjugates by n-propyl gallate. *Science, 217*, (2015), 1252–1255.

Fong, C. S., Kim, M., Yang, T. T., Liao, J.-C., & Tsou, M.-F. B. (2014). SAS-6 assembly templated by the lumen of cartwheel-less centrioles precedes centriole duplication. *Developmental Cell, 30*, 238–245.
Frank, J. (2009). Single-particle reconstruction of biological macromolecules in electron microscopy—30 years. *Quarterly Reviews of Biophysics, 42*, 139–158.
Fu, J., & Glover, D. M. (2012). Structured illumination of the interface between centriole and peri-centriolar material. *Open Biology, 2*, 120104.
Gao, L., Shao, L., Higgins, C. D., Poulton, J. S., Peifer, M., Davidson, M. W., et al. (2012). Noninvasive imaging beyond the diffraction limit of 3D dynamics in thickly fluorescent specimens. *Cell, 151*, 1370–1385.
Gao, L., Shao, L., Chen, B.-C., & Betzig, E. (2014). 3D live fluorescence imaging of cellular dynamics using Bessel beam plane illumination microscopy. *Nature Protocols, 9*, 1083–1101.
Garcia-Gonzalo, F. R., et al. (2011). A transition zone complex regulates mammalian ciliogenesis and ciliary membrane composition. *Nature Genetics, 43*, 776–784.
Gönczy, P. (2012). Towards a molecular architecture of centriole assembly. *Nature Reviews Molecular Cell Biology, 13*, 425–435.
Gopalakrishnan, J., et al. (2011). Sas-4 provides a scaffold for cytoplasmic complexes and tethers them in a centrosome. *Nature Communications, 2*, 359.
Guichard, P., et al. (2013). Native architecture of the centriole proximal region reveals features underlying its 9-fold radial symmetry. *Current Biology, 23*, 1620–1628.
Guichard, P., Chrétien, D., Marco, S., & Tassin, A.-M. (2010). Procentriole assembly revealed by cryo-electron tomography. *EMBO Journal, 29*, 1565–1572.
Gustafsson, M. G. (2000). Surpassing the lateral resolution limit by a factor of two using structured illumination microscopy. *Journal of Microscopy, 198*, 82–87.
Gustafsson, M. G., Shao, L., Carlton, P. M., Wang, C. J., Golubovskaya, I. N., Cande, W. Z., et al. (2008). Three-dimensional resolution doubling in wide-field fluorescence microscopy by structured illumination. *Biophysical Journal, 94*, 4957–4970.
Habermann, K., & Lange, B. M. (2012). New insights into subcomplex assembly and modifications of centrosomal proteins. *Cell Division, 7*, 17.
Hell, S. W. (2007). Far-field optical nanoscopy. *Science, 316*, 1153–1158.
Holden, S. J., Uphoff, S., & Kapanidis, A. N. (2011). DAOSTORM: an algorithm for high-density super-resolution microscopy. *Nature Methods, 8*, 279–280.
Holliger, P., & Hudson, P. J. (2005). Engineered antibody fragments and the rise of single domains. *Nature Biotechnology, 23*, 1126–1136.
Huang, B., Bates, M., & Zhuang, X. (2009). Super-resolution fluorescence microscopy. *Annual Review of Biochemistry, 78*, 993–1016.
Huang, F., Hartwich, T. M., Rivera-Molina, F. E., Lin, Y., Duim, W. C., Long, J. J., et al. (2013). Video-rate nanoscopy using sCMOS camera-specific single-molecule localization algorithms. *Nature Methods, 10*, 653–658.
Ishikawa, H., & Marshall, W. F. (2011). Ciliogenesis: building the cell's antenna. *Nature Reviews Molecular Cell Biology, 12*, 222–234.
Jakobsen, L., Schrøder, J. M., Larsen, K. M., Lundberg, E., & Andersen, J. S. (2013). Centrosome isolation and analysis by mass spectrometry-based proteomics. *Methods in Enzymology, 525*, 371–393. Elsevier Inc.
Jones, S. A., Shim, S.-H., He, J., & Zhuang, X. (2011). Fast, three-dimensional super-resolution imaging of live cells. *Nature Methods, 8*, 499–508.

Jord, A. Al, Lemaître, A. I., Delgehyr, N., Faucourt, M., Spassky, N., & Meunier, A. (2014). Centriole amplification by mother and daughter centrioles differs in multiciliated cells. *Nature*. http://dx.doi.org/10.1038/nature13770.

Juette, M. F., Gould, T. J., Lessard, M. D., Mlodzianoski, M. J., Nagpure, B. S., Bennett, B. T., et al. (2008). Three-dimensional sub-100 nm resolution fluorescence microscopy of thick samples. *Nature Methods, 5*, 527–529.

Juette, M. F., Terry, D. S., Wasserman, M. R., Zhou, Z., Altman, R. B., & Zheng, Q. (2014). The bright future of single-molecule fluorescence imaging. *Current Opinion in Chemical Biology, 20*, 103–111.

Kamiyama, D., & Huang, B. (2012). Development in the STORM. *Developmental Cell, 23*, 1103–1110.

Kanchanawong, P., Shtengel, G., Pasapera, A. M., Ramko, E. B., Davidson, M. W., Hess, H. F., et al. (2010). Nanoscale architecture of integrin-based cell adhesions. *Nature, 468*, 580–584.

Keller, L. C., Romijn, E. P., Zamora, I., Yates, J. R., & Marshall, W. F. (2005). Proteomic analysis of isolated *Chlamydomonas* centrioles reveals orthologs of ciliary-disease genes. *Current Biology, 15*, 1090–1098.

Kilmartin, J. V. (2014). Lessons from yeast: the spindle pole body and the centrosome. *Philosophical Transactions of the Royal Society of London B Biological Sciences, 369*.

Kitagawa, D., Vakonakis, I., Olieric, N., Hilbert, M., Keller, D., Olieric, V., et al. (2011). Structural basis of the 9-fold symmetry of centrioles. *Cell, 144*, 364–375.

Klar, T. A., Jakobs, S., Dyba, M., Egner, A., & Hell, S. W. (2000). Fluorescence microscopy with diffraction resolution barrier broken by stimulated emission. *Proceedings of the National Academy of Sciences of the United States of America, 97*, 8206–8210.

Klos Dehring, D. A., Vladar, E. K., Werner, M. E., Mitchell, J. W., Hwang, P., & Mitchell, B. J. (2013). Deuterosome-mediated centriole biogenesis. *Developmental Cell, 27*, 103–112.

Kner, P., Chhun, B. B., Griffis, E. R., Winoto, L., & Gustafsson, M. G. (2009). Super-resolution video microscopy of live cells by structured illumination. *Nature Methods, 6*.

Kobayashi, T., & Dynlacht, B. D. (2011). Regulating the transition from centriole to basal body. *Journal of Cell Biology, 193*, 435–444.

Kollman, J. M., Merdes, A., Mourey, L., & Agard, D. A. (2011). Microtubule nucleation by γ-tubulin complexes. *Nature Reviews Molecular Cell Biology, 12*, 709–721.

Lau, L., Lee, Y. L., Sahl, S. J., Stearns, T., & Moerner, W. E. (2012). STED microscopy with optimized labeling density reveals 9-fold arrangement of a centriole protein. *Biophysical Journal, 102*, 2926–2935.

Lawo, S., Hasegan, M., Gupta, G. D., & Pelletier, L. (2012). Subdiffraction imaging of centrosomes reveals higher-order organizational features of pericentriolar material. *Nature Cell Biology, 14*, 1148–1158.

Li, J. B., Gerdes, J. M., Haycraft, C. J., Fan, Y., Teslovich, T. M., May-Simera, H., et al. (May 14, 2004). Comparative genomics identifies a flagellar and basal body proteome that includes the BBS5 human disease gene. *Cell, 117*(4), 541–552.

Li, S., Fernandez, J.-J., Marshall, W. F., & Agard, D. A. (2012). Three-dimensional structure of basal body triplet revealed by electron cryo-tomography. *EMBO Journal, 31*, 552–562.

Mcdonald, K., & Morphew, M. K. (1993). Improved preservation of ultrastructure in difficult-to-fix organisms by high pressure freezing and freeze substitution: I. *Drosophila melanogaster* and *Strongylocentrotus purpuratus* embryos. *Microscopy Research and Technique, 473*, 465–473.

McGorty, R., Kamiyama, D., & Huang, B. (2013). Active microscope stabilization in three dimensions using image correlation. *Optical Nanoscopy, 2*, 1–7.

Mennella, V., Keszthelyi, B., McDonald, K. L., Chhun, B., Kan, F., Rogers, G. C., et al. (2012). Subdiffraction-resolution fluorescence microscopy reveals a domain of the centrosome critical for pericentriolar material organization. *Nature Cell Biology, 14*, 1159–1168.

Mennella, V., Agard, D. A., Huang, B., & Pelletier, L. (2013). Amorphous no more: subdiffraction view of the pericentriolar material architecture. *Trends in Cell Biology*, 1–10. http://dx.doi.org/10.1016/j.tcb.2013.10.001.

Miersch, S., & Sidhu, S. S. (2012). Synthetic antibodies: concepts, potential and practical considerations. *Methods, 57*, 486–498.

Mitchison, T. J., & Kirschner, M. (1986). Isolation of mammalian centrosomes. *Methods in Enzymology, 134*, 261–268.

Moisoi, N., Erent, M., Whyte, S., Martin, S., & Bayley, P. M. (2002). Calmodulin-containing substructures of the centrosomal matrix released by microtubule perturbation. *Journal of Cell Science, 115*, 2367–2379.

Moritz, M., Braunfeld, M. B., Fung, J. C., Sedat, J. W., Alberts, B. M., & Agard, D. A. (1995). Three-dimensional structural characterization of centrosomes from early embryos. *Journal of Cell Biology, 130*, 1149–1159.

Moritz, M., Zheng, Y., Alberts, B. M., & Oegema, K. (1998). Recruitment of the g-tubulin ring complex to *Drosophila* salt stripped centrosomes. *Journal of Cell biology, 142*, 775–786.

Mukamel, E. A., Babcock, H., & Zhuang, X. (2012). Statistical deconvolution for superresolution fluorescence microscopy. *Biophysical Journal, 102*, 2391–2400.

Müller, H., Schmidt, D., Steinbrink, S., Mirgorodskaya, E., Lehmann, V., Habermann, K., et al. (2010). Proteomic and functional analysis of the mitotic *Drosophila* centrosome. *EMBO Journal, 29*, 3344–3357.

Nido, G. S., Méndez, R., Pascual-García, A., Abia, D., & Bastolla, U. (2012). Protein disorder in the centrosome correlates with complexity in cell types number. *Molecular Biosystems, 8*, 353–367.

Nigg, E. A., & Raff, J. W. (2009). Centrioles, centrosomes, and cilia in health and disease. *Cell, 139*, 663–678.

Olivier, N., Keller, D., Gönczy, P., & Manley, S. (2013). Resolution doubling in 3D-STORM imaging through improved buffers. *PLoS One, 8*, e69004.

Paintrand, et al. (1992). Centrosome organization their sensitivity and centriole architecture: to divalent cations. *Journal of Structural Biology, 108*, 107–128.

Parthasarathy, R. (2012). Rapid, accurate particle tracking by calculation of radial symmetry centers. *Nature Methods, 9*, 724–726.

Pelletier, L., O'Toole, E., Schwager, A., Hyman, A. A., & Müller-Reichert, T. (2006). Centriole assembly in *Caenorhabditis elegans*. *Nature, 444*, 619–623.

Pertsinidis, A., Zhang, Y., & Chu, S. (2010). Subnanometre single-molecule localization, registration and distance measurements. *Nature, 466*, 647–651.

Puchner, E. M., Walter, J. M., Kasper, R., Huang, B., & Lim, W. A. (2013). Counting molecules in single organelles with superresolution microscopy allows tracking of the endosome maturation trajectory. *Proceedings of the National Academy of Sciences of the United States of America, 110*, 16015–16020.

Rogers, S. L., & Rogers, G. C. (2008). Culture of *Drosophila* S2 cells and their use for RNAi mediated loss-of-function studies and immunofluorescence microscopy. *Nature Protocols, 3*, 606–611.

Rust, M. J., Bates, M., & Zhuang, X. (Oct, 2006). Imaging by stochastic optical reconstruction microscopy (STORM). *Nat Methods, 3*(10), 793–795. Epub 2006 Aug 9.

Ryan, C. J., Cimermančič, P., Szpiech, Z. A., Sali, A., Hernandez, R. D., & Krogan, N. J. (2013). High-resolution network biology: connecting sequence with function. *Nature Reviews Genetics, 14*, 865–879.

Schermelleh, L., Carlton, P. M., Haase, S., Shao, L., Winoto, L., Kner, P., et al. (2008). Subdiffraction multicolor imaging of the nuclear periphery with 3D structured illumination microscopy. *Science, 320*, 1332–1336.

Schermelleh, L., Heintzmann, R., & Leonhardt, H. (2010). A guide to super-resolution fluorescence microscopy. *Journal of Cell Biology, 190*, 165–175.

Schnackenberg, B. J., Hull, D. R., Balczon, R. D., & Palazzo, R. E. (2000). Reconstitution of microtubule nucleation potential in centrosomes isolated from *Spisula solidissima* oocytes. *Journal of Cell Science, 953*, 943–953.

Sengupta, P., Van Engelenburg, S., & Lippincott-Schwartz, J. (2012). Visualizing cell structure and function with point-localization superresolution imaging. *Developmental Cell, 23*, 1092–1102.

Shcherbakova, D. M., Sengupta, P., Lippincott-Schwartz, J., & Verkhusha, V. V. (2014). Photocontrollable fluorescent proteins for superresolution imaging. *Annual Review of Biophysics, 43*, 303–329.

Shroff, H., Galbraith, C. G., Galbraith, J. A., & Betzig, E. (2008). Live-cell photoactivated localization microscopy of nanoscale adhesion dynamics. *Nature Methods, 5*, 417–423.

Shtengel, G., Wang, Y., Zhang, Z., Goh, W. I., Hess, H. F., & Kanchanawong, P. (2014). Imaging cellular ultrastructure by PALM, iPALM, and correlative iPALM-EM. *Methods in Cell Biology, 123C*, 273–294. Elsevier Inc.

Sir, J. H., Barr, A. R., Nicholas, A. K., Carvalho, O. P., Khurshid, M., Sossick, A., et al. (2011). A primary microcephaly protein complex forms a ring around parental centrioles. *Nature Genetics, 43*, 1147–1153.

Song, K., Awata, J., Tritschler, D., Bower, R., Witman, G. B., Porter, M. E., et al. (2015). In situ localization of N- and C-termini of subunits of the flagellar nexin–dynein regulatory complex (N-DRC) using SNAP-tag and cryo-electron tomography. *Journal of Biological Chemistry*, 0–27. http://dx.doi.org/10.1074/jbc.M114.626556.

Sonnen, K. F., Schermelleh, L., Leonhardt, H., & Nigg, E. A. (2012). 3D-structured illumination microscopy provides novel insight into architecture of human centrosomes. *Biology Open, 1*, 965–976.

Stinchcombe, J. C., & Griffiths, G. M. (2014). Communication, the centrosome and the immunological synapse. *Philosophical Transactions of the Royal Society of London B Biological Sciences, 369*.

Szymborska, A., de Marco, A., Daigle, N., Cordes, V. C., Briggs, J. A., & Ellenberg, J. (2013). Nuclear pore scaffold structure analyzed by super-resolution microscopy and particle averaging. *Science*. http://dx.doi.org/10.1126/science.1240672.

Tang, N., & Marshall, W. F. (2012). Centrosome positioning in vertebrate development. *Journal of Cell Science, 125*, 4951–4961.

Wallace, W., Schaefer, L. H., & Swedlow, J. R. (2001). A working person's guide to deconvolution in light microscopy. *Biotechniques, 31*, 1076–1078, 1080, 1082 passim.

Wang, S., Moffitt, J. R., Dempsey, G. T., Xie, X. S., & Zhuang, X. (2014). Characterization and development of photoactivatable fluorescent proteins for single-molecule-based superresolution imaging. *Proceedings of the National Academy of Sciences of the United States of America, 111*, 8452–8457.

Winey, M., & O'Toole, E. (2014). Centriole structure. *Philosophical Transactions of the Royal Society of London B Biological Sciences, 369.*

Woodruff, J. B., Wueseke, O., & Hyman, A. A. (2014). Pericentriolar material structure and dynamics. *Philosophical Transactions of the Royal Society of London B Biological Sciences, 369.*

Wueseke, O., Bunkenborg, J., Hein, M. Y., Zinke, A., Viscardi, V., Woodruff, J. B., et al. (2014). The *Caenorhabditis elegans* pericentriolar material components SPD-2 and SPD-5 are monomeric in the cytoplasm before incorporation into the PCM matrix. *Molecular Biology of the Cell, 25*, 2984–2992.

York, A. G., Chandris, P., Nogare, D. D., Head, J., Wawrzusin, P., Fischer, R. S., et al. (2013). Instant super-resolution imaging in live cells and embryos via analog image processing. *Nature Methods, 10*, 1122–1126.

Zhao, H., Zhu, L., Zhu, Y., Cao, J., Li, S., Huang, Q., et al. (2013). The Cep63 paralogue Deup1 enables massive de novo centriole biogenesis for vertebrate multiciliogenesis. *Nature Cell Biology, 15*, 1434–1444.

Zhu, L., Zhang, W., Elnatan, D., & Huang, B. (2012). Faster STORM using compressed sensing. *Nature Methods, 9*, 721–723.

CHAPTER

Probing mammalian centrosome structure using BioID proximity-dependent biotinylation

9

Elif Nur Firat-Karalar*, Tim Stearns[§,1]

**Department of Molecular Biology and Genetics, Koç University, Istanbul, Turkey*

[§]Department of Biology and Department of Genetics, Stanford University, Stanford, CA, USA

[1]Corresponding author: E-mail: stearns@stanford.edu

CHAPTER OUTLINE

Introduction ... 154

1. Materials ... 157

1.1 Equipment ... 157

1.2 Recipes ... 158

1.2.1 Biotin stock (2000X): 10 mM ... 158

1.2.2 Polyethylenimine (1 mg/mL) ... 158

1.2.3 HB buffer ... 158

1.2.4 BioID lysis buffer ... 158

1.2.5 Wash buffer 1 ... 158

1.2.6 Wash buffer 2 ... 158

1.2.7 Wash buffer 3 ... 158

1.2.8 Wash buffer 4 ... 159

1.2.9 SDS-PAGE sample buffer ... 159

2. Methods ... 159

2.1 Generation of BirA*-Fusion Protein Expression Plasmid ... 159

2.2 Choice of Cell Line and Expression Method ... 160

2.3 Validation of the BirA*-Fusion Protein in Cells ... 160

2.3.1 Expression of BirA-fusion protein in cells for large-scale pulldowns... 162*

2.3.2 Transient expression (PEI) ... 162

2.3.3 Stable cell lines ... 163

3. Large-Scale BioID Pulldown of BirA*-Fusion Proteins ... 163

3.1 Centrosome Enrichment before Streptavidin Affinity Purification ... 164

3.2 Whole-Cell Lysis before the Streptavidin Affinity Purification ... 165

3.3 Streptavidin Affinity Purification Experiments ... 165

3.4 Preparation of Sample for MS ... 166

Methods in Cell Biology, Volume 129, ISSN 0091-679X, http://dx.doi.org/10.1016/bs.mcb.2015.03.016

4. Data Analysis 166
Conclusion 167
Acknowledgments 168
References 168

Abstract

Understanding the structure and function of the centrosome will require identification of its constituent components and a detailed characterization of the interactions among these components. Here, we describe the application of proximity-dependent biotin identification (BioID) to identify spatial and temporal relationships among centrosome proteins. The BioID method relies on protein fusions to a promiscuous mutant of the *Escherichia coli* biotin ligase BirA, which biotinylates proteins that are in a ~10 nm labeling radius of the enzyme. The biotinylated proteins are captured by affinity and are identified by mass spectrometry. Proteins identified in this way are referred to as "proximity interactors." Application of BioID to a set of centrosome proteins demonstrated the utility of this approach in overcoming inherent limitations in probing centrosome structure. These studies also demonstrated the potential of BioID for building large-scale proximity interaction maps among centrosome proteins. In this chapter, we describe the work flow for identification of proximity interactions of centrosome proteins, including materials and methods for the generation and characterization of a BirA*-fusion protein expression plasmid, expression of BirA*-fusion proteins in cells, and purification and identification of proximity partners by mass spectrometry.

INTRODUCTION

Understanding the structure and function of the centrosome will require the identification of its constituent components and a detailed characterization of the interactions among these components. A combination of proteomic, bioinformatics, and comparative genomic studies has identified many, likely most, of the components of mammalian centrosomes (Alves-Cruzeiro, Nogales-Cadenas, & Pascual-Montano, 2014; Andersen et al., 2003; Firat-Karalar, Sante, Elliott, & Stearns, 2014; Hoh, Stowe, Turk, & Stearns, 2012; Jakobsen et al., 2011; Li et al., 2004). However, our understanding of the interactions between these components has been limited because the centrosome presents two challenges for such studies. First, there is only one centrosome in most nondividing cells, limiting the amount of material for biochemical studies; second, the centrosome is an insoluble structure, making the relevant interactions inaccessible to standard techniques such as co-immunoprecipitation. Efforts to overcome these limitations have included salt extraction of centrosome proteins from isolated centrosomes (Moritz et al., 1995; Schnackenberg, Khodjakov, Rieder, & Palazzo, 1998; Schnackenberg & Palazzo, 1999), and use of *Xenopus* egg extracts, which lack centrosomes but contain all of the components, as the starting material for immunoprecipitation (Hatch, Kulukian, Holland, Cleveland, & Stearns, 2010).

Recently, superresolution microscopy techniques have been applied to a subset of centrosome proteins, yielding new information on the spatial organization of centrosome proteins within subdomains of the organelle (Fu & Glover, 2012; Lawo, Hasegan, Gupta, & Pelletier, 2012; Mennella et al., 2012; Sonnen, Schermelleh, Leonhardt, & Nigg, 2012). Other approaches that have some promise to address this problem include fluorescence resonance energy transfer (Lukinavicius et al., 2013; Muller et al., 2005) and chemical cross-linking (Lukinavicius et al., 2013). Here, we describe the application of proximity-dependent biotin identification (BioID) (Roux, Kim, & Burke, 2013; Roux, Kim, Raida, & Burke, 2012) to the study of the mammalian centrosome. BioID has unique advantages in overcoming the limitations of studying interactions in the centrosome and has the potential for building large-scale interaction maps among centrosome proteins.

In the BioID approach, the protein of interest is tagged with a mutant form of *E. coli* biotin ligase BirA (R118G) (hereafter BirA*) (Roux et al., 2012) (Figure 1(A)). Wild-type BirA normally only transfers a biotin to a substrate bearing a specific recognition sequence. BirA* is promiscuous in that it activates biotin for transfer in the absence of a substrate peptide, and the activated biotinoyl-5′-AMP is free to diffuse away from the enzyme and covalently modify primary amines of nearby proteins. Since biotinoyl-5′-AMP is highly reactive and short-lived, the zone of modification in BioID is thought to extend only about 10 nm in radius around the BirA*-tagged protein. We will refer to biotin-labeled proteins in a BioID experiment as proximity partners. Biotinylated proximity partner proteins are purified by streptavidin-binding and identified by mass spectrometry (MS). It is important to note that the BioID biotinylation is a mark of potential proximity and not an evidence for physical interactions. In this respect, BioID is a screen for proteins that are proximal to the protein of interest at some point during labeling; a positive result might reflect direct or indirect physical interaction, or simple proximity without physical contact. Thus, the proximity interactions identified by BioID should be tested by subsequent experiments to, for example, determine which reflect physical interactions, and whether the identified proteins colocalize within the cell. Comparison of results from BioID and immunoprecipitation experiments for a number of proteins demonstrated that the two approaches yield an overlapping set of proteins (Couzens et al., 2013; Lambert, Tucholska, Go, Knight, & Gingras, 2014).

BioID has several advantages over traditional approaches for studying interactions at the centrosome, or other "solid-state" cell structures. First, due to the strong affinity of the biotin–streptavidin interaction, purification of biotinylated proteins can take place under denaturing conditions. This allows one to solubilize the centrosome while preserving the proximity relationship information, which is encoded in covalent protein modification. Second, proteins that are associated with the target protein only transiently can be more robustly identified by BioID, since such proteins are marked by biotinylation before cell lysis. These advantages of BioID have been successfully exploited in screens for proximity interactions in the mammalian centrosome (Comartin et al., 2013; Firat-Karalar, Rauniyar,

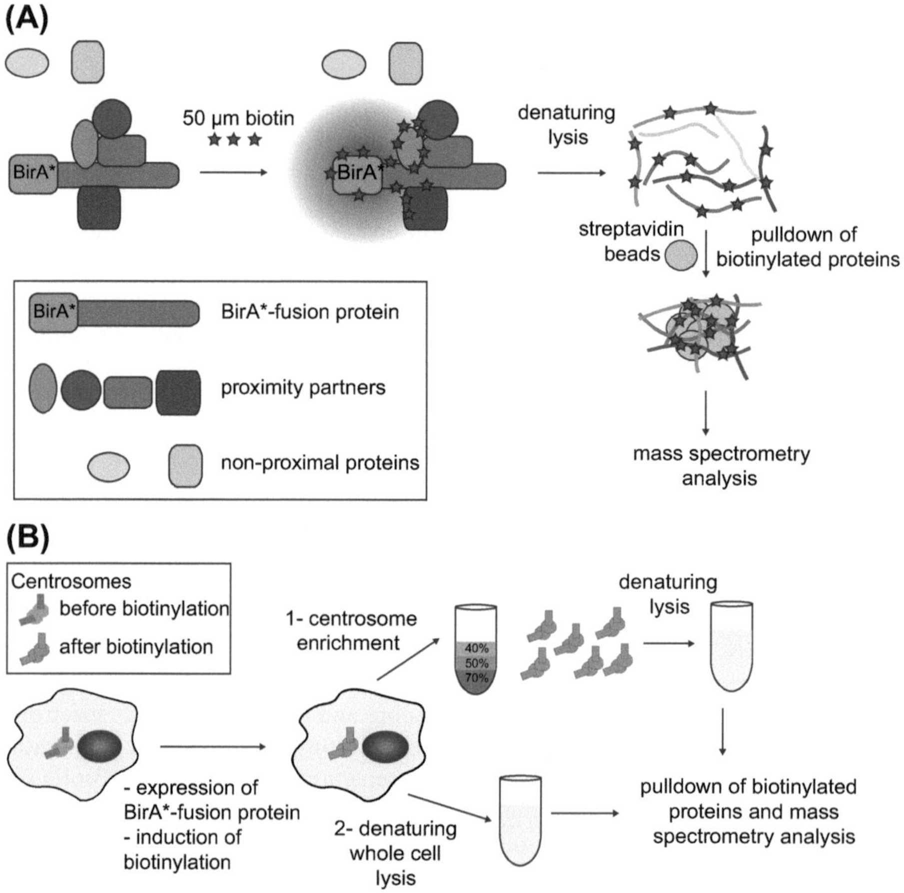

FIGURE 1 Work flow for the application of the BioID approach to the centrosome.

(A) In the BioID approach, the protein of interest is tagged with a mutant form of *E. coli* biotin ligase BirA (R118G), denoted as BirA*. Following incubation of cells with 50 μM of biotin for 18–24 h, BirA* biotinylates proteins that are in the proximity of the BirA*-fusion protein and we refer to such proteins as "proximity interactors." The labeling radius of BirA* is estimated to be around 10 nm. The proteins outside the proximity radius are denoted as "nonproximal proteins." Following incubation of cells with biotin, proteins are solubilized and biotinylated proteins are precipitated by streptavidin beads for subsequent mass spectrometry analysis. (B) Cell line of choice is transfected with the expression plasmid for the BirA*-fusion protein. Cells transiently or stably expressing the BirA*-fusion protein are incubated with excess biotin to induce biotinylation of the proximity interactors. These cells are then processed for identification of the biotinylated proteins by the following two methods: (1) Centrosome-enriched fractions are prepared from these cells by sucrose gradient fractionation, and centrosomes are solubilized under denaturing conditions. (2) Cells are lysed under denaturing conditions. Following lysis, the biotinylated proteins are affinity captured and are identified by mass spectrometry. (See color plate)

Yates, & Stearns, 2014), nuclear lamina and envelope (Kim et al., 2014; Roux et al., 2012), chromatin (Lambert et al., 2014), bilobe in *Trypanosoma bruce* (Morriswood et al., 2013) and Hippo signaling pathway (Couzens et al., 2013).

In this chapter, we describe the application of BioID to identify spatial and temporal proximity relationships among centrosome proteins. The basis for this description is our application of BioID to a set of core centrosome duplication and maturation proteins (Firat-Karalar, Rauniyar, et al., 2014). This work identified most of the known interactions between these proteins, and also identified previously uncharacterized relationships and proteins. We will describe the work flow for identification of proximity interactions of centrosome proteins, including materials and methods for the generation and characterization of a BirA*-fusion protein expression plasmid, expression of BirA*-fusion proteins in cells, and purification and identification of proximity partners by MS (Figure 1). In addition, we describe methods for combining BioID labeling with standard techniques for centrosome purification to enrich for centrosomal interactions (Figure 1(B)).

1. MATERIALS

Expression plasmid for generating BirA*-fusion proteins
Cells of choice for validation experiments and growth media for those cells
Biotin (2000X stock): 10 mM
Fixative (i.e., methanol, paraformaldehyde)
Antibodies specific to the epitope tag of the BirA* (i.e., myc, FLAG)
Antibodies specific to biotinylated proteins (Alexa Fluor-streptavidin, horseradish peroxidase (HRP)-streptavidin)
Antibodies that mark the centrosome (i.e., gamma-tubulin antibody)
Secondary antibodies that detect the chosen primary antibodies
DNA dyes (i.e., DAPI)
Streptavidin agarose resin
Standard immunofluorescence and immunoblotting reagents.

1.1 EQUIPMENT

Cell culture incubator, humidified, set at 37 °C, 5% CO_2
Refrigerated clinical centrifuge
Microcentrifuge
Ultracentrifuge
Sonicator with pulse capacity (i.e., Branson Digital Sonifier)
Rotator for microcentrifuge tubes
Sodium dodecyl sulfate polyacrylamide gel electrophoresis (SDS-PAGE) and transfer unit.

1.2 RECIPES

1.2.1 Biotin stock (2000X): 10 mM

Dissolve 122 mg biotin (Sigma #B4501) in 50 mL of dimethyl sulfoxide (DMSO). Aliquot to the desired volume and store at −20 °C.

1.2.2 Polyethylenimine (1 mg/mL)

Dissolve 50 mg polyethylenimine (PEI) 25 kD (Polysciences # 23966-2) in 50 mL purified water and adjust the pH to 7.0. Filter and sterilize the solution through 0.22 μm membrane, then aliquot to the desired volume, and store the PEI at −80 °C.

1.2.3 HB buffer

HEPES: 20 mM, pH 7.8
K-acetate: 5 mM
$MgCl_2$: 0.5 mM
DTT: 0.5 mM
Protease inhibitors.

1.2.4 BioID lysis buffer

Tris: 50 mM, pH 7.4
NaCl: 500 mM
SDS: 0.4%
Ethylene diamine tetraacetic acid (EDTA): 5 mM
DTT: 1 mM
Triton X-100: 2%
Protease inhibitors (10 μg/mL each of aprotinin, leupeptin, pepstatin, and 1 mM phenyl-methylsulfonyl fluoride).

1.2.5 Wash buffer 1

SDS (2%) in dH_2O.

1.2.6 Wash buffer 2

Deoxycholate: 0.2%, Triton X-100: 1%
NaCl: 500 mM
EDTA: 1 mM
HEPES: 50 mM, pH 7.5.

1.2.7 Wash buffer 3

Tris: 10 mM, pH 8.1
LiCl: 250 mM, 0.5% NP-40
Deoxycholate: 0.5%
Triton X-100: 1%
NaCl: 500 mM
EDTA: 1 mM.

1.2.8 Wash buffer 4

Tris: 50 mM, pH 7.4
NaCl: 50 mM.

1.2.9 SDS-PAGE sample buffer

Tris: 50 mM, pH 6.8
SDS: 2%
Glycerol: 10%
β-mercaptoethanol: 1%
Bromophenol-blue 0.005%.

2. METHODS

2.1 GENERATION OF BirA*-FUSION PROTEIN EXPRESSION PLASMID

The first step in the BioID approach is constructing an expression plasmid in which the gene of interest is fused with BirA*. To increase the likelihood that the BirA*-fusion protein will localize and function correctly, several points must be considered in the cloning strategy. BirA* is a 33.5 kDa protein, about 5 kDa larger than green fluorescent protein (GFP), and for GFP, the BirA*-fusion is usually made at either the N or C terminus of the target protein. Ideally, this fusion would not impair localization, stability, or function of the target protein, but this must be empirically determined for each protein. Our experience is that most proteins that are functional when tagged with GFP are also functional when tagged in the same site with BirA*; thus, prior evidence on successful GFP-fusion proteins should be used in deciding on BirA*-fusion orientation. We routinely make fusions to both termini for proteins about which we have no prior knowledge of fusion behavior. Second, if available, the protein structure or the spatial organization of the protein can also provide useful information; for example, superresolution imaging studies demonstrated that CEP152 has an elongated structure within the pericentriolar material, such that its N and C termini are in different spatial domains (Sonnen et al., 2012). Accordingly, analysis of both N- and C-terminal CEP152 fusions with BirA* yielded subprotein-level spatial information (Firat-Karalar, Rauniyar, et al., 2014).

In addition to choosing the orientation of the BirA*-fusion, one must also consider other desired properties of the construct. For example, most BirA*-fusion constructs also include an epitope tag (i.e., myc, FLAG) that allows both microscopic and biochemical detection of the fusion protein. One should also consider the promoter for the fusion protein, the nature of the vector, and the means for introducing that vector into the desired cells. Since these are all standard considerations for any mammalian cell expression experiment, we will not consider in depth here, except as they relate to the specifics of the experiments described.

2.2 CHOICE OF CELL LINE AND EXPRESSION METHOD

The centrosome is a small, single-copy organelle in most cultured cells. Therefore, limiting material is an ever-present challenge in studying centrosome protein interactions in situ. One can maximize the amount of relevant material by choosing cell lines that have small surface area-to-volume ratio. One such cell line, HEK293T, is highly transfectable, and has been successfully used for large-scale BioID pulldown of centrosome proteins. As is often the case with mammalian cells, there are trade-offs associated with the choice of cell line. Although HEK293T cells are easy to manipulate, they typically do not make a primary cilium, which limits their utility for some centrosome-related studies, for example. In contrast, RPE1 cells make a primary cilium at a high frequency, but are more difficult to grow to high density, and have a higher area-to-volume ratio than HEK293T cells. Thus, the cell line should be chosen based on the specific experimental requirements and constraints.

Once the cell line is chosen, the means of introducing and expressing the BirA*-fusion protein must also be chosen, based on the experimental objectives and the cell line. The BirA*-fusion proteins can be expressed either transiently or stably in cells using various means of introducing the vector (i.e., viral infection, lipid- or polymer-based transfection). Transient expression, which is usually accomplished by transfection of a plasmid vector, results in heterogeneous expression of the BirA*-fusion protein; depending on the vector used, this can range from lower to much higher than the endogenous protein is expressed. This is acceptable in some cases, but in others, particularly for the proteins that mislocalize upon overexpression, it would be preferable to have the BirA*-fusion protein expressed more homogeneously, and at a level more similar to the endogenous protein level. This can be accomplished by making stable cell lines and identifying those with the appropriate expression level. There are many methods for generating stable cell lines, including random integration by transient transfection of plasmid vectors or infection with integrating viral vectors (i.e., lentivirus), or integration at a specific genome location (i.e., Flip-in or homologous recombination). In the case for proteins that are toxic when stably expressed, inducible expression can be employed. Perhaps most ideally, the ability to target genomic loci by CRISPR/Cas9 endonuclease-mediated genome editing allows the creation of knock-in alleles in which BirA* is integrated in frame at the genomic locus of the gene of interest.

2.3 VALIDATION OF THE BirA*-FUSION PROTEIN IN CELLS

As the goal of BioID is to identify proximity interactions in the normal context of a target protein, the expression, localization, function, and biotinylation activity of the BirA*-fusion protein should be validated. Localization of the fusion protein can be compared to the endogenous protein by immunofluorescence microscopy (note that the BirA*-fusion vectors typically contain an epitope tag) (Figure 2(A)). Expression of a protein with the correct molecular weight can be determined by immunoblotting (Figure 2(B)). Function can be tested by complementation of a loss-of-function

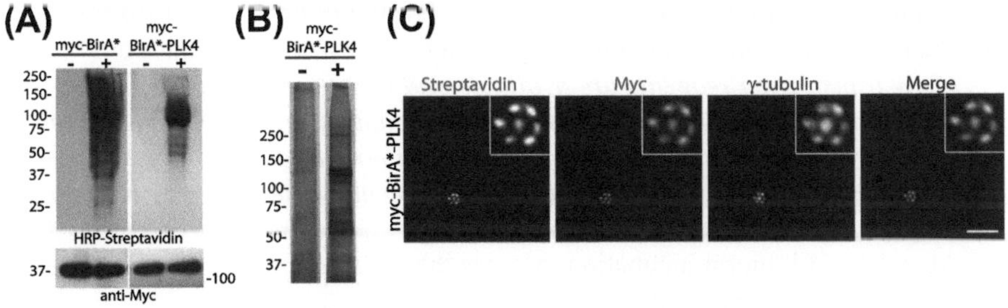

FIGURE 2 Localization and activity of myc-BirA*-PLK4 in mammalian cells.

(A) HEK293T cells were transfected with Myc-BirA* or Myc-BirA*-PLK4. After 18 h incubation with 50 μM biotin or the vehicle control DMSO, cells were lysed. Cell lysates were ran on an SDS-PAGE gel and processed for Western blotting using the following antibodies: (1) anti-Myc antibody to detect expression of the BirA*-fusion protein, (2) HRP-streptavidin to detect biotinylation of the proximity interactors of the BirA*-fusion protein. (B) HEK293T cells expressing Myc-BirA*-PLK4 were incubated with 50 μM biotin or DMSO for 18 h, lysed, ran on an SDS-PAGE gel, and stained by silver-staining. (C) U2OS cells were transfected with Myc-BirA*-PLK4. After 18 h incubation with biotin, cells were fixed, and stained for Myc-BirA*-PLK4 with anti-Myc antibody, for biotinylated proteins with fluorescent streptavidin and centrosomes with anti-γ-tubulin antibody. Scale bars = 10 μm, all insets show 4× enlarged centrosomes.

phenotype, or by presence of an overexpression phenotype that is the same as that observed with untagged protein (Figure 2(A)). However, this is not possible in many cases, as it requires that the function of the protein be known, and that there be an assay for that function. Lastly, the biotinylation activity of the BirA*-fusion protein, which is essential to the success of the method, can be assessed in cells by immunofluorescence with fluorescent-streptavidin and on blots with streptavidin-HRP. Although we have not characterized this in detail, most fusion proteins that localize to the structure of interest yield detectable biotinylation at that structure, either due to self-biotinylation of the BirA*-fusion protein, biotinylation of proximal proteins, or both. The experimental outline for validation experiments in a typical transfection-based experiment is described in this section.

1. For each experimental condition, place a single glass coverslip in one well of a standard six-well plate and plate cells to the wells the day before transfection. For maximum transfection efficiency, cells should be 80–90% confluent at the time of transfection.
   To monitor the induction of biotinylation, we recommend the following two experimental conditions: (1) transfected cells incubated with biotin (i.e., plus biotin condition) and (2) transfected cells incubated with vehicle (DMSO) (i.e., minus biotin condition).
2. Express the BirA*-fusion protein in cells using the choice of the transfection reagent. To induce biotinylation, change the medium 6–8 h after transfection to

medium supplemented with 50 μm biotin. For the minus biotin condition, change the medium to medium supplemented with DMSO (vehicle).
Many transfection reagents are available and the reagent must be chosen depending on the cell line. The ideal method should have high transfection efficiency, low toxicity, and minimal effects on normal physiology of the cells. For most cell lines, Lipofectamine LTX (Life Technologies) works well with the manufacturer's suggested protocol.

3. After 12–24 h biotin incubation, process the cells for immunofluorescence and immunoblotting experiments.
Incubation time with biotin can be adjusted depending on the protein of interest and the objective of the experiment. We note that levels of biotinylated proteins increase in parallel with biotin exposure time, reaching saturation within 12–24 h.
4. For immunofluorescence experiments, fix the coverslips with the choice of the fixation solution, block and stain the cells with the appropriate antibodies. After mounting the coverslips, image the cells by fluorescence microscopy.
We recommend the following primary antibodies for staining: (1) antibody against the epitope tag antibody to confirm the expression of the fusion protein, (2) fluorescent-streptavidin antibody to confirm biotinylation of proximity partners, and (3) antibody to mark the centrosome (i.e., gamma-tubulin antibody).
The signal form the staining of the epitope tag and the biotinylated peptides are expected to overlap given that the biotinylation radius of BirA was estimated to be 10 nm.*
5. For immunoblotting experiments, wash the cells twice with PBS. Add 1 mL of PBS +2 mM EDTA to the cells and incubate for 5 min at 37° to detach the cells from the wells. Transfer the cells to a microcentrifuge tube and pellet the cells at 15,000 rpm for 1 min. Following the centrifugation, resuspend the pellet in a volume of SDS-PAGE sample buffer appropriate for the number of cells to be lysed and boil the lysate at 95 °C for 5 min. Run the lysate on an SDS-PAGE gel, block and blot the membrane with primary antibodies against the epitope tag of the BirA*-fusion protein. Once the expression of the fusion protein is confirmed, strip the blot using standard methods, block and blot with streptavidin-HRP antibody to confirm biotinylation of the proximity partners of the BirA*-fusion protein.

2.3.1 Expression of BirA*-fusion protein in cells for large-scale pulldowns

Both transient and stable expressions of the BirA*-fusion protein in cells have been successfully used to identify proximity partners of centrosome proteins (Comartin et al., 2013; Firat-Karalar, Rauniyar, et al., 2014). The choice of transient or stable transfection depends on the properties of the protein of interest and objective of the experiment.

2.3.2 Transient expression (PEI)

For the transient transfection approach, it is important to use a cell line that has high transfection efficiency. Since untransfected cells are a source of contaminating background proteins, low transfection efficiencies will decrease the sensitivity of identifying proteins specific to the BirA*-fusion protein. Given their high transfection efficiency and availability of cost-effective transfection reagents, HEK293T cells are ideal for such experiments and they were successfully used for identifying proximity interactions at the centrosome (Firat-Karalar, Rauniyar, et al., 2014).

1. One day before transfection, plate the cells to five 15 cm dishes so that they are 70–90% at the time of transfection next day.
2. Transfect the cells using standard PEI-based transfection approach. For each 15 cm plate, mix 30 μg expression plasmid and the optimized PEI amount (3:1 dilution of 1 mg/mL PEI stock) in serum-free medium, incubate the transfection mixture for 30 min at room temperature, and add it dropwise to the cells. Then 4–6 h after transfection, change the medium with medium supplemented with 50 μm biotin.
   Other transfection reagents of choice can be used for transient expression of the BirA-fusion protein for large-scale experiments. We chose PEI-based transfection approach given its high transfection efficiency and the low cost for HEK293T cells.*
   We found 3:1 ratio of PEI to DNA (w/w) to be optimal for most genes we expressed in HEK293T cells. However, this ratio should be screened for each gene tested. Ratios between 1:1 and 5:1 can be used for screening.
3. As described in Section 3, 24 h after transfection, process the transfected cells for large-scale BioID pulldown experiments.

2.3.3 Stable cell lines

Stable cell lines can be generated using random integration or integration at a specific genome location. Inducible Flip-In T-Rex system (Life Technologies) was successfully used to generate isogenic stable mammalian cell lines that express BirA*-fusion protein from a specific genomic location under a tetracycline-inducible promoter (Comartin et al., 2013; Lambert et al., 2014). Once the correct localization and biotinylation activity of the BirA*-fusion protein is validated by the experiments described in Section 2.3, these cell lines could be used for large-scale BioID pulldown experiments.

3. LARGE-SCALE BioID PULLDOWN OF BirA*-FUSION PROTEINS

Experimental design: Appropriate controls are essential to the success of a BioID experiment. The chosen controls might differ based on the nature of the experiment, but at the least should include cells transfected with BirA* itself expressed in a manner similar to the BirA*-fusion protein, and a cell population that is not expressing any BirA* protein, that is, mock-transfected cells for a transient expression experiment, or the parental cell line for a stable cell line experiment. These control samples should be processed in parallel to the experimental samples for both validation (above) and large-scale experiments. Finally, at least three biological replicates for each experimental condition are required for the statistical analysis of the MS data, as described in Section 4.

3.1 CENTROSOME ENRICHMENT BEFORE STREPTAVIDIN AFFINITY PURIFICATION

1. Grow five 15 cm dishes of cells transiently or stably expressing BirA*-fusion protein at 37 °C, 5% CO_2. Once they are about 70–80% confluent, replace the medium with medium containing 50 μm biotin and incubate the cells further for 18–14 h to induce biotinylation.
2. Replace the medium with complete medium containing 5 μg/mL nocodazole and 5 μg/mL cytochalasin B and incubate the cells at 37 °C, 5% CO_2 for 1 h. *During this incubation time, prepare discontinuous sucrose gradient by progressively layering 5 mL of 70%, 3 mL of 50%, and 3 mL of 40% sucrose solutions upon one another in a Thinwall, Ultra-Clear Beckman tube (max volume 38.5 mL).*
3. Wash the cells twice with ice-cold HB buffer and incubate with 20 mL HB buffer at 4 °C for 10 min.
4. Scrape the cells from the plate gently with a cell scraper, and transfer the cells to a 15 mL dounce homogenizer and homogenize the cells by applying 10 strokes of the tight-fitting pestle of the dounce homogenizer.
5. Transfer the lysate to a 15 mL conical tube and centrifuge at 3000 rpm at 4 °C for 5 min to pellet nuclei.
6. Transfer the supernatant to a 50 mL conical tube and wash the pellet with 10 mL ice-cold HB buffer to remove centrosomes that pelleted with nuclei. Centrifuge the lysate at 3000 rpm at 4 °C for 5 min to pellet nuclei.
7. Collect the supernatant and combine with the first supernatant in the 50 mL conical tube. Add 0.1% Triton X-100 and centrifuge the lysate again at 3000 rpm at 4 °C for 5 min.
8. Collect the supernatant and load onto the sucrose gradient in Ultra-Clear Beckman tubes and centrifuge at 26,000 rpm in SW32Ti at 4 °C for 1 h.
9. Collect 0.5 mL fractions either by fractionating the gradient from the bottom of the tube by poking a hole or from the top using an automated fraction collector. *The centrosome peak should be approximately in fractions 9–11, starting from the bottom of the tube. To determine the peak centrosome fractions, collect 10 μL of each fraction in the 40–60% sucrose range, run on an SDS-PAGE gel, and process for immunoblotting using antibodies against gamma-tubulin (marker for the centrosomes).*
10. Pool the centrosome-containing fractions in a 15 mL conical tube and resuspend in lysis buffer (1.5 mL resuspension volume per 0.5 mL fraction pooled from the sucrose gradient). *Given that 0.5 mL of each fraction will be resuspended in a final volume of 1.5 mL lysis buffer, add 1.5X lysis buffer for a final resuspension concentration of 1X lysis buffer.*

11. Process the samples for streptavidin affinity purification experiments, as described in Section 3.3.

3.2 WHOLE-CELL LYSIS BEFORE THE STREPTAVIDIN AFFINITY PURIFICATION

1. Grow five 15 cm dishes of cells transiently or stably expressing BirA*-fusion protein at 37 °C, 5% CO_2. Once they are about 70–80% confluent, replace the medium with medium containing 50 μm biotin and incubate the cells further for 18–14 h to induce biotinylation.
2. Wash the cells twice with 15 mL of room temperature 1X PBS.
3. Add 1.5 mL of room temperature lysis buffer to each plate and scrape the cells gently with a cell scraper.
   High concentration of detergents and salt in the lysis buffer are required to denature and solubilize the centrosome proteins.
   High concentration of detergents and salt in the lysis buffer cause it to precipitate if the lysis buffer is stored at 4 °C.
4. Transfer the cell lysate to a 15 mL conical tube and process for streptavidin affinity purification experiments, as described in Section 3.3.

3.3 STREPTAVIDIN AFFINITY PURIFICATION EXPERIMENTS

1. Sonicate the sample in a Branson Digital Sonifier using the following parameters: (1) 30% amplitude for 1 min in pulses (15 s on, 45 s off), then wait for 2 min to allow sample to chill, and repeat the entire sequence two times; then (2) 40% amplitude for 30 s in pulses (15 s on, 45 s off), wait for 2 min to allow sample to chill, and repeat the entire sequence two times.
   Efficiency of lysis can be monitored by change in the viscosity of the sample or more quantitatively by the Bradford Colorimetric Assay for protein concentration. On different sonication machines, adjust the sonication regime for optimal lysis.
2. Add equal volume of ice-cold 50 mM of Tris, pH 7.4 to the lysate and mix well.
   Dilution of the lysis buffer twofold at this step is important for making the conditions more favorable for streptavidin–biotin binding.
3. Aliquot the lysate to microcentrifuge tubes and pellet the insoluble material by spinning the lysate at 15,000 rpm for 10 min at 4 °C.
   During the centrifugation step, wash streptavidin agarose resin (60 μL per plate) twice with 0.5X lysis buffer (lysis buffer diluted in 50 mM of Tris, pH 7.4) in a microcentrifuge tube, centrifuging between washes at 1000 rpm for 1 min.
4. Transfer the supernatant to a 15-mL conical tube on ice and add the bead suspension to the supernatant. Rotate the bead-lysate mixture at 4 °C overnight.
   If time is constrained during the experiment, the affinity capture time can be shortened. Due to the strong binding affinity between streptavidin and biotin, the binding will be saturated within several hours of incubation at 4 °C.

5. Pellet the beads at 1000 rpm for 1 min and carefully remove the supernatant by pipetting without disturbing the beads.
6. Resuspend the beads in 2 mL wash buffer one and transfer the beads to a microcentrifuge tube. Rotate the tube at room temperature for 10 min and pellet the beads as described in Step 5.
7. Wash the beads once with 2 mL wash buffer 2 as described in Step 6.
8. Wash the beads once with 2 mL wash buffer 3 as described in Step 6.
9. Wash the beads once with 2 mL wash buffer 4 as described in Step 6.
10. Pellet the beads at 1000 rpm for 1 min and resuspend the beads in 100 μL of 50 mM ammonium bicarbonate.
    Save 10 μL of beads (10% total) for immunoblot analysis. Prior to sending samples to MS, it is important to confirm successful pulldown of biotinylated proteins. Immunoblot analysis of the final beads using HRP-streptavidin antibody will detect biotinylated proteins and using an antibody against the epitope tag fused to BirA will detect BirA*-fusion protein.*

3.4 PREPARATION OF SAMPLE FOR MS

The protocol for releasing biotinylated peptides from the streptavidin beads should be carefully considered. The biotin–streptavidin interaction is one of the strongest noncovalent interactions known, and therefore, it is difficult to release biotinylated proteins from the streptavidin beads without losing significant material or without interfering with subsequent MS analysis. To circumvent this problem, we recommend releasing the peptides for analysis by MS using on-bead tryptic digestion. This method also prevents removal of the streptavidin itself from the beads that might interfere with MS analysis. If, instead, elution of the proteins followed by tryptic digest is the method of choice, elution conditions must be optimized to ensure efficient release of biotinylated proteins from the beads. Note that the method by which proteins are released from the streptavidin beads will affect the expected results: digestion of proteins from the beads will leave the biotinylated peptide on the beads, whereas release by elution will include the biotinylated peptide in the sample, and thus might allow identification of the site of biotinylation on peptides.

4. DATA ANALYSIS

In most cases, the identification of proteins in the BioID precipitates will be determined by MS followed by filtering of the data to generate a list of significant "hits." The MS methodology itself is not the subject of this article, and is subject to change as methods improve. However, various aspects of the data analysis are important and will be covered here.

Standard liquid chromotography-MS/MS analysis of the BioID samples will generate data on peptide masses, which are then compared to a database of virtually digested predicted proteins from the species in use. This will ultimately result in a list of proteins with associated information for each protein, including the number and identity of peptides, as well as the spectral counts, a measure of the abundance of that protein in the sample. We recommend that only proteins that were identified in at least two of the three recommended replicates, and that have a spectral count greater than four, be considered for further analysis.

Once the MS data from the experimental replicates and control samples are obtained, the next step in analysis is the normalization of data across experimental runs and among different samples. There are several normalization methods developed to quantitatively analyze label-free MS data (Neilson et al., 2011); we have used normalized spectral abundance factor (NSAF) analysis (Zybailov et al., 2006). The normalization in the NSAF approach takes into account the number of residues in each protein, as well as their spectral counts. As larger proteins are expected to generate more spectral counts than smaller proteins present in the same molar amount, the number of spectral counts for each protein is divided by number of residues, defining the spectral abundance factor (SAF). To account for variability between independent experimental runs, individual SAF values are normalized against the sum of all SAF values for a particular run, yielding the NSAF, which is a unitless, arbitrary value that can be used to compare the relative abundance of proteins across samples and experiments. The NSAF values are used to rank the mass spec hits by relative abundance.

To distinguish specific proximity interactions from nonspecific interactions, the ratio of the NSAFs of each protein in control and experimental datasets is calculated. A threshold for this NSAF ratio must be determined empirically based on the experimental objectives and knowledge of the system. For BioID analysis of centriole duplication proteins, a protein was considered a specific proximity interactor if the NSAF ratio was higher than 2.5 (Firat-Karalar, Rauniyar, et al., 2014). Information about specific proteins that are common contaminants in mass spec experiments can, if available, also be used to filter the data. For example, for some human and mouse cell lines, there is a publicly available repository of such contaminants maintained for this purpose (Mellacheruvu et al., 2013). In the case of proteins with multiple isoforms, the largest isoform should be chosen to assign spectral count for these proteins. It should be noted that the abundance, or lack thereof, does not necessarily correlate with biological relevance, and subsequent interaction and functional experiments are required to identify proximity interactors specific to the protein of interest.

CONCLUSION

Here we described the work flow of the application of BioID to the centrosome. So far, BioID has only been used to identify proximity interactions of a set of centrosome duplication and maturation proteins (Comartin et al., 2013; Firat-Karalar, Rauniyar, et al., 2014) and the results of these studies demonstrated the utility of the BioID approach in identifying relevant interactions at the centrosome. Future work to expand this approach to other centrosome proteins will be invaluable to ultimately build a proximity interaction map for the centrosome. It is important to note that the proximity information derived from BioID experiments is different from the physical information derived from biochemical approaches typically used to map protein–protein interactions. Therefore, BioID should be used complementary to such approaches. Together these studies can ultimately lead to the generation of an interactome for the centrosome that is essential for our understanding of the structure and function of the centrosome.

Proximity-labeling techniques have advanced significantly in recent years. In addition to the BioID approach, a new proximity-labeling approach that is capable of much higher temporal resolution has been developed recently (Hung et al., 2014; Lam et al., 2014; Rhee et al., 2013). This approach makes use of an engineered ascorbate peroxidase (APEX) and its substrate biotin-phenol to biotinylate proteins proximal to the fusion protein. Biotinylation in this case is achieved after only 1 min induction as opposed to 18–24 h induction in the BioID approach. Interactions among centrosome proteins are dynamic during the cell cycle and currently this is not a good approach available to identify temporal interactions among centrosome proteins. Given its short-labeling time, APEX-based proximity labeling has strong potential for identifying such interactions. This approach so far has only been used to identify the proteome of mitochondria (Hung et al., 2014; Rhee et al., 2013) and has not been tested for its feasibility in identifying temporal proximity interactions among proteins. Thus, the application of this approach in the context of the centrosome should be investigated by future studies.

ACKNOWLEDGMENTS

This work was supported by NRSA grant 5 F32 GM106620 to ENF and NIH grant R01 GM52022 to TS.

REFERENCES

Alves-Cruzeiro, J. M., Nogales-Cadenas, R., & Pascual-Montano, A. D. (2014). CentrosomeDB: a new generation of the centrosomal proteins database for Human and *Drosophila melanogaster. Nucleic Acids Research, 42*(Database issue), D430–D436.

Andersen, J. S., Wilkinson, C. J., Mayor, T., Mortensen, P., Nigg, E. A., & Mann, M. (2003). Proteomic characterization of the human centrosome by protein correlation profiling. *Nature, 426*(6966), 570–574.

Comartin, D., Gupta, G. D., Fussner, E., Coyaud, E., Hasegan, M., Archinti, M., et al. (2013). CEP120 and SPICE1 cooperate with CPAP in centriole elongation. *Current Biology, 23*(14), 1360–1366.

Couzens, A. L., Knight, J. D., Kean, M. J., Teo, G., Weiss, A., Dunham, W. H., et al. (2013). Protein interaction network of the mammalian Hippo pathway reveals mechanisms of kinase-phosphatase interactions. *Science Signaling, 6*(302), rs15.

Firat-Karalar, E. N., Rauniyar, N., Yates, J. R., 3rd, & Stearns, T. (2014). Proximity interactions among centrosome components identify regulators of centriole duplication. *Current Biology, 24*(6), 664–670.

Firat-Karalar, E. N., Sante, J., Elliott, S., & Stearns, T. (2014). Proteomic analysis of mammalian sperm cells identifies new components of the centrosome. *Journal of Cell Science, 127*, 4128–4133.

Fu, J., & Glover, D. M. (2012). Structured illumination of the interface between centriole and peri-centriolar material. *Open Biology, 2*(8), 120104.

Hatch, E. M., Kulukian, A., Holland, A. J., Cleveland, D. W., & Stearns, T. (2010). Cep152 interacts with Plk4 and is required for centriole duplication. *The Journal of Cell Biology, 191*(4), 721–729.

Hoh, R. A., Stowe, T. R., Turk, E., & Stearns, T. (2012). Transcriptional program of ciliated epithelial cells reveals new cilium and centrosome components and links to human disease. *PLoS One, 7*(12), e52166.

Hung, V., Zou, P., Rhee, H. W., Udeshi, N. D., Cracan, V., Svinkina, T., et al. (2014). Proteomic mapping of the human mitochondrial intermembrane space in live cells via ratiometric APEX tagging. *Molecular Cell, 55*(2), 332–341.

Jakobsen, L., Vanselow, K., Skogs, M., Toyoda, Y., Lundberg, E., Poser, I., et al. (2011). Novel asymmetrically localizing components of human centrosomes identified by complementary proteomics methods. *EMBO Journal, 30*(8), 1520–1535.

Kim, D. I., Birendra, K. C., Zhu, W., Motamedchaboki, K., Doye, V., & Roux, K. J. (2014). Probing nuclear pore complex architecture with proximity-dependent biotinylation. *Proceedings of the National Academy of Sciences of the USA, 111*(24), E2453–E2461.

Lambert, J. P., Tucholska, M., Go, C., Knight, J. D., & Gingras, A. C. (2014). Proximity biotinylation and affinity purification are complementary approaches for the interactome mapping of chromatin-associated protein complexes. *Journal of Proteomics, 118*, 81–94.

Lam, S. S., Martell, J. D., Kamer, K. J., Deerinck, T. J., Ellisman, M. H., Mootha, V. K., et al. (2014). Directed evolution of APEX2 for electron microscopy and proximity labeling. *Nature Methods, 12*, 51–54.

Lawo, S., Hasegan, M., Gupta, G. D., & Pelletier, L. (2012). Subdiffraction imaging of centrosomes reveals higher-order organizational features of pericentriolar material. *Nature Cell Biology, 14*(11), 1148–1158.

Li, J. B., Gerdes, J. M., Haycraft, C. J., Fan, Y., Teslovich, T. M., May-Simera, H., et al. (2004). Comparative genomics identifies a flagellar and basal body proteome that includes the BBS5 human disease gene. *Cell, 117*(4), 541–552.

Lukinavicius, G., Lavogina, D., Orpinell, M., Umezawa, K., Reymond, L., Garin, N., et al. (2013). Selective chemical crosslinking reveals a Cep57-Cep63-Cep152 centrosomal complex. *Current Biology, 23*(3), 265–270.

Mellacheruvu, D., Wright, Z., Couzens, A. L., Lambert, J. P., St-Denis, N. A., Li, T., et al. (2013). The CRAPome: a contaminant repository for affinity purification-mass spectrometry data. *Nature Methods, 10*(8), 730–736.

Mennella, V., Keszthelyi, B., McDonald, K. L., Chhun, B., Kan, F., Rogers, G. C., et al. (2012). Subdiffraction-resolution fluorescence microscopy reveals a domain of the centrosome critical for pericentriolar material organization. *Nature Cell Biology, 14*(11), 1159–1168.

Moritz, M., Braunfeld, M. B., Fung, J. C., Sedat, J. W., Alberts, B. M., & Agard, D. A. (1995). Three-dimensional structural characterization of centrosomes from early Drosophila embryos. *The Journal of Cell Biology, 130*(5), 1149–1159.

Morriswood, B., Havlicek, K., Demmel, L., Yavuz, S., Sealey-Cardona, M., Vidilaseris, K., et al. (2013). Novel bilobe components in *Trypanosoma brucei* identified using proximity-dependent biotinylation. *Eukaryotic Cell, 12*(2), 356–367.

Muller, E. G., Snydsman, B. E., Novik, I., Hailey, D. W., Gestaut, D. R., Niemann, C. A., et al. (2005). The organization of the core proteins of the yeast spindle pole body. *Molecular Biology of the Cell, 16*(7), 3341–3352.

Neilson, K. A., Ali, N. A., Muralidharan, S., Mirzaei, M., Mariani, M., Assadourian, G., et al. (2011). Less label, more free: approaches in label-free quantitative mass spectrometry. *Proteomics, 11*(4), 535–553.

Rhee, H. W., Zou, P., Udeshi, N. D., Martell, J. D., Mootha, V. K., Carr, S. A., et al. (2013). Proteomic mapping of mitochondria in living cells via spatially restricted enzymatic tagging. *Science, 339*(6125), 1328–1331.

Roux, K. J., Kim, D. I., & Burke, B. (2013). BioID: a screen for protein–protein interactions. *Current Protocols in Protein Science, 74*, Unit 19 23.

Roux, K. J., Kim, D. I., Raida, M., & Burke, B. (2012). A promiscuous biotin ligase fusion protein identifies proximal and interacting proteins in mammalian cells. *The Journal of Cell Biology, 196*(6), 801–810.

Schnackenberg, B. J., Khodjakov, A., Rieder, C. L., & Palazzo, R. E. (1998). The disassembly and reassembly of functional centrosomes in vitro. *Proceedings of the National Academy of Sciences of the USA, 95*(16), 9295–9300.

Schnackenberg, B. J., & Palazzo, R. E. (1999). Identification and function of the centrosome centromatrix. *Biology of the Cell, 91*(6), 429–438.

Sonnen, K. F., Schermelleh, L., Leonhardt, H., & Nigg, E. A. (2012). 3D-structured illumination microscopy provides novel insight into architecture of human centrosomes. *Biology Open, 1*(10), 965–976.

Zybailov, B., Mosley, A. L., Sardiu, M. E., Coleman, M. K., Florens, L., & Washburn, M. P. (2006). Statistical analysis of membrane proteome expression changes in *Saccharomyces cerevisiae*. *Journal of Proteome Research, 5*(9), 2339–2347.

CHAPTER

Purification of centrosomes from mammalian cell lines

10

Delphine Gogendeau*[,§], Paul Guichard[¶], Anne-Marie Tassin[||,1]

**Institut Curie, CNRS UMR144, Paris, France*

[§]*Institut Curie, Orsay, France*

[¶]*Swiss Institute for Experimental Cancer Research (ISREC), School of Life Sciences, Swiss Federal Institute of Technology (EPFL), Lausanne, Switzerland*

[||]*Institute for Integrative Biology of the Cell (I2BC), CEA, CNRS, Université Paris Sud, Gif sur Yvette, France*

[1]*Corresponding author: E-mail: anne-marie.tassin@cgm.cnrs-gif.fr*

CHAPTER OUTLINE

Introduction **172**
1. Methods **175**
1.1 Materials 175
1.2 Solutions and Reagents 176
1.3 Optimization of the $MgCl_2$ Concentration in the Cell Lysate Buffer 177
1.4 Centrosome Isolation 177
1.5 Immunolabeling of Centrosomal Fractions 179
1.6 Adaptation for Adherent Cells 180
1.6.1 Quality control/structural integrity of purified centrosome 181
1.7 Effect of $CaCl_2$ on the Centrosome 182
Conclusions **184**
Acknowledgments **184**
References **184**

Abstract

Centrosomes act as the main microtubule-organizing centre of animal cells and play critical roles in the cell, such as mitotic spindle organization, cell polarity, and motility. They are composed of two barrel-shaped structures, the centrioles, surrounded by the pericentriolar matrix. In mammalian cells, the two centrioles differ structurally due to generational difference, the oldest one bearing appendages which allow the transient docking of the centriole at the plasma membrane in order to grow a primary cilium.

Methods in Cell Biology, Volume 129, ISSN 0091-679X, http://dx.doi.org/10.1016/bs.mcb.2015.03.004

Centrosome components are highly conserved throughout evolution and several pathologies have been associated with centrosomal defects. The understanding of such a complex organelle has therefore been a challenge for many researchers and has led to the development of centrosomal purification procedures to assess molecular composition, biological function, and structural organization of centrosomes. In this paper, we detail a step-by-step procedure to generate high yield of purified centrosome obtained from various mammalian cell lines.

INTRODUCTION

Centrosomes are the major microtubule-organizing center (MTOC) of animal cells. At mitosis, hundreds of microtubules emanate from the centrosome in order to organize the mitotic spindle on which the chromosomes will segregate equally. But, centrosome function is not restricted to mitotic spindle organization, as it plays an essential role in the establishment of cell polarity and organelle positioning, cellular transport, and cell motility (Bornens, 2012). Centrosomes are composed of two barrel-shaped structures, the centrioles, surrounded by a meshwork of hundreds of proteins called the pericentriolar material (PCM). For a long time, PCM has been described as a mass of amorphous material; but recent studies using high resolution imaging have shown that pericentriolar proteins present a very precise organization (Fu & Glover, 2012; Lawo, Hasegan, Gupta, & Pelletier, 2012; Mennella et al., 2012; Sonnen, Schermelleh, Leonhardt, & Nigg, 2012). Centrioles are well conserved throughout evolution both in terms of structure and molecular composition (Carvalho-Santos, Azimzadeh, Pereira-Leal, & Bettencourt-Dias, 2011). Despite a few exceptions, their canonical structure consists in nine microtubule triplets that form a cylinder of approximately 200–250 nm of diameter. Interestingly, the two centrioles of the centrosome display structural and functional asymmetry due to their generational difference. The mother centriole is fully mature and bears distal and subdistal appendages at its distal end, whereas the daughter centriole remains immature until the next cell cycle (Nigg & Stearns, 2011). Both centrioles nucleate microtubules, but only the mother centriole anchors microtubules at its subdistal appendages (Piel, Meyer, Khodjakov, Rieder, & Bornens, 2000). In addition, in G0 phase, only the mother centriole can dock and anchor to the plasma membrane, via its distal appendages, in order to grow a primary cilium that is used as a cellular antenna essential for transduction pathways (Huangfu & Anderson, 2006; Oh & Katsanis, 2012). In proliferating cells, centrioles duplicate in a semiconservative manner once per cell cycle, where a procentriole is formed orthogonally to each preexisting centriole. This process is initiated in S phase and is tightly regulated. Interestingly, despite the fact that the centrosome is composed of hundreds of proteins, only a set of five core proteins, evolutionary and/or functionally conserved, are required for centriole duplication: Cep192/Spd2, Plk4/Sak/zyg1, Sas-6, Stil/ana2/Sas-5, and CPAP/Sas-4 (Arquint, Sonnen, Stierhof, & Nigg,

2012; Bettencourt-Dias et al., 2005; Dammermann et al., 2004; Delattre et al., 2004; Habedanck, Stierhof, Wilkinson, & Nigg, 2005; Kemp, Kopish, Zipperlen, Ahringer, & O'Connell, 2004; Kitagawa et al., 2011; Kirkham, Muller-Reichert, Oegema, Grill, & Hyman, 2003; Kleylein-Sohn et al., 2007; Leidel, Delattre, Cerutti, Baumer, & Gonczy, 2005; O'Connell et al., 2001; Pelletier, O'Tooe, Schwager, Hyman, & Muller-Reichert, 2006; Strnad et al., 2007; Tang et al., 2011; Vulprecht et al., 2012).

The understanding of the microtubule-organizing activity of the centrosome has led to the development of centrosome purification procedures (Mitchison & Kirschner, 1984), which would not have been possible without the identification and development of microtubule- and actin-filament-depolymerizing drugs (De Brabander, De May, Joniau, & Geuens, 1977; Tannenbaum, Tanenbaum, & Godman, 1977). To assess centrosome composition and function, this pioneering method has been modified to obtain high yields of purified centrosomes (Bornens, Paintrand, Berges, Marty, & Karsenti, 1987; Moudjou & Bornens, 1994). These purified centrosomes are functionally competent for microtubule nucleation both in vitro and in vivo (Bornens et al., 1987), enabling the testing of microtubule nucleating capacities of complete centrosomes and/or PCM fractions (Keating, Peloquin, Rodionov, Momcilovic, & Borisy, 1997; Mitchison & Kirschner, 1984; Wiese, Mayers, & Albee, 2009). The centrosome duplication activity of these purified centrosome has been tested in vivo using parthenogenesis assays in *Xenopus* eggs (Tournier, Karsenti, & Bornens, 1989) and in vitro in *Xenopus* egg extracts (Tournier, Cyrklaff, Karsenti, & Bornens, 1991). Chemical extraction of purified centrosomes by high salt or urea can cause changes in the structure and parthenogenetic capacity of the centrosome, although both are resistant to relatively high ionic strength (2 M KCl), but are sensitive to moderate concentrations of chaotropic agents (0.5 M KI) (Klotz et al., 1990). Finally, the identification of a protein as a core centrosomal protein can be achieved both by immunofluorescence as well as by determining its enrichment by western blot analysis on purified centrosome fractions (see, for example, Figure 1(A) and (B)) (Kellogg & Alberts, 1992, #283; Fry, Mayor, et al., 1998; Fry, Meraldi, & Nigg, 1998; Keryer et al., 2003; Moudjou, Bordes, Paintrand, & Bornens, 1996; Tassin, Celati, Moudjou, & Bornens, 1998). Isolated centrosomes from KE37 have been analyzed in 1D and 2D gels and revealed more than 100 proteins (Bornens & Moudjou, 1999). Later on, mass spectrometry studies enabled the identification of hundreds of centrosomal components in lymphoblastic cell lines (Andersen et al., 2003; Jakobsen, Schroder, Larsen, Lundberg, & Andersen, 2013), *Drosophila* (Muller et al., 2010), and mammalian sperm (Firat-Karalar, Sante, Elliott, & Stearns, 2014). Finally, 2D and 3D ultrastructural analyzes of isolated centrosomes, using both transmission electron microscopy and cryo-electron microscopy (cryo-EM) (Figure 1(C)), have enabled better characterization of the overall centriolar structure (Paintrand, Moudjou, Delacroix, & Bornens, 1992), the internal distal part of the centriole (Ibrahim, Messaoudi, Chichon, Celati, & Marco, 2009; Li, Fernandez, Marshall, & Agard, 2012), and the cartwheel (Guichard et al., 2012), as well as the first duplication steps of human centrioles (Guichard, Chretien, Marco, & Tassin, 2010).

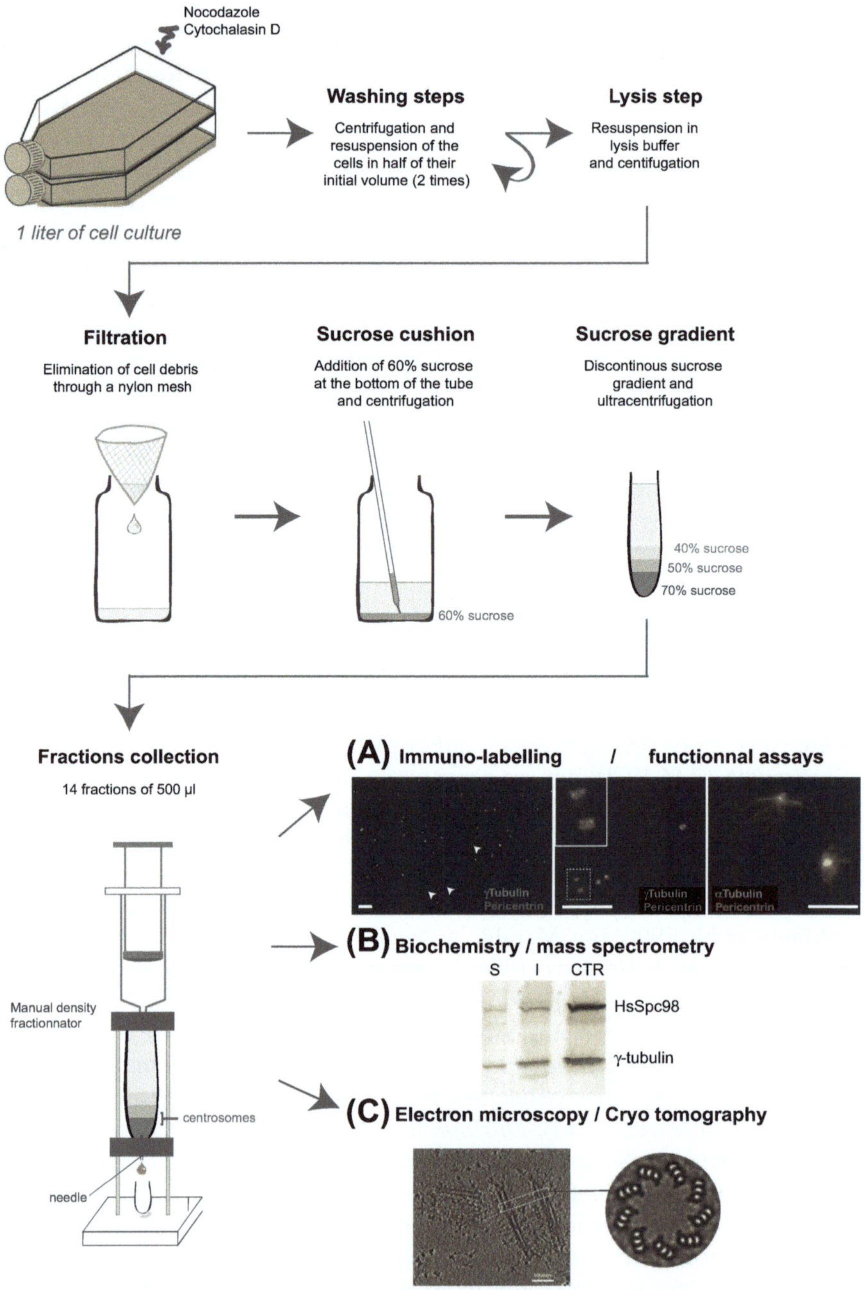

FIGURE 1 Overview of the centrosome purification procedure.

10^9 nonadherent cells growing in exponential phase, corresponding to approximately 1 L of cell culture, are treated with nocodazole and cytochalasin D. Cells are washed and

Here, we detail a step by step protocol for functional centrosome isolation from mammalian cultured cells. This method is adapted from Moudjou and Bornens (1994) and consists of cell lysis using hypotonic solutions, followed by separation of cellular compounds on sucrose gradients. We will insist on the importance of optimizing the $MgCl_2$ concentration during the lysis step according to the cell type. The consequences of changes in calcium ion concentration on the structure of the pericentriolar matrix will also be discussed.

1. METHODS

In our hands, centrosome purification from human (KE37) or murine (IIA1.6) lymphoblastic cell lines is similarly efficient: in both cell lines, the centrosome purification yield is about 25–30%.An overview of the technique is given in Figure 1.

1.1 MATERIALS

- Nylon mesh: SEFAR-NITEX 03-150/50 or cell strainers with a 40-μm pore size
- 250 mL centrifugation bottles

concentrated in half of their initial volume with either TBS or PBS and then resuspended in 20 mL of 0.1X TBS (or PBS)—8% sucrose. Cells are lysed and left 5 min on ice. The lysate in then centrifuged at 2500 g in a 50 mL tube and the supernatant filtered on the nylon mesh. The lysate is then concentrated on a 60% sucrose cushion before being treated by DNAse1 in order to remove the remaining DNA fragments. Lysates are then loaded on a discontinuous sucrose gradient. Several fractions of 0.5 mL are collected; usually centrosomes are enriched in fractions 5–8. Purified centrosomes can be used for (A) immunostainings (left and middle panels: IIA1.6 centrosomal fraction stained for γ-tubulin (red (gray in print versions)) and pericentrin (green (light gray in print versions)), a centrosome should be stained by the two markers) and functional assays (on the right: IIA1.6 centrosomal fraction incubated 5 min at 37 °C in the presence of tubulin, in nucleation buffer—80 mM K-Pipes pH 6.8, 1 mM $MgCl_2$, 1 mM EGTA, 1 mM GTP— and stained for α-tubulin and pericentrin. White arrowheads indicate pericentrin aggregates that are not centrosomes. Bars = 10 μm); (B) biochemistry (western blot of soluble (S), insoluble protein (I), and centrosomal fractions prepared as follows: KE37 cells were washed in PBS and lysed in a PHEM buffer containing 1% Triton X-100 and protease inhibitors. Insoluble proteins were pelleted at 300 g, solubilized in SDS-PAGE sample buffer (Laemmli, 1970), and then boiled for 5 min. Soluble proteins were precipitated with nine vol of methanol at 4 °C for 1 h and pelleted. The pellet was resuspended in the same amount of sample buffer that was used for the insoluble proteins. The membrane was probed for HsScp98 and γ-tubulin) and mass spectrometry experiments; or (C) ultrastructural studies (panels show the cryo-tomography of IIA1.6 isolated centrioles (left) and a symmetrization of a part of the mother centriole (right). Bar = 100 nm).

- 38 mL SW28 Beckman centrifugation polyallomer tubes
- Density gradient fractionator (ISCO 184 driven by hand)
- 15 mL glass Corex tubes with adapters for coverslips (see Figure 3) or Kimble adapters for round-bottom centrifuge tubes 15 mL.
- Rotor Beckman JS 7.5 and JS13 and SW28
- Slides and 12 mm coverslips for immunolabellings.
- Syringe with 21G needle.

1.2 SOLUTIONS AND REAGENTS

- RPMI-1640-Glutamax medium supplemented with fetal bovine serum (10%), Penicillin (100 U/mL), Streptomycin (100 μg/mL), Sodium pyruvate (1 mM), 2-mercaptoethanol (5 μM/mL) for IIA1.6 murine lymphoblasts cell line.
- DMEM/F12-Glutamax medium supplemented with fetal bovine serum (10%), Penicillin (100 U/mL), Streptomycin (100 μg/mL) for RPE1 cells
- Cytochalasin D 10 mM in DMSO (Sigma C8273)
- Nocodazole (10^{-3} M) in DMSO (Sigma M1404)
- DNAse1: Roche (10104159001)
- Tris-buffered saline solution TBS 1X: 10 mM Tris-HCl pH 7.4, 150 mM NaCl, or PBS 1X: 137 mM NaCl, 2.7 mM KCl, 10 mM Na_2HPO_4, 2 mM KH_2PO_4 can be used. For clarity, we will write only TBS in the protocol
- Saline buffered solution-sucrose, TBS-sucrose: TBS 0.1X, 8% glucose (w/v, i.e., 80 g of sucrose in a final volume of 1 L)
- Hepes 1 M, pH 7.2
- $MgCl_2$ 1 M
- β-mercaptoethanol
- *Lysis buffer*: 1 mM Hepes pH 7.2, 0.5% IGEPAL CA-630 (previously named NP-40), 0.5 mM $MgCl_2$, 0.1% β-mercaptoethanol. 1 mM of PMSF and protease inhibitors (aprotinin, leupeptin, pepstatin at 1 μg/mL each or complete protease inhibitor cocktail 1X tablet (Roche no. 11836153001)) are added extemporaneously. The concentration of $MgCl_2$ might be adjusted according to each cell type.
- *Gradient buffer*: 10 mM Pipes pH 7.2, 0.1% Triton X-100, 0.1% β-mercaptoethanol.
- *Sucrose solutions*:
  - 70% sucrose in gradient buffer (w/w, i.e., 70 g of sucrose in a final weight of 100 g of solution, which gives approximately 75 mL of solution).
  - 60% sucrose in gradient buffer (w/w)
  - 50% sucrose in gradient buffer (w/w)
  - 40% sucrose in gradient buffer (w/w)
- Antibodies for centrosome detection:
  - For murine centrosomes: GTU88 (Sigma) for γ-tubulin and Ab4448 (Abcam) for pericentrin.
  - For human centrosomes: Polyclonal γ-tubulin (Tassin et al., 1998) and CTR453 monoclonal antibody (Bailly, Doree, Nurse, & Bornens, 1989)

1.3 OPTIMIZATION OF THE $MgCl_2$ CONCENTRATION IN THE CELL LYSATE BUFFER

Before starting the centrosome purification, the optimum $MgCl_2$ concentration should be determined. This concentration can vary from one cell type to another and according to the culture conditions. Precise adjustment of the $MgCl_2$ concentration during the lysis step is a crucial point for the purity and the quality of the centrosomal preparation. As a matter of fact, $MgCl_2$ influences chromatin structure and solubility (Dixon & Burkholder, 1985). Figure 2 shows the influence of $MgCl_2$ concentration on the nuclear shape, the viscosity of the pellet, and the number of centrosomes obtained after the lysis step. At the right concentration, nuclei are swollen but the integrity of the nucleus is preserved and centrosomes are freed from nuclei. After centrifugation, the pellet is neither too compact nor too viscous: it appears slightly translucent, but neither white (a sign that too much $MgCl_2$ is present) nor too enlarged ($MgCl_2$ concentration too low). Concerning the lysate's appearance by immunofluorescence, if too little $MgCl_2$ is added, the nuclear envelope is disrupted and a lot of centrosomes become trapped in the DNA fragments during the discontinuous gradient centrifugation, even if DNAse is added in the preparation. Conversely, if too much $MgCl_2$ is present, nuclear structure is maintained, but the centrosome nucleus link is preserved. As a consequence, centrosomes are still connected with the nuclei and pellet during centrifugation (Figure 2).

1.4 CENTROSOME ISOLATION

1. Cells are grown in suspension at 37 °C and 5% CO_2 in RPMI-1640-Glutamax medium. Around 10^9 cells in exponential phase are required for the purification, which corresponds to approximately 1 L of culture at 10^6 cells/mL.
2. *Drug treatment*: Add 200 μL of 10^{-3} M nocodazole and 200 μL of 10 mM Cytochalasin D to the culture and incubate the cells for 1 h at 37 °C. This step allows depolymerization of the microtubules and actin filaments to facilitate the dissociation of centrosomes from the nuclei. After the treatment, perform all the following steps at 4 °C.
3. *Washing steps*: Sediment the cells by centrifugation at 1200 rpm (280 g) for 8 min. Resuspend the cells gently with half of the initial volume (500 mL) of TBS 1X, recentrifuge and then resuspend with 250 mL of TBS 0.1X, 8% sucrose (alternatively, PBS can replace TBS). During this step, the cells swell due to the hypotonic media. Afterward, they should be manipulated carefully.
4. *Lysis step*: Centrifuge the cells once more at 1200 rpm (280 g) for 8 min and resuspend them carefully in a total volume of 20 mL of TBS 0.1X, 8% sucrose. Then add the lysis buffer quickly, and all at once, using a volume of lysis buffer needed to obtain a final concentration of 10^7 cells/mL (i.e., 80 mL of lysis buffer to obtain a final volume of 100 mL). Carefully mix by inverting slowly the tube 10 times. Let the cell lysate sit for 5 min on ice. Take an aliquot of 200 μL to test the quality of the lysis later by immunofluorescence.

5. Remove the swollen nuclei, chromatin aggregates, and unlysed cells by centrifugation at 3500 rpm (2500 g) for 10 min. If the $MgCl_2$ concentration is correct, the pellet should be viscous but not too compact (see Figure 2).
6. Filter the lysis supernatant through a nylon mesh into a 250 mL centrifugation bottle and then add concentrated solution of 1 M Hepes to obtain a final concentration of 10 mM (1 mL for 100 mL lysis buffer). Take an aliquot to test the quality of the lysis supernatant later by immunofluorescence.

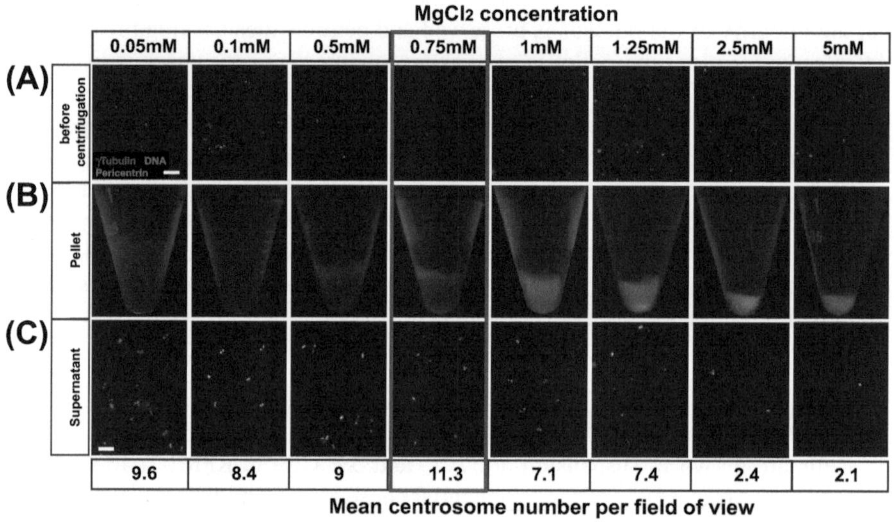

FIGURE 2 Influence of $MgCl_2$ during lysis step.

(A) Lysates of IIA1.6 murine cells before centrifugation labeled by γ-tubulin (red), pericentrin (green), and DNA (blue). Below 0.5 mM $MgCl_2$, nuclei are damaged and large chromatin aggregates are present in the lysates. Above 1.25 mM $MgCl_2$, nuclei are compact and more importantly, centrosomes are associated to the nuclei. (B) Appearance of the pellets following 10 min centrifugation at 2500 rpm. The volume of the nuclei pellet decreases as the $MgCl_2$ concentration increases (from 0.4 mL at 0.02 mM $MgCl_2$ to 0.1 mL at 10 mM $MgCl_2$). At low $MgCl_2$ concentrations, from 0.05 to 0.1 mM, the nuclei pellet appears translucent and highly viscous, whereas at higher concentrations, from 2.5 to 5 mM $MgCl_2$, the nuclei pellet is white and very compact. At the optimal concentration, the pellets should still appear viscous and whitish, but not too white. (C) Supernatants labeled by γ-tubulin (red), pericentrin (green), and DNA (blue). Under 0.1 mM $MgCl_2$, DNA is still found associated with centrosomes. Above 2.5 mM $MgCl_2$, supernatants contain very few centrosomes, still associated with the nuclei. Concentrations between 0.5 and 1.25 mM give a good purity of centrosomes. The number given under each panel gives the average number of centrosomes per field of view (at least 10 fields were counted for each condition). For IIA.6 cells, the optimal $MgCl_2$ concentration is 0.75 mM. Bars = 10 μm. (See color plate)

7. Add DNAse1 to a final concentration of 1 µg/mL (2 U/mL). Mix well and leave the centrosomal suspension sit for 30 min.
8. *Centrosome concentration*: At this step, the centrosomes are sedimented on a sucrose cushion, in order to be concentrated in a smaller volume. To perform this step, transfer the centrosome-containing supernatant to a 250 mL centrifuge bottle and underlay it with 12.5 mL of the 60% sucrose solution. Sediment the centrosomes by centrifugation at 7500 rpm (10,400 g) for 30 min in a J2-21 Beckman centrifuge using a JS 7.5 rotor. Carefully remove the supernatant by aspiration from the top until only about 1/3 of the initial volume (i.e., 25–30 mL) remains in the bottom of the tube. The concentrated solution of centrosomes now contains 20–25% sucrose.
9. *Final centrosome purification*: Prepare a discontinuous sucrose gradient in a 38 mL SW28 Beckman pollyallomer tube by filling the tube with 3 mL of 70% sucrose, 3 mL of 50% sucrose, and 3 mL of 40% sucrose solution. Delicately add the 20–25% sucrose solution containing the centrosomes on top of the gradient and centrifuge at 25,000 rpm for 1 h in a Beckman L8.50B ultracentrifuge with an SW28 rotor.

   Tip: To facilitate the preparation of the gradient and avoid the mixing of the sucrose solutions, use a piece of polystyrene and add each sucrose solutions on top of this polystyrene.

   Fraction Collection: Place the SW28 tube in the density gradient fractionator (ISCO 184 driven by hand). Carefully perforate the tube at the bottom using a 21G needle and collect 14 fractions of 500 µL. Keep an aliquot of each fraction for immunofluorescence experiments and flash freeze the rest in liquid nitrogen. Fractions can be stored at −80 °C for months.

1.5 IMMUNOLABELING OF CENTROSOMAL FRACTIONS

Immunolabeling of fractions: 100 µL of aliquots taken after the lysis step and after the filtration of the lysate on the nylon mesh, as well as 10 µL of each fraction, are sedimented by centrifugation at 10,000 g for 15 min in a swinging rotor (JS13.1 Beckman) on a glass coverslip, in 15 mL Corex tubes equipped with adaptors presented in Figure 3 and also described in (Evans, Mitchison, & Kirschner, 1985; Jakobsen et al., 2013). After centrifugation the coverslips are recovered and fixed for 6 min in methanol at −20 °C. Centrosomes are labeled using two antibodies directed against the centrosome, for example, CTR453 monoclonal antibody (Bailly, Doree, Nurse, & Bornens, 1989) and a polyclonal antibody directed against gamma-tubulin (Tassin et al., 1998) for human centrosomes or anti γ-tubulin (GTU88, Sigma) and anti pericentrin (Ab4448, Abcam) for murine centrosomes. A DAPI staining is also performed to assess the DNA content of the preparation.

Immunofluorescence results: After the lysis step, nuclei should still contain their membrane integrity even if appear swollen. The centrosomes should not be connected to the nuclei anymore (see Section 1.3 and Figure 3 for more details). In the filtered lysates, pairs of centrioles should be observed. Importantly, intact nuclei

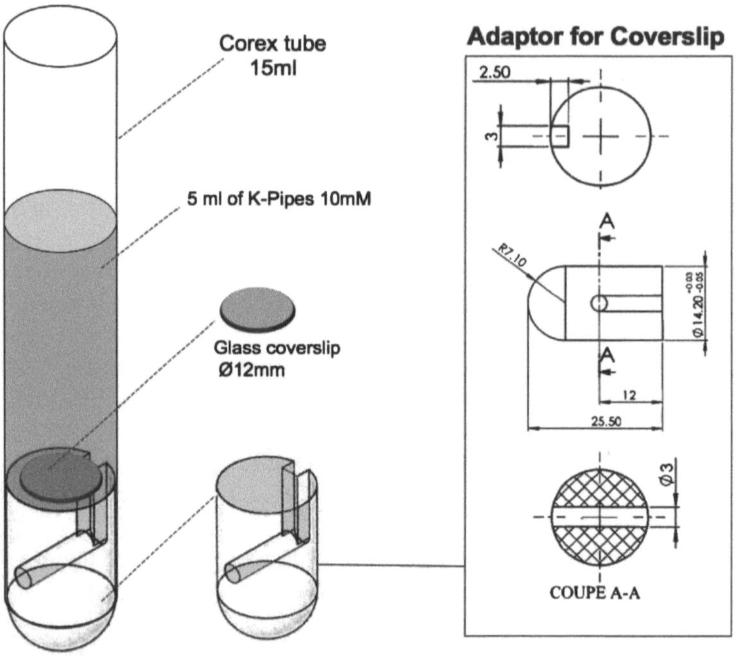

FIGURE 3 Coverslip adaptors to perform immunolabelings.

In order to sediment centrosomes on the coverslip, 10 μL of each fraction are resuspended in 5 mL of K-pipes 10 mM and centrifuged for 10 min at 10,000 g using 15 mL Corex tubes and a JS13.1 Beckman swinging rotor. To sediment uniformly the centrosomes and recover the coverslip after centrifugation, homemade adaptors are used: these adaptors are machined from plastic to fit the curved shape of the Corex tube and have a notch to facilitate the recovery of the adaptor and the coverslip after centrifugation using a hook.

should not be present and only low amounts of chromatin should be seen. In the purified fractions, centrosomes should be highly concentrated and DNA content should be very low.

Determination of purification yield: Centrosome number per fraction is determined using the following procedure: centrosomes are counted under an epifluorescence microscope in a given field, whose dimensions have been measured by a calibration slide micrometer. The number of centrosomes per fraction is then calculated following the total area on which the centrosomes have been sedimented and the fraction volume.

1.6 ADAPTATION FOR ADHERENT CELLS

This protocol can be adapted to adherent cells. According to the final use of the centrosomes and to the purity required for further experiments, the cell number may vary between 2×10^8 and 10^9 cells. To obtain 10^9 cells can be very demanding in term of costs and incubator space. It is known that cell number per flask varies according to cell type. For some cell lines, such as HEK293, it is easy to reach the 10^9 cells required since the cells can be easily grown to high density. The protocol detailed in the previous paragraphs can thus be applied.

However, for other cell types such as RPE1 cell line, the number of cells that can be grown per flask is greatly reduced, meaning that four flasks of 300 cm^2 at 80–90% confluence will yield approximately 2×10^8 cells. In this case, the procedure can be stopped at the sucrose concentration step and the centrosomes used directly, for example, in cryo-EM experiments. However, the purity of the preparation will be lower. Therefore the choice of the cell line should be taken in consideration according to the experiment requirement (EM, mass spectrometry, etc.).

Nevertheless, if a large volume of culture is grown to obtain 10^9 RPE1 cells, this also implies that a large amount of drugs have to be used at the first step of the procedure. In order to decrease the cost of the preparation, a simple trick is applied: instead of adding the drugs directly to the culture medium, the cells are mechanically scraped and resuspended in 250 mL of prewarmed medium containing cytochalasin D and nocodazole (2×10^{-6} and 2×10^{-7} M final concentrations, respectively) for 1 h. The centrosome purification is then carried out as previously described.

As for cells growing in suspension, the optimal $MgCl_2$ concentration should be determined. In our hands, for RPE1 cells, usually 0.5 mM of $MgCl_2$ is used in the lysis buffer. The yield of the centrosome purification for RPE1 cells is similar to the one of KE37 cells, i.e., about 25–30%.

1.6.1 Quality control/structural integrity of purified centrosome

Centrosomes from lymphoblastic lines (KE37, IIA1.6) and those from RPE1 cells were analyzed by cryo-microscopy. The overall structure is conserved with apparent intact microtubule triplets of the centriolar wall surrounded by the pericentriolar material (PCM). Interestingly, centrioles purified form KE37 and from RPE1 display a striking difference in density surrounding the distal part of the mature centrioles. Cryo-microscopy images of RPE1 centrioles show the presence of electron dense material corresponding to the localization of the appendages, whereas these densities are not clearly visible for KE37 centrioles (Figure 4). To understand this difference, cryo-tomography acquisition of KE37- and RPE1-isolated centrosomes revealed that the appendages are present in both centrosomes, but they are more electron dense for RPE1 centrioles as expected from the 2D view. This is an interesting finding that deserves full ultrastructural characterization regarding the function of subdistal appendages within the cell. However, up to now a global structural analysis of centrioles by cryo-microscopy in numerous cell lines has not yet been achieved. It is tempting to propose that subdistal appendages may differ in their overall structure according to their function. It is known that microtubules nucleated at the centrosome can be anchored at the subdistal appendages or released from the centrosome (Abal et al., 2002) and these activities vary from one cell type to another. Subdistal appendages are also proposed to turn into basal foot, during primary cilia growth (Seeley & Nachury, 2010; Tateishi et al., 2013). As a matter of

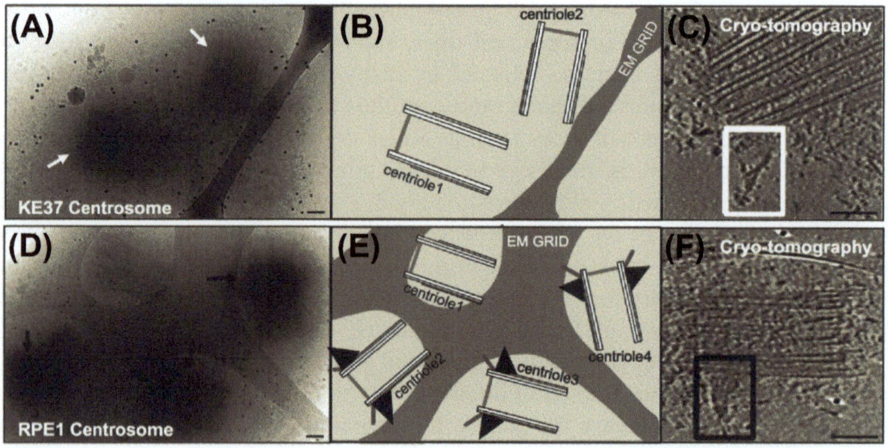

FIGURE 4 Cryo-microscopy of isolated centrosomes from KE37 and RPE1 cells lines.

(A) Cryo-microscopy of isolated KE37 centrosome and the schematic representation in (B). Note the appendages are not visible at the distal part of the centrioles in cryo-microscopy (white arrows), but visible in cryo-tomography (white box in (C)). (D) Cryo-microscopy of isolated RPE1 centrosome and the schematic representation in (E). Please note the highest density of the appendages at the distal part of the centrioles (red (white in print versions) arrows) compare to the centrioles isolated from KE37 cells. RPE1 centriole appendages seem to be more dense by cryo-tomography (red (white in print versions) box in (F)). Bars = 100 nm.

The image in Figure 4(C) is reprint with permission from Guichard et al. (2010).

fact, differences are noticed in the ability of lymphoblastic cells and epithelial cells to grow a cilium.

1.7 EFFECT OF $CaCl_2$ ON THE CENTROSOME

In 1992, an interesting study from Paintrand et al. (1992), showed that the centrosome structure is influenced by the environmental calcium concentration. Namely, in their study, the authors demonstrated that the presence of EDTA, which chelates cations and in particular Ca^{++}, influences several aspects of centrosome structure. First, it changes the conformation of the pericentriolar matrix, in particular the connection between mother and daughter centrioles. This connection appears narrow and long in the presence of EDTA and compact and surrounding the centrioles without EDTA. This conformational change induces a modification of the distance between the two centrosomes (Figure 5). Second, the appendages are also impacted in their density and orientation by the presence of calcium (Figure 5). Last but not least, the centriolar diameter itself is strikingly reduced by calcium addition, with a more compact distal end.

These effects are likely due to the presence of centrins (Moudjou, Paintrand, Vigues, & Bornens, 1991; Paoletti, Moudjou, Paintrand, Salisbury, & Bornens,

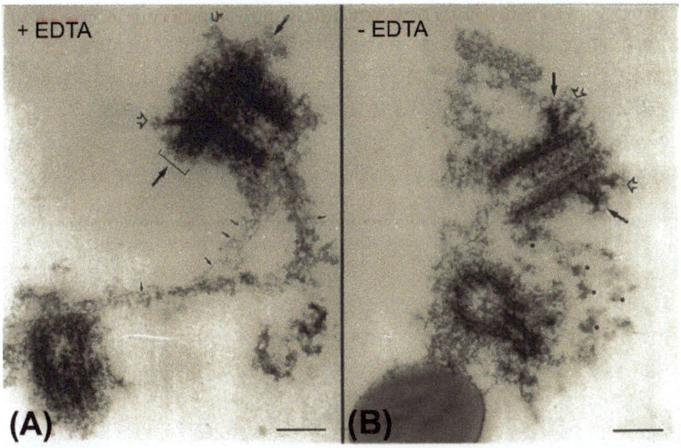

FIGURE 5 Effect of calcium on pericentriolar material (PCM) and distal appendages.

Centrosomes purified in the presence (A) or in the absence (B) of EDTA, a calcium chelator. The presence of calcium at the centrosome influences the PCM's shape (narrow linker vs a cloud surrounding the centrioles), the intercentriolar distance, the shape of the distal and subdistal appendages, and the centriole's diameter. Bars = 200 nm.

Adapted from Paintrand et al., 1992. Reprint with permission.

1996), small proteins possessing four Ca^{++} binding EF-hand domains, as well as centrin-binding proteins such as Sfi1 (Kilmartin, 2003) and POC5 (Azimzadeh et al., 2009) and C2 domain proteins (Shu et al., 2005; Ye, Zeng, Ning, Reiter, & Liu, 2014) within the centrosomes. In human and murine cells, centrin and POC5 have been described in the distal lumen of the centriole (Azimzadeh et al., 2009; Paoletti et al., 1996), explaining the contraction of the centriole after EDTA addition. Studies in ptK2 cells have shown that centrin also localized to the pericentriolar lattice (Baron, Suman, Nemeth, & Salisbury, 1994), which exhibits a calcium sensitive behavior (Baron et al., 1994). Even if such centrin-containing fibers have not been described in the PCM of murine or human cells, pericentriolar matrix contractibility observed in the presence of Ca^{++} can be explained by the presence of both centrins and Sfi1-domain proteins. Upon calcium binding, centrins change their conformation (reviewed in Dantas, Daly, & Morrison, 2012) and centrins have been involved in impressive structural changes in cytoskeletal structures that are particularly abundant in protists and green algae. For example, in *Chlamydomonas reinhardtii*, centrin-containing fibers are localized in the flagellar rootlet, in the nucleus basal body connector, and in the stellar fibrils present in the basal body transition zone (Geimer & Melkonian, 2005; Salisbury, Baron, Surek, & Melkonian, 1984; Sanders & Salisbury, 1994). Upon calcium influx, the contraction of all these structures leads to the deflagellation and repositioning of the basal body within the cell. Centrins are also the major component of the myoneme of many ciliates (Huang & Pitelka, 1973) and particularly the infraciliary lattice of *Paramecium* whose fibers result from centrin-Sfi1 association (Gogendeau et al., 2007). Even if such centrin-containing fibers have not been described in the PCM of murine or human cells, both Sfi1 and centrins are present at the centrosome, suggesting that this molecular contractility could participate in the structural changes due to calcium.

From their study, Paintrand et al. suggest that the condition most comparable to the native one is where EDTA is completely absent from the buffer, implying that centrins present at the centrosome are naturally bound to calcium. The fact that addition of calcium to this steady state induces a dramatic shrinking of the centriole wall, which could lead to a breakage in the ninefold symmetry, suggests that some centrins can still bind calcium and change their conformation. The addition of calcium or EDTA during the centrosome preparation should therefore be avoided or restricted to a very low amount.

CONCLUSIONS

Here, we describe an updated protocol, enabling the purification of centrosomes from either adherent cells or cells grown in suspension. This protocol can be used for any cell line after having adjusted the $MgCl_2$ concentration. These purified centrosomes can then be widely used for IF, biochemical studies, electron microscopy studies. In the future, it will be interesting to understand the structural differences concerning the mother centriole appendages observed in this study.

ACKNOWLEDGMENTS

We greatly acknowledge Renata Basto for her kind invitation to write this method paper. We are grateful to Michel Bornens for giving us the permission to use the figure on the function of Ca^{++} in centrosome structure. We thank the workshop of EPFL and especially Samuel Gex for the coverslip adaptor representation. We acknowledge James Sillibourne and Anne Aubusson-Fleury for proofreading and the critical comments on the manuscript. Figures 4(C) and 5 are reprint with permission (licence no.: 3513680640536 and 3490761008341, respectively). Delphine Gogendeau is financed by an ANR grant (no.ANR-11-BSV8-0016).

REFERENCES

Abal, M., Piel, M., Bouckson-Castaing, V., Mogensen, M., Sibarita, J. B., & Bornens, M. (2002). Microtubule release from the centrosome in migrating cells. *Journal of Cell Biology, 159*(5), 731–737. http://dx.doi.org/10.1083/jcb.200207076.

Andersen, J. S., Wilkinson, C. J., Mayor, T., Mortensen, P., Nigg, E. A., & Mann, M. (2003). Proteomic characterization of the human centrosome by protein correlation profiling. *Nature, 426*(6966), 570–574. http://dx.doi.org/10.1038/nature02166.

Arquint, C., Sonnen, K. F., Stierhof, Y. D., & Nigg, E. A. (2012). Cell-cycle-regulated expression of STIL controls centriole number in human cells. *Journal of Cell Science, 125*(Pt 5), 1342–1352. http://dx.doi.org/10.1242/jcs.099887.

Azimzadeh, J., Hergert, P., Delouvee, A., Euteneuer, U., Formstecher, E., Khodjakov, A., et al. (2009). hPOC5 is a centrin-binding protein required for assembly of full-length centrioles. *Journal of Cell Biology, 185*(1), 101–114. http://dx.doi.org/10.1083/jcb.200808082.

Bailly, E., Doree, M., Nurse, P., & Bornens, M. (1989). p34cdc2 is located in both nucleus and cytoplasm; part is centrosomally associated at G2/M and enters vesicles at anaphase. *EMBO Journal, 8*(13), 3985–3995.

Baron, A. T., Suman, V. J., Nemeth, E., & Salisbury, J. L. (1994). The pericentriolar lattice of PtK2 cells exhibits temperature and calcium-modulated behavior. *Journal of Cell Science, 107*(Pt 11), 2993–3003.

Bettencourt-Dias, M., Rodrigues-Martins, A., Carpenter, L., Riparbelli, M., Lehmann, L., Gatt, M. K., et al. (2005). SAK/PLK4 is required for centriole duplication and flagella development. *Current Biology, 15*(24), 2199–2207. http://dx.doi.org/10.1016/j.cub.2005.11.042.

Bornens, M. (2012). The centrosome in cells and organisms. *Science, 335*(6067), 422–426. http://dx.doi.org/10.1126/science.1209037.

Bornens, M., & Moudjou, M. (1999). Studying the composition and function of centrosomes in vertebrates. *Methods in Cell Biology, 61*, 13–34.

Bornens, M., Paintrand, M., Berges, J., Marty, M. C., & Karsenti, E. (1987). Structural and chemical characterization of isolated centrosomes. *Cell Motility and the Cytoskeleton, 8*(3), 238–249. http://dx.doi.org/10.1002/cm.970080305.

Carvalho-Santos, Z., Azimzadeh, J., Pereira-Leal, J. B., & Bettencourt-Dias, M. (2011). Evolution: Tracing the origins of centrioles, cilia, and flagella. *Journal of Cell Biology, 194*(2), 165–175. http://dx.doi.org/10.1083/jcb.201011152.

Dammermann, A., Muller-Reichert, T., Pelletier, L., Habermann, B., Desai, A., & Oegema, K. (2004). Centriole assembly requires both centriolar and pericentriolar material proteins. *Developmental Cell, 7*(6), 815–829. http://dx.doi.org/10.1016/j.devcel.2004.10.015.

Dantas, T. J., Daly, O. M., & Morrison, C. G. (2012). Such small hands: the roles of centrins/caltractins in the centriole and in genome maintenance. *Cellular and Molecular Life Sciences, 69*(18), 2979–2997. http://dx.doi.org/10.1007/s00018-012-0961-1.

De Brabander, M., De May, J., Joniau, M., & Geuens, G. (1977). Ultrastructural immunocytochemical distribution of tubulin in cultured cells treated with microtubule inhibitors. *Cell Biology International Reports, 1*(2), 177–183.

Delattre, M., Leidel, S., Wani, K., Baumer, K., Bamat, J., Schnabel, H., et al. (2004). Centriolar SAS-5 is required for centrosome duplication in *C. elegans*. *Nature Cell Biology, 6*(7), 656–664. http://dx.doi.org/10.1038/ncb1146.

Dixon, D. K., & Burkholder, G. D. (1985). The effects of sodium and magnesium-ion interactions on chromatin structure and solubility. *European Journal of Cell Biology, 36*(2), 315–322.

Evans, L., Mitchison, T., & Kirschner, M. (1985). Influence of the centrosome on the structure of nucleated microtubules. *Journal of Cell Biology, 100*(4), 1185–1191.

Firat-Karalar, E. N., Sante, J., Elliott, S., & Stearns, T. (2014). Proteomic analysis of mammalian sperm cells identifies new components of the centrosome. *Journal of Cell Science*. http://dx.doi.org/10.1242/jcs.157008.

Fry, A. M., Mayor, T., Meraldi, P., Stierhof, Y. D., Tanaka, K., & Nigg, E. A. (1998). C-Nap1, a novel centrosomal coiled-coil protein and candidate substrate of the cell cycle-regulated protein kinase Nek2. *Journal of Cell Biology, 141*(7), 1563–1574.

Fry, A. M., Meraldi, P., & Nigg, E. A. (1998). A centrosomal function for the human Nek2 protein kinase, a member of the NIMA family of cell cycle regulators. *EMBO Journal, 17*(2), 470–481. http://dx.doi.org/10.1093/emboj/17.2.470.

Fu, J., & Glover, D. M. (2012). Structured illumination of the interface between centriole and peri-centriolar material. *Open Biology, 2*(8), 120104. http://dx.doi.org/10.1098/rsob.120104.

Geimer, S., & Melkonian, M. (2005). Centrin scaffold in *Chlamydomonas reinhardtii* revealed by immunoelectron microscopy. *Eukaryotic Cell, 4*(7), 1253–1263. http://dx.doi.org/10.1128/EC.4.7.1253-1263.2005.

Gogendeau, D., Beisson, J., de Loubresse, N. G., Le Caer, J. P., Ruiz, F., Cohen, J., et al. (2007). An Sfi1p-like centrin-binding protein mediates centrin-based Ca2+ -dependent contractility in Paramecium tetraurelia. *Eukaryot Cell, 6*(11), 1992–2000. http://ec.asm.org/content/6/11/1992.long.

Guichard, P., Chretien, D., Marco, S., & Tassin, A. M. (2010). Procentriole assembly revealed by cryo-electron tomography. *EMBO Journal, 29*(9), 1565–1572. http://dx.doi.org/10.1038/emboj.2010.45.

Guichard, P., Desfosses, A., Maheshwari, A., Hachet, V., Dietrich, C., Brune, A., et al. (2012). Cartwheel architecture of Trichonympha basal body. *Science, 337*(6094), 553. http://dx.doi.org/10.1126/science.1222789.

Habedanck, R., Stierhof, Y. D., Wilkinson, C. J., & Nigg, E. A. (2005). The Polo kinase Plk4 functions in centriole duplication. *Nature Cell Biology, 7*(11), 1140–1146. http://dx.doi.org/10.1038/ncb1320.

Huang, B., & Pitelka, D. R. (1973). The contractile process in the ciliate, Stentor coeruleus. I. The role of microtubules and filaments. *Journal of Cell Biology, 57*(3), 704–728.

Huangfu, D., & Anderson, K. V. (2006). Signaling from Smo to Ci/Gli: conservation and divergence of Hedgehog pathways from Drosophila to vertebrates. *Development, 133*(1), 3–14. http://dx.doi.org/10.1242/dev.02169.

Ibrahim, R., Messaoudi, C., Chichon, F. J., Celati, C., & Marco, S. (2009). Electron tomography study of isolated human centrioles. *Microscopy Research and Technique, 72*(1), 42–48. http://dx.doi.org/10.1002/jemt.20637.

Jakobsen, L., Schroder, J. M., Larsen, K. M., Lundberg, E., & Andersen, J. S. (2013). Centrosome isolation and analysis by mass spectrometry-based proteomics. *Methods in Enzymology, 525*, 371–393. http://dx.doi.org/10.1016/B978-0-12-397944-5.00018-3.

Keating, T. J., Peloquin, J. G., Rodionov, V. I., Momcilovic, D., & Borisy, G. G. (1997). Microtubule release from the centrosome. *Proceedings of the National Academy of Sciences of the United States of America, 94*(10), 5078–5083.

Kellogg, D. R., & Alberts, B. M. (1992). Purification of a multiprotein complex containing centrosomal proteins from the Drosophila embryo by chromatography with low-affinity polyclonal antibodies. *Molecular Biology of the Cell, 3*(1), 1–11. http://www.molbiolcell.org/content/3/1/1.long.

Kemp, C. A., Kopish, K. R., Zipperlen, P., Ahringer, J., & O'Connell, K. F. (2004). Centrosome maturation and duplication in *C. elegans* require the coiled-coil protein SPD-2. *Developmental Cell, 6*(4), 511–523.

Keryer, G., Di Fiore, B., Celati, C., Lechtreck, K. F., Mogensen, M., Delouvee, A., et al. (2003). Part of Ran is associated with AKAP450 at the centrosome: involvement in microtubule-organizing activity. *Molecular Biology of the Cell, 14*(10), 4260–4271. http://dx.doi.org/10.1091/mbc.E02-11-0773.

Kilmartin, J. V. (2003). Sfi1p has conserved centrin-binding sites and an essential function in budding yeast spindle pole body duplication. *Journal of Cell Biology, 162*(7), 1211–1221. http://dx.doi.org/10.1083/jcb.200307064.

Kirkham, M., Muller-Reichert, T., Oegema, K., Grill, S., & Hyman, A. A. (2003). SAS-4 is a *C. elegans* centriolar protein that controls centrosome size. *Cell, 112*(4), 575–587.

Kitagawa, D., Kohlmaier, G., Keller, D., Strnad, P., Balestra, F. R., Fluckiger, I., et al. (2011). Spindle positioning in human cells relies on proper centriole formation and on the microcephaly proteins CPAP and STIL. *Journal of Cell Science, 124*(Pt 22), 3884–3893. http://dx.doi.org/10.1242/jcs.089888.

Kleylein-Sohn, J., Westendorf, J., Le Clech, M., Habedanck, R., Stierhof, Y. D., & Nigg, E. A. (2007). Plk4-induced centriole biogenesis in human cells. *Developmental Cell, 13*(2), 190–202. http://dx.doi.org/10.1016/j.devcel.2007.07.002.

Klotz, C., Dabauvalle, M. C., Paintrand, M., Weber, T., Bornens, M., & Karsenti, E. (1990). Parthenogenesis in Xenopus eggs requires centrosomal integrity. *Journal of Cell Biology, 110*(2), 405–415.

Laemmli, U. K. (1970). Cleavage of structural proteins during the assembly of the head of bacteriophage T4. *Nature, 227*(5259), 680–685.

Lawo, S., Hasegan, M., Gupta, G. D., & Pelletier, L. (2012). Subdiffraction imaging of centrosomes reveals higher-order organizational features of pericentriolar material. *Nature Cell Biology, 14*(11), 1148–1158. http://dx.doi.org/10.1038/ncb2591.

Leidel, S., Delattre, M., Cerutti, L., Baumer, K., & Gonczy, P. (2005). SAS-6 defines a protein family required for centrosome duplication in *C. elegans* and in human cells. *Nature Cell Biology, 7*(2), 115–125. http://dx.doi.org/10.1038/ncb1220.

Li, S., Fernandez, J. J., Marshall, W. F., & Agard, D. A. (2012). Three-dimensional structure of basal body triplet revealed by electron cryo-tomography. *EMBO Journal, 31*(3), 552–562. http://dx.doi.org/10.1038/emboj.2011.460.

Mennella, V., Keszthelyi, B., McDonald, K. L., Chhun, B., Kan, F., Rogers, G. C., et al. (2012). Subdiffraction-resolution fluorescence microscopy reveals a domain of the centrosome critical for pericentriolar material organization. *Nature Cell Biology, 14*(11), 1159–1168. http://dx.doi.org/10.1038/ncb2597.

Mitchison, T., & Kirschner, M. (1984). Microtubule assembly nucleated by isolated centrosomes. *Nature, 312*(5991), 232–237.

Moudjou, M., Bordes, N., Paintrand, M., & Bornens, M. (1996). gamma-Tubulin in mammalian cells: the centrosomal and the cytosolic forms. *Journal of Cell Science, 109*(Pt 4), 875–887.

Moudjou, M., & Bornens, M. (1994). Isolation of centrosomes from cultured animal cells. In C. J.E. (Ed.), *Cell biology: A laboratory handbook* (pp. 595–604). Academic Press.

Moudjou, M., Paintrand, M., Vigues, B., & Bornens, M. (1991). A human centrosomal protein is immunologically related to basal body-associated proteins from lower eucaryotes and is involved in the nucleation of microtubules. *Journal of Cell Biology, 115*(1), 129–140.

Muller, H., Schmidt, D., Steinbrink, S., Mirgorodskaya, E., Lehmann, V., Habermann, K., et al. (2010). Proteomic and functional analysis of the mitotic *Drosophila* centrosome. *EMBO Journal, 29*(19), 3344–3357. http://dx.doi.org/10.1038/emboj.2010.210.

Nigg, E. A., & Stearns, T. (2011). The centrosome cycle: centriole biogenesis, duplication and inherent asymmetries. *Nature Cell Biology, 13*(10), 1154–1160. http://dx.doi.org/10.1038/ncb2345.

O'Connell, K. F., Caron, C., Kopish, K. R., Hurd, D. D., Kemphues, K. J., Li, Y., et al. (2001). The *C. elegans* zyg-1 gene encodes a regulator of centrosome duplication with distinct maternal and paternal roles in the embryo. *Cell, 105*(4), 547–558.

Oh, E. C., & Katsanis, N. (2012). Cilia in vertebrate development and disease. *Development, 139*(3), 443–448. http://dx.doi.org/10.1242/dev.050054.

Paintrand, M., Moudjou, M., Delacroix, H., & Bornens, M. (1992). Centrosome organization and centriole architecture: their sensitivity to divalent cations. *Journal of Structural Biology, 108*(2), 107–128.

Paoletti, A., Moudjou, M., Paintrand, M., Salisbury, J. L., & Bornens, M. (1996). Most of centrin in animal cells is not centrosome-associated and centrosomal centrin is confined to the distal lumen of centrioles. *Journal of Cell Science, 109*(Pt 13), 3089–3102.

Pelletier, L., O'Toole, E., Schwager, A., Hyman, A. A., & Muller-Reichert, T. (2006). Centriole assembly in Caenorhabditis elegans. *Nature, 444*(7119), 619–623. http://dx.doi.org/10.1038/nature05318.

Piel, M., Meyer, P., Khodjakov, A., Rieder, C. L., & Bornens, M. (2000). The respective contributions of the mother and daughter centrioles to centrosome activity and behavior in vertebrate cells. *Journal of Cell Biology, 149*(2), 317–330.

Salisbury, J. L., Baron, A., Surek, B., & Melkonian, M. (1984). Striated flagellar roots: isolation and partial characterization of a calcium-modulated contractile organelle. *Journal of Cell Biology, 99*(3), 962–970.

Sanders, M. A., & Salisbury, J. L. (1994). Centrin plays an essential role in microtubule severing during flagellar excision in *Chlamydomonas reinhardtii*. *Journal of Cell Biology, 124*(5), 795–805.

Seeley, E. S., & Nachury, M. V. (2010). The perennial organelle: assembly and disassembly of the primary cilium. *Journal of Cell Science, 123*(Pt 4), 511–518. http://dx.doi.org/10.1242/jcs.061093.

Shu, X., Fry, A. M., Tulloch, B., Manson, F. D., Crabb, J. W., Khanna, H., et al. (2005). RPGR ORF15 isoform co-localizes with RPGRIP1 at centrioles and basal bodies and interacts with nucleophosmin. *Human Molecular Genetics, 14*(9), 1183–1197. http://dx.doi.org/10.1093/hmg/ddi129.

Sonnen, K. F., Schermelleh, L., Leonhardt, H., & Nigg, E. A. (2012). 3D-structured illumination microscopy provides novel insight into architecture of human centrosomes. *Biology Open, 1*(10), 965–976. http://dx.doi.org/10.1242/bio.20122337.

Strnad, P., Leidel, S., Vinogradova, T., Euteneuer, U., Khodjakov, A., & Gonczy, P. (2007). Regulated HsSAS-6 levels ensure formation of a single procentriole per centriole during the centrosome duplication cycle. *Developmental Cell, 13*(2), 203–213. http://dx.doi.org/10.1016/j.devcel.2007.07.004.

Tang, C. J., Lin, S. Y., Hsu, W. B., Lin, Y. N., Wu, C. T., Lin, Y. C., et al. (2011). The human microcephaly protein STIL interacts with CPAP and is required for procentriole formation. *EMBO Journal, 30*(23), 4790–4804. http://dx.doi.org/10.1038/emboj.2011.378.

Tannenbaum, J., Tanenbaum, S. W., & Godman, G. C. (1977). The binding sites of cytochalasin D. I. Evidence that they may be peripheral membrane proteins. *Journal of Cellular Physiology, 91*(2), 225–237. http://dx.doi.org/10.1002/jcp.1040910208.

Tassin, A. M., Celati, C., Moudjou, M., & Bornens, M. (1998). Characterization of the human homologue of the yeast spc98p and its association with gamma-tubulin. *Journal of Cell Biology, 141*(3), 689–701.

Tateishi, K., Yamazaki, Y., Nishida, T., Watanabe, S., Kunimoto, K., Ishikawa, H., et al. (2013). Two appendages homologous between basal bodies and centrioles are formed using distinct Odf2 domains. *Journal of Cell Biology, 203*(3), 417–425. http://dx.doi.org/10.1083/jcb.201303071.

Tournier, F., Cyrklaff, M., Karsenti, E., & Bornens, M. (1991). Centrosomes competent for parthenogenesis in *Xenopus* eggs support procentriole budding in cell-free extracts. *Proceedings of the National Academy of Sciences of the United States of America, 88*(22), 9929–9933.

Tournier, F., Karsenti, E., & Bornens, M. (1989). Parthenogenesis in Xenopus eggs injected with centrosomes from synchronized human lymphoid cells. *Developmental Biology, 136*(2), 321–329.

Vulprecht, J., David, A., Tibelius, A., Castiel, A., Konotop, G., Liu, F., et al. (2012). STIL is required for centriole duplication in human cells. *Journal of Cell Science, 125*(Pt 5), 1353–1362. http://dx.doi.org/10.1242/jcs.104109.

Wiese, C., Mayers, J. R., & Albee, A. J. (2009). Analysis of centrosome function and microtubule dynamics by time-lapse microscopy in Xenopus egg extracts. *Methods in Molecular Biology, 586*, 89–113. http://dx.doi.org/10.1007/978-1-60761-376-3_5.

Ye, X., Zeng, H., Ning, G., Reiter, J. F., & Liu, A. (2014). C2cd3 is critical for centriolar distal appendage assembly and ciliary vesicle docking in mammals. *Proceedings of the National Academy of Sciences of the United States of America, 111*(6), 2164–2169. http://dx.doi.org/10.1073/pnas.1318737111.

CHAPTER

Isolation, cryotomography, and three-dimensional reconstruction of centrioles

11

Paul Guichard, Virginie Hamel, Aitana Neves, Pierre Gönczy[1]

Swiss Institute for Experimental Cancer Research (ISREC), School of Life Sciences, Swiss Federal Institute of Technology (EPFL), Lausanne, Switzerland

[1]*Corresponding author: E-mail: pierre.gonczy@epfl.ch*

CHAPTER OUTLINE

Introduction **192**

1. Methods **194**

1.1 Materials and Reagents 194

1.1.1 Trichonympha *and centriole isolation* 194

1.1.2 Detection of Trichonympha *centrioles by immunofluorescence* 195

1.1.3 Electron microscopy 195

1.1.4 Software 196

1.2 *Trichonympha* Centriole Isolation 196

1.2.1 Trichonympha *isolation* 197

1.2.2 Isolation of centrioles from Trichonympha 197

1.3 Analysis of Isolated *Trichonympha* Centrioles 197

1.3.1 Immunofluorescence experiment 199

1.3.2 Analysis by transmission electron microscopy 200

1.4 Cryotomography of Centrioles 200

1.4.1 Sample preparation and transfer into a cryomicroscope 200

1.4.2 Tomogram acquisition 202

1.4.3 Tomogram reconstruction using IMOD 203

1.5 Subtomogram Averaging 203

Conclusions **206**

Acknowledgments **207**

References **207**

Methods in Cell Biology, Volume 129, ISSN 0091-679X, http://dx.doi.org/10.1016/bs.mcb.2015.04.003

Abstract

Centrioles and basal bodies (referred to hereafter as centrioles for simplicity) are microtubule-based cylindrical organelles that are typically ~450-nm long and ~250 nm in diameter. The centriole is composed of three distinct regions: the distal part characterized by microtubule doublets, the central core that harbors microtubule triplets, which are also present in the proximal part that also contains the cartwheel, a structure crucial for centriole assembly. The cartwheel was initially revealed by conventional electron microscopy of resin-embedded samples and is thought to impart the near universal ninefold symmetry of centrioles. Deciphering the native architecture of the cartwheel has proven challenging owing to its small dimensions and the difficulties in isolating it. Here, we present a method to purify and analyze the structure of the exceptionally long *Trichonympha* centriole by cryotomography and subtomogram averaging. Using this method, we revealed the native architecture of the proximal cartwheel-containing region at ~40 Å-resolution. This method can be applied as a general strategy for uncovering the structure of centrioles in other species.

INTRODUCTION

Centrioles are among the largest macromolecular complexes of eukaryotic cells and exhibit a remarkably conserved cylindrical structure containing nine microtubule blades (term used from here on to collectively refer to microtubule doublets and triplets). The correct assembly of centrioles is critical for the formation of centrosomes and cilia (reviewed in Nigg & Raff (2009)). Deciphering the architecture of centrioles is key for understanding overall centriole assembly and function (reviewed in Gönczy (2012)). Due to the typically small dimensions of centrioles (~450-nm long and ~250-nm diameter), electron microscopy of resin-embedded samples has been the method of choice over the last 60 years to study their ultrastructure (e.g., Allen, 1969; Bornens, Paintrand, Berges, Marty, & Karsenti, 1987; Cavalier-Smith, 1974; Dippell, 1968; Geimer & Melkonian, 2004; Vorobjev & Chentsov, 1982). These studies showed that the canonical centriole is not only composed of nine microtubule blades, but also displays other signature features. Among these are the distal and subdistal appendages, which are needed for anchorage to the plasma membrane during cilium formation. Another example is the central core that comprises periodic densities along the A-microtubule. Finally, the proximal region contains a remarkable structure called the cartwheel that is thought to impart the ninefold symmetry of the entire centriole (Figure 1(A)).

Despite these advances, analysis of resin-embedded specimens is limited by the steps necessary for preparation of samples for electron microscopy. The successive steps of chemical/cryofixation, dehydration in pure ethanol, as well as staining using heavy metals such as osmium tetroxide and uranyl acetate, induce artifacts of sample preparation and reduce overall resolution (Dubochet & Sartori Blanc, 2001). These limitations have been overcome by the development over the last 30 years of cryoelectron microscopy, which allows observation of biological samples in a hydrated

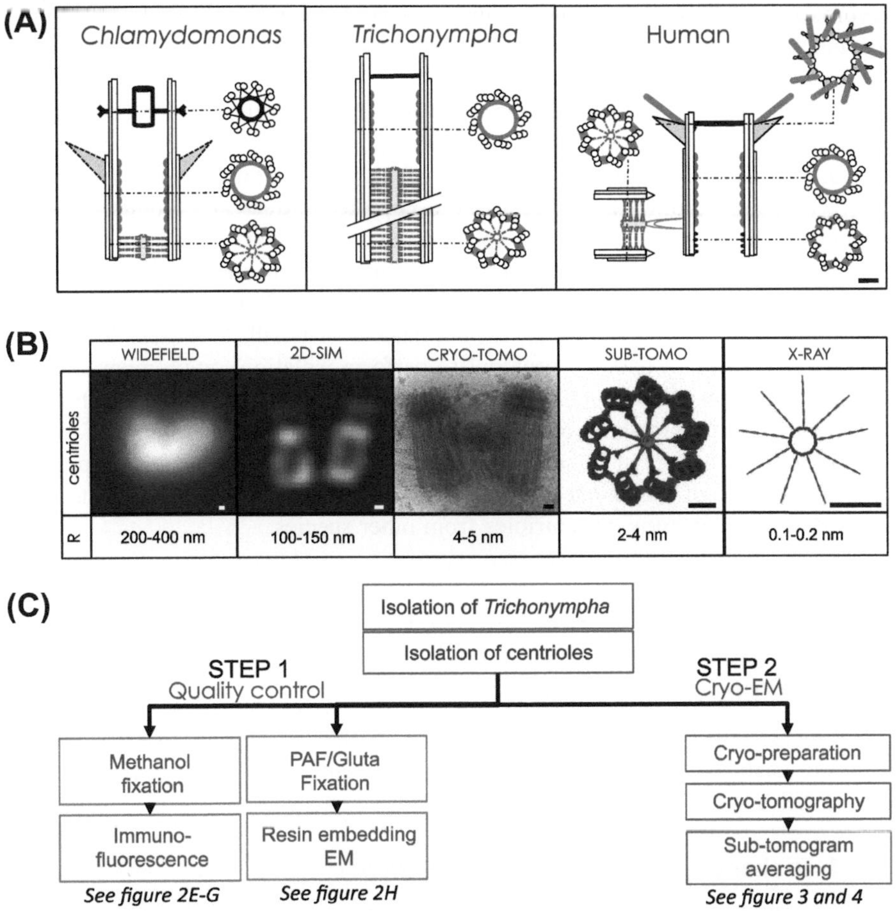

FIGURE 1 Analysis of centriolar architecture.

(A) Schematic representation of centriolar architecture in *Chlamydomonas*, *Trichonympha*, and human cells. Note that the overall organization of the centriole is conserved, with the ninefold symmetrical arrangement of outer microtubule blades (small circles), the central core (blue (dark gray in print versions)), and the cartwheel (red (gray in print versions)). (B) Methods utilized to analyze centriolar structure at different resolutions (R). (C) Summary of the steps taken to perform cryomicroscopy analysis of purified *Trichonympha* centrioles. Scale bar: 50 nm.

state, also dubbed native state. This method has been developed by the group of Jacques Dubochet and relies on the ability to create a state close to liquid water but frozen at −180 °C, the vitreous state (Dubochet, 2012; Dubochet & McDowall, 1981). Briefly, samples such as proteins, with an overall size not exceeding several hundreds of nanometers, are deposited directly on EM grids and quickly plunged in liquid ethane so as to trap them in this vitreous state prior to analysis by cryomicroscopy. In this condition, the sample does not need to be stained, as the inherent electron density of the sample generates the contrast on the image (Bouchet-Marquis & Hoenger, 2011).

Over the following decades, important developments in image processing and especially in tomography and particle averaging have brought forward electron microscopy as an alternative method to analyze protein complexes, filling the gap between superresolution fluorescent microscopy techniques and atomic resolution methods such as X-ray crystallography (Figure 1(B)). With these advances, one can now thoroughly analyze the native structure of centrioles by cryomicroscopy using cryotomography and image processing (Guichard, Chrétien, Marco, & Tassin, 2010; Guichard et al., 2012, 2013; Li, Fernandez, Marshall, & Agard, 2011).

Here, we report a detailed protocol to perform isolation, cryotomography, and three-dimensional reconstruction of centrioles. To resolve the native architecture of the centriolar proximal region at ~40-Å resolution, we take advantage of the centrioles found in the protist *Trichonympha*, which have a ~40 times longer proximal region than that of canonical centrioles, and exemplify a procedure that can be applied also to the study of centrioles from other species.

1. METHODS

1.1 MATERIALS AND REAGENTS

Listed below are specific material and reagents needed for the different steps of centriole analysis. Usual material for molecular biology is also needed (pipettes, tips, distilled water, etc.).

1.1.1 Trichonympha *and centriole isolation*

- *Zootermopsis nevadensis* termites
- Tweezer DUMONT n°5 (Dumont, ref 0101-5-PO)
- Ice bucket
- 10-cm-diameter petri dish
- K-PIPES 10 mM pH 7.2
- Differential interference contrast (DIC) microscope with a 20× objective
- Lysis buffer: 10-mM K-PIPES pH 7.2 + 0.5% NP-40 (IGEPAL) + protease inhibitors Complete (SIGMA; 1/1000)
- Tabletop centrifuge at 4 °C or in a cold room
- 12-mm round coverslip
- Glass slide
- 15-ml Kontes™ KIMAX™ Reusable high strength centrifuge tubes without cap (Fisher Scientific, ref 09-500-34)—"Corex" tubes
- 16-mm-diameter adapter rubber to put the Corex tubes in the rotor (Fisher Scientific, ref 05-569-11)
- JS-13 (Avanti) or HB6 (Beckman) centrifuge rotors (to be used for the Corex tubes)

- Adapters for coverslips
- Curved-tip needle
- Sucrose

1.1.2 *Detection of* Trichonympha *centrioles by immunofluorescence*

- Methanol (100%)
- Ceramic rack (Thomas Scientific, ref 8542E40)
- 6-well plate (used in cell culture)
- Parafilm
- Mouse monoclonal antiacetylated tubulin 6-11B-1 (Sigma–Aldrich, ref: T6793)
- Mouse centrin-2 antibody 20H5 (Merck Millipore, ref 04-1624)
- Secondary antibodies anti-mouse Alexa 568 (Life technologies, ref A11004)
- Secondary antibodies anti-rabbit Alexa 488 (Life technologies, ref A11034)
- PBS, Tween, BSA (PBT): Phosphate Buffered Saline (PBS) + Tween 0.05% + Bovine Serum Albumin (BSA) 1%
- Mounting medium: 4% n-Propyl-Gallate; 1× PBS; 90% glycerol
- Nail polish (transparent)
- Fluorescence microscope (epifluorescence, confocal, 63× oil objective, etc.)

1.1.3 Electron microscopy

1.1.3.1 Conventional electron microscopy

- Nunc™ Thermanox™ 13-mm-diameter coverslips (ThermoScientific, ref 174950)
- Paraformaldehyde (EMS, ref 15700)
- Glutaraldehyde (EMS, ref 16000)
- Fixative solution (PBS with 2% paraformaldehyde and 1.5% glutaraldehyde)
- Cacodylate buffer 0.1 M, pH 7.4 (EMS, ref 11650)
- Osmium and Uranyl acetate (EMS, ref 19100 and ref 22400, respectively)
- Graded ethanol series: 50%, 70%, 90%, 95%, and 100%
- Watch glass
- Durcupan resin (EMS, ref 14040)
- Vacuum oven (EMS, ref 3606)
- Razor blade
- Ultramicrotome for electron microscopy preparation
- Electron microscope operating at 80 KeV for observation of resin section

1.1.3.2 Cryomicroscopy

- Tweezer Style L5 with clamping rings (EMS, ref 0508-L4-PO)
- 10-nm gold beads (Aurion, ref 210.133)
- Cryo-EM grids Lacey 300 mesh (EMS, ref LC325-Cu)
- Homemade cryoplunger or FEI Vitrobot or Leica EM GP system
- Liquid nitrogen
- Bottle of ethane gas
- Whatman filter paper n°4
- Cryomicroscope equipped with field emission gun (FEG) operating at 200 KeV

1.1.4 Software

1.1.4.1 Tomogram acquisition

- FEI tomogram acquisition
- JEOL tomogram acquisition
- SerialEM http://bio3d.colorado.edu/SerialEM/

1.1.4.2 Tomogram reconstruction

- Input/ouput plug-in to convert stack in an .mrc format: http://u759.curie.fr/en/download/softwares/input-output/input-output-00741
- IMOD (etomo): http://bio3d.colorado.edu/imod/

1.1.4.3 Image processing and subtomogram averaging

- Matlab: mathworks.com/products/matlab/
- Coordinates interpolation: http://gonczy-lab.epfl.ch/resources
- Bsoft: http://lsbr.niams.nih.gov/bsoft/
- Spider: http://spider.wadsworth.org/spider_doc/spider/docs/spider.html
- PEET: http://bio3d.colorado.edu/PEET/
- Dynamo: https://dynamo.bioz.unibas.ch/
- Jsubtomo: http://www.opic.ox.ac.uk/wiki/index.php/Jsubtomo

Other software for tomography and image processing can be found at: http://en.wikibooks.org/wiki/Software_Tools_For_Molecular_Microscopy/Specific_packages.

1.1.4.4 Visualization

- ImageJ: http://imagej.nih.gov/ij/
- UCSF Chimera: http://www.cgl.ucsf.edu/chimera/

1.2 *TRICHONYMPHA* CENTRIOLE ISOLATION

The parabasalid protists *Trichonympha* are symbiotes of termites that live in their gut and break down the cellulose of the wood ingested by their host (Schmidt, 1956). To obtain a large amount of starting material, we used *Zootermopsis nevadensis* termites, in whose gut reside three species of *Trichonympha: Trichonympha campanula, Trichonympha collaris, and Trichonympha sphaerica*. *Trichonympha campanula* is the most abundant among these, such that the resulting isolated centrioles will be mostly from this species.

Note: After purification of the centrioles, we recommend not to freeze the sample because the slow formation of ice can affect sample integrity. In the following protocol, analyze the sample directly by immunofluorescence and conventional EM and in a second step, repeat the exact same protocol and directly perform cryo-EM (Figure 1(C)).

If the sample needs to be frozen, add sucrose with a final concentration of at least 30% and snap freeze the sample. In this condition, the sample can be still be analyzed by immunofluorescence and regular EM but not by cryo-EM due to the sucrose that prevents formation of good vitreous ice, thus leading to considerably reduced contrast (Grassucci, Taylor, & Frank, 2007).

1.2.1 Trichonympha *isolation (Figure 2(A)–(B))*

- Place a petri dish on ice.
- Collect one *Zootermopsis nevadensis* termite and put it in the petri dish.
- Once the termite is anesthetized by cold (around 5 min), remove its head with tweezers (Figure 2(A), steps 1 and 2).
- Pull the gut out from the termite's abdomen with the tweezers. The gut comes out easily but be gentle so as not to break it (Figure 2(A), steps 3 and 4).
- Tear the gut open using the tweezers as much as possible.
- Place the gut in an Eppendorf tube containing 1 mL of 10-mM K-PIPES pH 7.2. The gut contains a mix of bacteria and different types of flagellates, including *Trichonympha* (Figure 2(A), step 5).
- Let *Trichonympha* cells sediment for 10 min on ice; the bacteria and smaller protists will remain in the supernatant (Figure 2(B)).
- Carefully aspirate the supernatant with a pipette and then resuspend the sedimented material in 10-mM K-PIPES pH 7.2 (Figure 2(B)). Let the *Trichonympha* cells sediment for 10 min and repeat the entire procedure (sedimentation, removal of the supernatant, and resuspension of the pellet) 3 times.
- Visualize the integrity of the cells by DIC microscopy and verify that most bacteria are gone (Figure 2(C)).

1.2.2 Isolation of centrioles from Trichonympha

All steps hereafter should be performed at 4 °C.

- Lyse the cells by adding 1 mL of lysis buffer to the sedimented *Trichonympha* cells.
- Incubate 20 min (if you want to keep the rostrum) or 1 h (if you want to dissociate the rostrum) (Figure 2(D)). Most centriole-flagella are released after 20 min.
- Centrifuge at 500 g for 3 min in a tabletop centrifuge to remove the cellular debris, collect the supernatant, and centrifuge it at 1000 g for 5 min. The pellet contains isolated centrioles still attached to their flagella (Figure 2(E)).

1.3 ANALYSIS OF ISOLATED *TRICHONYMPHA* CENTRIOLES (FIGURE 2(E–H))

To determine the quality of the sample, an immunofluorescence experiment and an analysis using conventional transmission electron microscopy are performed (Figure 2(E)–(H)). The first method serves to evaluate whether select proteins have remained in place during isolation for instance by using antibodies that recognize centriolar proteins. The second method evaluates the purity of the sample and its structural integrity.

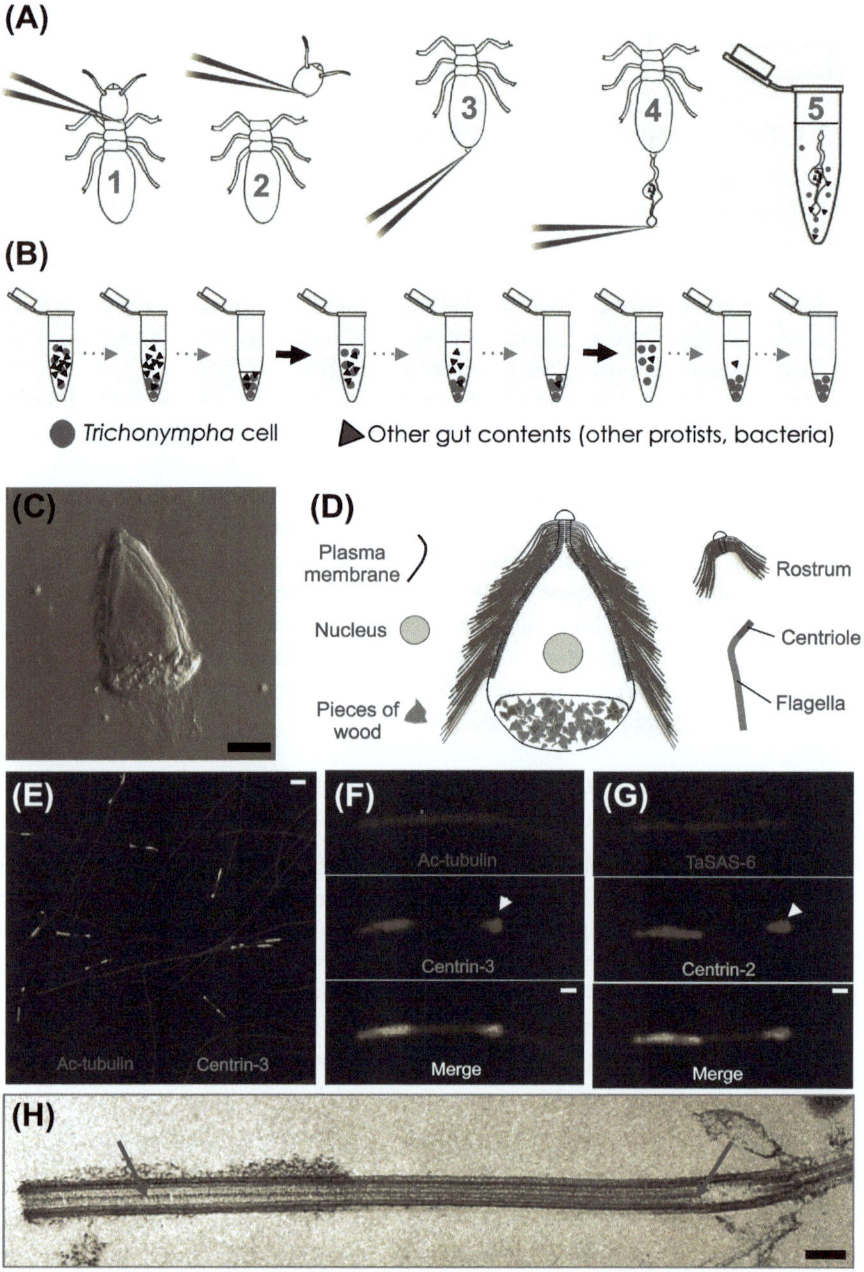

FIGURE 2 Isolation of *Trichonympha* centrioles.

(A) Dissection of *Zootermopsis nevadensis* termite and extraction of its gut. (B) *Trichonympha* cells are isolated from the rest of the gut contents by three rounds of washes

1.3.1 Immunofluorescence experiment

- Place an adapter inside a 15-mLCorex tube containing 5 mL of 10-mM K-PIPES buffer and insert a 12-mm diameter round glass coverslip (For details, see chapter on purification of centrosomes from mammalian cell lines).
- Load 50 μL of isolated centrioles preparation into the K-PIPES buffer solution in the Corex tube.
- Centrifuge at 11,000 g for 10 min at 4 °C using a JS-13 (Avanti) or HB6 (Beckman) swinging rotor with the corresponding rubber adapter.
- Gently pull out the adapter from the Corex tube using a curved-tip needle and carefully remove the coverslip with tweezers (For details, see chapter on purification of centrosomes from mammalian cell lines).
- Fix the coverslip by putting it for 4 min in precooled pure methanol at −20 °C in a ceramic rack.
- Place the coverslip in a well of a 6-well plate filled with PBS and rinse for 5 min.
- Centrioles are subsequently stained with antibodies against acetylated tubulin (1/1000 dilution in PBT) to mark centriolar microtubules and flagella, as well as centrin-2 (1/1000 dilution in PBT) to mark only centrioles. Practically speaking, the coverslip is deposited on a strip of Parafilm and covered by 50−100 μL of antibody, incubating for 1 h at room temperature with a lid to prevent dehydration.
- Wash the coverslip by putting it in a well of a 6-well plate filled with PBS for 5 min.
- Incubate with secondary antibodies diluted in PBT (anti-mouse Alexa-568 1/1000 and anti-rabbit Alexa 488 1/1000) for 45 min at room temperature as for the primary antibodies.
- Wash the coverslip in PBS for 5 min.
- Mount the coverslip on a glass slide. Pipette 6 μL of mounting medium onto a glass slide. Take the coverslip with tweezers to invert it (sample facing down) and gently deposit it onto the drop of mounting medium. Seal the coverslip with nail polish.

and sedimentation. (C) Intact isolated *Trichonympha* cells under DIC microscope. Scale bar: 20 μm. (D) Schematic representation of a *Trichonympha* cell. Note the rostrum, which dissociates later than the cell body. (E−G) Isolated *Trichonympha* centrioles connected to flagella stained with antibodies against centrin (green) (E−G) and acetylated tubulin (red, E, F) or TaSAS-6 (red, G). Note that in the *Trichonympha* centriole, the pattern of centrin (centrin-2 or centrin-3) staining is different from that observed for *Chlamydomonas* or human centrioles. Indeed, beside the evolutionary conserved localization of the proteins at the distal region of centrioles (arrowheads, F, G), centrin antibodies also decorate the proximal-most third of the proximal region. Scale bars represent 2 μm in (E) and 500 nm in (F−G). (H) Conventional transmission electron microscopy of the exceptionally long *Trichonympha* centriole. Note that the central hub of the cartwheel is visible along the centriole (red arrow). Scale bar: 200 nm. (See color plate)

- Analyze the stained centrioles under an epifluorescence microscope or a confocal microscope as shown in Figure 2(E)–(G).

In addition to this analysis by fluorescent microscopy, the structural integrity of isolated centrioles and flagella is analyzed by conventional electron microscopy (Figure 2(H)). To do so:

1.3.2 Analysis by transmission electron microscopy

- Place an adapter inside a 15-mL Corex tube containing 5 mL of 10-mM K-PIPES buffer and insert a 13-mm diameter Thermanox coverslip (ThermoScientific).
- Load 50 μL of isolated centrioles directly into the K-PIPES buffer in the Corex tube.
- Centrifuge at 11,000 g for 10 min at 4 °C using a JS-13 (Avanti) or HB6 (Beckman) swing rotor with the corresponding rubber adapter.
- Gently pull out the adapter from the Corex tube using a curved-tip needle and carefully remove the coverslip with tweezers. Note that the following steps are conducted at room temperature.
- Fix the coverslip by putting it in a well of a 6-well plate containing 2 mL of fixative solution (PBS with 2% paraformaldehyde and 1.5% glutaraldehyde) for 60 min.
- Wash the coverslips 3 times in cacodylate buffer (0.1 M, pH 7.4) for 5 min.
- Postfix with 1% osmium tetroxide in cacodylate buffer for 60 min.
- Wash coverslip with distilled water (2 times for 5 min).
- Stain with 2% uranyl acetate for 30 min.
- Wash once coverslips in distilled water.
- In watch glasses, dehydrate in a graded alcohol series, 3 min each (2 × 50%, 1 × 70%, 1 × 90%, 1 × 95%, 2 × 100%).
- Add 1:1 EtOH (100%): Durcupan for 30 min.
- Embed in 100% Durcupan for at least 4 h at room temperature.
- Incubate in oven at 65 °C overnight.
- Separate the Thermanox coverslip from the resin using a razor blade.
- Perform serial 60-nm thick sections using an ultramicrotome.

Note: By keeping the resin surface parallel to the knife in the ultramicrotome, the centrioles are cut longitudinally, thus allowing observation of the structure through its entire length (Figure 2(H)).

1.4 CRYOTOMOGRAPHY OF CENTRIOLES (FIGURE 3)

Once the centriole isolation procedure is robust as judged by immunofluorescence and electron microscopy, proceed to repeat the centriole isolation procedure and perform analysis by cryomicroscopy (Figure 3).

1.4.1 Sample preparation and transfer into a cryomicroscope

- Mix the sample with 10-nm gold beads, so as to reach a concentration of at least 10 gold beads per field of view in cryo-EM at the magnification used for

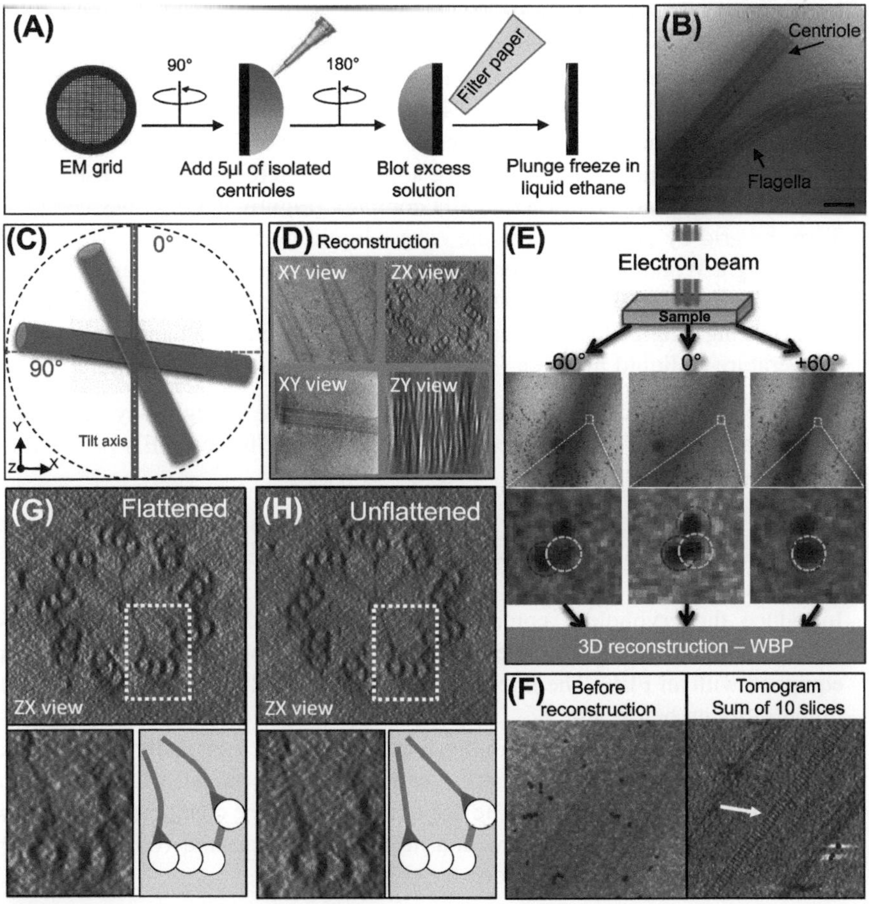

FIGURE 3 Cryotomography of isolated centrioles.

(A) Preparation of EM grids for cryomicroscopy. (B) Image of a centriole and a cilium in cryomicroscopy. Scale bar: 200 nm. (C) Sample orientation compared to the tilt axis (red (dark gray in print versions)). In blue (light gray in print versions), the sample is almost parallel to the tilt axis and in orange (gray in print versions), perpendicular to it. (D) Effects of the missing wedge depending on sample orientation. The sample in blue (light gray in print versions) exhibits a good reconstruction in the ZX view (blue (light gray in print versions)), whereas the sample in orange (gray in print versions) is not reconstructed correctly in the ZY view. (E) Tomogram acquisition and reconstruction: fiducial markers (here three gold beads are highlighted) can be followed at different tilts. The tomogram is reconstructed using a weighted back projection (WBP). (F) Before tomogram reconstruction (0° tilt), structures inside the centriole are not visible, but they appear after tomogram reconstruction (Tomogram), as exemplified by the central hub of the cartwheel (White arrow (dark gray in print versions)). (G and H) ZX views of tomogram reconstructions exemplifying a flattened tomogram with bent radial spokes (G) or unflattened with characteristic straight radial spokes (H).

tomography. This concentration should be calibrated before use by analyzing gold beads on their own at different dilutions by cryo-EM.
- Use L5 style tweezers with a clamping ring to catch the extremity of a cryo-EM carbon film grid (Lacey 300 microMesh, EMS) (Figure 3(A)).
- Deposit 5 μL of the purified centrioles on the surface of the cryo-EM grid using a 20-μL tip cut at the extremity to have a larger diameter and thus preserve the structure of the long centriole/flagella complex (Figure 3(A)).
- Turn the grid by 180°. Remove the excess liquid by blotting the surface of the grid using a Whatman n°4 filter paper in order to form a thin layer of liquid (approximately 50–300 nm, this cannot be controlled precisely). Immediately after blotting, plunge the grid with the tweezers into liquid ethane cooled at the temperature of liquid nitrogen (−180 °C) to vitrify the sample (Figure 3(A)).

Note: To perform this vitrification step, a homemade plunging system can be used; alternatively, use a commercial setup such as a Vitrobot (FEI) or the Leica EM GP system.

- Transfer the grid to an electron microscope cryoholder. Ensure that the grid always stays in liquid nitrogen (or in nitrogen gas above the liquid nitrogen) during the transfer.
- Introduce the cryoholder containing the grid into the cryomicroscope. For optimal imaging, it is preferable to use a microscope operating at 200 keV and equipped with an FEG. The microscope should be cooled at the temperature of liquid nitrogen before use.
- Select a zone in the grid containing a centriole, which is easily distinguished from a flagellum owing to its shape, centrioles being straight, whereas flagella are usually bent (Figure 3(B)). The observation should be conducted at very low magnification (i.e., 6000×) but with a strong defocus (around 100 μm) in order to see the organelle in the vitreous ice (Figure 3(B)).

Note: After detection of a centriole at low magnification, ensure that the orientation of the centriole axis is close to parallel to the tilt axis of the microscope; this will be critical for the quality of the reconstruction due to the "missing wedge" effect. As exemplified in Figure 3(C)–(D), if the sample axis is close to the tilt axis (blue cylinder), the reconstruction will be suitable for further analysis by subtomogram averaging. In contrast, if the sample axis is close to 90° compared with the tilt axis (orange cylinder), a strong artifact will be present in the tomogram reconstruction, precluding further analysis. For more details, regarding the missing wedge effect, see Guesdon, Blestel, Kervrann, and Chrétien (2013).

1.4.2 Tomogram acquisition

- After selection of a centriole at low magnification, use the low dose mode to acquire an image at high magnification in a region close to the centriole. Set the defocus at ~2–4 μm. Note that this calibration step should not be conducted on the centriole itself, the structure of which would be affected by the electron beam (Baker & Rubinstein, 2010).

- Perform the acquisition of the tomogram. Images are recorded at 29,000× magnification and a final pixel size of 0.748 nm. Note that images can be collected using a 2k charge-coupled device (CCD) camera, but a 4k camera is preferable due to the possibility of using binning 2 during acquisition, thus reducing sample degradation by the electron beam.
- Acquire the tilt series covering an angular range from −60° to +60° with 2° increments.

Note: Acquisition parameters (binning, number of images, angle increment) can be changed, but the final dose of electrons per surface area should not exceed 70 $e^-/Å^2$ in order to preserve the integrity of the sample and thus the final resolution (Kudryashev, Castaño-Díez, & Stahlberg, 2012).

1.4.3 Tomogram reconstruction using IMOD

To start the tomogram reconstruction using IMOD, a stack of image files containing all images of the tomogram acquisition and the parameters used should be created in an .mrc format. The FEI software generates this file automatically, but the JEOL software does not. To generate an .mrc file in this case, an ImageJ plugin can be used to transform a stack with .tif format to an .mrc format. This plugin is mentioned in the Material and reagent section for the tomogram reconstruction software.

Note: This part will not explain in detail how to use the software etomo (IMOD package, Boulder, Colorado, (Kremer, Mastronarde, & McIntosh, 1996)). For a detailed tutorial, please use this link:

http://bio3d.colorado.edu/imod/doc/etomoTutorial.html.

- Rename the extension ".mrc" of the stack to ".st," which indicates to the software that the tomogram is a single tilt acquisition.
- Load the tomogram and the angles file in etomo.
- Start the alignment of images using the fiducial mode in etomo. Use the 10-nm gold fiducial markers to precisely align the images (Figure 3(E)).
- Carefully verify that the software tracked the beads correctly in all tilt-angle images. As exemplified in Figure 3(E), beads can overlap sometimes at different angles (see colored dashed lines monitoring three gold beads in the Figure 3(E)).
- After correct alignment, reconstruct the tomogram using R-weighted back projection in etomo. As shown in Figure 3(F), after reconstruction internal structures such as the central hub (red arrow) can be visualized.

1.5 SUBTOMOGRAM AVERAGING (FIGURE 4)

In tomography, the final 3D reconstruction is known to be imperfect due to the missing wedge effect. This is due to the impossibility of acquiring images at high

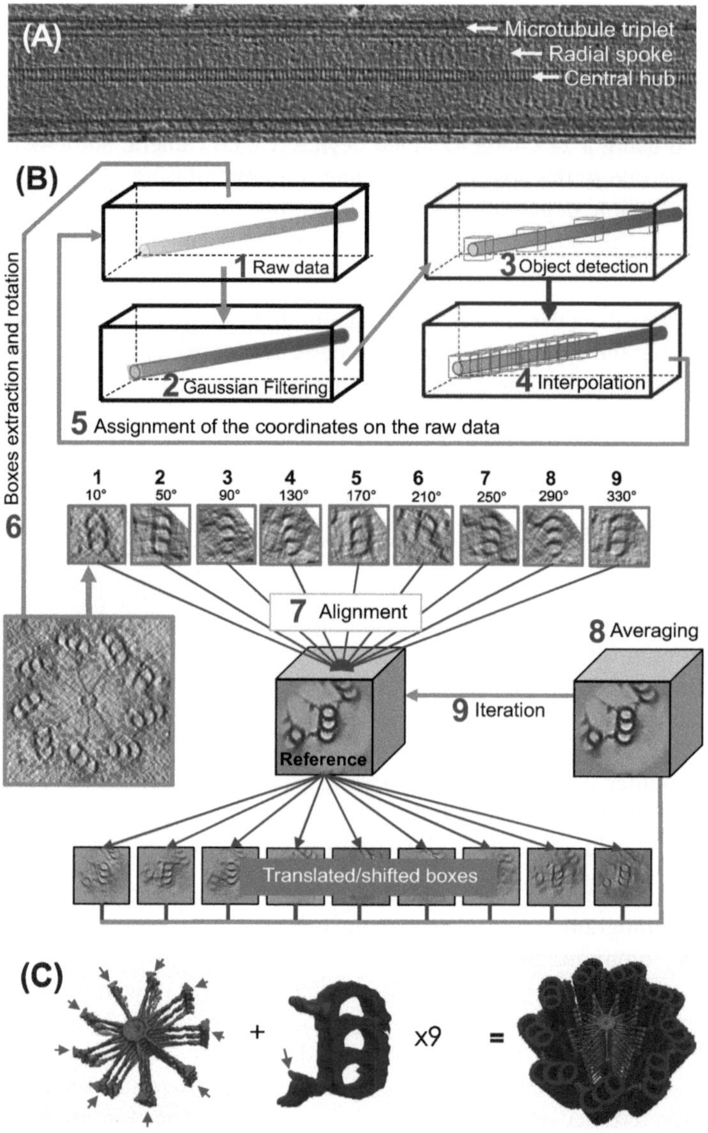

FIGURE 4 Subtomogram averaging.

(A) Cryotomogram section of a *Trichonympha* centriole. Microtubule triplets, the central hub of the cartwheel, and the radial spokes are highlighted. The radial spokes are difficult to visualize due to their low density. (B) Steps of image processing. The tomogram is first duplicated to preserve a version of the raw data (1) and then filtered using a Gaussian Filter (2). Detection of a specific microtubule triplet is done by picking five different regions along the structure (3). Those coordinates are saved and used to calculate a curve that corresponds to the trajectory of the microtubule triplet. This curve is used to calculate new angles, limiting the information during tomogram reconstruction and resulting in low information in the z-axis. The missing wedge effect is compensated in the case of the centriole by taking advantage of the ninefold symmetry of the structure during subtomogram averaging (Bui & Ishikawa, 2013; Guichard et al., 2013; Li et al., 2011).

To perform subtomogram averaging, it is crucial to use tomograms of centrioles that have not been compressed. Such compression can appear during sample preparation and especially during formation of the thin ice layer before freezing. If the ice layer is thinner than the sample (i.e., less than ~250 nm), the centrioles will be affected. Moreover, the centriole can be compressed by touching the carbon edges of the cryo-EM grid. Sometimes the compression is not visible before reconstruction. As exemplified in Figure 3(G) and (H), compression can affect only a subset of the centrioles, which are probably less rigid than others (blue, radial spokes of the cartwheel). In the case of structural analysis of the radial spokes for instance, it is very important to find centrioles without compression to avoid problem of averaging by merging straight radial spokes and abnormal curved radial spokes.

Figure 4(A) shows a cryoelectron tomogram with the different regions of the centriole, whereas Figure 4(B) illustrates the steps applied to the raw data to obtain a 3D structure: Duplication of the tomogram stack to preserve a version of the raw data (1), Gaussian filtering (2), object detection (3), interpolation (4), assignment of the coordinates on the raw tomogram (5), box extraction (6), alignment (7), and averaging (8). The steps of alignment and averaging are an iterative procedure (9).

- Due to the low contrast and high noise, the reconstructed tomogram is first duplicated to keep the initial raw volume and then filtered using a Gaussian filter (or a low-pass filter) in order to help visualization of the centriolar microtubules in the tomogram (Figure 4(B), steps 1 and 2). This step can be done using ImageJ, Bsoft, or Spider softwares.
- Using Bsoft in a Bshow mode, continuous microtubule triplets are detected in the tomogram and picked (micrograph/pick particles) in five different regions along the microtubule blade (Figure 4(B), step 3). Then the coordinates are saved using the item "micrograph/write parameters" in Bshow. This step creates a .star file. This step is repeated for each microtubule blade.

Note: The picked boxes along the blade are used as landmarks to calculate a curve that simulates the trajectory of the microtubule triplets. This curve is generated

coordinates spaced by 17 nm (4). Using these new coordinates, subtomograms are extracted from the raw tomogram (5). This step is repeated for the nine microtubule triplets (6). After extraction, boxes from each triplet are rotated to be in the same orientation as the reference map and processed to align them with the reference (7). The resulting translated boxes are used for the averaging (8), and the output is used as a new reference iteratively (9). (C) 3D reconstruction views of the cartwheel and the microtubule triplet. These two regions have been reconstructed separately. Using common features between the two maps (i.e., the pinhead structure, red (dark gray in print versions) arrows) and by duplicating 9 times the triplet map, the complete architecture of the centriole proximal region is reconstructed.

by interpolating the spline passing through the points, applying iteratively the centripetal scheme described in Lee (1989). After curve determination, 3D coordinates of 17 nm-spaced boxes are calculated along the microtubules. We have written a Matlab® script, freely available on our Web site (http://gonczy-lab.epfl.ch/resources), to perform this step. Using the .star file with manually selected coordinates as an input, the script generates a new .star file (called filename I.STAR) with the coordinates of 17-nm-spaced boxes along the interpolated cubic spline curve.

- Use the coordinates saved previously for each .star file to generate the new interpolated coordinates of 17-nm-spaced boxes for each selected microtubule blade (Figure 4(A) and (B) step 4).
- Using the command-line mode of Bsoft (Heymann, Cardone, Winkler, & Steven, 2008), extract the subtomogram based on the new coordinates file using the command "bpick" on the raw tomogram (Figure 4(B), steps 5 and 6).
- Because of the ninefold symmetry, each microtubule triplet should be rotated with an increment angle of 40°, from the first microtubule blade to the last microtubule blade (Figure 4(B), step 6).
- Align all extracted and rotated subtomograms using PEET (Nicastro et al., 2006), Jsubtomo (Huiskonen et al., 2010) or Dynamo (Castaño-Díez, Kudryashev, Arheit, & Stahlberg, 2012). As for IMOD, theses pieces of software are very well described, and tutorials are available elsewhere (Figure 4(B), steps 7 and 8).
  To perform this alignment, a reference map is needed. Three maps are available for centriole in the Electron microscopy Data Bank:
  - *Chlamydomonas* basal body microtubule triplet: EMD-5252 (Li et al., 2011)
  - *Trichonympha* basal body microtubule triplet: EMD-2330 (Guichard et al., 2013)
  - *Trichonympha* cartwheel: EMD-2329 (Guichard et al., 2013)
- After alignment, the new averaged map can be visualized using UCSF Chimera (Pettersen et al., 2004).

CONCLUSIONS

We presented a step-by-step protocol to purify and analyze the proximal part of *Trichonympha* centrioles in their native state with unprecedented detail using cryotomography (Figure 4(C)). This method can be applied more generally to study centrioles from other species. By taking advantage of repetitive structures, it is possible to use such periodicities to extract multiple copies of the same structural unit to perform subtomogram averaging. However, this method requires a large number of copies to obtain a robust 3D reconstruction (2000–5000 copies, (Guichard et al., 2012, 2013; Li et al., 2011)). This represents a prohibitive limitation for small or nonperiodical structures. For instance, the canonical height of the cartwheel is ~100 nm, with about 5 repeats, meaning that only 5 copies can be extracted per centriole (using the ninefold symmetry, these 5 copies can be turned into 45 final copies.). To circumvent this limitation, we took advantage of the exceptionally long proximal region of the *Trichonympha* centriole, which enabled us to extract about 900 copies per tomogram. Using only 6 tomograms, we could thus obtain the ~5000 copies needed for correct subtomogram averaging. For the analysis of a small or nonperiodic structure, the strategy would then be to acquire a much larger number of tomograms and extract few boxes from each tomogram before averaging. For instance, in the case of the canonical cartwheel, ~50–100 tomograms of unflattened and well-oriented centrioles should be acquired. Although this number of tomograms was difficult to reach until recently, software and hardware development that allow automatic tomogram acquisition open the possibility to perform this task. Moreover, the development of new camera detectors with better contrast and resolution (Grigorieff, 2013; Ruskin, Yu, & Grigorieff, 2013) allows one to decrease the number of particles needed for subtomogram averaging. These advances will bring the architecture of centrioles from many species within reach and thus are expected to yield exciting new discoveries.

ACKNOWLEDGMENTS

We thank Veronika Nemcikova Villimova and Nicola J. Brown for reading the manuscript. We thank Capucine Hachet for help in preparing Figure 2(A). Work on centriole formation in the Gönczy laboratory is funded notably by the European Research Council (ERC) (AdG 233335 and AdG 340227). A.N. was supported by a SystemsX.ch Transition Postdoc Fellowship (SXFSI0_141995).

REFERENCES

Allen, R. D. (1969). The morphogenesis of basal bodies and accessory structures of the cortex of the ciliated protozoan *Tetrahymena pyriformis*. *The Journal of Cell Biology, 40*, 716–733.

Baker, L. A., & Rubinstein, J. L. (2010). *Radiation damage in electron cryomicroscopy.* Elsevier Masson SAS.

Bornens, M., Paintrand, M., Berges, J., Marty, M. C., & Karsenti, E. (1987). Structural and chemical characterization of isolated centrosomes. *Cell Motility and the Cytoskeleton, 8*, 238–249.

Bouchet-Marquis, C., & Hoenger, A. (2011). Cryo-electron tomography on vitrified sections: a critical analysis of benefits and limitations for structural cell biology. *Micron, 42*, 152–162.

Bui, K. H., & Ishikawa, T. (2013). *3D structural analysis of flagella/cilia by cryo-electron tomography* (1st ed.). Elsevier Inc.

Castaño-Díez, D., Kudryashev, M., Arheit, M., & Stahlberg, H. (2012). Dynamo: a flexible, user-friendly development tool for subtomogram averaging of cryo-EM data in high-performance computing environments. *Journal of Structural Biology, 178*, 139–151.

Cavalier-Smith, T. (1974). Basal body and flagellar development during the vegetative cell cycle and the sexual cycle of *Chlamydomonas reinhardtii*. *Journal of Cell Science, 16*, 529.

Dippell, R. V. (1968). The development of basal bodies in paramecium. *Proceedings of the National Academy of Sciences of the United States of America, 61*, 461.

Dubochet, J. (2012). Cryo-EM-the first thirty years. *Journal of Microscopy, 245*, 221–224.

Dubochet, J., & McDowall, A. W. (1981). Vitrification of pure water for electron microscopy. *Journal of Microscopy, 124*, RP3–RP4.

Dubochet, J., & Sartori Blanc, N. (2001). The cell in absence of aggregation artifacts. *Micron (Oxford, England: 1993), 32*, 91–99.

Geimer, S., & Melkonian, M. (2004). The ultrastructure of the *Chlamydomonas reinhardtii* basal apparatus: identification of an early marker of radial asymmetry inherent in the basal body. *Journal of Cell Science, 117*, 2663–2674.

Gönczy, P. (2012). Towards a molecular architecture of centriole assembly. *Nature Reviews Molecular Cell Biology, 13*, 425–435.

Grassucci, R. A., Taylor, D. J., & Frank, J. (2007). Preparation of macromolecular complexes for cryo-electron microscopy. *Nature Protocols, 2*, 3239–3246.

Grigorieff, N. (2013). Direct detection pays off for electron cryo-microscopy. *eLife, 2013*, 2–4.

Guesdon, A., Blestel, S., Kervrann, C., & Chrétien, D. (2013). Single versus dual-axis cryo-electron tomography of microtubules assembled in vitro: limits and perspectives. *Journal of Structural Biology, 181*, 169–178.

Guichard, P., Chrétien, D., Marco, S., & Tassin, A.-M. (2010). Procentriole assembly revealed by cryo-electron tomography. *EMBO Journal, 29*, 1565–1572.

Guichard, P., Desfosses, A., Maheshwari, A., Hachet, V., Dietrich, C., Brune, A., et al. (2012). Cartwheel architecture of *Trichonympha* basal body. *Science, 337*, 2012.

Guichard, P., Hachet, V., Majubu, N., Neves, A., Demurtas, D., Olieric, N., et al. (2013). Native architecture of the centriole proximal region reveals features underlying its 9-fold radial symmetry. *Current Biology: CB, 23*, 1620–1628.

Heymann, J. B., Cardone, G., Winkler, D. C., & Steven, A. C. (2008). Computational resources for cryo-electron tomography in Bsoft. *Journal of Structural Biology, 161*, 232–242.

Huiskonen, J. T., Hepojoki, J., Laurinmäki, P., Vaheri, A., Lankinen, H., Butcher, S. J., et al. (2010). Electron cryotomography of Tula hantavirus suggests a unique assembly paradigm for enveloped viruses. *Journal of Virology, 84*, 4889–4897.

Kremer, J. R., Mastronarde, D. N., & McIntosh, J. R. (1996). Computer visualization of three-dimensional image data using IMOD. *Journal of Structural Biology, 116*, 71–76.

Kudryashev, M., Castaño-Díez, D., & Stahlberg, H. (2012). Limiting factors in single particle cryo electron tomography. *Computational and Structural Biotechnology Journal, 1*.

Lee, E. T. Y. (1989). Choosing nodes in parametric curve interpolation. *Computer-Aided Design, 21*, 363–370.

Li, S., Fernandez, J.-J., Marshall, W. F., & Agard, D. A. (2011). Three-dimensional structure of basal body triplet revealed by electron cryo-tomography. *EMBO Journal*, 1–11.

Nicastro, D., Schwartz, C., Pierson, J., Gaudette, R., Porter, M. E., & McIntosh, J. R. (2006). The molecular architecture of axonemes revealed by cryoelectron tomography. *Science (New York, N.Y.), 313*, 944–948.

Nigg, E. A., & Raff, J. W. (2009). Centrioles, centrosomes, and cilia in health and disease. *Cell, 139*, 663–678.

Pettersen, E. F., Goddard, T. D., Huang, C. C., Couch, G. S., Greenblatt, D. M., Meng, E. C., et al. (2004). UCSF Chimera—a visualization system for exploratory research and analysis. *Journal of Computational Chemistry, 25*, 1605–1612.

Ruskin, R. S., Yu, Z., & Grigorieff, N. (2013). Quantitative characterization of electron detectors for transmission electron microscopy. *Journal of Structural Biology, 184*, 385–393.

Schmidt, H. (1956). Studies on Flagellata in intestine of termites. II. Wood particles as cell inclusions and nutrition in Trichonympha and Joenia. *Zeitschrift fur Parasitenkunde (Berlin, Germany), 17*, 269–275.

Vorobjev, I. A., & Chentsov, Yu S. (1982). Centrioles in the cell cycle. I. Epithelial cells. *Journal of Cell Biology, 93*, 938–949.

CHAPTER

New insights into centrosome imaging in *Drosophila* and mouse neuroepithelial tissues

12

Maria A. Rujano[*,§], **Renata Basto**[*,1], **Véronique Marthiens**[*,1]

*Institut Curie, CNRS UMR144, Paris, France

[§]Imagine Institute, Paris, France

[1]Corresponding authors: E-mail: renata.basto@curie.fr; veronique.marthiens@curie.fr

CHAPTER OUTLINE

Introduction ... **212**

1. Imaging Centrosomes in *Drosophila* Neuroepithelial Cells ... **214**

1.1 Materials ... 215

1.1.1 Reagents ... 215

1.1.2 Equipment ... 215

1.2 Methods ... 217

1.2.1 Larval staging ... 217

1.2.2 Dissection ... 217

1.2.3 Immunolabeling of fixed brains ... 218

1.2.4 Live imaging ... 219

2. High Spatial Resolution Imaging of Centrosomes in Neural Stem Cells of the Mouse Neocortex ... **220**

2.1 Material ... 220

2.1.1 Reagents ... 220

2.1.2 Equipment ... 222

2.2 Methods ... 222

2.2.1 Preparation of dorsal telencephalon explants ... 222

2.2.2 Immunolabeling of whole-mount cortical explants ... 223

2.2.3 Imaging of centrosomes and mitotic spindles from an en face perspective ... 225

Conclusion ... **226**

Acknowledgments ... **226**

References ... **227**

Methods in Cell Biology, Volume 129, ISSN 0091-679X, http://dx.doi.org/10.1016/bs.mcb.2015.04.005

Abstract

The centrosome is the main microtubule-organizing center in animal cells. It participates in the assembly of a bipolar spindle that ensures accurate segregation of chromosomes during mitosis. Recently, mutations in centrosome genes have been identified in patients affected by neurodevelopmental disorders. In fact, the etiology of several neurodevelopmental pathologies seems to be linked to defects in the assembly of the mitotic spindle in the neural stem cell compartment during neurogenesis. Therefore, getting better insights into the structure and function/dysfunction of the mitotic spindle apparatus in an intact tissue environment is of utmost importance. However, imaging nanometer-scale structures like centrosomes and microtubule bundles within the depth of a tissue is still challenging. Here we describe two procedures to acquire high-resolution images on fixed tissues and to perform live imaging of microtubule-based structures in the neuroepithelia of the *Drosophila* brain and of the mouse neocortex. We take advantage of the accumulation of centrosomes and mitotic figures at the apical surface of these polarized tissues to improve the quality of staining and imaging. Both *Drosophila* and mouse models with centrosome dysfunction showed abnormalities in the neuroepithelium reminiscent of the ones described in brains of human patients. These observations have highlighted their value as model organisms to study the etiology of human neurodevelopmental pathologies.

INTRODUCTION

Centrioles are cylindrical structures of 200 nm diameter and 500 nm length formed by triplets of microtubules arranged in ninefold symmetry (Bornens, 2002; Firat-Karalar & Stearns, 2014). Centrioles are embedded in and organize the pericentriolar material (PCM), the site of microtubule nucleation and anchoring (Fu & Glover, 2012; Lawo, Hasegan, Gupta, & Pelletier, 2012; Mennella, Agard, Huang, & Pelletier, 2014). A key step to investigate the function and dysfunction of centrosomes under normal and pathological conditions is to perform an in-depth characterization of their architecture and dynamics at different stages of the cell cycle (Chavali, Putz, & Gergely, 2014; Godinho & Pellman, 2014; Nigg & Raff, 2009). Nonetheless, imaging nanoscale centrosomal structures and the associated microtubule network in 3-dimensions (3D) within a tissue has remained challenging.

The *Drosophila* neuroepithelium in the larval optic lobe is composed of symmetrically dividing neuroepithelial cells that generate asymmetrically dividing neural stem cells (NSCs) (Egger, Boone, Stevens, Brand, & Doe, 2007; Yasugi, Umetsu, Murakami, Sato, & Tabata, 2008) (Figure 1(A)). These cells form a monolayered pseudostratified columnar epithelial tissue (Rujano, Sanchez-Pulido, Pennetier, le Dez, & Basto, 2013) that displays typical apicobasal polarity within the larval optic lobe, and gives rise to all the neurons and glia of the adult optic lobe (Meinertzhagen & Hanson, 1993). The possibility of studying the mechanisms involved in cell fate transition from a proliferative (neuroepithelial) to a neurogenic (neuroblast) state, underscore the relevance of the optic lobe to study neurogenesis. Furthermore, the

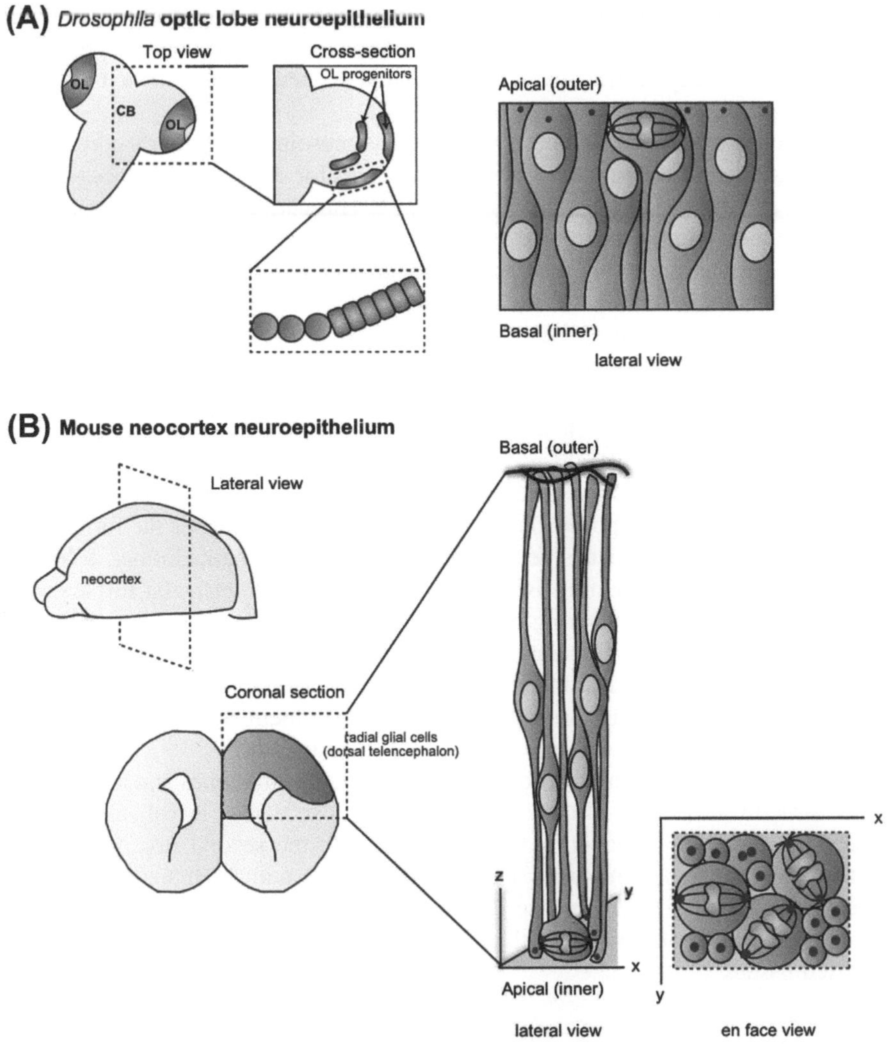

FIGURE 1 ***Drosophila*** **and mouse neuroepithelial tissue architecture.**

Drawings depicting the organization of embryonic neural stem cells within the neuroepithelium of *Drosophila* optic lobe (A) and mouse neocortex (B). Nuclei undergo interkinetic nuclear migration. This gives a pseudostratified appearance to the neuroepithelia. Centrosomes (red (gray in print versions) dots) and mitotic spindles always accumulate at the apical side of the tissue. (A) *Drosophila* neuroepithelial cells display a columnar shape and apicobasal polarity, with the apical side facing the outer part of the larval optic lobe and the basal side facing the inner part. Tissue width lies between 15 and 20 μm. (B) Mouse radial glial cells display an elongated bipolar morphology and apicobasal polarity, with processes keeping attachment to both basal lamina (outer) and ventricular lining (inner). Tissue width is increasing over neurogenesis, while the size of the ventricular lining is decreasing. En face view of the ventricular surface on the right-hand drawing illustrates the accumulation of mitotic spindles and interphase centrosomes from an apical point of view.

Drosophila neuroepithelium is amenable to live imaging with high spatial and temporal resolution from both the apical and the lateral side and as such, it represents an ideal system to study centrosome architecture and dynamics in the context of a living developing tissue.

The mouse embryonic neuroepithelium of the presumptive cerebral cortex is one of the best-documented model systems in mammals for studying the process of neurogenesis and related defects (Taverna, Gotz, & Huttner, 2014). Embryonic NSCs, also called radial glial cells, divide symmetrically to amplify the pool of progenitors and then asymmetrically to give rise to all neurons and glial cells populating functional adult brains (Temple, 2001). NSCs display an elongated shape spanning the whole thickness of the tissue. During the cell cycle, their nucleus moves up and down the depth of the tissue while their processes keep attachment to both basal and apical sides (Figure 1(B)) (Noctor, Flint, Weissman, Dammerman, & Kriegstein, 2001). Centrosomes remain closely attached to the apical side by an unknown mechanism. The apical surface of the neuroepithelium is in contact with the cerebrospinal fluid circulating in the third ventricle and is therefore often referred to as the ventricular surface. During interphase, one of the two centrioles templates the formation of a primary cilium at the apical side (Dubreuil, Marzesco, Corbeil, Huttner, & Wilsch-Brauninger, 2007). Before mitosis, the nucleus reaches the ventricular surface where centrosomes accumulate. During mitosis, the two apical centrosomes organize the two poles of the spindle apparatus that participates in the faithful segregation of chromosomes in the two daughter cells. We take advantage of the apical centrosome location in NSCs all along the cell cycle to visualize them with high resolution in mouse cerebral cortical explants using en face view imaging.

In the first part, we provide protocols for imaging centrosomes and microtubule-based structures in NSCs of the *Drosophila* optic lobe during larval stages by live imaging and/or immunolabeling of fixed tissues. In the second part, we describe methods to analyze centrosomes and mitotic spindles with high spatial resolution in embryonic NSCs of the mouse dorsal telencephalon. In particular, we explain how to perform whole-mount cortical explant analysis from an en face perspective on fixed tissues.

1. IMAGING CENTROSOMES IN *DROSOPHILA* NEUROEPITHELIAL CELLS

The protocol described thereafter works for immunolabeling or live imaging of brains in larval and pupal stages, but NSCs are actively dividing only during 3rd instar larvae. Furthermore, the protocol is also suitable for immunolabeling and live imaging of cellular structures other than centrosomes. First, we describe how to collect larvae at the desired stages and how to dissect the tissue. Then, we explain how to fix and label centrosomes with antibodies and finally, we describe how to mount the tissue for live imaging (Figure 2).

1.1 MATERIALS

1.1.1 Reagents

- *Drosophila* larvae at developmental stage of interest
- Phosphate-buffered saline (PBS) (1X, pH 7.4)
- PBST 0.1% (PBS containing 0.1% Triton X-100)
- PBST 0.3% (PBS containing 0.3% Triton X-100)
- Normal goat serum (NGS, Invitrogen, PCN5000)
- Blocking solution (5% NGS v/v in PBST 0.3%)
- Paraformaldehyde 16% (Electron Microscopy Sciences, 15710)
- Paraformaldehyde 4% (diluted from paraformaldehyde 16% in PBST 0.1%)
- Primary antibodies diluted in PBST 0.3%
- One or more appropriate secondary antibodies diluted in PBST 0.3%. We recommend Alexa Fluor® 488, 546, and 647 (Molecular Probes-Life technologies)
- Hoechst 33342 diluted at 0.5 μg/mL in PBST 0.3% (Hoechst 33342 trihydrochloride trihydrate, Molecular Probes-Life technologies, H1399; stock solution prepared at 10 mg/mL in DMSO and stored at 4 °C)
- Mounting medium (1.25% n-propyl gallate (Sigma, P3130), 75% glycerol (bidistilled, 99.5%, VWR, 24388-295), 25% H_2O). Prepared by mixing ingredients overnight (O/N) at 4 °C and stored at −20 °C
- Schneider's *Drosophila* medium (Gibco, 21720-024)
- Heat-inactivated fetal bovine serum (FBS) (Gibco, 10500)
- Penicillin/streptomycin (Gibco, 15140)
- Live-imaging medium (Schneider's *Drosophila* medium supplemented with 10% FBS, 100 units/mL Penicillin and 100 μg/mL streptomycin)
- Oil 10S Voltalef (VWR BDH Prolabo, 24627-188)

1.1.2 Equipment

- Vials with cornmeal medium
- Dissecting microscope
- Silicon pad for dissection (plastic dish filled with Sylgard® 184 silicone elastomer kit, Down Corning)
- Glass dishes (Nine-well Pyrex plate, Electron Microscopy Sciences, 71563)
- Forceps (two pairs; Dumont #5; Fine Science Tools, 11200-14)
- Micropipettes and tips
- Minutien pins (0.1-mm diameter, FST, 26002-10)
- Microdissecting pin holder (FST, 26018-17)
- Microscope slides
- Round cover glasses (12 mm ø, N°1, Marienfeld Lab Glassware, 01-115-20)
- Nail polish
- Orbital shaker
- Paintbrush
- Slide holder

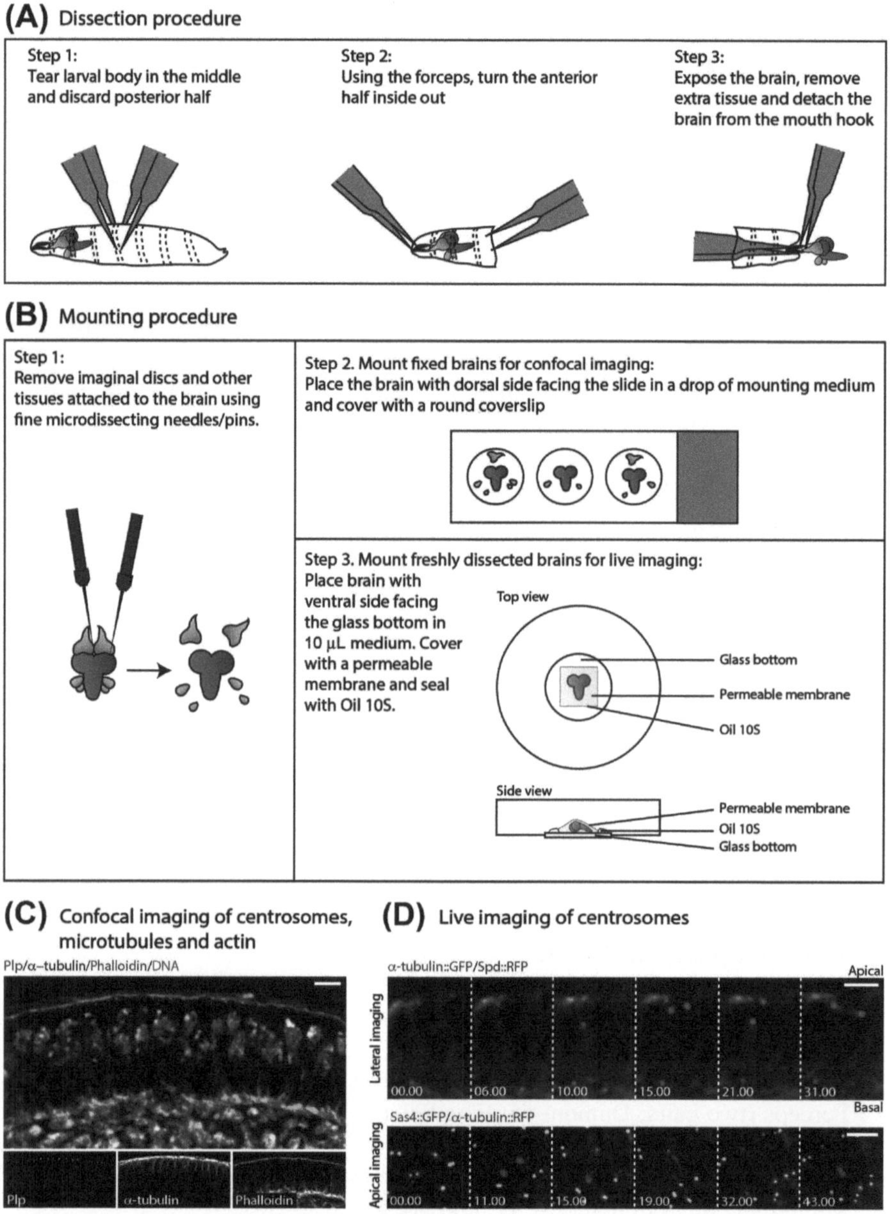

FIGURE 2 Imaging of *Drosophila* neuroepithelial cells.

(A) Schematic representations of the successive steps of dissection required for isolating the *Drosophila* larval brain. (B) Larval brain mounting for imaging. The left-side panel shows a scheme for cleaning the brain before imaging. Right panel shows schemes of mounted samples for confocal imaging of fixed brains (top) and live imaging of fresh samples (bottom). (C) Example of confocal acquisition for centrosomes, microtubules, and actin staining.

- Glass bottom 35 mm dish (P35G-1.5-14-C, MatTek Corporation)
- Permeable membrane (Standard membrane kit, YSI)

1.2 METHODS

1.2.1 Larval staging

As the developmental period in *Drosophila* is temperature-dependent, to facilitate reproducible staging of larvae, we recommend raising fly stocks and performing the staging at 25 °C.

In order to analyze centrosomes and cytoskeletal structures at different developmental stages in neuroepithelial cells, larvae are staged and the tissue is collected and analyzed at early, mid-, and late-third instar larvae (L3).

1. Place at least 20 adult flies (males and females) in vials with cornmeal medium at 25 °C.
2. Collect eggs by flipping the flies from the vial after 2–4 h and keep the vial with eggs at 25 °C.
3. Collect as many times as necessary depending on the fertility and viability of parental lines.
4. Select larvae for dissection of brains 72, 96, or 120 h after collection of eggs for early, mid-, and late-L3 larval stages, respectively.

Optional: Collect only freshly hatched larvae in a 2–4 h time window and stage on cornmeal medium to early, mid-, and late-L3 larval stages (48, 72, or 96 h after larval hatching).

1.2.2 Dissection

This is the general procedure for dissecting larval brains (Figure 2(A)). *Critical*: If dissection is followed by immunolabeling of fixed tissues, the dissection is performed in PBS. Alternatively, when preparing brains for live imaging, the dissection should be performed in the medium for live imaging.

Centrosomes are depicted in red and immunostained with *Drosophila* pericentrin-like protein (Plp), microtubules are in blue and immunostained with α-tubulin, actin is in red and stained with phalloidin, and DNA is in gray labeled with Hoechst. (D) Example of spinning disc confocal live imaging of centrosomes and microtubules. Centrosomes are visualized with SPD::RFP (top panel) or Sas4::GFP (bottom panel) and microtubules are visualized with α-tubulin::GFP (top panel) or α-tubulin::RFP (bottom panel). Time is given in minutes. The neuroepithelium is imaged from the lateral side (top panel) and from the apical side (bottom panel). Four-dimensional z-stacks of 5–10 μm at 0.3 μm intervals were acquired every 30 or 60 s using an X60, NA 1.4 oil immersion objective. Scale bars: 5 μm. (See color plate)

1. Collect the larvae at the stage of interest from the medium or the walls of the vials with a PBS-soaked paintbrush or forceps and place them in one of the wells in a glass dish filled with PBS. Remove the fly food from the surface of the larvae carefully with a paintbrush. Subsequently, transfer the larvae into another well filled with fresh PBS to complete cleaning.
2. Transfer the larvae individually into drops of cold PBS or live-imaging medium placed on a silicon pad for dissection.
3. Tear larval body in the middle using the forceps and discard the posterior half.
4. Expose the brain, by holding the anterior half of the larvae by the mouth hook with one pair of forceps while turning the anterior inside out with the other pair of forceps.
5. Once the brain is located and while still holding the mouth hook, gently remove all other tissues from around the brain leaving only some imaginal discs attached. These ones are used to manipulate the sample without touching/damaging the brain.
6. Gently detach the brain from the mouth hook with the forceps and transfer to a clean drop of PBS or live-imaging medium by grabbing the tissue from one of the imaginal discs.

A typical immunolabeling experiment requires 10–20 brains. *Critical*: To avoid tissue degradation, the time between dissection and fixation should not exceed 15 min. If necessary, dissect and fix brains in groups of 5–10.

For live imaging, 1–4 brains can be mounted immediately after dissection.

1.2.3 Immunolabeling of fixed brains

1.2.3.1 Fixation

1. Using a P-20 micropipette, transfer the brains from the PBS drop, into a well of a glass dish that contains 200–500 μL of freshly prepared 4% paraformaldehyde. *Tip*: to avoid the tissue sticking to the microtip, before pipetting the tissue, pipet up and down on a solution of PBST or NGS.
2. Incubate the brains in the fixative on an orbital shaker with gentle agitation for 20 min at room temperature (RT, 20 °C). *Critical*: the tissue should be completely submerged in fixative. If necessary, remove air bubbles with the help of forceps and/or fine mounted pins.

1.2.3.2 Staining

1. After fixation, wash the tissue in PBST 0.3% for 10 min with gentle agitation. Repeat this step 2 more times.

Optional: Some antibodies may require a blocking step to minimize background. To block, add 100–500 μL blocking solution and incubate the fixed brains on an orbital shaker with gentle agitation for 30 min at RT.

2. Replace the PBST 0.3% with 50–200 μL primary antibody solution and incubate O/N at 4 °C. *Tip*: to avoid evaporation, place the dish containing the brains on top of a water soaked Kimwipe inside a plastic box.

3. Remove antibody solution and wash in PBST 0.3% for 10 min with gentle agitation at RT. Repeat the washing step 2 more times.
4. Replace the PBST 0.3% with 50–200 μL secondary antibody solution and incubate O/N at 4 °C. *Optional*: The incubation time in this step can be reduced to 3 h at RT but the antibody penetration greatly increases when performed O/N.
5. Remove antibody solution and wash in PBST 0.3% for 10 min with gentle agitation at RT. Repeat the washing step 2 more times.
6. Replace the PBST 0.3% with Hoescht solution and incubate for 3 h at RT with gentle agitation.
7. Remove the nuclear staining solution and wash with PBS for 10 min with gentle agitation at RT. Repeat the washing step 2 more times.

1.2.3.3 Mounting (Figure 2(B) and (C))

1. Gently transfer one brain to an ~7 μL drop of mounting medium on a microscope slide grabbing the tissue by one of the imaginal discs using the forceps.
2. Use fine mounted pins to remove the remaining imaginal discs and any other tissue attached to the brain (If not too big and is placed well separated from the brain, this excess tissue can be left on the slide).
3. Place the brain with the dorsal part facing the slide and ventral up. Use the fine mounted pins for this step.
4. Gently cover the brain with a round coverslip.
5. Soak excess mounting medium from the borders of the coverslip using a Kimwipe.
6. Seal the edges carefully with nail polish and place the samples on a dark slide holder. *Optional*: Up to three brains can be mounted on one slide.
7. Proceed to imaging.

Perform imaging using a confocal microscope. For imaging of centrosomes on L3 larval brain cells, we recommend a 63× or 100× oil objective and a z-step of 0.2 μm *Critical*: Imaging should be done as soon as possible to get the best signal. After imaging, return the samples to a dark slide holder and keep them at 4 °C.

1.2.4 Live imaging (Figure 2(B) and (D))

1.2.4.1 Mounting

1. Use fine-mounted pins to remove imaginal discs and any other tissue attached to the brain.
2. Using a P-20 micropipette, gently transfer 1–4 freshly dissected brains to a 10 μL drop of medium on the center of a glass bottom 35 mm dish. Avoid transferring more than 5 μL of medium together with the brains. *Tip*: to avoid the tissue sticking to the microtip, before pipetting the tissue, pipet up and down live-imaging medium.
3. To image the neuroepithelium laterally, place the brain with the dorsal part facing the glass bottom. If imaging is to be done from the apical side, place the

brain with the ventral part facing the glass bottom. Use the fine mounted pins for this step.

4. Cover the brains with a piece of permeable membrane small enough to fit on the space of the round glass bottom.
5. Seal the membrane edges with 10S Voltalef oil.
6. Proceed to imaging.

Perform imaging using a spinning disc confocal microscope. For imaging of centrosomes in L3 larval brain cells, we recommend a 63× or 100× oil objective and a z-step of 0.2 μm. For centrosome dynamics studies, acquire images at intervals of 5–30 s maximum.

2. HIGH SPATIAL RESOLUTION IMAGING OF CENTROSOMES IN NEURAL STEM CELLS OF THE MOUSE NEOCORTEX

First we provide a detailed protocol on how to dissect and fix the embryonic cortical wall of the presumptive cerebral cortex (Figure 3). Then, we provide detailed explanations on how to perform immunofluorescence staining of whole-mount cortical explants and how to image en face views of mitotic figures within the apical progenitor compartment.

2.1 MATERIAL

2.1.1 Reagents

- Mouse embryos collected between embryonic day 11.5 and 16.5 (referred to as E11.5 and E16.5, respectively)

Nota Bene: All the solutions are prepared fresh the day of the experiment.

- PBS (1×, pH 7.4)
- Paraformaldehyde 4%, prepared from a stock of PFA 16% (Electron Microscopy Sciences, 15710), diluted in PBS 1×
- PBST 1%, PBS 1× containing 1% Triton X-100 v/v (Sigma, T9284) and sodium azide 0.02%
- PBST 0.3%: PBS 1×, 0.3% Triton X-100 v/v, sodium azide 0.02%
- Bovine Serum Albumin (BSA, FractionV-pH7, Euromedex, 04-100-811C)
- Blocking solution: 3% BSA w/v diluted in PBST 0.3%
- Primary and secondary antibodies are diluted in the blocking solution. We recommend the highly cross-adsorbed Alexa Fluor® 488, Alexa Fluor® 568, and Alexa Fluor® 647 raised in goat for confocal imaging (Molecular Probes-Life technologies)
- DNA is stained with Dapi (4′,6′-diamidino-2-phenylindole, Thermo Scientific, 46190). Dapi is diluted in the solution of secondary antibodies at a 3 μg/mL final concentration

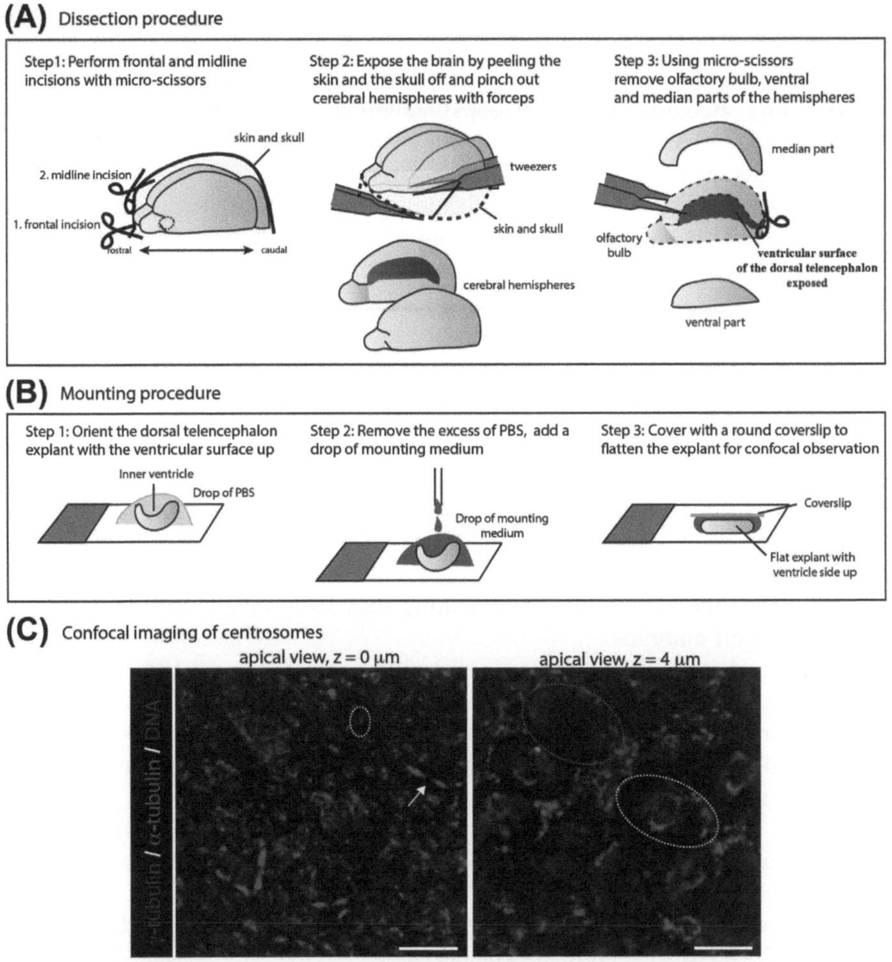

FIGURE 3 En face views imaging of centrosomes in the mouse neocortex.

(A) Schematic representations of the successive steps of dissection required for isolating dorsal telencephalon explants from mouse embryonic brains before fixation and immunolabeling procedures. (B) Drawings explaining the mounting procedure to obtain apical side up dorsal telencephalon explants to image centrosomes in interphase and mitotic figures from an en face point of view. (C) Examples of en face view confocal acquisitions showing centrosomes (γ-tubulin in red), microtubules (α-tubulin in green), and DNA in blue. The same region was imaged at different z-positions along the apicobasal axis. At the uttermost apical position ($z = 0$ μm) (left-hand panel), the interphase centrosomes within the apical processes can be visualized (dashed white circle). In addition, the cleavage furrows are also visualized (white arrow). Just 4 μm above ($z = 4$ μm) (right-hand panel) mitotic cells are visualized in different stages (white dashed circle surrounds one metaphase and red dashed circle surrounds one anaphase). In the left-hand panels, note that centrosomes in interphase belong to cells at different stages of the cell cycle, i.e., one or two centrosomes can be visualized per apical process. In the right-hand panels, note that the parallel orientation of the mitotic spindle in relation to the ventricular lining enables the visualization of the two mitotic centrosomes at the same z-position. Scale bars: 8 μm. (See color plate)

- Mounting medium (1.25% n-propyl gallate (Sigma, P3130), 75% glycerol (bidistilled, 99.5%, VWR, 24388-295), 25% H_2O). Prepared by mixing ingredients overnight (O/N) at 4 °C and stored at −20 °C
- *Optional*: For tissue pre-permeabilization before fixation, prepare a fresh solution of PBST 0.1%: PBS 1× containing 0.1% Triton X-100 v/v, respectively, and sodium azide 0.02%

2.1.2 Equipment

- Dissecting microscope
- One pair of scissors 8.5 cm straight (Fine Science Tools, FST, 14084-08), to collect uterine horns containing embryos by caesarean
- Two pairs of forceps (FST, Dumont #5, 11200-14)
- One pair of forceps angled up 7 mm (FST, Dumont #7, Dumoxel 11271-30), to hold embryos before chopping their heads off and pinch out brains and cerebral hemispheres
- One pair of fine scissors angled up, cutting edge 10 mm (FST, 15017-10), to chop the heads off embryos
- Silicon pad for dissection (plastic dish filled with Sylgard® 184 silicone elastomer kit, Down Corning)
- One pair of fine scissors straight, cutting edge 3 mm (FST, 15000-00), to cut the skin and the skull of the heads and microdissect the median and ventral parts of the telencephalon
- Spatula type Chattaway (L:100 × 1:4 mm, Dutscher, 442180) or plastic dispensable micropipette (VWR, 612-1747), to handle and/or transfer dorsal telencephalon explants
- Microknife (Corneal/Scleral V-Lance Knives, Alcon Ophthalmic, 20G, 1.3 mm, 8065912001) to slice cortical explants
- 24-well plates for cell culture (TPP, 92024)
- Orbital shaker
- Micropipettes and tips
- Microscope slides
- Square cover glasses (18 mm diameter, No.1.5H, Marienfeld, 0117580)
- Nail polish

2.2 METHODS

2.2.1 Preparation of dorsal telencephalon explants

2.2.1.1 Collection of cerebral hemispheres

1. The observation of a vaginal plug is considered as day 0.5 of mating. Collect cerebral hemispheres when neurogenesis occurs between embryonic day 11.5 (E11.5) and embryonic day 16.5 (E16.5) for analysis.
2. Pregnant females are anaesthetized with isoflurane before being sacrificed by cervical dislocation. Collect embryos by caesarean with the help of 8.5 cm

straight scissors. Remove them from the uterine horns and yolk sacs with a pair of forceps (#5). Hold the embryos with 7 mm angled-up forceps and chop heads off embryos with angled up fine scissors. Place the heads in a drop of PBS on a silicon plate for further dissection at RT (20 °C).

Note: Dissections are always performed at RT in order to prevent microtubule depolymerization before fixation.

2.2.1.2 Dissection of dorsal telencephalon explants (Figure 3(A))

The dorsal part of the telencephalon is isolated from other brain regions using the following dissection steps.

1. Use a pair of fine straight scissors to make a horizontal incision of the skin at the base of the forebrain.
2. Use the same scissors to cut the skin and the forming skull at the midline along the rostrocaudal axis. Perform the incisions superficially to prevent any damage of the underlying cerebral hemispheres. Care should be taken while dissecting E15.5–E16.5 brains since the skull is getting harder at these stages. Push away the skin and the skull laterally with a pair of forceps (#5) to uncover the cerebral hemispheres.
3. Pinch out each cerebral hemisphere from the rest of the brain with the #5 tweezers and orient the neocortical explant with the ventricular surface up in a drop of clean PBS.
4. Use tweezers (#5) to gently hold the cerebral hemisphere while cutting away the olfactory bulb, the ventral and median parts of the telencephalon with the fine straight scissors. *Critical*: Expose the ventricle surface of the dorsal telencephalon as much as possible at this step to enable a better access of the antibodies during immunolabeling. In that way, we can observe homogeneous staining of the whole surface area and a high reproducibility of our immunolabeling results.
   Notes:
   a. Perform the dissection steps as quickly as possible for a better preservation of the tissues. The whole dissection procedure should not exceed 45 min.
   b. The size of the ventricular surface is expected to decrease during late stages of neurogenesis as a consequence of depletion of the NSC pool.
   c. If the regional variability along the rostrocaudal and mediolateral axes is not an issue for the analysis, each cortical explant can be further sliced in four pieces maximum with the help of a microknife (V-Lance knife) on the silicon pad.

2.2.2 Immunolabeling of whole-mount cortical explants

2.2.2.1 General recommendations

1. Prepare all fixative solutions and buffers freshly the day of the experiment.
2. Perform the following steps by incubating each free-floating dorsal telencephalon explant in a well filled with the appropriate working solution. The use of

24-well plates is recommended for the whole procedure, but 96-well plates can be used to reduce the amount of antibodies.

3. Fill and remove solutions from the wells with the help of a 1 mL blue pipet tip, or alternatively transfer explants from one well to another one with the help of a microspatula or a dispensable plastic micropipette.
4. Perform the whole procedure at RT. Incubate explants under gentle rotation (without the pre-extraction and fixation steps).

2.2.2.2 Fixation of dorsal telencephalon explants

1. *Optional*: Pre-extraction of the cytoplasmic pool of proteins before fixation improves the intensity of staining on centrosomes and other microtubule-based structures. For example, the organization of microtubule bundles within the spindle apparatus appears more clearly after pre-extraction of the pool of soluble α-tubulin.

To perform this step, incubate the explants in a solution of PBST 0.1% for 2 min and wash them quickly 2 times in PBS.

2. Fix tissues in −20 °C-precooled methanol for 10 min or alternatively in PFA 4% for 1 h at RT, depending on the primary antibody to be used. Wash the explants 3 times in PBS for at least 10 min before proceeding with immunolabeling.

Note: It is highly recommended to perform the immunostainings immediately after methanol fixation. PFA-fixed explants can be stored at 4 °C for one month in PBS 1× containing 0.02% azide before immunolabeling.

2.2.2.3 Immunofluorescent staining and mounting (Figure 3(B))

1. Incubate fixed explants with 1 mL of PBST 1% for 15 min.
2. Perform the blocking step by incubating the explants in 1 mL blocking solution (3% BSA w/v diluted in PBST 0.3%) for 1 h at RT under gentle rotation. Use the same blocking solution to proceed for staining with primary and secondary antibodies.
3. Incubate the explants with primary antibodies (250 μL per well for 24-well plates, 50 μL per well for 96-well plates) O/N at RT under gentle agitation. Wash them 3 times with 1 mL blocking solution for at least 1 h.
4. Incubate the explants with secondary antibodies along with Dapi (staining of the DNA) O/N at RT, protected from light. Wash them 3 times with 1 mL blocking solution for at least 1 h and perform a last wash in PBS before mounting.
5. Place the explant in a drop of PBS on a glass slide for ease of manipulation and further orientation under a binocular microscope (Figure 3(B)). To obtain an en face view of the ventricular side of a cerebral cortex, place the explant as flat as possible with the ventricular surface up. Remove the excess of PBS with a 100 μL pipet yellow tip and add a drop of mounting

medium onto the explant. Place a coverslip on the sample and seal it with nail polish.

6. Place the mounted samples at 4 °C in a dark slide holder before imaging. Wait at least 24 h before imaging. The samples can be imaged until 1 month after immunolabeling.

2.2.3 *Imaging of centrosomes and mitotic spindles from an en face perspective (Figure 3(C))*

2.2.3.1 Confocal imaging of centrosomes and mitotic figures

1. Image the ventricular side of dorsal telencephalon explants with a laser scanning confocal microscope. We can image centrosomes and mitotic spindles with the highest resolution using the Leica SP8, equipped with hybrid detectors (PICT-IBISA Imaging Platform, Institut Curie) (Figure 3(C)). A 63× oil immersion lens (Leica × HCX PL APO 1.4 oil CS2) is used for all the acquisitions. If mitotic figures can be analyzed without any further zooming, we can get the best resolution for analyzing centrosomes and microtubule bundles within the mitotic spindle apparatus with an electronic zoom of 2, at a 1024 × 1024 pixels image resolution.
2. Interphase and mitotic centrosomes of NSCs remain at the ventricular surface and can be imaged at the same time using three-dimensional z-stacks of 8–10 μm at 0.2 μm intervals starting from the ventricular lining. Indeed, mitotic cells display a diameter between 8 and 10 μm depending on the mitotic stage and the embryonic age of the tissue. The ventricular lining can be easily identified by the accumulation of centrosomes at the apical surface. Centrioles are around 200 nm in diameter. Therefore, images taken using steps of 0.2 μm along the z-axis provide the best resolution of centrosomes (Figure 3(C)).

Note: Regional variability can be observed along the rostrocaudal and mediolateral axes within the same cortical explant. For example, we have already noticed variability in the mitotic indexes (number of NSC in mitosis over total number of NSC in contact with the ventricle through their apical process). To check for and/or get rid of such variability in our analysis, we perform each combination of staining using one whole cortical explant and image different area within the same explant (8–10 images acquired per explant).

2.2.3.2 Choice of antibodies for centrosomes and mitotic spindles staining

1. Centrosomes appear as small dots after fluorescent immunolabeling using antibodies, against centriole or PCM proteins (Figure 3(C)). In addition, mammalian cells display some satellite structures—considered as reserve pools of centriole or PCM—which are spread around the centrosome and can be easily misidentified as the centrosome. Therefore, centrosomes should always be

visualized by a combination of at least two centrosome markers. We recommend the following primary antibodies:

a. γ-tubulin antibody at 1/500 dilution (mouse monoclonal, clone GTU88, Sigma, T5326). This antibody stains the centriole wall and the PCM of centrosomes in methanol-fixed or PFA-fixed explants.
b. pericentrin antibody at 1/500 dilution (rabbit polyclonal, Abcam, ab4448). This antibody stains the PCM of centrosomes in methanol-fixed or PFA-fixed explants.
c. cdk5rap2 antibody at 1/500 dilution (rabbit polyclonal, Bethyl Laboratories, IHC-00063). This antibody stains the PCM of centrosomes in PFA-fixed explants.

Alternatively, since centrosomes are microtubule-organizing centers during both interphase and mitosis, centrosome markers can be used in combination with α-tubulin.

2. To visualize the mitotic spindle apparatus, we recommend the following primary antibody:
   a. α-tubulin antibody at 1/1000 dilution (mouse monoclonal, clone DM1α, Sigma, T6199). This antibody stains microtubule bundles within the mitotic spindle apparatus, and especially microtubules nucleated at the centrosome, as well as the microtubule network below the plasma membrane.

CONCLUSION

Here we described different approaches to study centrosomes within native tissue environments. These procedures were initially settled in our laboratory to analyze centrosome/spindle pole integrity and/or behavior in the NSCs of *Drosophila* and mouse mutant models for microcephaly-associated disorders (Marthiens et al., 2013; Rujano et al., 2013). They can now be used to get further insights into centrosome/spindle pole characteristics in relation to other neurodevelopmental disorders related to NSC dysfunctions.

ACKNOWLEDGMENTS

Image acquisitions were performed on workstations of the PICT-IBiSA Lhomond Imaging facility of Institut Curie. The authors wish to thank V.Fraisier, L.Sengmanivong, and F.Waharte from the PICT-IBiSA Lhomond Imaging facility of Institut Curie for technical assistance. Work in our lab is supported by a grant from AICR (13–0170) and an ERC starting grant (Centrostemcancer 242598), as well as by the Institut Curie, the CNRS and INSERM. Our laboratory is a member of the CelTisPhyBio labex.

REFERENCES

Bornens, M. (2002). Centrosome composition and microtubule anchoring mechanisms. *Current Opinion in Cell Biology, 14*(1), 25–34.

Chavali, P. L., Putz, M., & Gergely, F. (2014). Small organelle, big responsibility: the role of centrosomes in development and disease. *Philosophical Transactions of the Royal Society of London, B: Biological Sciences, 369*(1650).

Dubreuil, V., Marzesco, A. M., Corbeil, D., Huttner, W. B., & Wilsch-Brauninger, M. (2007). Midbody and primary cilium of neural progenitors release extracellular membrane particles enriched in the stem cell marker prominin-1. *Journal of Cell Biology, 176*(4), 483–495.

Egger, B., Boone, J. Q., Stevens, N. R., Brand, A. H., & Doe, C. Q. (2007). Regulation of spindle orientation and neural stem cell fate in the *Drosophila* optic lobe. *Neural Development, 2*, 1.

Firat-Karalar, E. N., & Stearns, T. (2014). The centriole duplication cycle. *Philosophical Transactions of the Royal Society of London, B: Biological Sciences, 369*(1650).

Fu, J., & Glover, D. M. (2012). Structured illumination of the interface between centriole and peri-centriolar material. *Open Biology, 2*(8), 120104.

Godinho, S. A., & Pellman, D. (2014). Causes and consequences of centrosome abnormalities in cancer. *Philosophical Transactions of the Royal Society of London, B: Biological Sciences, 369*(1650).

Lawo, S., Hasegan, M., Gupta, G. D., & Pelletier, L. (2012). Subdiffraction imaging of centrosomes reveals higher-order organizational features of pericentriolar material. *Nature Cell Biology, 14*(11), 1148–1158.

Marthiens, V., Rujano, M. A., Pennetier, C., Tessier, S., Paul-Gilloteaux, P., & Basto, R. (2013). Centrosome amplification causes microcephaly. *Nature Cell Biology, 15*(7), 731–740.

Meinertzhagen, I. A., & Hanson, T. E. (1993). In A. Martinez-Arias, & M. Bate (Eds.), *The development of* Drosophila melanogaster (pp. 1363–1491). Cold Spring Harbor Laboratory Press.

Mennella, V., Agard, D. A., Huang, B., & Pelletier, L. (2014). Amorphous no more: subdiffraction view of the pericentriolar material architecture. *Trends in Cell Biology, 24*(3), 188–197.

Nigg, E. A., & Raff, J. W. (2009). Centrioles, centrosomes, and cilia in health and disease. *Cell, 139*(4), 663–678.

Noctor, S. C., Flint, A. C., Weissman, T. A., Dammerman, R. S., & Kriegstein, A. R. (2001). Neurons derived from radial glial cells establish radial units in neocortex. *Nature, 409*(6821), 714–720.

Rujano, M. A., Sanchez-Pulido, L., Pennetier, C., le Dez, G., & Basto, R. (2013). The microcephaly protein Asp regulates neuroepithelium morphogenesis by controlling the spatial distribution of myosin II. *Nature Cell Biology, 15*(11), 1294–1306.

Taverna, E., Gotz, M., & Huttner, W. B. (2014). The cell biology of neurogenesis: toward an understanding of the development and evolution of the neocortex. *Annual Reviews of Cell and Developmental Biology, 30*, 465–502.

Temple, S. (2001). The development of neural stem cells. *Nature, 414*(6859), 112–117.

Yasugi, T., Umetsu, D., Murakami, S., Sato, M., & Tabata, T. (2008). *Drosophila* optic lobe neuroblasts triggered by a wave of proneural gene expression that is negatively regulated by JAK/STAT. *Development, 135*(8), 1471–1480.

CHAPTER 13

Microinjection techniques for studying centrosome function in *Drosophila melanogaster* syncytial embryos

Paul T. Conduit*[,1], Daniel Hayward[§], James G. Wakefield[§,1]

*Department of Zoology, University of Cambridge, Cambridge, UK

[§]Biosciences, College of Life and Environmental Sciences, University of Exeter, Exeter, UK

[1]Corresponding authors: E-mail: ptc29@cam.ac.uk; j.g.wakefield@exeter.ac.uk

CHAPTER OUTLINE

Introduction **230**

1. Protocol **231**

2. Collection and Preparation of *Drosophila* Embryos **232**

2.1 Preparation of Fruit Juice Agar Plates 232

2.2 Preparation of Yeast Paste 232

2.3 Setting Up Fly Collection Chambers 233

3. Preparing Samples for Microinjection **233**

3.1 Drugs or Lyophilized Peptides 233

3.2 Unlabeled Affinity-Purified Antibodies or Other Proteins 233

3.3 Fluorescently Coupling Affinity-Purified Antibodies, Proteins, or Peptides. 234

3.4 In vitro Transcribed mRNA 235

4. Mounting Embryos for Spinning Disc Confocal Microscopy **235**

4.1 Preparing the Heptane-Glue Solution 235

4.2 Preparing the Imaging Surface: MatTek Dishes 236

4.3 Preparing the Imaging Surface: Coverslips 236

4.4 Mounting the Embryos 236

5. Microinjection of Embryos **238**

5.1 Pulling Glass Microinjection Needles (if Necessary) 238

5.2 Loading the Sample in the Needle 239

Methods in Cell Biology, Volume 129, ISSN 0091-679X, http://dx.doi.org/10.1016/bs.mcb.2015.03.007

5.3 Breaking the Needle Tip (if Using Home-Pulled Needles)........................ 239
5.4 Embryo Injection.. 240
6. Perturbation via Cold Treatment 241
6.1 Cold Treatment Assay 241
6.2 Cold Treatment Assay with Microinjection 242
7. Imaging and Data Analysis.. 242
7.1 Example 1 242
7.2 Example 2 244
7.3 Example 3 245
References 247

Abstract

Microinjection is a powerful technique that can be used to study protein function. Early *Drosophila* embryos are particularly amenable to microinjection due to their large size and their single cell status. Here, we report methods to microinject these embryos with various reagents to study the function of proteins at centrosomes and centrosome function more generally. Although precise details vary between laboratories, many aspects of the process are conserved. We describe the process from setting up a fly cage to imaging the injected embryos on a spinning disk confocal microscope and use specific examples to highlight the potency of this technique.

INTRODUCTION

Centrosomes comprise a pair of centrioles that recruit a network of proteins known as the pericentriolar material (PCM) in order to facilitate the controlled nucleation of microtubules. Microtubules have many important cellular roles, which differ depending on cell cycle stage, cell differentiation, and the context of a cell within a multicellular environment, such as forming the spindle structure that separates duplicated chromosomes during mitosis and establishing cell polarity. Until relatively recently, it was widely believed that centrosomes were indispensable for mitotic progression in most animal cell types, but it is now known that alternative spindle assembly pathways exist and can compensate for a lack of centrosomes (Duncan & Wakefield, 2011). These pathways rely on the nucleation of microtubules from other sources, such as from the chromatin and the sides of preexisting microtubules. Nonetheless, centrosomes can still justifiably be defined as the main microtubule organizing centers in mitotic animal cells.

Over the last decade, many of the proteins required for centrosome assembly and function have been identified, but how most of these proteins carry out their functions remains unclear. The precise interplay between different spindle assembly pathways is also under intense investigation. In this report, we detail how microinjection experiments in *Drosophila* syncytial embryos can be used to study both the function of centrosome proteins and the role of noncentrosomal pathways in spindle assembly.

Microinjection experiments are an extremely powerful way of examining protein function. This can be achieved by (1) injecting interfering antibodies, peptides, or drugs that sterically interfere with endogenous protein:protein interactions (Brust-Mascher, Sommi, Cheerambathur, & Scholey, 2009; Buttrick et al., 2008; Conduit et al., 2010; Conduit, Richens, et al., 2014; Gergely, Kidd, Jeffers, Wakefield, & Raff, 2000; Hayward & Wakefield, 2014; Lucas & Raff, 2007; Novak, Conduit, Wainman, & Raff, 2014; Oliveira, Hamilton, Pauli, Davis, & Nasmyth, 2010; Raff & Glover, 1988, 1989; Sharp et al., 2000; Sharp, Yu, Sisson, Sullivan, & Scholey, 1999; Silverman-Gavrila & Wilde, 2006; Sisson, Field, Ventura, Royou, & Sullivan, 2000); (2) injecting fluorescently tagged wild-type or mutant proteins (or mRNA) to observe protein localization and/or the function of mutant proteins (Conduit, Feng, et al., 2014; Kellum, Raff, & Alberts, 1995; Oegema, Marshall, Sedat, & Alberts, 1997; Oliveira et al., 2010); (3) injecting proteins or exogenous enzymes that artificially induce cell cycle events at specific times (Oliveira et al., 2010; Oliveira & Nasmyth, 2013; Royou, McCusker, Kellogg, & Sullivan, 2008). Microinjection can also be combined with noninvasive perturbations, such as temperature shifts, which transiently disrupt microtubule organization (Hayward, Metz, Pellacani, & Wakefield, 2014).

Such experiments are challenging in many eukaryotic cell types, due to the need to inject or perturb single cells with diameters of ~10–50 μm. The *Drosophila* early embryo, in contrast, begins development as a single cell, a syncytium, of approximately 150 μm wide and 500 μm long. A continual and rapid division of nuclei within the cytosol results in hundreds of nuclei that occupy the embryonic cortex ~1.5 h after fertilization, creating a syncytial blastoderm. These nuclei and associated centrosomes continue to divide in near synchrony a further four times at the cortex, allowing easy visualization of mitotic events. By microinjecting transgenic embryos expressing GFP-fusion proteins that localize to centrosomes (centrioles and/or the PCM) or microtubules, and by imaging the embryos using spinning disc confocal microscopy, high spatiotemporal information can be gained about protein function and spindle assembly. Such powerful techniques are becoming increasingly accepted and are shedding new light on the acute response of biological systems to the loss or gain of protein function.

1. PROTOCOL

Below, we describe protocols that allow the manipulation and perturbation of *Drosophila melanogaster* syncytial embryos expressing GFP-tagged proteins, in order to assess centrosomal function through the cell cycle. We provide methods for the microinjection of drugs, fluorescently tagged antibodies, proteins, or mRNA; for controlled temperature shifts; and for the analysis of the perturbations using spinning disc confocal microscopy. As specific protocols tend to vary between laboratories and are determined by, for example, scarcity of available injection reagents or micromanipulation apparatus, we detail a range of specific protocols currently in use in the Wakefield, Raff, and Conduit laboratories.

2. COLLECTION AND PREPARATION OF *DROSOPHILA* EMBRYOS

Collection of 1–2 h *Drosophila* embryos is standard procedure and is based on the preference of *Drosophila* adult females to lay fertilized embryos on fruit juice agar plates, containing yeast paste, within a specific collection chamber. The optimal composition of the fruit source for the collection plates varies, and is usually determined within the specific research environment. The three most common sources are apple juice, grape juice, and cranberry juice, or a combination of these. *Drosophila* are sensitive to environmental fluctuations and females will retain fertilized embryos upon perturbation, leading to variations in aging of laid embryo populations. As such, it is good practice to set up the collection chamber 2 or 3 days in advance of the first collections, to allow acclimatization.

2.1 PREPARATION OF FRUIT JUICE AGAR PLATES

1. Using a microwave, boil 9 g of agar in 300 mL of distilled water and 10 g sucrose in 100 mL apple/cranberry/grape juice[1] in separate 500 mL glass bottles until the agar and sucrose have dissolved.
2. Combine the two solutions and mix well.
3. Pour to a depth of ~15 mm into Petri dishes that have the appropriate diameter to fit the bottom of a fly cage[2] (see below; usually 60 mm diameter).
4. To sterilize the plates and to remove surface air bubbles, flame briefly with a Bunsen burner set to blue flame. Replace the lids and leave to dry for ~1 h at room temperature. Store at 4 °C for up to 2 weeks, or until mold develops.

2.2 PREPARATION OF YEAST PASTE

1. Mix dried yeast[3] and tap water in a 50 mL sterile tube using a metal spatula until a paste forms. Aim for an extra thick double cream consistency. Store at 4 °C.
2. Add a spot of yeast to a fruit juice agar plate and allow the yeast paste to dry at room temperature. For a 60 mm diameter plate, when collecting 1–2 h old embryos for injection, this should roughly equate to the size of a raisin. When maintaining flies overnight on a 60 mm plate, the yeast spot size should be roughly the size of a grape. Only add yeast on the day the plate is required.[4]

[1]Obtained at local supermarkets.
[2]Perspex collection chambers can be obtained from Flystuff Inc (Ref no. 59-100 and 59-101). The two sizes (small and large) fit petri dishes of 60 and 100 mm, respectively. Alternatively, collection chambers can be manufactured in-house.
[3]We generally use fast-action dried baking yeast, obtained at local supermarkets.
[4]Fruit juice agar plates can be washed in tap water and reused ~3 times, depending on the time the plate is in use.

2.3 SETTING UP FLY COLLECTION CHAMBERS

1. Transfer flies from vials/bottles to a perspex chamber of appropriate size.[4] The bottom of the cage should be sealed with a fruit juice agar plate that has a dried yeast paste spot in the center. An ideal number of flies will cover the surface of the fruit juice agar plate when the flies are banged down to the bottom.
2. Maintain flies at 25 °C throughout their occupancy of the chamber.
3. Exchange the plate for a fresh one at the end of the first day, ensuring the fresh plate has enough yeast paste to sustain the flies through the night.
4. Change the plate approximately every 3 h on subsequent days leading up to the experiment.

3. PREPARING SAMPLES FOR MICROINJECTION

The choice of sample type to be injected will be specific for the experiment. Typically, drugs are injected at 1000-fold higher concentrations than in cell culture applications, reflecting the dilution that occurs within the embryo cytosol once injected. Peptides, proteins, and affinity-purified antibodies should be injected at 2–10 mg/mL, and mRNA at between 1 and 2 mg/mL. Fluorescent labeling provides a useful way to monitor injection efficiency and to correlate concentrations of injected sample with observed phenotypes. In Section 3.3, we provide a detailed protocol for labeling large amounts of affinity-purified antibodies with a specific fluorescent dye.

3.1 DRUGS OR LYOPHILIZED PEPTIDES

Dilute the appropriate compound or peptide to an appropriate concentration in injection buffer:

- 50 mM HEPES
- 50 mM KCL
- 1 mM $MgCl_2$
- 1 mM EGTA, pH 7.5

  Adjusted to pH 7.5 with K-OH

3.2 UNLABELED AFFINITY-PURIFIED ANTIBODIES OR OTHER PROTEINS

1. Buffer exchange the antibody or protein into injection buffer using an appropriate desalting cartridge,[5] as per the manufacturer's instructions. Ensure the choice of column is appropriate for the volume of initial sample size (e.g., 10 mL column for volumes of 1–2 mL; 0.5 mL for volumes of 50–100 μL).

[5]For example, P-6 (BioRad) or Zeba (Pierce).

2. Measure the protein concentration on a nanodrop. If the concentration is below 2 mg/mL, concentrate the sample as follows.
3. Wash an appropriate ultracentrifugation filter device with injection buffer.
4. Add the antibody/protein solution and centrifuge at 2700 rcf until the appropriately calculated volume has been spun through the column (i.e., for a 1 mL protein solution of initial concentration 1 mg/mL, centrifuge until total volume of 900 μL has been spun through). If necessary, discard the solution at the bottom of the collecting tube and recentrifuge again until the desired amount remains in the filter.
5. Measure the concentration of the protein left on the filter side using a nanodrop to ensure the appropriate concentration of the reagent. Flash-freeze 3 μL aliquots in liquid nitrogen. Store at −80 °C.

3.3 FLUORESCENTLY COUPLING AFFINITY-PURIFIED ANTIBODIES, PROTEINS, OR PEPTIDES

1. Make, sterilize, and degas injection buffer and a modified injection buffer at pH 8.6.
2. Dissolve Texas Red-X succinimidyl ester (Life Technologies, T-6134) in DMSO to final concentration of 50 mg/mL. Mix well and flash-freeze 3 μL aliquots in liquid nitrogen. Store at −80 °C.
3. Thaw 2 mg (2 mL of 1 mg/mL) affinity-purified antibody/protein.
4. Use a 10 mL P-6 desalting cartridge (BioRad, 732-5304) to buffer exchange the antibody into injection buffer pH 8.6, as per the manufacturer's instructions.
5. Collect 2 mL from the peak fractions.
6. Thaw an aliquot of Texas Red-X succinimidyl ester and add 2 μL to the 2 mL antibody/protein solution. Rotate 1 h at room temperature.
7. Use the 10 mL P-6 desalting cartridge to separate the uncoupled Texas Red-X succinimidyl ester from the coupled antibody/protein and collect 2 mL of the peak coupled-antibody/protein fractions.
8. Wash an ultra centrifugation concentration filter[6] with injection buffer pH 7.5.
9. Add the conjugated antibody/protein solution and spin 2700 rcf for 2 min. Discard the solution at the bottom of the collecting tube and spin again until ~50 μL remains in the filter.
10. Measure the protein concentration on a nanodrop.[7] Flash-freeze 3 μL aliquots in liquid nitrogen. Store at −80 °C.

Labeling of smaller amounts of proteins or affinity-purified antibodies (100 μg at 1 mg/mL) with a variety of fluorescent dyes can alternatively be achieved using

---

[6]For example, Amicon Ultra-15 Centrifugal Filter Unit, 50 KDa cut off.
[7]It should be close to 10 mg/mL.

DyLight Microscale Antibody Labelling Kits (Life Technologies) following manufacturer's instructions.

3.4 IN VITRO TRANSCRIBED mRNA

1. Clone the desired template DNA into a suitable plasmid adjacent to a 5′ T3 or T7 promoter.
2. Linearize the plasmid DNA using a suitable restriction enzyme (aim to cut opposite to the location of the template DNA sequence).
3. Use either a T3 or T7 mMESSAGE mMACHINE capped in vitro *RNA* transcription kit (Life Technologies) to generate capped mRNA, as per the manufacturer's instructions.
4. Use an RNAeasy Mini Kit (Qiagen) to purify the mRNA, as per the manufacturer's instructions.
5. Measure the mRNA concentration and adjust to between 1 and 2 mg/mL if necessary. Flash-freeze 3 μL aliquots and store at −80 °C.

4. MOUNTING EMBRYOS FOR SPINNING DISC CONFOCAL MICROSCOPY

The principle underlying embryo mounting is that the embryos should be dechorionated, glued to a glass coverslip surface, and allowed to partially dessicate prior to injection. These steps ensure firstly that the embryos do not move when the needle touches them and, secondly, that the internal osmotic pressure is relieved to the point that puncture with a needle, under pressure, results in microinjection of the desired solution, rather than fatal embryo lysis, or "popping." As with previous steps, there is no single standard protocol between laboratories. Below, we describe two such approaches that differ in their use of standard coverslips or MatTek dishes, and their approach to removing the chorion of the embryos. Dechorionation via bleach treatment has the advantage that dessication time will be identical for all embryos aligned on the glass surface. However, manual dechorionation using double-sided sticky tape minimizes the loss of viable, unpopped embryos, should only a small number of embryos be available.

4.1 PREPARING THE HEPTANE-GLUE SOLUTION

1. Fill a 50 mL screw-cap plastic tube with exposed double-sided sticky tape (e.g., Scotch Tape). Add heptane up to the 20 mL line.
2. Rotate at room temperature overnight.
3. Remove 1 mL heptane-glue solution into a 1.5 mL microcentrifuge tube and centrifuge for 1 min at 15,000 rcf.
4. Remove supernatant into a fresh 1.5 mL microcentrifuge tube. If the glue is too thick, add some heptane; if the glue is too thin, open the lid and allow it to

evaporate until the desired thickness is reached (see Section 4.2). Store at room temperature.[8]

4.2 PREPARING THE IMAGING SURFACE: MatTek DISHES

1. Use a razor blade to cut away a section of plastic from the side of the MatTek dish; this will allow the needle to enter the dish at an acute angle. To do this, align the razor blade with one side of the square coverslip and carefully press down until the cut reaches the bottom of the plastic dish. Align the razor blade with the opposite side of the coverslip and repeat the cut. Place two fingertips inside the MatTek dish adjacent to the two cuts and hold the dish firmly in place, then use your other hand to bend and break the plastic section off.
2. Dip a 20–200 μL pipette tip into the heptane-glue solution and allow the heptane-glue to enter by capillary action. Press the narrow end of the pipette tip against the coverslip on the MatTek dish and move the tip across the coverslip to draw a "glue line" parallel to the side of the dish that has been cut away. The line should be ~2 mm thick.
3. Place the lid of the MatTek dish underneath the dish so that the coverslip is raised close to the height of a glass microscope slide; this will facilitate the transfer of embryos from a microscope slide to the coverslip.

4.3 PREPARING THE IMAGING SURFACE: COVERSLIPS

1. Place a 22 × 22 mm glass coverslip under a stereomicroscope.
2. Dip a 20–200 μL pipette tip into the heptane-glue solution and allow the heptane-glue to enter by capillary action. Press the narrow end of the pipette tip against the coverslip and move the tip across the coverslip to draw a "glue line." The line should be ~2 mm thick.

4.4 MOUNTING THE EMBRYOS

1. Allow flies to lay embryos on fruit juice agar plates for 2 h (to obtain a batch of 0–2 h embryos), or for 1 h, followed by incubation at 25 °C for 1 h (to obtain a batch of 1–2 h embryos), or for 15 min (to obtain a batch of 0–15 min embryos specifically for mRNA injection).

   For manual dechorionation:

   a. Attach some clear, double-sided sticky tape (e.g., Scotch Tape) to a glass microscope slide.[9]

---

[8]The heptane in the 50 mL plastic tube will evaporate, but the tube can be topped up and reused for years.
[9]The double-sided sticky tape should be sticky enough to attach to the chorion of an embryo, but not so sticky that a dechorionated embryo cannot roll freely over the tape. Scotch Tape works well, but other brands may be just as suitable.

b. Using a paint brush, transfer the embryos onto the double-sided sticky tape (this can be done under a stereomicroscope if necessary, especially if the flies have laid only a few embryos); try to get an even spread.
c. Under a stereomicroscope, use one half of a blunted pair of forceps to carefully roll an embryo on the tape until its chorion pops open. Position the dechorionated embryo on top of the empty chorion casing so that it is not in contact with the sticky tape and then use the forceps to pick up the embryo and transfer it to the glue line on the coverslip/MatTek dish.
d. Repeat the process, placing the embryos lengthways on the glue line and leaving a small gap between each embryo (~1/5th length of an embryo). Line up as many embryos as possible within 20 min to avoid uneven desiccation of the embryos (see end of the section).

For bleach dechorionation:

a. Add a few milliliters of embryo wash (dH_20 + 0.05% Tween 20) to the fruit juice agar plate containing the embryos.
b. Agitate gently using a paint brush to free embryos from the surface of the plate.
c. Pour the resuspended embryos into a glass filtration funnel device, attached to a Buchner flask.[10] An appropriately sized square piece of PTFE/Teflon membrane should be inserted between the filter chamber and the fret.
d. Remove the embryo wash from the filtration device, using a vacuum pump attached to the Buchner flask.
e. Add 5 mL of thin bleach[11] to the filtration device. Briefly agitate with a paint brush to ensure the embryos mix evenly with the bleach. Leave for 2 min, with occasional reagitation.
f. Remove the bleach using vacuum aspiration and replace with 10 mL of embryo wash.
g. Agitate embryos with a paint brush to ensure even washing and remove solution using vacuum aspiration.
h. Repeat a further wash with embryo wash.
i. Unclamp the filtration device and carefully remove the PTFE/Teflon membrane. The top surface of the membrane should contain all the dechorionated embryos.
j. The embryos can now be taken to the stereomicroscope, picked up with a pair of blunted forceps, and lined up on the glue line of the coverslip/MatTek dish.

[10]The device should be similar to the 15 mL Millipore device consisting of glass filter/fret, filter clamp, and stopper, attached to a buchner flask. An appropriately sized square piece of PTFE/Teflon membrane should be inserted between the filter chamber and the fret.
[11]Thin bleach refers either to neat "value" supermarket brands or to a 60:40 ratio of standard bleach to water.

2. Place the coverslip/MatTek dish with embryos in a 25 °C incubator to desiccate the embryos. Monitor their desiccation by removing the dish and observing the embryos under a stereomicroscope with reflected light. The light produces lines along the surface of the embryos; the straighter the lines, the lesser the embryos are desiccated.
3. Once desiccated to the desired level,[12] add a line of mineral oil or a 1:1 mixture of Halocarbon oil 27:700 (Sigma) over the embryos to prevent further desiccation. The embryos are now ready for microinjection.

5. MICROINJECTION OF EMBRYOS

The precise methodology of microinjection will vary, depending on the micromanipulation apparatus available. One crucial aspect of the injection is the angle of the pipette holder to the microscope stage. This is usually determined by the physical restrictions of the microinjection apparatus in relation to the microscope. The ideal angle of the microinjector and needle is ~15°. However, angles up to 30° are possible. If the angle is larger than this then, upon the movements associated with injection, the tip of the needle is likely to pass through the embryo and embed itself in the glue or coverslip beneath, damaging both embryo and needle. Below, we outline methods associated with either a manually operated syringe or an Eppendorf Injectman II. The former uses needles that have been made from glass capillary tubes in-house, while the latter requires Eppendorf Femtotip II needles. The advantage of the Femtotip II needles is their standardized, premade opening size (0.5 μm inner diameter), while the advantage of homemade needles is their insubstantial cost. Both methods, however, use a micromanipulator to initially align needle tips with the focal plane of the sample, the microscope stage to move the embryos in relation to the needles and to inject, and manual syringe pressure to expel the drug, antibody, protein, peptide, or mRNA into the embryo. This manual approach is superior to automated expelling systems in that it allows fine-tuning of the injection between embryos, due to slight variations of dessication.

5.1 PULLING GLASS MICROINJECTION NEEDLES (IF NECESSARY)

1. Use a vertical micropipette puller (P-30 Sutter, or similar) to make microinjection needles from 1 mm O.D. × 0.58 mm I.D. glass capillary tubes, as per the manufacturer's instructions.

[12]For manual dechorionation, the line of embryos will vary in their level of dessication due to the time difference between dechorionating the first and last embryo. If the time required for optimal dessication has been determined (i.e., through earlier injections that day), this can be corrected for by adding mineral/halocarbon oil to the top, middle, and bottom areas of the coverslip/MatTek dish at different times. For bleach dechorionation, all embryos on a coverslip/MatTek dish will be dessicated to the same extent.

2. Store the needles on a line of blue-tac in a suitable Petri dish making sure their tips do not touch the sides of the dish. Note: the tips of the needles can be checked using a stereomicroscope.

5.2 LOADING THE SAMPLE IN THE NEEDLE

1. Remove an aliquot of the sample to be injected from the −80 °C freezer and spin in a bench top centrifuge at 15,000 rcf for 20 min to ensure that any particles sediment at the bottom of the tube.
2. Transfer ~2 μL of injection solution into a needle using a microcapillary pipette tip. Push the microcapillary pipette tip into the wide end of the needle and eject the solution as close to the fine end as possible.
3. Attach the needle to a pipette holder connected to an injection syringe. For FemtoTip II needles, this requires the correct screwtop connector.
4. Mount the pipette holder onto a micromanipulator positioned to the side of the confocal microscope. Crudely adjust the position of the needle so that the fine end is in line with the light path of the 10X or 20X lens.

5.3 BREAKING THE NEEDLE TIP (IF USING HOME-PULLED NEEDLES)

1. Make a needle breaker by sticking a 22 × 22 mm glass coverslip to the center of a glass microscope slide; breathe on the coverslip and press it down onto the slide until a firm connection is made and the coverslip does not move.
2. Add a drop of mineral/halocarbon oil on the left edge of the coverslip so that the oil covers a portion of both the slide and the coverslip.
3. Place the slide under the microscope to which the micromanipulator is attached in transmitted light mode and focus on the junction between the coverslip and the slide (below the oil). Move the focus up slightly, so that the focal plane is above the surface of the slide but below the level of the coverslip surface.
4. Carefully bring the needle down into the oil using the micromanipulator controls while looking down the eyepieces—it should be clear when the needle enters the oil. As you lower the needle, move it from side to side in order to keep track of its position. Make sure the tip of the needle comes down to the left of the coverslip so that it points towards the left-hand side of the coverslip. Continue lowering the needle until the tip is in focus. It should now be positioned above the surface of the slide but below the level of the coverslip surface.
5. To break open the tip of the needle, gently move the coverslip towards the tip of the needle using the microscope stage controls and touch the side of the coverslip against the tip of the needle. Move the coverslip back and try to expel some antibody solution from the needle by pressing on the syringe (air may come out first). Repeat the process until the desired amount of antibody solution can be expelled. Leave the tip of the needle a safe distance from the surface of the slide and from the side of the coverslip, but still within the oil.

5.4 EMBRYO INJECTION

1. Briefly raise the needle.
2. Place the MatTek dish or coverslip containing the line of embryos on the microscope stage sample holder, making sure the embryo line is perpendicular to the orientation of the needle.
3. Focus on the embryos using either a 10X or 20X lens.
4. Lower the needle into the oil using the micromanipulator, until the tip also comes into focus. Position the needle so that the tip is at a safe distance from the side of the embryos.
5. Select an embryo for injection using brightfield. The desired stage of embryonic development depends on the specific experiment. When injecting samples that are designed to immediately perturb protein function, such as antibodies or drugs, it is normally best to inject into embryos in nuclear cycle 10–13. Under brightfield, these embryos will have a uniform, brown cytoplasm with a uniform perimeter of clearing around the cortex. It should be possible to see a series of regular invaginations within this perimeter, corresponding to the deformations brought about by the actin cytoskeleton surrounding the cortical nuclei. In addition, there should be clear invaginations at the posterior pole, corresponding to the already-formed pole cells. Embryos younger than cycle 10 will show little or no perimeter clearing, or pole cells. Embryos older than cycle 13 (i.e., cellularized embryos) will show a clear monolayer of cells at the cortex, or invaginations throughout the cortex, corresponding to gastrulation.[13] When injecting mRNA samples designed to generate proteins of interest to be analyzed in cycles 10–13, injection must be carried out as soon as possible after egg laying, and at least 1 h prior to the syncytial blastoderm stage (see Section 7.2). In these cases, all embryos that lack pole cells are injected sequentially.
6. Using the micromanipulator, position the needle so that its tip is at the correct height to contact the side of the embryo.[14] Gently insert the needle into the embryo by moving the embryo towards the needle using the microscope stage adjustment (i.e., keep the needle stationary while moving the embryos into the needle); the tip of the needle should insert ~1/4th width of the embryo. Eject the desired amount of sample by pressing firmly on the

[13]Cellularization occurs ~3 h postfertilization. Gastrulation begins approximately 1 h later. Although collection of 0–2 h embryos should ensure no cellularized or gastrulating embryos within the population, a proportion of embryos are retained by the mother after fertilization. Thus, it is not unusual to observe some embryos at this stage of development. However, if the majority of 0–2 h embryos are consistently cellularized prior to microinjection, it is likely the density of flies in the collection chamber, in relation to the available yeast paste, is too high.

[14]It is preferable to be slightly higher than the middle of the embryo, given the 15°–30° angle of the needle.

syringe and then move the embryo away from the needle so that the needle is back in its original position.[15]

6. PERTURBATION VIA COLD TREATMENT

The Wakefield laboratory has recently described a noninvasive protocol based on the cold-labile properties of microtubules. The method leads to rapid microtubule depolymerization during metaphase that can be maintained for over 90 min without affecting regular centrosome spacing or centrosome–cortex interactions.[16] Upon rewarming, mitotic microtubule organization reinitiates, but does so predominantly from chromatin, rather than centrosomes. Not only does this procedure allow the dynamic relocalization of centrosomal and microtubule-associated proteins to be explored, but it can also be combined with microinjection techniques, in order to probe the functional relationships between microtubule-generating pathways and specific proteins.

6.1 COLD TREATMENT ASSAY

1. At least 30 min prior to initial imaging, stick two 18 mm round coverslips together by breathing on them, then stick these to the middle of a 60 mm Petri dish. Repeat for at least two additional dishes.[17]
2. Place the Petri dishes, with lids, and a 1.5 mL microcentrifuge tube filled with a 1:1 mixture of 700:27 halocarbon oil on ice.
3. Mount 1–2 h single embryos on heptane-glue each on a 22 mm^2 coverslip, using the manual dechorionation technique. Repeat for five to eight embryos.
4. Cover each embryo in a thin layer of halocarbon oil.
5. Image each embryo using the 60X lens on a spinning disc confocal, keep the laser exposure low, monitoring the GFP fluorescence until an embryo at cycle 11 is identified.
6. Image the embryo until 30 s following NEB, after spindle bipolarity has clearly been achieved.
7. Immediately take the coverslip off the objective and place on one of the cold Petri dishes, directly on top of the 18 mm coverslip stack.[18]

[15]Sometimes cytoplasm can spool out and remain attached to the side of the embryo; this should be removed by dragging it away with the needle, otherwise cytoplasm continues to leak out through the hole in the embryo.

[16]Wild-type embryos have been observed to maintain centrosome spacing and the ability to reinitiate spindle formation capable of accurate chromosome segregation after 4 h on ice. We have not attempted longer time periods.

[17]There is a high probability of being unable to focus and begin reimaging before microtubule nucleation reinitiates. For standard cold treatment we routinely place 2–3 times more embryos on ice than we plan to image.

[18]If the embryos enter anaphase before they are sufficiently chilled on ice, they will undergo endomitosis. This can easily be identified, as centrosomes will have duplicated upon reinitiation of microtubule nucleation.

8. Cover the embryo with ~100 μL of ice-cold halocarbon oil. Put a lid on the Petri dish and completely bury in ice. Be very careful not to get any ice in the Petri dishes.
9. After 90 min, place the coverslip back on the microscope. As quickly as possible, find the embryo using brightfield microscopy, switch to fluorescence, focus the centrosomes, and begin to image. Microtubule nucleation is visible approximately 20–30 s after the coverslip is removed from ice.

6.2 COLD TREATMENT ASSAY WITH MICROINJECTION

1. Follow Steps 1–6 as above in Section 6.1.[19]
2. After 75 min on ice, place a coverslip on the objective. Image under brightfield and inject, as per Step 5 in Section 5.4. It is vital that the injection is achieved within 30 s of placing on the objective.
3. Put the coverslip back on ice for at least another 15 min. Image.

7. IMAGING AND DATA ANALYSIS

Imaging conditions depend upon the specific experiment and the microscope setup available. Below, we describe three examples that examine the dynamic behavior of GFP-tagged centrosomal proteins. In Example 1, embryos were injected with antibodies to disrupt centrosomes during the M-phase to S-phase transition; in Example 2, mutant embryos were injected with mRNA encoding a rescue construct tagged with GFP; in Example 3, the dynamic localization of a centrosomal protein is monitored before, and after, cold-induced microtubule depolymerization.

7.1 EXAMPLE 1

The embryo expressed Spd-2-GFP and had been injected with Texas-red-labeled anti-Asl antibodies. The imaging was carried out on a Perkin Elmer ERS spinning disk confocal system mounted on a Zeiss Axiovert microscope, using a 63X 1.4NA oil immersion objective, an ion argon laser (to produce 488 nm light), and 440 and 568 nm diode-pumped solid state lasers. The microscope was run using Volocity software, which allows photobleaching to be carried out when and where the user wants. The experiment analyzes the effect of inhibiting Asl on Spd-2-GFP recruitment to centrosomes (Figure 1).

[19]The quality of the heptane-glue is especially important for this procedure. It should always be prepared 24 h in advance and the brand of double-sided sticky tape makes a difference. We find Scotch Tape to be suitable. If the glue is too thick it will be difficult to inject and conditions may become anoxic; if it is too thin the embryos will become unstuck when microinjection is attempted.

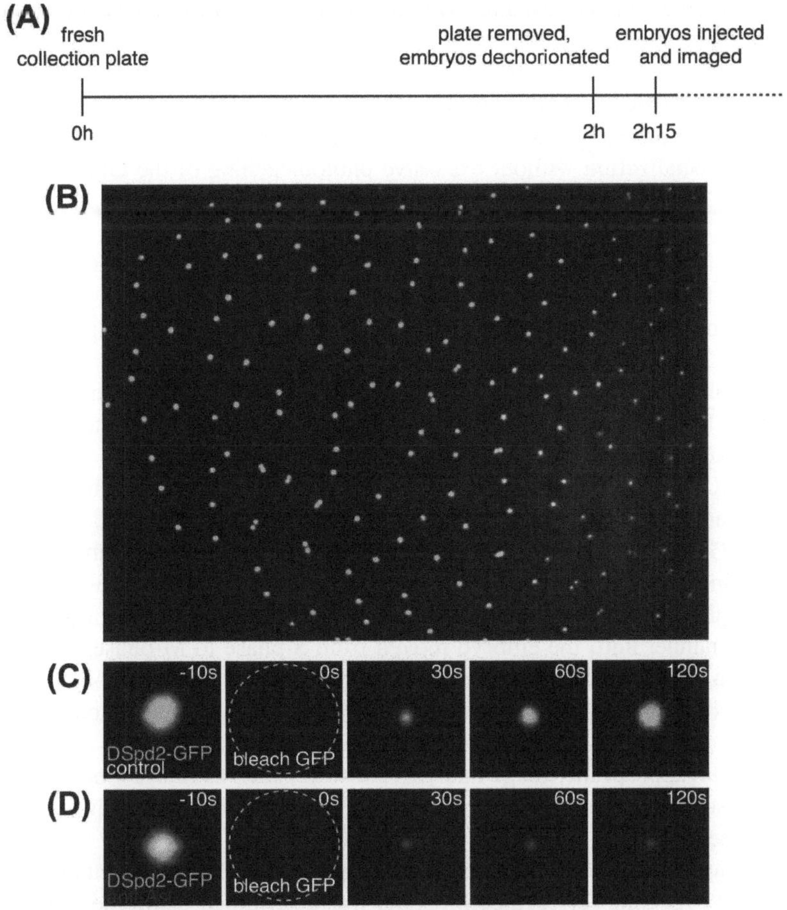

FIGURE 1 Antibody-mediated inhibition combined with FRAP analysis.

(A) A timeline of the experiment. (B) An image of a syncytial embryo expressing Spd-2-GFP (green) shortly after having been injected with Texas-red-labeled anti-Asl antibodies (red). Note how the antibody is concentrated on one side of the field of view, allowing a comparison between centrosomes that have been bound by the antibody (bound centrosomes) and centrosomes that have not been bound by the antibody (control centrosomes). (C, D) Examples from an FRAP experiment where the Spd-2-GFP signal was photobleached from a control centrosome (C) or from a bound centrosome (D). Time before and after photobleaching (t = 0 s) is indicated in the top right of each panel. Note how the Spd-2-GFP signal recovers robustly at the control centrosome, but not at the bound centrosome, showing that Asl inhibition perturbs the incorporation of new Spd-2-GFP molecules into the centrosome. (See color plate)

Modified from Conduit, et al. (2014) – eLife.

1. A 63X objective was used to view the embryo with the 488 and 568 nm lasers. The labeled antibody was strongly concentrated in a localized area of the embryo and the microscope stage was adjusted so that the field of view contained centrosomes both bound and not bound by antibody (Figure 1b).
2. The argon laser and diode laser powers and exposure times were adjusted to allow visualization without excessive photobleaching of the GFP signal (~50% power; 500 ms exposure) and the Texas-red signal (~20% power; 500 ms exposure).
3. The software was programmed to take 7 × 0.5 μm Z stacks every 10 s and the FRAP module to photobleach a circular ROI of 5 μm diameter using the 440 nm laser (the 440 nm laser avoids excessive photobleaching of the antibodies fluorescent signal).
4. A prebleached frame was taken and then the ROI was moved over a centrosome bound by antibody and the FRAP module was activated to photobleach the Spd-2-GFP signal. The software then took a bleached image.
5. The ROI was immediately moved to a control centrosome not bound by antibody and the Spd-2-GFP signal was photobleached.
6. Both GFP signals were allowed to recover for 2 min, (see Figure 1C and 1D) enabling accurate measurements of the initial fluorescence recovery rate.
7. Multiple pairs of bound and unbound centrosomes were photobleached in the same manner until the embryo moved from S-phase to M-phase (centrosome protein dynamics change on entry into M-phase).
8. The whole process was repeated on multiple embryos until enough data was collected to enable adequate statistical comparisons.
9. Using an appropriate image analysis software, such as image J or Volocity, the fluorescent signal above background at each centrosome at each time point was measured.
10. The data were plotted in Microsoft Excel and the change in fluorescence intensity over time for each centrosome was determined using the "LINEST" function.
11. The values from bound and unbound centrosomes were compared using an appropriate statistical test, such as a T-test or Mann–Whitney U test.

7.2 EXAMPLE 2

The Raff lab has recently described an assay to quickly screen different mutant versions of the Centrosomin (Cnn) protein. Mothers mutant for the *cnn* gene lay eggs that develop until they display severe mitotic defects during the syncytial stage and arrest prior to cellularization. If mRNA encoding GFP-tagged Cnn protein is injected shortly after fertilization (at least 1 h prior to the syncytial blastoderm stage), the resulting GFP-Cnn protein localizes normally to centrosomes and rescues the mitotic defects otherwise observed in the syncytial blastoderm. mRNA encoding mutant versions of GFP-Cnn can be injected to quickly test the localization and function of different mutants, in a *cnn* null mutant background. This is particularly useful when screening large numbers of mutants, as it avoids the time-consuming generation of multiple transgenic lines. In theory, this technique can be used to screen mutations in other centrosomal proteins, although it may not always be possible to do this in a null mutant background, as many centrosomal mutants arrest very early in development prior to injection or mRNA translation.

In the following example, the embryo had a cnn^{f04547}/cnn^{HK21} genotype and had been injected with mRNA expressing wild-type GFP-Cnn. The imaging was carried out on the same system as in Example 1.

1. The injected embryos was allowed to develop for at least 1 h at 25 °C.
2. The MatTek dish was placed on the microscope stage and a 63X lens was used to scan through the line of embryos with the 488 nm laser. An embryo in cycle 10 that was expressing GFP-Cnn protein was selected for imaging.
3. The field of view was moved to the area of the cortex where the GFP-Cnn protein was being expressed and the embryo was imaged from the start of nuclear cycle 11 until early in nuclear cycle 13 (Figure 2).

7.3 EXAMPLE 3

Syncytial embryos expressing EB1-GFP were cold-treated and injected with antibodies raised against the Augmin component Dgt6 (Figure 3). A Femtotips® II needle was loaded with anti-Dgt6 antibody in injection buffer at 6 mg/mL and attached to the CellTram® oil microinjector and Transferman NK2 (Eppendorf) injection apparatus. Imaging was performed on a Visitron Systems Olympus IX81 microscope with a CSOX1 spinning disk using an UPlanS APO 1.3 NA (Olympus) 60X objective and Piezo Flexure Objective Scanner with an argon 488 nm laser.

In this example the effects of Augmin perturbation on microtubule regrowth from centrosomes and chromatin are observed.

1. Single coverslips were imaged using a 60X objective and brightfield illumination to identify cycle 10 or 11 embryos, and then laser fluorescence (with low exposure) was used to identify prophase embryos.
2. Once embryos had reached metaphase (typically 20 s following the influx of EB1 comets into the nuclear region characterizing nuclear envelope breakdown) the coverslips were immediately placed on an ice-cold Petri dish and covered in ice-cold halocarbon oil. The Petri dish was then buried in ice.
3. During this time, a few nonexperimental embryos on a separate coverslip were injected to test the microinjection injection needle. This ensured the needle could be immediately used to inject the experimental embryos once they had been removed from the ice.
4. After 75 min the embryos were removed from the petri dish while still on ice using a pair of fine forceps and a needle. The coverslip was quickly mounted on the microscope system and the embryo identified under brightfield using a 10X objective. Injection was performed within 30 s and the coverslip was then quickly placed back onto an ice-cold Petri dish and covered with cold halocarbon oil.

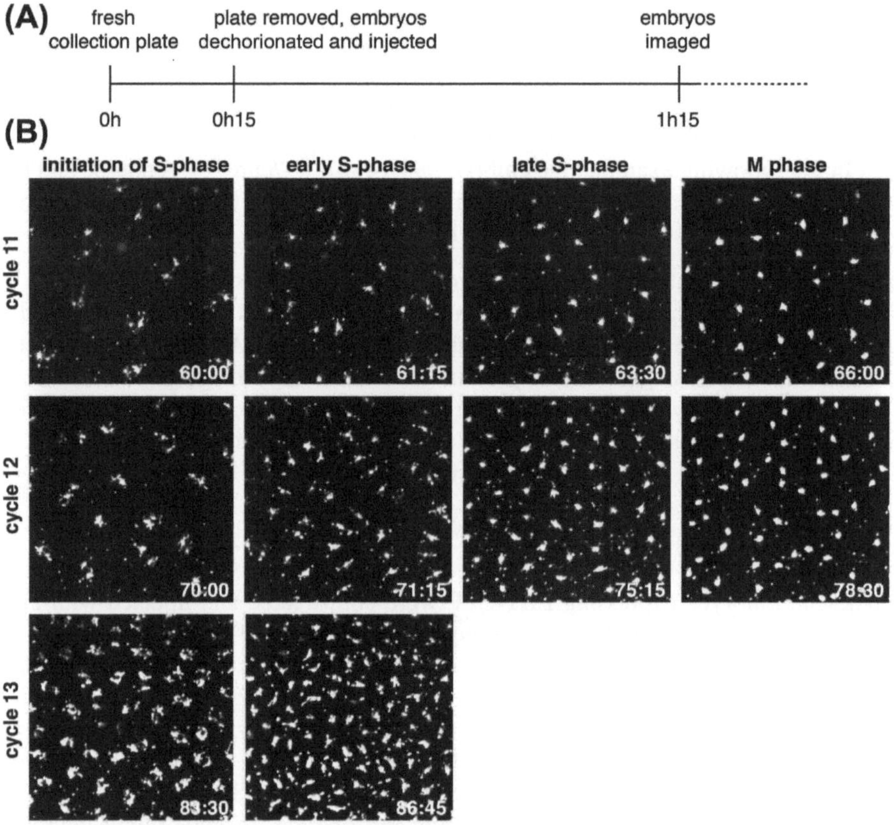

FIGURE 2 Injection of mRNA encoding GFP-Cnn into a *cnn* null embryo.

(A) A timeline of the experiment. (B) Sequential images of a *cnn* null mutant syncytial embryo that has been injected with in vitro transcribed mRNA encoding GFP-Cnn. Time after injection (min:s) is shown in the bottom right of each panel. The injected mRNA is translated into GFP-Cnn protein within the embryo. The GFP-Cnn protein localizes to centrosomes in a manner that is indistinguishable from genomically encoded GFP-Cnn protein. The portions of the embryo that express GFP-Cnn do not display the mitotic phenotypes typically associated with *cnn* null mutants. The embryo was filmed from the start of nuclear cycle 11 until the early stages of nuclear cycle 13.

5. After a further 15 min, the coverslip was removed from the Petri dish and the same embryo was quickly identified using brightfield with a 60X objective.
6. GFP was visualized using a 60X objective and 488 nm argon laser with an exposure time of 400 ms and laser power of 25%. Images of five focal planes, 0.2 μm apart, were acquired every 2 s. Image acquisition was initiated as soon as possible, often before the exact focal plane had been found.

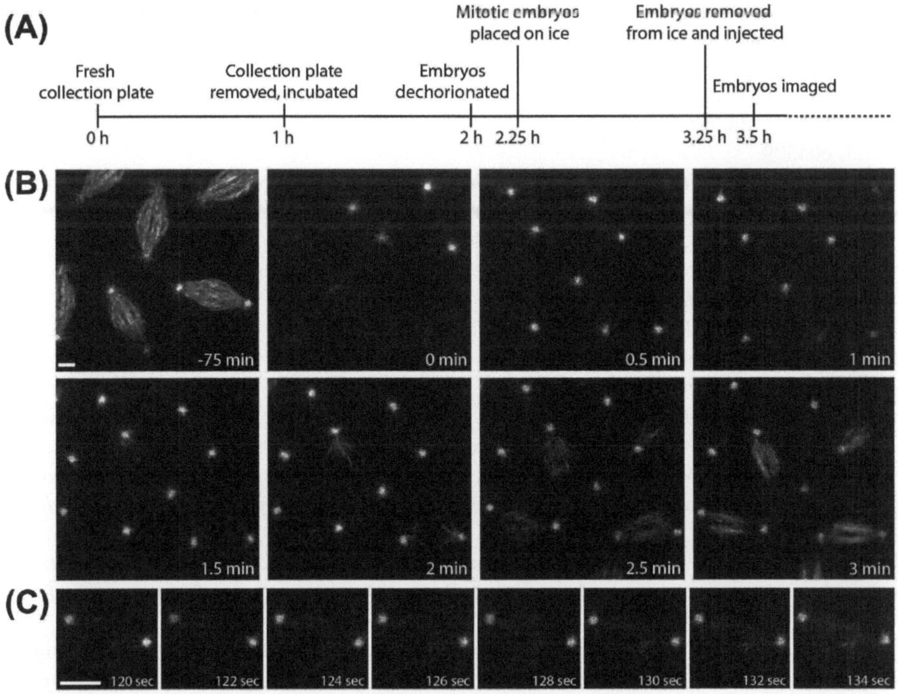

FIGURE 3 Combining embryo cold treatment with antibody injection.

(A) A timeline of the experiment. (B, C) Stills from a time-lapse movie showing an EB1-GFP expressing embryo before cold treatment (−75 min) and following cold treatment and anti-Dgt6 antibody injection. Time after removal from ice and refocusing on objective (min/s) is shown in the bottom right of each panel. Note how microtubule growth is predominantly initiated from the centrosomes, unlike in a noninjected embryo where the major site of microtubule growth is from the around the chromatin. (C) A close-up of a single spindle from 120−134 s, the predominant period of spindle formation from centrosomal microtubules. Scale bar—5 μm.

REFERENCES

Brust-Mascher, I., Sommi, P., Cheerambathur, D. K., & Scholey, J. M. (2009). Kinesin-5-dependent poleward flux and spindle length control in *Drosophila* embryo mitosis. *Molecular Biology of the Cell, 20*, 1749−1762.

Buttrick, G. J., Beaumont, L. M., Leitch, J., Yau, C., Hughes, J. R., & Wakefield, J. G. (2008). Akt regulates centrosome migration and spindle orientation in the early *Drosophila melanogaster* embryo. *The Journal of Cell Biology, 180*, 537−548.

Conduit, P. T., Brunk, K., Dobbelaere, J., Dix, C. I., Lucas, E. P., & Raff, J. W. (2010). Centrioles regulate centrosome size by controlling the rate of Cnn incorporation into the PCM. *Current Biology, 20*, 2178−2186.

Conduit, P. T., Feng, Z., Richens, J. H., Baumbach, J., Wainman, A., Bakshi, S. D., et al. (2014). The centrosome-specific phosphorylation of Cnn by Polo/Plk1 drives Cnn scaffold assembly and centrosome maturation. *Developmental Cell, 28*, 659–669.

Conduit, P. T., Richens, J. H., Wainman, A., Holder, J., Vicente, C. C., Pratt, M. B., et al. (2014). A molecular mechanism of mitotic centrosome assembly in *Drosophila*. *Elife, 3*, e03399.

Duncan, T., & Wakefield, J. G. (2011). 50 Ways to build a spindle: the complexity of microtubule generation during mitosis. *Chromosome Research, 19*, 321–333.

Gergely, F., Kidd, D., Jeffers, K., Wakefield, J. G., & Raff, J. W. (2000). D-TACC: a novel centrosomal protein required for normal spindle function in the early *Drosophila* embryo. *EMBO Journal, 19*, 241–252.

Hayward, D., Metz, J., Pellacani, C., & Wakefield, J. G. (2014). Synergy between multiple microtubule-generating pathways confers robustness to centrosome-driven mitotic spindle formation. *Developmental Cell, 28*, 81–93.

Hayward, D., & Wakefield, J. G. (2014). Chromatin-mediated microtubule nucleation in *Drosophila* syncytial embryos. *Communicative and Integrative Biology, 7*, e28512.

Kellum, R., Raff, J. W., & Alberts, B. M. (1995). Heterochromatin protein 1 distribution during development and during the cell cycle in *Drosophila* embryos. *Journal of Cell Science, 108*, 1407–1418.

Lucas, E. P., & Raff, J. W. (2007). Maintaining the proper connection between the centrioles and the pericentriolar matrix requires *Drosophila* centrosomin. *The Journal of Cell Biology, 178*, 725–732.

Novak, Z. A., Conduit, P. T., Wainman, A., & Raff, J. W. (2014). Asterless licenses daughter centrioles to duplicate for the first time in *Drosophila* embryos. *Current Biology, 24*, 1276–1282.

Oegema, K., Marshall, W. F., Sedat, J. W., & Alberts, B. M. (1997). Two proteins that cycle asynchronously between centrosomes and nuclear structures: *Drosophila* CP60 and CP190. *Journal of Cell Science, 110*, 1573–1583.

Oliveira, R. A., Hamilton, R. S., Pauli, A., Davis, I., & Nasmyth, K. (2010). Cohesin cleavage and Cdk inhibition trigger formation of daughter nuclei. *Nature Cell Biology, 12*, 185–192.

Oliveira, R. A., & Nasmyth, K. (2013). Cohesin cleavage is insufficient for centriole disengagement in *Drosophila*. *Current Biology, 23*, R601–R603.

Raff, J. W., & Glover, D. M. (1988). Nuclear and cytoplasmic mitotic cycles continue in *Drosophila* embryos in which DNA synthesis is inhibited with aphidicolin. *The Journal of Cell Biology, 107*, 2009–2019.

Raff, J. W., & Glover, D. M. (1989). Centrosomes, and not nuclei, initiate pole cell formation in *Drosophila* embryos. *Cell, 57*, 611–619.

Royou, A., McCusker, D., Kellogg, D. R., & Sullivan, W. (2008). Grapes(Chk1) prevents nuclear CDK1 activation by delaying cyclin B nuclear accumulation. *The Journal of Cell Biology, 183*, 63–75.

Sharp, D. J., Brown, H. M., Kwon, M., Rogers, G. C., Holland, G., & Scholey, J. M. (2000). Functional coordination of three mitotic motors in *Drosophila* embryos. *Molecular Biology of the Cell, 11*, 241–253.

Sharp, D. J., Yu, K. R., Sisson, J. C., Sullivan, W., & Scholey, J. M. (1999). Antagonistic microtubule-sliding motors position mitotic centrosomes in *Drosophila* early embryos. *Nature Cell Biology, 1*, 51–54.

Silverman-Gavrila, R. V., & Wilde, A. (2006). Ran is required before metaphase for spindle assembly and chromosome alignment and after metaphase for chromosome segregation and spindle midbody organization. *Molecular Biology of the Cell, 17*, 2069–2080.

Sisson, J. C., Field, C., Ventura, R., Royou, A., & Sullivan, W. (2000). Lava lamp, a novel peripheral golgi protein, is required for *Drosophila melanogaster* cellularization. *The Journal of Cell Biology, 151*, 905–918.

CHAPTER

14

A yeast two-hybrid approach for probing protein—protein interactions at the centrosome

Brian J. Galletta[1], Nasser M. Rusan

Cell Biology and Physiology Center, National Heart Lung and Blood Institute, National Institutes of Health, Bethesda, MD, USA

[1]*Corresponding author: E-mail: brian.galletta@nih.gov*

CHAPTER OUTLINE

Introduction **252**

1. Identifying Direct Protein Interactions Have Had a Profound Impact on Understanding Centrosome Function **253**

2. Challenges to Understanding Protein—protein Interactions in the Centrosome **255**

3. Dissection of Complex, Multicomponent Protein Machines Using Y2H **256**

3.1 General Principle of the Y2H 256

3.2 Advantages of the Y2H System for Complex, Multicomponent Protein Machines 257

3.3 Limitations of the Y2H System 258

4. Performing an Array Based Screen for Centrosome Protein Interactions **259**

4.1 Selecting a Y2H System 259

4.2 Dividing Proteins into Smaller Fragments 260

4.3 Generating the Y2H Library 261

4.4 Autoactivation and False Positive Rate Identification 262

4.5 Prescreen Planning 262

4.6 Screening for Interactions 263

4.6.1 Protocol 264

4.7 Interpreting Screening Results 267

5. Generating Specific, Separation of Function Mutations by Reverse Y2H **267**

5.1 Rationale 268

5.2 Generating Mutant Library and Screening for Loss of Interaction 268

5.2.1 Protocol 270

Methods in Cell Biology, Volume 129, ISSN 0091-679X, http://dx.doi.org/10.1016/bs.mcb.2015.03.012

Summary ... 272
Acknowledgments ... 273
References ... 273

Abstract

As a large, nonmembrane bound organelle, the centrosome must rely heavily on protein–protein interactions to assemble itself in the cytoplasm and perform its functions as a microtubule-organizing center. Therefore, to understand how this organelle is built and functions, one must understand the protein–protein interactions made by each centrosome protein. Unfortunately, the highly interconnected nature of the centrosome, combined with its predicted unstructured, coil-rich proteins, has made the use of many standard approaches to studying protein–protein interactions very challenging. The yeast-two hybrid (Y2H) system is well suited for studying the centrosome and is an important complement to other biochemical approaches. In this chapter we describe how to carry out a directed Y2H screen to identify the direct interactions between a given centrosome protein and a library of others. Specifically, we detail using a bioinformatics-based approach (structure prediction programs) to subdivide proteins and screen for interactions using an array-based Y2H approach. We also describe how to use the interaction information garnered from this screen to generate mutations to disrupt specific interactions using mutagenic-PCR and a "reverse" Y2H screen. Finally, we discuss how information from such a screen can be integrated into existing models of centrosome assembly and how it can initiate and guide extensive in vitro and in vivo experimentation to test these models.

INTRODUCTION

The centrosome is the major microtubule-organizing center (MTOC) of many cells (Figure 1(A)). As an MTOC, the centrosome functions in a variety of cellular processes including cell migration, neuronal path finding, axon selection, immune

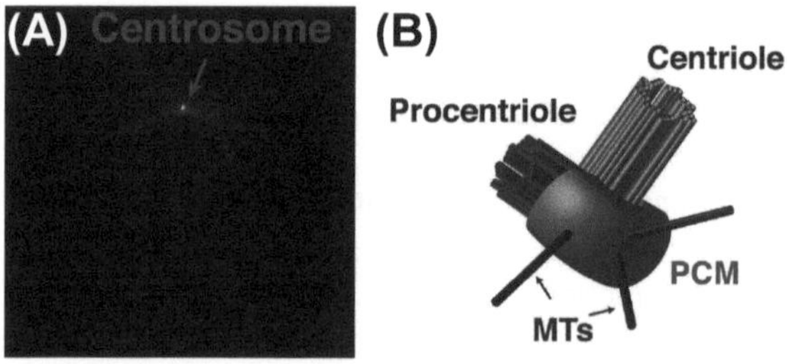

FIGURE 1 The centrosome is the major MTOC of many cells.

(A) Immnuofluorescent micrograph of an interphase *Drosophila* S2 cell showing the centrosome (green (gray in print versions)) organizing the microtubules (red (dark gray in print versions)). (B) Schematic of a centrosome, showing the centriole, procentriole, PCM, and MTs.

cell function, cell polarity, and mitosis (reviewed in Angus & Griffiths, 2013; Bettencourt-Dias, Hildebrandt, Pellman, Woods, & Godinho, 2011; Bornens, 2012; Elric & Etienne-Manneville, 2014; Sakakibara, Ando, Sapir, & Tanaka, 2013). Centrosomes consist of two major components, the core centriole pair and the pericentriolar material (PCM), both identifiable by light and electron microscopy (EM). Centrioles are composed of nine triplet microtubules (MTs) and many associated proteins arranged as a barrel (Figure 1(B)). Centrioles play many critical roles, such as ensuring proper centrosome duplication, serving as an anchor and organizer of PCM, and functioning as the basal body responsible for nucleating cilia and flagella. The PCM surrounds the centrioles and is the region from which MTs are nucleated and anchored (Figure 1(A) and (B)). While the PCM appears relatively unstructured by EM, superresolution microscopy has revealed some amount of spatial order to several PCM proteins (Fu & Glover, 2012; Lawo, Hasegan, Gupta, & Pelletier, 2012; Mennella et al., 2012; Sonnen, Schermelleh, Leonhardt, & Nigg, 2012).

The protein composition of centrosomes is increasingly well understood. Multiple groups have succeeded in purifying the entire organelle from cells and have revealed several 100 proteins associated with the centrosome (Andersen et al., 2003; Jakobsen et al., 2011; Muller et al., 2010). As is the case with other large molecular assemblies, a detailed understanding of the function of any centrosome protein must include an understanding of how it physically interacts with others. Only with this level of understanding can one begin to probe the true consequences of perturbing protein function by mutagenesis. In this chapter, we will discuss some of the challenges to studying the protein–protein interactions of the centrosome. We will then discuss the benefits of studying these interactions by Y2H and describe how to carry out an array-based screen to identify the interactions within a collection of centrosome proteins. Finally, we will discuss how the information gained from this method of Y2H screening can guide further experimentation, including the generation of highly specific, separation of function mutants using a reverse-Y2H approach. In combination with other powerful genetic, biochemical, and cell biological approaches, these Y2H techniques should help propel our understanding of the biology of the centrosome.

1. IDENTIFYING DIRECT PROTEIN INTERACTIONS HAVE HAD A PROFOUND IMPACT ON UNDERSTANDING CENTROSOME FUNCTION

A few direct interactions among centrosome proteins have been successfully identified and the conclusions drawn from these studies have truly advanced our understanding of centrosome biogenesis. A particularly insightful set of interactions are those identified among the core centriole proteins, Sas6, SCL/TAL1 Interrupting Locus (STIL)/Ana2/Sas5, Centrosomal protein 135 kDa (Cep135)/Bld10, and Centrosomal P4.1-Associated Protein (CPAP)/Sas4 (Figure 2(A)). For this set of proteins the addition of direct protein–protein interaction data to the genetic and

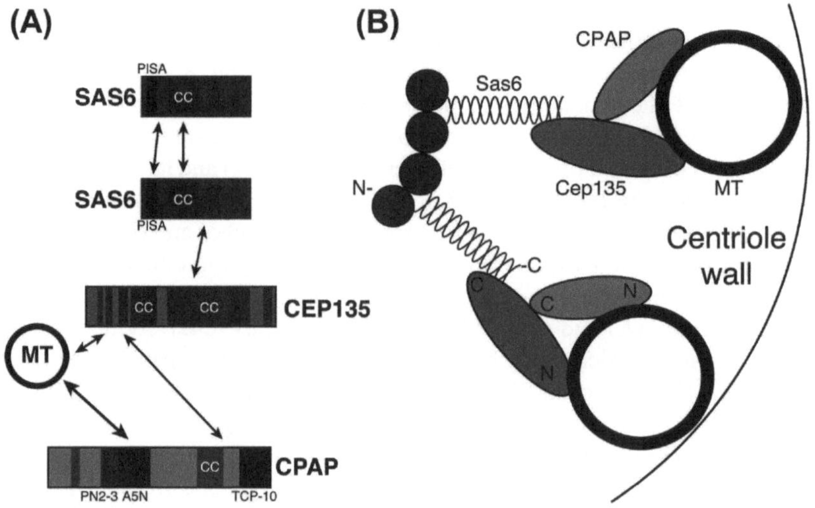

FIGURE 2 The interactions among core centriole proteins dictate the organization of the structure.

(A) Known direct protein–protein interactions among core centriole proteins. Sas6 is included twice to illustrate its homotypic interactions. Shaded regions denote known or predicted protein motifs, PISA (present in Sas6), CC (coiled-coil), PN2-3 (MT destabilizing motif), A5N (MT binding motif), and TCP-10 (T complex protein 10). (B) Cartoon of the proteins that compose two of the nine arms of the cartwheel of a centriole and how they might interact to connect the symmetry of Sas6 to the MTs of the centriole. Approximate locations of the amino-terminal (N) and carboxy-terminal (C) ends of the proteins are indicated. See Section 2 for details and references.

structural information has begun to crystallize a view of the centriole architecture. The interaction between Sas6 and STIL/Ana2/Sas5, which in some systems is regulated by the master centriole duplication kinase Plk4, is likely one of the earliest events in the construction of a new centriole, termed a procentriole (Dzhindzhev et al., 2014; Leidel, Delattre, Cerutti, Baumer, & Gonczy, 2005; Ohta et al., 2014). The interactions that Sas6, and its *Chlamydomonas reinhardtii* ortholog Bld12, can make with itself seem likely to help establish the stereotypic centriole symmetry. Sas6 homodimerizes via its C-terminal tails and oligomerizes via its globular heads. Together, these interactions drive the formation of higher order structures that likely help establish the ninefold radial symmetry of the procentriole's cartwheel (van Breugel et al., 2011; Kitagawa et al., 2011). In this higher order structure, the C-termini of Sas6 dimers radiate out from a central hub (Figure 2(B), two of nine Sas6 dimers are shown). The C-terminal end of Sas6 can interact with Cep135, which in turn, interacts with CPAP/Sas4. Since both Cep135 and CPAP/Sas4 can interact with MTs, an attractive model is that these interactions link the spokes of the Sas6 cartwheel to the MTs of the centriole wall, thus connecting the ninefold symmetry of Sas6 tails to the triplet MTs (Hiraki,

Nakazawa, Kamiya, & Hirono, 2007; Lin et al., 2013; Roque et al., 2012). Therefore, the identification of direct interactions, in combination with other approaches, has helped shape this fundamental model of the centriole core.

Interactions between centrosome proteins have offered insight into other centrosomal processes, including the regulation of centriole duplication (Dzhindzhev et al., 2014; Hatch, Kulukian, Holland, Cleveland, & Stearns, 2010; Kim et al., 2013; Ohta et al., 2014; Sonnen, Gabryjonczyk, Anselm, Stierhof, & Nigg, 2013) and centriole length control (Spektor, Tsang, Khoo, & Dynlacht, 2007). Insight provided from these interactions bodes extremely well for the success of future endeavors to define more interactions among centrosome proteins.

2. CHALLENGES TO UNDERSTANDING PROTEIN–PROTEIN INTERACTIONS IN THE CENTROSOME

As illustrated by the examples above, understanding how centrosomes are assembled, regulated, and perform their cellular functions will require a detailed understanding of how its proteins physically relate to each other. Loss-of-function and other genetic studies in vivo have been extremely fruitful in identifying proteins critical for major aspects of centrosome biology, such as centriole duplication and MTOC activity. In fact, much of our understanding of the initial steps of centriole duplication stems from pioneering genetic work in *Caenorhabditis elegans* (Dammermann et al., 2004; Delattre et al., 2004; Kemp, Kopish, Zipperlen, Ahringer, & O'Connell, 2004; Leidel et al., 2005; Leidel & Gonczy, 2003; O'Connell et al., 2001; Pelletier, O'Toole, Schwager, Hyman, & Muller-Reichert, 2006) and later from Ribonucleic Acid Interference (RNAi)-based screens in cultured cells (Balestra, Strnad, Fluckiger, & Gonczy, 2013; Dobbelaere et al., 2008; Goshima et al., 2007). However, these genetic studies have only provided limited insight into the interconnectedness of the centrosome. Some of this stems from the fact that disrupting many of the most important proteins leads to loss of the entire organelle, making it difficult to assess how the absence of one protein impacts the behaviors of others.

Understanding the nature of the protein–protein interactions within the centrosome, which are critical for its function, is challenging for a number of reasons. One major obstacle is the centrosome's size—it is a micron-scale, supramolecular machine. It consists of hundreds of proteins, 50–100 of which are important for its centriole and MTOC functions. This predicts a very large number of possible interactions. Investigating each in a pairwise fashion using biochemical methods such as in vitro binding assays of purified components would be prohibitively laborious. Furthermore, it is becoming more evident that centrosome proteins are an unusually challenging class of proteins to study in vitro. More than half of the residues in human centrosome proteins are predicted to be in disordered structures and many are predicted to be part of coiled-coils (Dos Santos et al., 2013). It is possible that many proteins only adopt a stable state in the context of their binding partners and/or the local environment of the centrosome. In a high-throughput screen for soluble centrosome proteins, only 32.5% of centrosome proteins were soluble in lysates of *E. coli* expressing recombinant forms of the proteins. This success rate was not improved when only using putative globular domains (Dos Santos et al., 2013). These unusual characteristics of centrosome proteins may contribute to the reason why only a few centrosome proteins have been purified for crystallography to date, all of which are relatively small portions of the protein. Furthermore, several have required complex, or unconventional procedures to ensure protein behavior (van Breugel et al., 2011; Cottee et al., 2013; Qiao, Cabral, Lettman, Dammermann, & Dong, 2012; Zheng et al., 2014). The need for such protein-specific treatment, although fruitful as in the case of the core centriole components, is a low-throughput approach and is not conducive to elucidating the large numbers of interaction that occur in the centrosome. Furthermore, many techniques for studying interactions in vitro are biased toward high-affinity interactions (Bruckner, Polge, Lentze, Auerbach, & Schlattner, 2009) and could miss lower-affinity interactions that might be critical for centrosome protein dynamics, especially within the PCM. As an alternative to using purified component, one can probe centrosome protein interactions in vivo using a co-immunoprecipitation (Co-IP) approach. However, Co-IP experiments of any centrosome protein leave open the possibility that the detected interaction is not direct. Thus, such an approach should be paired up with other methods such as direct in vitro binding assays (if possible) or a Y2H analysis as discussed below.

3. DISSECTION OF COMPLEX, MULTICOMPONENT PROTEIN MACHINES USING Y2H

Given how little we know about the protein connections within the centrosome, there is a great need for a method that would uncover all the potential interactions in a quick and high-throughput manner. Here, we suggest that a Y2H approach is part of the solution serving as a critical first step to solve the complex interactions within the centriole and PCM.

3.1 GENERAL PRINCIPLE OF THE Y2H

The Y2H system was originally described by Fields and Song (1989). The original system relied on the fact that the DNA-binding domain (BD) and transcriptional activation domain (AD) of the *Saccharomyces cerevisiae* transcription factor GAL4 can be separated. When separated neither can drive transcription from Gal4 responsive promoters on their own. However, a functional transcription factor can be reconstituted and drive transcription when a protein fused to a BD interacts with a protein fused to an AD. In turn, reporter gene activity allows one to infer a direct interaction between the proteins fused to the AD and BD (Figure 3(A) and (B)). Since its introduction, many modifications and variations of the Y2H system have been developed to make it easier to use and improve its efficacy, but the general principle of Y2H systems remains the same.

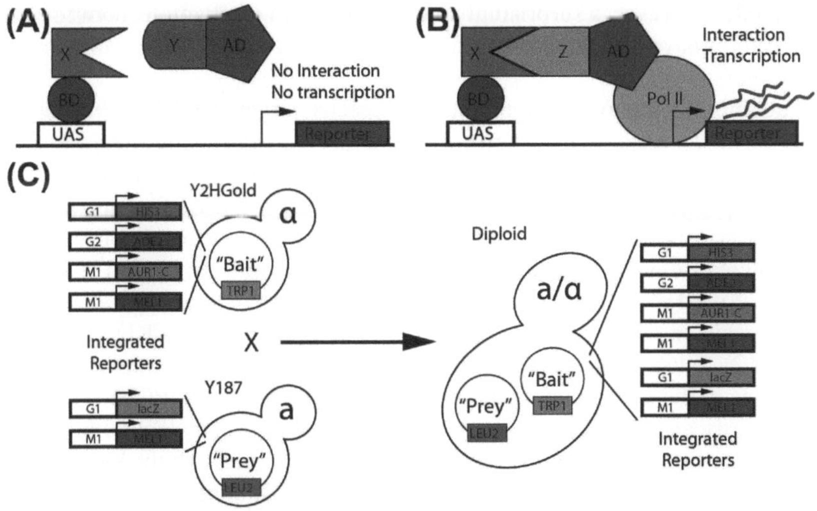

FIGURE 3 Schematic of the Y2H system.

(A) Protein X is fused to the Gal4 BD. Protein Y is fused to the Gal4 AD. Since protein X does not interact with protein Y, no transcription is driven from the reporter gene. (B) Protein X is fused to the Gal4 BD protein Z is fused to the Gal4 AD. Since protein X and Z interact, a functional transcription factor is generated and reporter gene transcription occurs. (C) Diagram of the mating based Matchmaker Gold Y2H system (Clonetech) showing the integrated reporter genes in the indicated yeast strains. Refer to Section 4.6.1, Step 10 for details.

3.2 ADVANTAGES OF THE Y2H SYSTEM FOR COMPLEX, MULTICOMPONENT PROTEIN MACHINES

Testing interactions by the Y2H system has several advantages. Y2H requires little specialized equipment. Any lab capable of growing yeast cultures can perform it in its simplest forms. It does not require proteins to be purified to perform the assay, an advantage for studying challenging proteins, like those of the centrosome. The test for interaction by Y2H occurs in the nucleus of the yeast cells, so unless one is studying a protein normally native to the yeast nucleus, there is a reduced likelihood of detecting indirect interactions. Finally, with some modifications, Y2H is amenable to being used in a high-throughput setting, allowing a protein of interest to be tested for interactions with many proteins.

The Y2H system is well suited for studying complex multiprotein systems, because of its modularity and ease of use. In fact, several groups have used the Y2H to elucidate the interactions among proteins of whole organisms from viruses to yeast (reviewed in Roberts, Parrish, Mangiola, & Finley, 2012). In *Drosophila melanogaster*, our model system of choice, three large scale Y2H-based interactomes have been generated (Formstecher et al., 2005; Giot et al., 2003; Stanyon et al., 2004). However, a surprisingly small number of interactions between centrosome proteins have been identified. Below, we discuss a possible explanation for this low number.

In addition to genome-scale interaction screens, targeted Y2H screens have been critical for understanding the interactions that occur within multiprotein assemblies involved in a variety of cellular processes. Of significant interest to those who study the centrosome are the interactions determined among proteins of complex structures related in form or function to the centrosome. For example, a large proportion of centrosome proteins are predicted to form coiled-coils (Dos Santos et al., 2013). Therefore, studies that used a Y2H system to map coiled-coil protein interactions in *Saccharomyces cerevisiae* could be very insightful (Newman, Wolf, & Kim, 2000; Wang et al., 2012). Interestingly these screens identified a number of interactions among the proteins of the kinetochore, responsible for connecting MTs to the chromosomes during mitosis, and also among the proteins of spindle pole body (SPB), the functional equivalent of the centrosome in yeast. Like the centrosome, both the kinetochore and the SPB are coiled-coil rich organelles involved in regulating MT attachments. The protein–protein interactions of the kinetochore (Ikeuchi, Nakano, Kamiya, Yamane, & Kawarasaki, 2010; Ikeuchi, Sasaki, Kawarasaki, & Yamane, 2003; Shang et al., 2003; Wong et al., 2007) and the SPB (reviewed in Adams & Kilmartin, 1999; Elliott, Knop, Schlenstedt, & Schiebel, 1999; Schramm, Elliott, Shevchenko, & Schiebel, 2000; Schramm, Janke, & Schiebel, 2001) have been extensively probed by Y2H, leading to important discoveries about protein function, as well as providing critical information for understanding the role of the larger protein assembly. In *Drosophila*, the Y2H system has been successfully used to uncover extensive direct interactions between the PCM proteins Cnn, Asl, and Spd2 (Conduit et al., 2014). Extrapolating from this work, we predict that a massive number of centrosome interactions have yet to be discovered; investigating the importance of the Cnn–Asl–Spd2 and other interactions will be a challenging, yet an exciting endeavor.

3.3 LIMITATIONS OF THE Y2H SYSTEM

Like all approaches, there are several limitations to the Y2H method. The system generates false negatives that arise for a variety of reasons. In some cases the interaction between test proteins might place the BD and AD in a spatial configuration where they cannot form a functional unit, for example, due to steric hindrances. False negatives also arise when one or both of the test proteins are not present in the yeast nucleus due to protein instability or its failure to enter the nucleus. Finally, many protein–protein interactions are regulated by post-translational modification. These modifications will likely be absent in an Y2H assay and interactions that require them will be missed (reviewed in Bruckner et al., 2009; Rajagopala & Uetz, 2009).

The system can also produce false positives. Some proteins, when fused to either the AD or BD can activate transcription in the absence of an interaction partner. Fortunately, as discussed below, these proteins can be easily identified and the results discarded. Several approaches have been utilized to further reduce the number of false positive interactions that rely on increasing the stringency of the interaction. A widely used method is the addition of 3-Amino-1,2,4-triazole (3-AT), a competitive inhibitor of HIS3, to the yeast growth media. Increasing concentrations of 3-AT increases the amount of HIS3 that must be produced to support growth, thus reducing the background of spurious false positives (Durfee et al., 1993; Fields, 1993). Another method to reduce false positives is to use multiple reporters under the control of different promoters. The ability of the interaction to initiate transcription on multiple promoters increases the confidence in the detected interaction (Rajagopala & Uetz, 2009). As discussed below, we have found the use of multiple reporters greatly reduces the positive interaction hit rate. However, the inability to activate transcription on one promoter does not mean the interaction is necessarily false. Therefore, in the protocol described below, we suggest screening for interactions using different combinations of reporters. Testing an interaction using multiple stringencies can provide several layers of information to help eliminate false positives, while not missing weak, but significant interactions. We also recommend screening against a selection of noncentrosomal proteins, especially coiled-coil proteins, to serve as negative controls; this can help eliminate false positives due to "sticky" proteins. We stress here that the Y2H should be viewed as an initial guide to further secondary experimentation that could support, or refute a direct interaction. In the case of the centrosome, we strongly believe that the Y2H is a necessary first step because of the obstacles mentioned above. Once a Y2H interaction is identified, the proteins of interest should be tested for interaction in vitro (if possible) using purified components, or in vivo using Co-IPs. These studies can then guide genetic studies in vivo to truly show functional relevance of interactions. High standards and experimental rigor should naturally eliminate false positives.

4. PERFORMING AN ARRAY BASED SCREEN FOR CENTROSOME PROTEIN INTERACTIONS

In this chapter we will describe how to generate an arrayed library of centrosome proteins using protein subfragments, and how to then screen a single protein or fragment against the entire library. However, the principles described can easily be modified to scale down to testing single interactions, or scaled up with automation to test a larger set of proteins. While we use the proteins of the centrosome as an example, it is by no means limited to this organelle. Our approach should be useful to examine the interactions among any interconnected protein complex. To allow for the greatest degree of flexibility, we assume the reader has already generated their own list of proteins from which they wish to construct their library.

4.1 SELECTING A Y2H SYSTEM

Since its original development, many variations on the principle of the Y2H system have been developed, utilizing different fusion proteins, different reporters, and different host cell types. We direct the reader to a comprehensive review of Y2H systems that includes detailed discussion of their advantages and disadvantages (Bruckner et al., 2009). Some thought should be given to the system selected prior to beginning the screen. Of particular note is the finding that different systems used to probe the same set of proteins did not yield a completely overlapping set of interactions (Rajagopala, Hughes, & Uetz, 2009; Stanyon et al., 2004). This work also highlights that an interaction might only be detected when protein X is used as a bait and protein Y is used as a prey, but not vice versa. This is consistent with our experience as well. We therefore test as many protein–protein pairs as possible in both directions. We describe a procedure to test interactions using an approach centered on the Matchmaker Gold system (Figure 3(C), Clonetech, Mountain View, CA), but this protocol could be easily modified to accommodate other Y2H systems.

Since we are specifically interested in identifying how one centrosome protein can interact with other centrosome proteins, and not how it can interact with any protein per se, screening using an array-based Y2H system is perfectly suited for this application. In this type of screen, a collection of clones of known sequence is constructed in Y2H vectors. Bait and prey plasmids are separately transformed into yeast of opposite mating types. A collection of yeast strains of one mating type, each carrying a plasmid encoding a unique centrosome protein, are systematically arrayed in a manageable format, like that of a multiwell plate. Then, yeast of the opposite mating type, carrying the test Y2H vector, are mixed with the entire array. Yeast mating then brings the bait and prey plasmids together in the same diploid yeast cell. As a result each colony on the array contains a unique pair of proteins, which can be tested for interaction using reporter genes carried by the parental strains (reviewed in Cagney & Uetz, 2001; Rajagopala & Uetz, 2009; Uetz, 2002). This method accommodates testing a large number of combinations of potential interacting proteins in a systematic fashion and, as such, is ideally suited to understanding the interactions among proteins of a complex, an organelle and even a whole genome.

There are several advantages to performing a Y2H screen in an array-based format. Arrays allow for the immediate identification of the interacting proteins based on their position in the array. Unlike random library screening, plasmids from positive interactions do not need to be recovered and sequenced. This system can also be automated, allowing large sets of proteins to be systematically assayed. Finally, the array based format reduces the occurrence of certain types of false positives and makes the elimination of others easier due to the ease of retesting the interaction via a simple yeast mating (reviewed in Uetz, 2002).

4.2 DIVIDING PROTEINS INTO SMALLER FRAGMENTS

Several large scale Y2H studies have indicated that it is advantageous to examine subregions of proteins in addition to full-length proteins (Boxem et al., 2008; Flajolet et al., 2000; Flores et al., 1999; Formstecher et al., 2005; Fromont-Racine et al., 2000). There are several advantages to including smaller protein fragments in a

Y2H screen. One is that it can separate regions of proteins problematic for use in Y2H from the remainder of the protein. This might include regions that contain elements toxic to the yeast or that can activate transcription of the Y2H reporter genes in the absence of an interaction (autoactivation). When separated, regions of the protein that do not have these issues can still be used to test for interactions. Another advantage is that subfragments might uncover interactions not revealed by full-length proteins. The studies referenced above indicate that some full-length proteins, while expressed, do not yield interactions, while subfragments of the same proteins do. Our experience confirms this observation. A fragment might outperform full-length proteins for a number of reasons. For example, full-length proteins might not fold correctly in yeast, precluding proper interaction with its partner. It is also possible that the BD or AD may be placed in a position causing steric hindrance to protein–protein interaction or to transcriptional activation. The failure of full-length proteins to interact may also be reflective of their biology. For example, many proteins are regulated via intramolecular autoinhibition mechanisms that restrict access to portions of themselves until a specific cellular event occurs. Thus, dividing a protein into smaller fragments might alleviate this autoinhibition and expose a necessary binding domain or surface. The use of full-length proteins might be a factor contributing to the limited number of centrosome protein interactions identified in genome wide Y2H screens mentioned above. We therefore recommend that proteins be assayed for interaction as both full-length and as small protein fragments, if possible.

We suggest a rational, structure-based (existing or predicted) approach to subdividing proteins prior to use in Y2H screens. For each centrosome protein we first determined if any structures of the protein has been solved. In the absence of existing structural information, we perform secondary and tertiary protein structure predictions utilizing two available structure prediction servers, Jpred3 and Phyre2 (Cole, Barber, & Barton, 2008; Kelley & Sternberg, 2009). We then screen the protein for known structural or functional motifs using the Simple Modular Architecture Research Tool (SMART) web server (Letunic, Doerks, & Bork, 2014). Finally, since centrosome proteins are rich in sequences predicted to participate in the formation of coiled-coils, we use the COILS web server to predict such regions (Lupas, Van Dyke, & Stock, 1991). With this information in hand we divide these proteins into smaller fragments with the least disruption to the above features. As an alternative, several groups referenced above describe screening protocols where a protein of interest is screened against a collection of protein fragments that have been randomly generated prior to screening.

4.3 GENERATING THE Y2H LIBRARY

Commercial Y2H systems provide vectors that contain multiple cloning sites allowing for restriction enzyme based cloning. To reduce the labor in producing an array of protein fragments, bait and prey vectors modified to accommodate cloning techniques more conducive for use in high-throughput circumstances can be used. One such modification was to make the Y2H vectors pGBKT7 and pGADT7 compatible with the Gateway cloning system (Rossignol, Collier, Bush, Shaw, & Doonan, 2007; Life Technologies, Grand Island, NY). Our lab has further modified the Gateway compatible pGBKT7 vector by replacing the kanamycin resistance cassette with one providing resistance against ampicillin so that it could be used with Gateway Entry clones (Galletta et al., 2014). Sequences encoding the fragments should be generated by PCR and then cloned into Entry vectors. After verification by DNA sequencing, Gateway recombination reactions are performed to transfer these sequences into bait (pGBKT7) and prey (pGADT7) vectors. Other cloning systems can also be used, such as plasmid construction by homologous recombination in yeast.

As discussed above, bait and prey plasmids carried in yeast of opposite mating type are used to introduce pairs of proteins into the same yeast by mating. For this procedure, bait plasmids (pGBKT7) are transformed into the Y2HGold yeast strain, a MAT**a** strain, and prey (pGADT7) into the Y187 yeast strain, a MATα strain. Single colonies of each are selected, propagated, and stocks of each bait in Y2HGold and each prey in Y187 are generated.

4.4 AUTOACTIVATION AND FALSE POSITIVE RATE IDENTIFICATION

A common limitation to testing protein interactions by Y2H is that some protein fragments, when introduced into the system, can activate the Y2H reporters in the absence of any binding partner. While this is more commonly a problem with fragments fused to the GAL4-BD (bait), this can occur in GAL4-AD (prey) fusions as well (Serebriiskii & Golemis, 2001). Prior to use in testing interactions, all strains carrying Y2H vectors should be tested for autoactivation by first generating "empty strains" (prey-empty = Y187 containing empty pGADT7; bait-empty = Y2H; gold containing empty pGBKT7). Then, all newly generated stocks should be crossed to these "empty stains" and tested for reporter activity as described below. If the process is performed in a more high-throughput fashion, these "empty stains" should be placed into the array. Plates showing a positive signal in these control wells can be eliminated from the analysis. Protein fragments that cause autoactivation cannot be used in interaction testing. It is unlikely, however, that a protein will autoactivate as both prey and bait, and it is also unlikely that all subfragments of the same proteins cause autoactivation. Therefore, in a properly designed screen, little information is lost or missed due to autoactivation events. If this proves to be a major problem, we suggest repeating the process by designing, generating, and testing new protein subfragments. Another useful control is to test for interactions against random, noncentrosome proteins, to serve as a control for particularly sticky proteins. Since so many proteins in the centrosome contain coiled-coils and disordered regions, proteins with these types of regions should be included.

4.5 PRESCREEN PLANNING

Depending on the number of interactions that one will test, screening using an array Y2H based system will involve growing a large number of yeast strains, moving and mixing a large number of cultures in a very specific order, dealing with many yeast plates and keeping track of a large data set on the backend. Prescreen planning to ensure the availability of the correct amount of media, number of tubes and plates, etc. can help ensure the fewest missteps on the day of the screen. Tables 1 and 2 list the yeast media and equipment required for the screen we describe.

Some thought should also be given to the design of the array. We recommend including blank wells on each half of the 96-well array, in different positions, to aid in plate orientation (see below). We also recommend including a well for each of the "empty stains" to function as negative controls on each plate; they will serve as a readout of any unexpected autoactivation that might occur on the day of the experiment. Finally, this type of screen will yield a large data set that must be tracked and organized. We recommend building a database, using software such as Filemaker Pro (FileMaker, Inc., Santa Clara, CA) or Access (Microsoft, Redmond, WA), to track all the data.

4.6 SCREENING FOR INTERACTIONS

Here we describe the details for how to screen a single protein or protein fragment, against the collection of centrosome proteins generated above. As discussed, some protein–protein interactions only occur when protein X serves as the bait and

Table 1 Yeast Media Recipes

SD media base	5 g ammonium sulfate, 3.4 g yeast nitrogen base without amino acids and ammonium sulfate, 20 g of D-glucose, appropriate DO mix, water to 1 L. Add 20 g agar/L for plates.
Complete amino acid mix	800 mg adenine, 2 g arginine, 800 mg histidine, 2 g isoleucine, 4 g leucine, 2 g lysine, 800 mg methionine, 2 g phenylalanine, 8 g threonine, 2 g tryptophan, 2 g tyrosine, 800 mg uracil, 6 g valine
-leu drop out (DO) mix	CAA, omit leucine—use 730 mg/L
-trp DO mix	CAA, omit tryptophan—use 780 mg/L
-leu -trp DO mix	CAA, omit leucine and tryptophan—use 680 mg/L
-ade -leu -trp -ura DO mix	CAA, omit adenine, leucine, tryptophan, and uracil—use 640 mg/mL
SD-leu	SD media base, -leu DO mix
SD-trp	SD media base, -trp DO mix
DDO	SD media base, -leu -trp DO mix
QDO	SD media base, -ade -leu -trp -ura DO mix
DDOXA	SD -leu -trp DO mix, Aureobasidin A (Clonetech, Mountain View, CA), X-α-Gal (Clonetech, Mountain View, CA and Gold Biotechnology, St Louis, MO)
QDOXA	SD -ade -leu -trp -ura DO mix, Aureobasidin A, X-α-Gal
2× YPAD	20 g tryptone, 10 g yeast extract, 80 mg adenine, 20 g D-glucose, water to 1 L

Table 2 Materials Needed

Item	Commercial Example (Company, Catolog#)
Culture tubes	Falcon—352059
96-well flat bottom dishes	Olympus—25-104
96 deep well plate	Axygen—P-2ML-SQ-C-S
Reagent reservoirs	Corning—4870
Multichannel pipettor	Gilson—F14404
48 pin multiblot replicator	V&P Scientific—VP 407AH
Petri dishes	Fisher Scientific—FB0875712
Replica-plating tool	Bel Art Products—378480000
Sterile velvet squares	Bel Art Products—378480002

protein Y as the prey, and not vice versa. Therefore, one must test any protein of interest against a bait array and a prey array to maximize the number of interactions identified. For simplicity, we will describe testing the centrosome protein encoded by **Y**our **F**avorite **G**ene in a prey plasmid (YFG-prey) against a collection of strains carrying bait plasmids encoding a variety of centrosome proteins in an array format (ARRAY-bait). However, we strongly suggest that the tests also be done in the opposite direction (YFG-bait against ARRAY-prey). Figure 4 shows the steps involved in the screening procedure.

4.6.1 Protocol

1. Grow fresh cultures of all yeast strains to be tested. Inoculate liquid cultures of yeast carrying Y2H plasmids for the array (ARRAY-bait), as well as for the protein or fragment to be tested (YFG-prey), at 30 °C with shaking in synthetic defined (SD)-leu media or SD-trp media, as appropriate to maintain plasmid selection. This can be done in individual culture tubes or directly in a 96-well format using a deep well plate, although the latter may not be optimal for yeast growth. Grow to optical density $(OD)_{600} \approx 0.5$. Some strains may grow faster than others. Generally this takes
   1–3 days. It may be useful to estimate the growth rate of the strains prior to starting. Then the time of growth for individual strains can be adjusted so that all strains reach the desired OD_{600} at approximately the same time.
2. Array the ARRAY-bait cultures by transferring 20 μL of each into a single well of a 96-well, flat bottom plate. If more than one YFG-prey strain is to be tested against the array, it is useful to set up the ARRAY-bait in a master plate (using a deep well, 96-well plate if necessary) and then use a multichannel pipette to transfer the array to multiple, identical ARRAY-bait plates.
3. In a sterile reagent reservoir, mix 2 mL of YFG-prey culture with 10 mL of 2× yeast-extract peptone adenine dextrose (YPAD) media.
4. Using a multichannel pipette, transfer 120 μL of the YFG-prey/2× YPAD mixture into each well of the 96-well ARRAY-bait plate. Mix by pipetting up and down a few times. This is now referred to as the mating-plate.

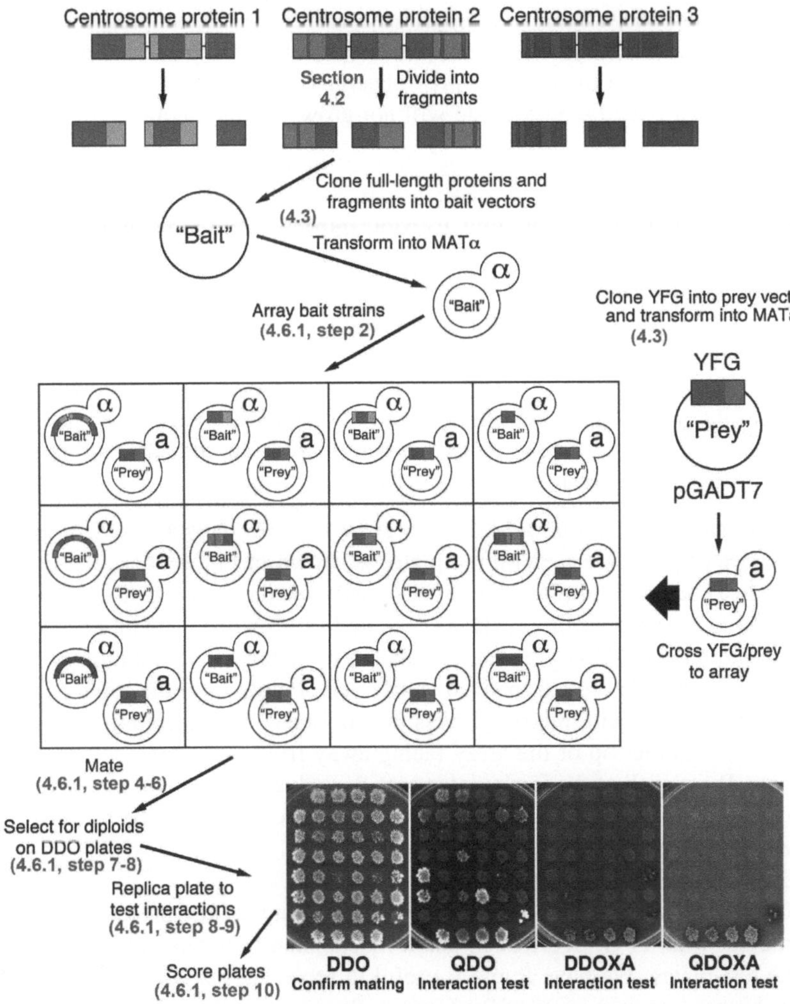

FIGURE 4 Workflow of an array-based Y2H screen for centrosome–protein interactions.

Refer to sections indicated on the figure for details describing each step.

5. Repeat Steps 3–4 until all YFG-prey samples have been crossed with the ARRAY-bait.
6. Grow mating-plates for 20–24 h at 30 °C with shaking to allow the yeast to mate. The success of the mating reaction can be assayed by examining a small sample of the culture for the presence of zygotes by phase contrast microscopy, although this is usually not necessary.
7. Transfer approximately 3 μL of each mating culture from the mating-plate onto double dropout (DDO) plates. This can be facilitated using a 48-pin multiblot replicator (VP 407AH, V&P Scientific, San Diego, CA). In this case, the

cultures from one 96-well mating-plate are transferred as two 48-sample halves to each of two DDO plates. These plates will select for growth of diploids that have received both the bait and prey plasmids from their parents. Parental haploids that have failed to mate will not grow on this media.

Sterilize the replicator before each use by immersing the pins into a dish of ethanol or isopropanol. Gently shake off excess and place the pins in the flame of a Bunsen burner. Allow the pins to cool. Introduce the replicator into one half of the 96-well mating-plate and swirl it in the media to ensure the yeast is evenly suspended. Remove the replicator from the mating-plate, taking care not to touch the sides of the wells. Gently set the replicator down onto the surface of a DDO plate, taking care to not let the replicator slide laterally. Lift the replicator off the plate, leaving ~3 μL of culture behind. Place the replicator back in the dish with alcohol. Repeat for the other half of the 96-well mating-plate. Mark each DDO plate so that the orientation relative to the array can be determined. These plates will be referred to as diploid-plates. Repeat for all mating-plates.

8. Allow the yeast on diploid-plates to grow for 3–5 days at 30 °C until robust patches of yeast are seen on the plate, as shown in Figure 4.
9. Replica each diploid-plate onto DDO, quadruple dropout (QDO), DDO with X-α-Gal and Aureobasidin A (DDOXA), and QDO with X-α-Gal and Aureobasidin A (QDOXA) plates, all labeled to match the orientation of the diploid-plate. To replica, place a sterile velvet cloth onto the replica plating tool and secure with the ring. Press the surface of the diploid-plate onto the velvet, with the top of the array facing away from you. Remove the diploid-plate. Press each of the fresh plates onto the velvet and remove to make a copy. These new DDO, QDO, DDOXA, and QDOXA plates will be referred to as test-plates. Repeat for all diploid-plates.
10. Grow test-plates for 5 days at 30 °C.
11. Test-plates can now be scored to determine if any of the proteins in the array interact with YFG. Score each patch independently for its growth on each of the test-plates. We have found it useful to score the result of protein pair on each test plate on a scale of 0–3, where 0 = no growth, 1 = minimal growth/color, 2 = moderate growth/color, and 3 = robust growth/color. The plates are scored as follows.

DDO—Media lacks leucine and tryptophan, which selects diploids carrying both bait and prey plasmids. Ensures that replica plating was successful at all positions.

QDO (two growth interaction reporters, Figure 3(C))—Scored for growth. Media lacks leucine and tryptophan, which maintains selection for the bait and prey plasmids. Growth on this media, which lacks histidine and adenine indicates activation of the *HIS3* and *ADE2* Y2H reporters, respectively, and indicates a bait–prey interaction.

DDOXA (two drug interaction reporters, Figure 3(C))—Scored for growth and development of blue colony color. Media lacks leucine and tryptophan, which maintains selection for the bait and prey plasmids. Growth on this media, which contains the antibiotic agent Aureobasidin A indicates the activation of *AUR1-C* Y2H reporter. Development of a blue color on this media, which contains X-α-Gal, indicates the activation of the *MEL1* Y2H reporter. Activation of both these reporters indicates a bait–prey interaction.

QDOXA (two growth interaction reporters, two drug reporters, Figure 3(C))—Scored for growth and development of blue colony color. Media lacks leucine and tryptophan, which maintains selection for the bait and prey plasmids. This media lacks histidine and adenine, and contains Aureobasidin A and X-α-Gal. The growth and development of the blue color requires activation of the *ADE2*, *HIS3*, *AUR1-C*, and *MEL1* Y2H reporters and indicates an interaction under the most stringent conditions.

4.7 INTERPRETING SCREENING RESULTS

As discussed above, the yeast strains used in this Y2H system carry multiple reporters driven by different promoters. Each of these reporters should have subtle differences in the false positives they yield and when used in combination they should reduce the incidence of false positives. The plates used in the protocol test for activity of these reporters in different combinations. QDO plates are similar to the plates used historically in many Y2H screens. We have found that these plates show a much greater number of interactions than the other plates. In our experience, of the centrosome protein pairs that show an interaction on QDO, only about 60% of these pairs show growth on DDOXA and only 50% show growth on QDOXA (Galletta & Rusan, unpublished observation). This is consistent with an increased stringency with additional promoters and likely a significant elimination of false positives.

Once the initial screen is scored, all pairs showing an interaction should be retested by taking the original yeast stocks and preforming small scale mating assays to validate positive interactions. This simple retesting will eliminate a significant number of false positives (Rajagopala & Uetz, 2009; Uetz, 2002). The interactions identified can then be used in combination with biochemical, cellular biological, and other approaches to truly determine protein function. One particularly powerful use of the information gained in this type of screen is to guide a genetic approach to identify mutations to disrupt specific protein–protein interactions.

5. GENERATING SPECIFIC, SEPARATION OF FUNCTION MUTATIONS BY REVERSE Y2H

Mutations are powerful tools for elucidating protein function. Even more powerful are mutations that specifically disrupt the interaction between a protein and only one of its binding partners. It is critical to note that any mutation, even a single point mutation, has the potential to disrupt more than one interaction. This is especially a concern in a complex, multiprotein structure like the centrosome, which is highly interconnected. However, with the knowledge obtained from the interaction studies described in the previous section it is possible to generate mutations that disrupt specific subsets of interactions, and possibly exclusively a single interaction. In this section we describe how to generate such a mutant by a reverse-Y2H approach.

5.1 RATIONALE

The reverse two-hybrid approach used here is based on the method described by Bennett, Shern, and Kahn (2004) with significant modifications. This method utilizes low-fidelity PCR to introduce random mutations into DNA encoding a protein of interest. The mutagenized DNA is then cloned into the Y2H vectors directly in the Y2H strains by homologous recombination mediated repair. These mutant alleles can then be screened to identify ones that disrupt a known interactor.

The major modification we have made is to adapt the procedure for use in a mating-based, arrayed format. Similar to Bennett et al. (2004), we generate random mutations in the sequence encoding YFG by low-fidelity PCR and use homologous recombination mediated repair to clone the mutated YFG fragments. However, instead of cotransforming the mutated-YFG with a plasmid encoding the interaction partner being tested against, we perform the recombination in a haploid Y2H strain without its interaction counterpart. The YFG mutants are then clonally collected and put into an array. Once the YFG mutant array is generated, it can be tested for the loss of interactions by mating the array to Y2H strains carrying plasmids encoding the interacting protein of interest to identify mutations that abolish the interaction.

Performing the screen in the fashion described below has several advantages over cotransforming random mutants with their interaction partner. Most significantly, to ensure that the generated mutation only disrupts a specific protein–protein interaction of interest, a candidate clone can easily be pulled from the master array and tested for its ability to interact with all interaction partners. There is no need to first isolate the mutant plasmid from yeast, a labor-intensive process. This secondary screen allows the investigator to eliminate nonspecific mutants simply by performing additional yeast matings. The investigator would only recover the few mutants that fit the desired criteria. This method saves a significant amount of time and effort. A workflow diagram of the mutagenesis and screen is shown in Figure 5.

5.2 GENERATING MUTANT LIBRARY AND SCREENING FOR LOSS OF INTERACTION

To facilitate the use of this system with any protein or fragment of interest we have designed universal primers that allow amplification from the Y2H vectors (pGADT7 and pGBKT7) generated in Section 4.3 above (Table 3). PCR products of putative YFG mutants are cloned by cotransforming them into the Y2H yeast strains with

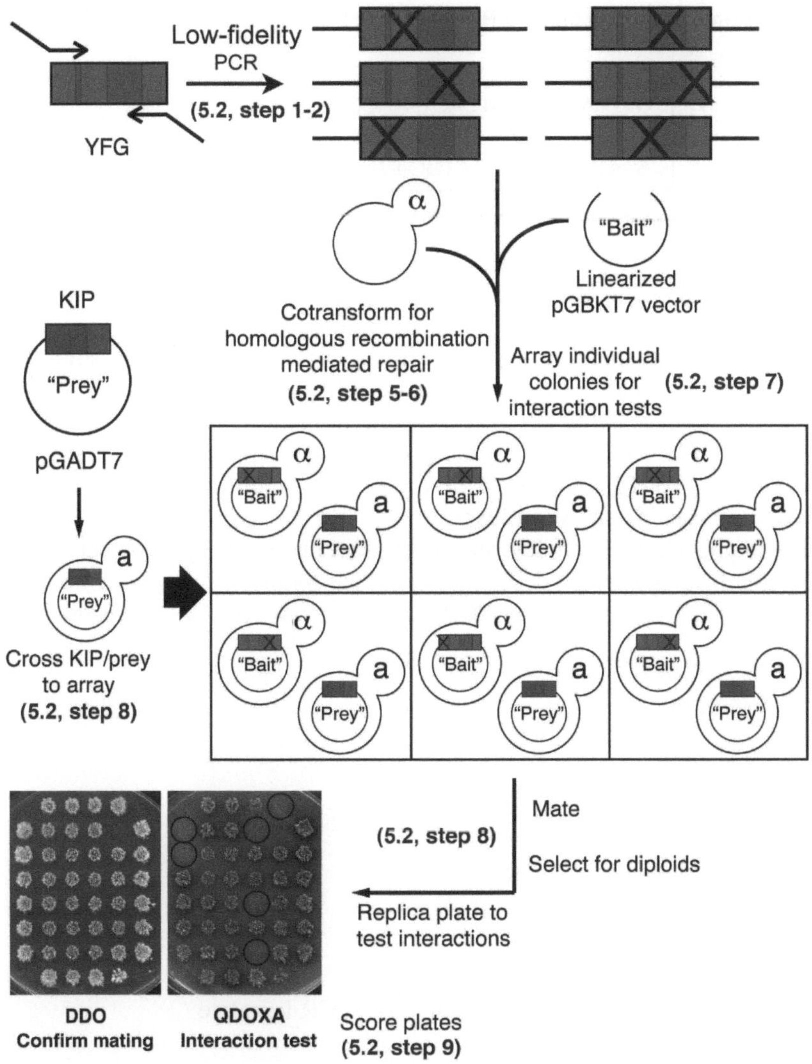

FIGURE 5 Workflow of an array-based reverse Y2H screen to generate and identify mutations that disrupt protein–protein interactions.

Refer to sections indicated on the figure for details describing each step. Red (gray in print versions) X's represent random point mutations introduced by mutagenic PCR. Yeast colonies carrying mutations that disrupt the protein–protein interaction are circled in red (gray in print versions).

Table 3 Primers for Low Fidelity PCR

Primer Name	Sequence
T7 sequencing	5′-TAATACGACTCACTATAGGGCG-3′
pGBKT7 Mut	5′-CGGAATTAGCTTGGCTGC-3′
pGADT7 Mut	5′-ATGGTGCACGATGCACAG-3′

linearized Y2H vectors and then selecting for the plasmid. For simplicity, we describe mutagenizing YFG and cloning it into the bait vector (pGBKT7) in the bait Y2H strain (Y2HGold). An array of YFG mutants is then mated to a Known Interacting Protein (KIP) in a prey vector (pGADT7) in the prey strain (Y187) and screened to identify mutations that disrupt the YFG/KIP interaction. While we describe mutagenizing a bait and testing it against a prey, this process works equally well when mutagenizing the prey. Simply replace the primer "pGBKT7 Mut" with "pGADT7 Mut" listed in Table 3 for amplification and switch to the appropriate plasmids and yeast hosts.

The mutagenic PCR we describe generates a mutation approximately every 250 base pairs. If mutations are desired more or less frequently, we direct the reader to studies focusing on low-fidelity PCR (Cadwell & Joyce, 1992; Wilson & Keefe, 2001).

5.2.1 Protocol

1. Mutagenic PCR mix:
   a. Taq polymerase
   b. 1× Taq polymerase buffer (supplied buffer by the manufacturer)
   c. 0.05 mM $MnCl_2$
   d. 0.06 mM dATP
   e. 0.25 mM dCTP
   f. 0.25 mM dGTP
   g. 0.25 mM dTTP
   h. YFG in pGBKT7 (from Section 4.3)—PCR template
   i. T7 Sequencing Primer
   j. pGBKT7 Mut Primer
2. The following conditions for PCR were used for the pGBKT7 primers to amplify a product of ~1 kb. Adjust conditions as necessary.
   a. 95 °C 2 min
   b. 95 °C 30 s
   c. 54 °C 30 s
   d. 72 °C 1 min
   e. Repeat Steps 2–4 for 30 cycles.
3. Gel purify mutant YFG PCR product.
4. Linearize pGBKT7 by restriction codigestion with *Eco*RI and *Pst*I (If mutagenizing prey clones, pGADT7 can be linearized by codigesting with *Eco*RI and *Xho*I.).

Gel purify linearized vector to ensure that there is no uncut plasmid present, as any will increase the background of clones that appear to lose interaction.

5. Cotransform equimolar amounts of the mutant YFG PCR product with the linearized pGBKT7 vector, for a total of 0.5–1 μg DNA, into Y2HGold. The exact amount of DNA needed will have to be determined empirically to yield optimal results. The goal is to find amounts that yield a plate full of colonies with adequate separation to allow individual colonies to be picked.
6. Plate on SD-trp plates to select for repaired plasmids containing mutant versions of YFG.
7. Load a 96-well plate with 100 μL/well SD-trp liquid media. Inoculate individual colonies from the plate in Step 6 into each well. Each well now contains a unique mutant version of YFG in pGBKT7 in Y2HGold. Grow at 30 °C with shaking for 1–2 days until $OD_{600} = \sim 0.5$. Pin a copy of the array onto solid SD-trp media, as in Section 4.6.1, Step 7; this will serve as a duplicate of the array.

Note: The number of mutant colonies that must be screened to identify a desired mutant cannot be known a priori and must be determined empirically. We have found that the number can vary dramatically. In some cases we have identified the desired mutation after screening only a few 100 mutants, others have taken several 1000, and others we have never been able to generate.

8. Transfer the YFG mutants from Step 7 into an array as described in Section 4.6.1 above. Follow Steps 2–9 above and mate the array with the KIP in pGADT7 in Y187.
9. Score plates. Ensure the presence of both the YFG mutant plasmid and the KIP plasmid by growth on DDO. Score colonies for interactions using QDO, DDOXA, and QDOXA plates. In this application, the experimenter is looking for colonies that grow on DDO, but show reduced or no growth on QDO, DDOXA, and QDOXA plates as compared with the results of unmutagenized YFG and KIP.
10. Recover all clones that displayed a loss of the YFG/KIP interaction from the YFG mutant array. Retest these against KIP. This should help eliminate certain types of false hits.
11. Screen the selected mutant YFG clones from Step 10 for their ability to interact with all other proteins known to interact with YFG. This is accomplished by crossing the YFG mutant clones to the known interactors in pGADT7 in Y187 generated in the original screen and testing them as described above. An ideal mutant allele would disrupt only the interaction with the desired KIP and maintain interaction with all other interacting proteins.
12. Once a clone harboring the desired mutation is found, the insert contained in the clone can be recovered by performing colony PCR (Sambrook & Russell, 2006; Sathe, O'Brien, McLaughlin, Watson, & Livi, 1991) using the same primers used for mutagenesis. This PCR product can be sequenced to identify the mutations it harbors. These mutations can be engineered back into the sequence encoding the fragment to confirm they are causative of the loss of interaction.

SUMMARY

Many important cellular functions rely on large, multiprotein assemblies. In order to truly understand the function of these complexes, and the functions of their constituent parts, an understanding of the connections among these proteins is critical. This is especially true for the centrosome, which as a nonmembrane bound organelle is, in many respects, a truly enormous and highly interconnected protein complex. As discussed above, there are numerous challenges to understanding of the protein–protein interactions within a complex like the centrosome. The Y2H system is a powerful tool for probing direct protein–protein interactions within complexes. It allows the experimenter to identify interactions within the structure that might not be accessible using other techniques, such as low-affinity and transient interactions.

On the simplest level, interaction information can provide an understanding of how the proteins of the complex fit together. But beyond this, interaction information can be critical to direct experiments to probe function. Mutagenesis is one of the most powerful tools used to understand protein function in the cell. However, multiprotein complexes present special challenges to interpreting the results of these studies. The potential interconnectedness means that complete loss-of-function mutations might alter many protein–protein interactions within the complex. However, once the interactions within a complex are known, the consequence of a given mutation on all of the interactions can be assessed. This can allow increased confidence that the consequence of a mutation is the result of disruption of a specific interaction. Interaction data can also guide reconstitution and structure determination experiments. As discussed above, the proteins of the centrosome are frequently challenging to working with in vitro. Proteins functioning in complexes can sometimes not fold correctly in the absence of their partners. The Y2H data from a screen like the one described will provide an excellent resource to identify pairs or groups of proteins with which to begin copurifying proteins together with their partners. The data from this screen could also help identify interactions critical for building a structure, but not necessarily present in the final structure. In addition, Y2H of fragments can also reveal regulated interactions, for example, interactions involving domains that are masked in the context of the full-length protein until some event uncovers them. Finally, Y2H screens can reveal interactions that are important in certain contexts that are not important in others. For example, it might identify interactions that take place in mitotic centrosomes, but not in interphase centrosomes; or in basal bodies, but not in centrioles.

Y2H screens like this have the potential to yield large sets of interaction data. This leaves the task of understanding how the interactions we identify contribute to how the centrosome assembles and performs its functions. This will certainly involve integrating the Y2H interaction data with data generated by other techniques. For example, combining Y2H interaction data with proximity-mapping techniques, like the BirA biotin ligase system (Firat-Karalar, Rauniyar, Yates, & Stearns, 2014), and cross-linking studies, like those done by selective covalent crosslinking (S-CROSS) (Lukinavicius et al., 2013), will be a powerful approach to probe the landscape of protein assemblies within the centrosome. We also envision the Y2H method playing an important role in understanding the growing number of human congenital disease linked to mutations in centrosome and cilia proteins. Testing these mutants against arrays of centrosome proteins might reveal the interactions altered by these mutations. This type of detailed interaction information allows for the generation of separation of function mutations that only affect a single protein–protein interaction. This combination has the promise of becoming a powerful reverse genetic approach to understand the molecular details of protein function in the centrosome.

ACKNOWLEDGMENTS

We thank Alexander Kelly and Colleen Skau for critical reading of the manuscript. NMR is supported by the division of intramural research at the National Institutes of Health/NHLBI (1ZIAHL006104).

REFERENCES

Adams, I. R., & Kilmartin, J. V. (1999). Localization of core spindle pole body (spb) components during spb duplication in *Saccharomyces cerevisiae*. *The Journal of Cell Biology, 145*, 809–823.

Andersen, J. S., Wilkinson, C. J., Mayor, T., Mortensen, P., Nigg, E. A., & Mann, M. (2003). Proteomic characterization of the human centrosome by protein correlation profiling. *Nature, 426*, 570–574.

Angus, K. L., & Griffiths, G. M. (2013). Cell polarisation and the immunological synapse. *Current Opinion in Cell Biology, 25*, 85–91.

Balestra, F. R., Strnad, P., Fluckiger, I., & Gonczy, P. (2013). Discovering regulators of centriole biogenesis through sirna-based functional genomics in human cells. *Developmental Cell, 25*, 555–571.

Bennett, M. A., Shern, J. F., & Kahn, R. A. (2004). Reverse two-hybrid techniques in the yeast *Saccharomyces cerevisiae*. *Methods in Molecular Biology, 261*, 313–326.

Bettencourt-Dias, M., Hildebrandt, F., Pellman, D., Woods, G., & Godinho, S. A. (2011). Centrosomes and cilia in human disease. *Trends in Genetics, 27*, 307–315.

Bornens, M. (2012). The centrosome in cells and organisms. *Science, 335*, 422–426.

Boxem, M., Maliga, Z., Klitgord, N., Li, N., Lemmens, I., Mana, M., et al. (2008). A protein domain-based interactome network for *C. elegans* early embryogenesis. *Cell, 134*, 534–545.

van Breugel, M., Hirono, M., Andreeva, A., Yanagisawa, H.-, Yamaguchi, S., Nakazawa, Y., et al. (2011). Structures of sas-6 suggest its organization in centrioles. *Science (New York, NY), 331*, 1196–1199.

Bruckner, A., Polge, C., Lentze, N., Auerbach, D., & Schlattner, U. (2009). Yeast two-hybrid, a powerful tool for systems biology. *International Journal of Molecular Sciences, 10*, 2763–2788.

Cadwell, R. C., & Joyce, G. F. (1992). Randomization of genes by pcr mutagenesis. *PCR Methods and Applications, 2*, 28–33.

Cagney, G., & Uetz, P. (2001). High-throughput screening for protein–protein interactions using yeast two-hybrid arrays. *Current Protocols in Protein Science* (Chapter 19), Unit 19.6.

Cole, C., Barber, J. D., & Barton, G. J. (2008). The jpred 3 secondary structure prediction server. *Nucleic Acids Research, 36*, W197–W201.

Conduit, P. T., Richens, J. H., Wainman, A., Holder, J., Vicente, C. C., Pratt, M. B., et al. (2014). A molecular mechanism of mitotic centrosome assembly in drosophila. *Elife, 3*, e03399.

Cottee, M. A., Muschalik, N., Wong, Y. L., Johnson, C. M., Johnson, S., Andreeva, A., et al. (2013). Crystal structures of the cpap/stil complex reveal its role in centriole assembly and human microcephaly. *Elife, 2*, e01071.

Dammermann, A., Muller-Reichert, T., Pelletier, L., Habermann, B., Desai, A., & Oegema, K. (2004). Centriole assembly requires both centriolar and pericentriolar material proteins. *Developmental Cell, 7*, 815–829.

Delattre, M., Leidel, S., Wani, K., Baumer, K., Bamat, J., Schnabel, H., et al. (2004). Centriolar sas-5 is required for centrosome duplication in *C. elegans*. *Nature Cell Biology, 6*, 656–664.

Dobbelaere, J., Josue, F., Suijkerbuijk, S., Baum, B., Tapon, N., & Raff, J. (2008). A genome-wide RNAi screen to dissect centriole duplication and centrosome maturation in drosophila. *PLoS Biology, 6*, e224.

Dos Santos, H. G., Abia, D., Janowski, R., Mortuza, G., Bertero, M. G., Boutin, M., et al. (2013). Structure and non-structure of centrosomal proteins. *PLoS One, 8*, e62633.

Durfee, T., Becherer, K., Chen, P. L., Yeh, S. H., Yang, Y., Kilburn, A. E., et al. (1993). The retinoblastoma protein associates with the protein phosphatase type 1 catalytic subunit. *Genes and Development, 7*, 555–569.

Dzhindzhev, N. S., Tzolovsky, G., Lipinszki, Z., Schneider, S., Lattao, R., Fu, J., et al. (2014). Plk4 phosphorylates ana2 to trigger sas6 recruitment and procentriole formation. *Current Biology, 24*, 2526–2532.

Elliott, S., Knop, M., Schlenstedt, G., & Schiebel, E. (1999). Spc29P is a component of the spc110p subcomplex and is essential for spindle pole body duplication. *Proceedings of the National Academy of Sciences of the USA, 96*, 6205–6210.

Elric, J., & Etienne-Manneville, S. (2014). Centrosome positioning in polarized cells: common themes and variations. *Experimental Cell Research, 328*, 240–248.

Fields, S. (1993). The two-hybrid system to detect protein–protein interactions. *Methods, 5*, 116–124.

Fields, S., & Song, O. (1989). A novel genetic system to detect protein–protein interactions. *Nature, 340*, 245–246.

Firat-Karalar, E. N., Rauniyar, N., Yates, J. R. r., & Stearns, T. (2014). Proximity interactions among centrosome components identify regulators of centriole duplication. *Current Biology, 24*, 664–670.

Flajolet, M., Rotondo, G., Daviet, L., Bergametti, F., Inchauspe, G., Tiollais, P., et al. (2000). A genomic approach of the hepatitis c virus generates a protein interaction map. *Gene, 242*, 369–379.

Flores, A., Briand, J. F., Gadal, O., Andrau, J. C., Rubbi, L., Van Mullem, V., et al. (1999). A protein–protein interaction map of yeast RNA polymerase iii. *Proceedings of the National Academy of Sciences of the USA, 96*, 7815–7820.

Formstecher, E., Aresta, S., Collura, V., Hamburger, A., Meil, A., Trehin, A., et al. (2005). Protein interaction mapping: a drosophila case study. *Genome Research, 15*, 376–384.

Fromont-Racine, M., Mayes, A. E., Brunet-Simon, A., Rain, J. C., Colley, A., Dix, I., et al. (2000). Genome-wide protein interaction screens reveal functional networks involving sm-like proteins. *Yeast, 17*, 95–110.

Fu, J., & Glover, D. M. (2012). Structured illumination of the interface between centriole and peri-centriolar material. *Open Biology, 2*, 120104.

Galletta, B. J., Guillen, R. X., Fagerstrom, C. J., Brownlee, C. W., Lerit, D. A., Megraw, T. L., et al. (2014). Drosophila pericentrin requires interaction with calmodulin for its function at centrosomes and neuronal basal bodies but not at sperm basal bodies. *Molecular Biology of the Cell, 25*, 2682–2694.

Giot, L., Bader, J. S., Brouwer, C., Chaudhuri, A., Kuang, B., Li, Y., et al. (2003). A protein interaction map of drosophila melanogaster. *Science, 302*, 1727–1736.

Goshima, G., Wollman, R., Goodwin, S. S., Zhang, N., Scholey, J. M., Vale, R. D., et al. (2007). Genes required for mitotic spindle assembly in drosophila s2 cells. *Science, 316*, 417–421.

Hatch, E. M., Kulukian, A., Holland, A. J., Cleveland, D. W., & Stearns, T. (2010). Cep152 interacts with plk4 and is required for centriole duplication. *The Journal of Cell Biology, 191*, 721–729.

Hiraki, M., Nakazawa, Y., Kamiya, R., & Hirono, M. (2007). Bld10P constitutes the cart-wheel-spoke tip and stabilizes the 9-fold symmetry of the centriole. *Current Biology, 17*, 1778–1783.

Ikeuchi, A., Nakano, H., Kamiya, T., Yamane, T., & Kawarasaki, Y. (2010). A method for reverse interactome analysis: high-resolution mapping of interdomain interaction network in dam1 complex and its specific disorganization based on the interaction domain expression. *Biotechnology Progress, 26*, 945–953.

Ikeuchi, A., Sasaki, Y., Kawarasaki, Y., & Yamane, T. (2003). Exhaustive identification of interaction domains using a high-throughput method based on two-hybrid screening and pcr-convergence: molecular dissection of a kinetochore subunit spc34p. *Nucleic Acids Research, 31*, 6953–6962.

Jakobsen, L., Vanselow, K., Skogs, M., Toyoda, Y., Lundberg, E., Poser, I., et al. (2011). Novel asymmetrically localizing components of human centrosomes identified by complementary proteomics methods. *EMBO Journal, 30*, 1520–1535.

Kelley, L. A., & Sternberg, M. J. (2009). Protein structure prediction on the web: a case study using the phyre server. *Nature Protocols, 4*, 363–371.

Kemp, C. A., Kopish, K. R., Zipperlen, P., Ahringer, J., & O'Connell, K. F. (2004). Centrosome maturation and duplication in *C. elegans* require the coiled-coil protein spd-2. *Developmental Cell, 6*, 511–523.

Kim, T. S., Park, J. E., Shukla, A., Choi, S., Murugan, R. N., Lee, J. H., et al. (2013). Hierarchical recruitment of plk4 and regulation of centriole biogenesis by two centrosomal scaffolds, cep192 and cep152. *Proceedings of the National Academy of Sciences of the USA, 110*, E4849–E4857.

Kitagawa, D., Vakonakis, I., Olieric, N., Hilbert, M., Keller, D., Olieric, V., et al. (2011). Structural basis of the 9-fold symmetry of centrioles. *Cell, 144*, 364–375.

Lawo, S., Hasegan, M., Gupta, G. D., & Pelletier, L. (2012). Subdiffraction imaging of centrosomes reveals higher-order organizational features of pericentriolar material. *Nature Cell Biology, 14*, 1148–1158.

Leidel, S., Delattre, M., Cerutti, L., Baumer, K., & Gonczy, P. (2005). Sas-6 defines a protein family required for centrosome duplication in *C. elegans* and in human cells. *Nature Cell Biology, 7*, 115–125.

Leidel, S., & Gonczy, P. (2003). Sas-4 is essential for centrosome duplication in *C. elegans* and is recruited to daughter centrioles once per cell cycle. *Developmental Cell, 4*, 431–439.

Letunic, I., Doerks, T., & Bork, P. (2014). Smart: recent updates, new developments and status in 2015. *Nucleic Acids Research, 43*, D257–D260.

Lin, Y. C., Chang, C. W., Hsu, W. B., Tang, C. J., Lin, Y. N., Chou, E. J., et al. (2013). Human microcephaly protein cep135 binds to hsas-6 and cpap, and is required for centriole assembly. *EMBO Journal, 32*, 1141–1154.

Lukinavicius, G., Lavogina, D., Orpinell, M., Umezawa, K., Reymond, L., Garin, N., et al. (2013). Selective chemical crosslinking reveals a cep57-cep63-cep152 centrosomal complex. *Current Biology, 23*, 265–270.

Lupas, A., Van Dyke, M., & Stock, J. (1991). Predicting coiled coils from protein sequences. *Science, 252*, 1162–1164.

Mennella, V., Keszthelyi, B., McDonald, K. L., Chhun, B., Kan, F., Rogers, G. C., et al. (2012). Subdiffraction-resolution fluorescence microscopy reveals a domain of the centrosome critical for pericentriolar material organization. *Nature Cell Biology, 14*, 1159–1168.

Muller, H., Schmidt, D., Steinbrink, S., Mirgorodskaya, E., Lehmann, V., Habermann, K., et al. (2010). Proteomic and functional analysis of the mitotic drosophila centrosome. *EMBO Journal, 29*, 3344–3357.

Newman, J. R., Wolf, E., & Kim, P. S. (2000). A computationally directed screen identifying interacting coiled coils from *Saccharomyces cerevisiae*. *Proceedings of the National Academy of Sciences of the USA, 97*, 13203–13208.

O'Connell, K. F., Caron, C., Kopish, K. R., Hurd, D. D., Kemphues, K. J., Li, Y., et al. (2001). The *C. elegans* zyg-1 gene encodes a regulator of centrosome duplication with distinct maternal and paternal roles in the embryo. *Cell, 105*, 547–558.

Ohta, M., Ashikawa, T., Nozaki, Y., Kozuka-Hata, H., Goto, H., Inagaki, M., et al. (2014). Direct interaction of plk4 with stil ensures formation of a single procentriole per parental centriole. *Nature Communications, 5*, 5267.

Pelletier, L., O'Toole, E., Schwager, A., Hyman, A. A., & Muller-Reichert, T. (2006). Centriole assembly in *Caenorhabditis elegans*. *Nature, 444*, 619–623.

Qiao, R., Cabral, G., Lettman, M. M., Dammermann, A., & Dong, G. (2012). Sas-6 coiled-coil structure and interaction with sas-5 suggest a regulatory mechanism in *C. elegans* centriole assembly. *EMBO Journal, 31*, 4334–4347.

Rajagopala, S. V., Hughes, K. T., & Uetz, P. (2009). Benchmarking yeast two-hybrid systems using the interactions of bacterial motility proteins. *Proteomics, 9*, 5296–5302.

Rajagopala, S. V., & Uetz, P. (2009). Analysis of protein–protein interactions using array-based yeast two-hybrid screens. *Methods in Molecular Biology, 548*, 223–245.

Roberts, G. G., 3rd, Parrish, J. R., Mangiola, B. A., & Finley, R. L., Jr. (2012). High-throughput yeast two-hybrid screening. *Methods in Molecular Biology, 812*, 39–61.

Roque, H., Wainman, A., Richens, J., Kozyrska, K., Franz, A., & Raff, J. W. (2012). Drosophila cep135/bld10 maintains proper centriole structure but is dispensable for cartwheel formation. *Journal of Cell Science, 125*, 5881–5886.

Rossignol, P., Collier, S., Bush, M., Shaw, P., & Doonan, J. H. (2007). Arabidopsis pot1a interacts with tert-v(i8), an n-terminal splicing variant of telomerase. *Journal of Cell Science, 120*, 3678–3687.

Sakakibara, A., Ando, R., Sapir, T., & Tanaka, T. (2013). Microtubule dynamics in neuronal morphogenesis. *Open Biology, 3*, 130061.

Sambrook, J., & Russell, D. W. (2006). Analyzing yeast colonies by pcr. *Cold Spring Harbor Protocols, 2006*.

Sathe, G. M., O'Brien, S., McLaughlin, M. M., Watson, F., & Livi, G. P. (1991). Use of polymerase chain reaction for rapid detection of gene insertions in whole yeast cells. *Nucleic Acids Research, 19*, 4775.

Schramm, C., Elliott, S., Shevchenko, A., & Schiebel, E. (2000). The bbp1p-mps2p complex connects the spb to the nuclear envelope and is essential for spb duplication. *EMBO Journal, 19*, 421–433.

Schramm, C., Janke, C., & Schiebel, E. (2001). Molecular dissection of yeast spindle pole bodies by two hybrid, in vitro binding, and co-purification. *Methods in Cell Biology, 67*, 71–94.

Serebriiskii, I. G., & Golemis, E. A. (2001). Two-hybrid system and false positives. Approaches to detection and elimination. *Methods in Molecular Biology, 177*, 123–134.

Shang, C., Hazbun, T. R., Cheeseman, I. M., Aranda, J., Fields, S., Drubin, D. G., et al. (2003). Kinetochore protein interactions and their regulation by the aurora kinase ipl1p. *Molecular Biology of the Cell, 14*, 3342–3355.

Sonnen, K. F., Gabryjonczyk, A. M., Anselm, E., Stierhof, Y. D., & Nigg, E. A. (2013). Human cep192 and cep152 cooperate in plk4 recruitment and centriole duplication. *Journal of Cell Science, 126*, 3223–3233.

Sonnen, K. F., Schermelleh, L., Leonhardt, H., & Nigg, E. A. (2012). 3D-structured illumination microscopy provides novel insight into architecture of human centrosomes. *Biology Open, 1*, 965–976.

Spektor, A., Tsang, W. Y., Khoo, D., & Dynlacht, B. D. (2007). Cep97 and cp110 suppress a cilia assembly program. *Cell, 130*, 678–690.

Stanyon, C. A., Liu, G., Mangiola, B. A., Patel, N., Giot, L., Kuang, B., et al. (2004). A drosophila protein-interaction map centered on cell-cycle regulators. *Genome Biology, 5*, R96.

Uetz, P. (2002). Two-hybrid arrays. *Current Opinion in Chemical Biology, 6*, 57–62.

Wang, Y., Zhang, X., Zhang, H., Lu, Y., Huang, H., Dong, X., et al. (2012). Coiled-coil networking shapes cell molecular machinery. *Molecular Biology of the Cell, 23*, 3911–3922.

Wilson, D. S., & Keefe, A. D. (2001). Random mutagenesis by pcr. *Current Protocols in Molecular Biology* (Chapter 8), Unit 8.3.

Wong, J., Nakajima, Y., Westermann, S., Shang, C., Kang, J. S., Goodner, C., et al. (2007). A protein interaction map of the mitotic spindle. *Molecular Biology of the Cell, 18*, 3800–3809.

Zheng, X., Gooi, L. M., Wason, A., Gabriel, E., Mehrjardi, N. Z., Yang, Q., et al. (2014). Conserved tcp domain of sas-4/cpap is essential for pericentriolar material tethering during centrosome biogenesis. *Proceedings of the National Academy of Sciences of the USA, 111*, E354–E363.

CHAPTER

Genome-wide RNAi screens in S2 cells to identify centrosome components

15

Jeroen Dobbelaere

Max F. Perutz Laboratories, University of Vienna, Vienna Biocenter (VBC), Vienna, Austria

E-mail: jeroen.dobbelaere@univie.ac.at

CHAPTER OUTLINE

Introduction **280**
Centrosome Function and Structure **280**
Early Characterization of the Molecular Composition of Centrosomes **281**
Genetic Identification of Centriole Components in *Drosophila* and *C. elegans* and other Model Systems **281**
The First Genome-Wide RNA Interference Screen for Centrosomes **282**
Identification of the Complete Proteome of Purified Centrosomes **282**
Use of RNAi in *Drosophila* Cells to Study Genes Involved in MT Formation and Centrosome Function **283**
1. Methods **284**
1.1 Screen Design 284
1.1.1 Interphase versus mitotic centrosomes 284
1.1.2 Cell line selection and growing conditions 286
1.1.3 Cell staining to visualize centrosomes 286
1.1.4 Analysis 287
2. Genome-Wide Screen **288**
2.1 Library Selection 288
2.2 Controls 288
2.3 High-Throughput Cell Assay 288
2.4 Staining 289
3. Image Acquisition **291**
4. Scoring Phenotypes and Image Analysis **292**
5. Validation of Potential Hits **294**

Methods in Cell Biology, Volume 129, ISSN 0091-679X, http://dx.doi.org/10.1016/bs.mcb.2015.03.011

Discussion 294
Acknowledgments 296
References 297

Abstract

Centrosomes act as the major microtubule organizing centers in animal cells. To fully understand how the centrosome functions, a detailed analysis of its principal structural components and regulators is needed. Genome-wide RNA interference (RNAi) allows for comprehensive screening of all components. *Drosophila* tissue culture cells provide an attractive model for such screens. First, *Drosophila* centrosomes are similar to their human counterparts, but less complex. Thus, all major centrosome components are conserved and fewer redundancies apply. Second, RNAi is highly efficient in *Drosophila* tissue culture cells and, compared to RNAi in human cells, it is cost-effective. Finally, the availability of comprehensive libraries permits easy genome-wide screening of most of *Drosophila*'s 14,000 protein coding genes. In this paper, we present detailed instructions for designing, performing, and analyzing a genome-wide screen in *Drosophila* tissue culture cells to identify centrosome components using a microscopy-based approach.

INTRODUCTION

CENTROSOME FUNCTION AND STRUCTURE

In most of the higher eukaryotic cells, the centrosome functions as the major microtubule (MT) organizing center. As such, the centrosome contributes to diverse cellular processes, including vesicle transport, cell polarity, cell migration, and cell division (Arquint, Gabryjonczyk, & Nigg, 2014; Nigg & Raff, 2009). Furthermore, many regulators of cell signaling concentrate at the centrosome (Arquint et al., 2014). This suggests that besides its role in MT nucleation, the centrosome also functions as a scaffold, coordinating and controlling key cellular decisions.

Centrosomes consist of a cloud of MT organizing pericentriolar material (PCM) organized around a pair of centrioles. Centriole duplication is tightly regulated to ensure centrosomes duplicate only once per cell cycle. At the end of mitosis, the two centrioles disengage (Tsou & Stearns, 2006). Subsequently, during the S-phase of the following cell cycle, new centrioles form adjacent to the two mother centrioles. As cells enter mitosis, centrioles recruit additional PCM, a process called centrosome maturation. Mature centrosomes drive MT nucleation and form the poles of the mitotic spindle.

In nondividing cells, centrioles function as basal bodies at the base of cilia (Kim & Dynlacht, 2013). During ciliogenesis, centrioles migrate to the plasma membrane where they act as a template for the ciliary axoneme. Docking to the membrane occurs via centriolar appendages. While cilia are present in most vertebrate cells, *Drosophila melanogaster* and *Caenorhabditis elegans* possess cilia exclusively in sensory neurons and in the case of *Drosophila* also in sperm. Cilia defects only affect sensory functions (worms and flies) and male fertility (flies) but not the development of the organism (Nigg & Raff, 2009). Thus, *Drosophila* and *C. elegans* are ideal model systems to dissect the role of centrioles as centrosome organizers independent of their role as basal bodies (Vincensini, Blisnick, & Bastin, 2011).

To further elucidate centrosome structure and better understand how the centrosome functions at a molecular level building MT networks, a complete catalog of its structural and regulatory components is needed.

EARLY CHARACTERIZATION OF THE MOLECULAR COMPOSITION OF CENTROSOMES

In the early 1980s, several labs succeeded in purifying centrosomes from cell lines and analyzing their MT-nucleation capabilities (Hamaguchi & Kuriyama, 1982; Kuriyama & Borisy, 1981; Mitchison & Kirschner, 1984). Ultrastructural analysis of isolated centrosomes showed the presence of various substructures of the centriole and the presence of PCM. Although the raw material for a molecular characterization of the centrosome was therefore available, no complete genome had yet been sequenced to allow for conclusive identification of novel centrosome components. To circumvent this issue, the Alberts Lab injected gel-separated purified centrosomes into rabbits to obtain antibodies against components of the *Drosophila* centrosome (Kellogg, Field, & Alberts, 1989). This resulted in the identification of several novel centrosome components, although none of the core centriole or PCM components were identified.

Purification of centrosomes from cultured cells yields mostly naked centrioles or centrioles with little PCM because these cells are mostly in interphase. To obtain centrioles with full-sized mitotic PCM, the Agard Lab purified centrosomes from the *Drosophila* syncytial embryo, which undergo synchronous waves of mitosis (Moritz & Alberts, 1999). Biochemical characterization showed that these centrosomes are potent MT organizers. Furthermore, the researchers were able to demonstrate that MTs are nucleated within the PCM and not from centrioles. Although this analysis provided invaluable insights into how centrosomes nucleate MTs, only a few molecular components could be identified using this approach.

GENETIC IDENTIFICATION OF CENTRIOLE COMPONENTS IN *DROSOPHILA* AND *C. ELEGANS* AND OTHER MODEL SYSTEMS

Forward genetic screens in *Drosophila* and *C. elegans* were central to the identification of many critical centrosomal components and regulators. Mutational screens using Ethylmethansulfonat (EMS) or P-elements in *Drosophila* by the Glover Lab revealed Polo and Aurora-A kinase as key regulators of centrosome maturation and mitotic spindle assembly (Glover, Leibowitz, McLean, & Parry, 1995; Sunkel & Glover, 1988). Asp (Gonzalez et al., 1990) and later Cnn (Li et al., 1998; Megraw,

Li, Kao, & Kaufman, 1999), two components of the PCM organizing the MT cytoskeleton, were later discovered and have since been shown to cause neurological defects when mutated in human patients (Nigg, Cajanek, & Arquint, 2014). Two other components of the centrosome, pericentrin and γ-tubulin, were also identified using classical methods (Doxsey, Stein, Evans, Calarco, & Kirschner, 1994; Oakley & Oakley, 1989). Pericentrin was identified using antisera from scleroderma patients; it localizes to the interface between centrioles and PCM (Mennella et al., 2012). γ-Tubulin, originally identified in the fungus *Aspergillus nidulans* (Oakley & Oakley, 1989), is a key MT nucleator and component of the PCM (Moritz & Agard, 2001); it is also important for centriole duplication (Dammermann, Maddox, Desai, & Oegema, 2008; Kleylein-Sohn et al., 2007).

The key breakthrough in the centrosome field came with work performed in *C. elegans* in the late 1990s. Using the stereotypical division pattern of the *C. elegans* embryo and live microscopy, the White Lab was able to identify mutants with easily scored defects in mitotic spindle assembly due to centrosome dysfunction (Hamill, Severson, Carter, & Bowerman, 2002; Nigg et al., 2014; O'Connell, Leys, & White, 1998; O'Connell, Maxwell, & White, 2000). Most informative was the phenotype of a failure of centriole duplication. A centriole duplication defect is characterized by a normal first embryonic mitosis followed by monopolar spindle formation in both daughter cells during the second mitosis. This phenotype led to the identification of the kinase that drives centriole duplication, ZYG-1 (O'Connell et al., 2001). Such mutant analysis also yielded several other important centrosomal proteins, including SAS-5, SPD-2, and SPD-5 (for a review, see Fu, Hagan, and Glover (2015), Nigg et al. (2014), and Nigg and Raff (2009)).

THE FIRST GENOME-WIDE RNA INTERFERENCE SCREEN FOR CENTROSOMES

Combining the stereotypical division pattern in *C. elegans* with the emergence of tools for genome-wide RNA interference (RNAi), the Hyman Lab performed the first comprehensive screen for all factors required for embryonic cell division (Gonczy et al., 2000). This screen led to the identification of further key centrosomal proteins, including the machinery underlying centriole duplication (Fu et al., 2015; Nigg et al., 2014). Subsequent studies showed this centriole duplication pathway to be largely conserved across evolution (Strnad & Gonczy, 2008), highlighting the power of genome-wide approaches.

IDENTIFICATION OF THE COMPLETE PROTEOME OF PURIFIED CENTROSOMES

The availability of sequenced genomes since 1998 (Consortium, 1998) also allowed for the biochemical characterization of purified centrosomes. Mass spectrometry of purified centrosomes from vertebrate cultured cells, *Chlamydomonas* and *Drosophila*, resulted in the identification of numerous novel centrosome components, many of which remain yet to be characterized (Andersen et al., 2003; Jakobsen et al., 2011; Keller, Romijn, Zamora, Yates, & Marshall, 2005; Muller et al., 2010). This comprehensive dataset obtained in several model systems has enabled researchers to analyze the function of these centrosome components using genetics, RNAi, microscopy, and other techniques.

USE OF RNAi IN *DROSOPHILA* CELLS TO STUDY GENES INVOLVED IN MT FORMATION AND CENTROSOME FUNCTION

The ease of conducting RNAi in *Drosophila* tissue culture cells (Clemens et al., 2000) and the availability of a fully sequenced genome made it possible to carry out genome-wide screens in *Drosophila*. Using this technology to identify novel centrosome components and regulators in *Drosophila*, it was further facilitated by the large array of tools available to study centrosomes from previous research. To allow a clear identification of cell cycle, mitotic spindle, morphology, and centrosome defects, microscopy-based assays were developed that were compatible with RNAi. Using a visual cell-based assay allows both identification of novel genes and an initial functional characterization in the primary screen. The first microscopy-based screens in S2 cells used a subset of genes, including all kinases (Bettencourt-Dias et al., 2004) or all MT motors (Goshima & Vale, 2003) to screen for cell cycle and mitotic spindle defects or changes in cell morphology in a set of $\approx$1000 genes (Kiger et al., 2003). The first genome-wide screen (>90% of the *Drosophila* genes) in S2 cells analyzed cell growth and viability using a luciferase-based assay (Boutros et al., 2004). This paved the way for microscopy-based approaches to examine a variety of cellular processes (for an overview, see http://www.genomernai.org or http://www.flyrnai.org).

Cell-based visual screening was used to identify novel centrosome components and regulators in *Drosophila* (Dobbelaere et al., 2008; Goshima et al., 2007). To obtain a sufficient number of cells to analyze, both screens arrested cells in mitosis (PCM present) and used a microscopy-based screening method to analyze defects in centrosome formation. The PCM markers γ-tubulin or centrosomin (Cnn) were used to identify both PCM and, indirectly, centriole defects. Although the markers used were different, similar results were obtained. Thus, proteomic and RNAi screening enabled us to obtain a near complete list of centrosome components and regulators in *Drosophila*. RNAi screens identified novel components like Ana1, Ana2, and Ana3 and yielded Plk4/Sak and other homologs of *C. elegans* proteins that could not be identified by homology (Bettencourt-Dias et al., 2004; Dobbelaere et al., 2008; Goshima et al., 2007; Stevens, Dobbelaere, Wainman, Gergely, & Raff, 2009). Recently, a similar screen in human cells focusing on centriole duplication was completed using the centriole marker centrin (Balestra, Strnad, Fluckiger, & Gonczy, 2013). Thus, RNAi screening in *C. elegans*, *Drosophila*, and human cells was instrumental in creating a comprehensive list of centrosomal components and regulators.

This Chapter provides an overview of how to set up a genome-wide microscopy-based screen. Detailed protocols are provided and important intermediate steps and potential pitfalls are highlighted. Protocols in this review are based on previous centrosome screen (Dobbelaere et al., 2008).

1. METHODS

1.1 SCREEN DESIGN

While the goal of a genome-wide screen is to identify novel components in a particular molecular pathway, detailed prior knowledge is invaluable to set up an assay that is easy to perform and will be successful in identifying novel components. From design and setting up the assay, to performing the screen, analyzing and confirming results takes two to three years of work. A robust assay and careful planning are therefore key for the success of a genome-wide screen.

1.1.1 Interphase versus mitotic centrosomes

The centrosome cycle in *Drosophila* tissue culture is similar to that in vertebrate cells (Nigg & Raff, 2009). After cell division, each daughter cell inherits one pair of centrioles. These centrioles separate and each parental centriole promotes the formation of one new daughter in S-phase to once again end up with two pairs of centrioles in mitosis. One difference to vertebrates is that PCM accumulates around centrioles in substantial quantities only in mitosis (Dobbelaere et al., 2008). PCM in *Drosophila* tissue culture cells only forms around centrioles. Thus, the presence of PCM foci can be used to screen for PCM formation and also serves as an indicator for the presence of centrioles. It should be noted that the number of centrioles cannot be accurately determined in this way. To obtain sufficient protein knockdown by RNAi, depletion in Drosophila cells is usually done for 3 or 4 days. This means that during an RNAi experiment most cells will undergo three to four rounds of cell divisions (S2 and S2R+ cells divide every 20–24 h at 25 °C). Depletion of proteins required for centriole duplication typically leads to a reduction in the number of centrioles to ~25% in this time frame (Bettencourt-Dias et al., 2005; Rodrigues-Martins et al., 2007). When visualizing a PCM marker, a reduced number of centrioles is shown by monopolar spindles with one PCM dot only. The advantage of using PCM proteins compared to centriolar markers is that the PCM is much bigger and therefore easier to stain and visualize. Two PCM components, γ-tubulin (Goshima et al., 2007) and Cnn (Dobbelaere et al., 2008), were used to visualize the PCM in *Drosophila* cells and enable the identification of centriole and PCM components.

Under normal conditions only 1–2% of the *Drosophila* tissue cells are in mitosis. To be able to assess all aspects of centrosome defects after RNAi, therefore, requires mitotic cells to be enriched. To obtain reliable automated scoring, 100–300 mitotic

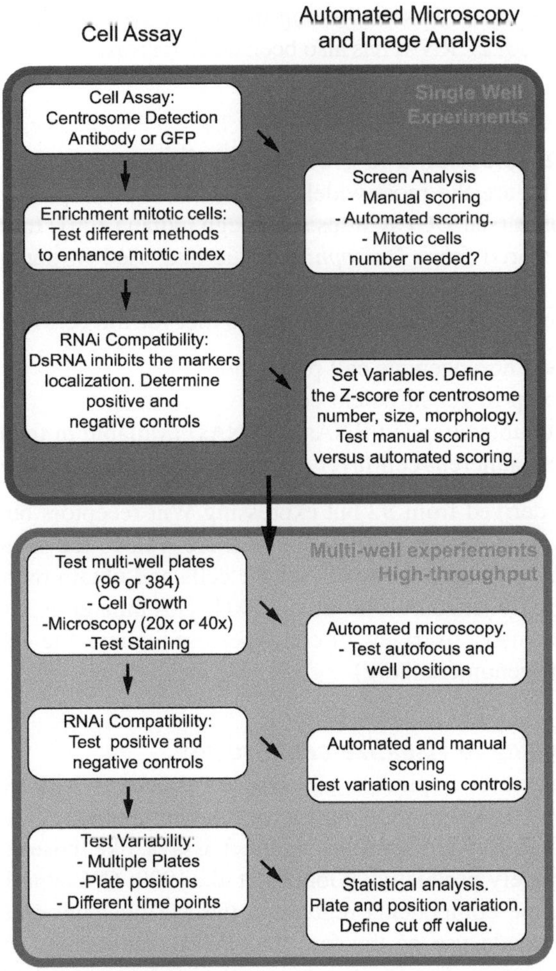

FIGURE 1 Overview of screen design.

Strategy to develop a cell-based assay for genome-wide RNAi screening starting with experiments on coverslips (green (dark gray in print versions)) to optimizing high-throughput screening in multiwell plates (orange (light gray in print versions)). Conditions for cell growth and RNAi-mediated depletion and fluorescence microscopy and image analysis are developed concurrently.

cells are needed per experiment to ensure a reliable measurement (see Figure 1 and Section 1.1.4). *Drosophila* cells are difficult to arrest using the standard protocols employed in human tissue culture (double thymidine block) or specific *Drosophila* drugs like the molting hormone ecdysone. We opted for an enrichment method using the MT depolymerizing drug colchicine to obtain a mitotic index of 20–30%. Colchicine also had the interesting sideeffect that the PCM became bigger (1.5 to

2-fold the original size), thus facilitating the microscopy process. Inhibition of the proteasome using *cdc27* RNAi has also been used to arrest cells in mitosis (Goshima et al., 2007).

1.1.2 Cell line selection and growing conditions

Although S2 cells are the most widely used cells in *Drosophila*, many other cell types are available (see DGCR https://dgrc.cgb.indiana.edu). In addition, primary cells can be prepared from *Drosophila* tissue and used in genome-wide screens (Bai et al., 2008; Bai, Sepp, & Perrimon, 2009). For genome-wide high-content screening, the following conditions should be met for the cell line selected:

- Easy to grow a homogenous cell population
- Easy to scale up to get sufficient cells
- Sensitive to double-stranded RNAs (dsRNAs) available in the existing libraries
- Growing as an adherent cell layer to enable easy staining and visualization.

S2R+ cells (derived from S2 but expressing Wnt receptors on the cell surface) were our cell type of choice because they adhere well to the substrate and form a uniform monolayer. Other labs use S2 or Kc cells and obtain monolayers by using precoated dishes (Bettencourt-Dias et al., 2004; Goshima et al., 2007; Kiger et al., 2003). For an extensive overview of cell types used, check the genomeRNAi database (http://www.genomernai.org)

1.1.3 Cell staining to visualize centrosomes

Centrosomes and mitotic spindles are complex structures with multiple functions (Nigg & Raff, 2009). Although using a molecular component of the centrosome appears to be a straightforward approach, indirect assays, like those used in *C. elegans*, can also be extremely productive (Gonczy et al., 1999; O'Connell et al., 1998). To identify centrosome components regulating the MT-nucleation capabilities of the centrosome, we made use of the fact that PCM components in *Drosophila* are only present in mitosis. The PCM component Cnn is one of the first components to be recruited and makes up the bulk of the *Drosophila* PCM (Lucas & Raff, 2007; Megraw, Kao, & Kaufman, 2001). The availability of a specific antibody (Lucas & Raff, 2007) and the possibility to indirectly identify centriole components outweighed the fact that potential components downstream of Cnn could be missed (proteins localizing to the PCM but not influencing Cnn recruitment would thus not be picked up by this readout). A similar approach was used by the Vale Lab using γ-tubulin as their centrosome marker (Goshima et al., 2007).

RNAi can be combined with both live and fixed analysis of samples. Although technically feasible even in human cells with multiple siRNAs as demonstrated by the MitoCheck project (Neumann et al., 2010), live microscopy is technically challenging and requires extensive image analysis. Tagged proteins in *Drosophila* cell lines also tend to be overexpressed, which can result in the formation of extranumerary centrioles (Dobbelaere et al., 2008), reducing their suitability for genome-wide screening. The easier alternative is therefore to combine a genome-wide screen with fixing cells and performing immunofluorescence staining for multiple markers.

A key criterion for a successful cell-based screen is reliable staining with a good signal-to-noise ratio. Staining in multiwell plates is very different from single experiments and antibodies and, under these conditions, other reagents should be tested for suitability early on. Of the 12 centriole and PCM antibodies tested under screening conditions, only the Cnn antibody used in the final screen (Dobbelaere et al., 2008) and the commercial γ-tubulin antibody (Sigma #T6557) resulted in reproducible staining detectable by automated microscopy at 20× magnification. After testing several protocols for immunofluorescence, a harsh permeabilization step to reduce cytoplasmic background was also included to avoid false signals in the automated analysis (Figure 1).

1.1.4 Analysis

Genome-wide screens, and microscopy-based screens in particular, create enormous amounts of data. Therefore, an automated pipeline for image analysis is needed to allow reliable and unbiased scoring of the phenotypes. In general, two approaches are used.

- A primary screen with a simple assay is used to identify potential candidates. Detailed phenotypic analysis is then performed in a secondary screen on a much smaller set of genes. This allows rapid screening of the genome-wide data and reduces the need for extensive image analysis or grouping (Dobbelaere et al., 2008). This approach can also be performed manually to enhance or complement the scoring method (Dobbelaere et al., 2008; Goshima et al., 2007).
- Many phenotypes can already be scored in the primary screen. If many known genes are included in the dataset, these can help assigning a function to newly identified genes. This approach is often used in smaller screens (Green et al., 2011) but also employed in genome-wide screens (Neumann et al., 2010; Neumuller et al., 2011).

Automated scoring of phenotypes in genome-wide screens brings several challenges. A number of publications provide details about image analysis (Jones et al., 2009; Singh, Carpenter, & Genovesio, 2014; Zwolinski, Kozak, & Kozak, 2013). For identifying centrosomal defects in the primary screen, two sets of considerations are key (Figure 3):

- How many cells are needed for unambiguous scoring of centrosomal defects? Manual scoring is often done on picture libraries (Rohn et al., 2011), with phenotypes scored from a few cells. Potential artifacts are easily identified and eliminated compared to automated methods. Automatic scoring, on the other hand, has the big advantage that it is unbiased. By necessity, automatic scoring uses thresholds, with ambiguous cells excluded from further analysis. Therefore, more cells need to be analyzed compared to manual methods. Often a hybrid of manual and automated scoring is used to avoid false negatives (Dobbelaere et al., 2008; Goshima et al., 2007).

- Does staining allow identification of the selected objects and can individual cells be identified using image segmentation? Object detection and image segmentation to detect cell boundaries should be reproducibly achieved to allow for identification of cellular defects. Several image processing tools exist to enhance the image but a good primary image is needed in all cases.

2. GENOME-WIDE SCREEN

2.1 LIBRARY SELECTION

Several genome-wide libraries are available for *Drosophila*, both in DNA and in RNA form (see http://www.nextrnai.org/). Most cover >90% of the annotated *Drosophila* genes. We chose a commercially available dsRNA library from Ambion (AM85000). Since *Drosophila* libraries rely on long dsRNAs (300–500 base pair (bp)) targets are mostly well depleted and few off-target effects have been observed, leading to strong correlation when different dsRNAs are used (Horn & Boutros, 2010).

2.2 CONTROLS

To avoid day-to-day and well-to-well variation, controls should be carefully chosen to compensate for external variation. Cells in outer wells tend to grow more slowly than those in wells at the center of the plate. Having both negative and positive controls on the inside and outside of the plate can help to control for this effect. The different categories of controls are as follows:

- Negative controls: We used wells without dsRNA and a dsRNA against Discosoma red fluorescent protein (DsRED)
- Positive controls: dsRNAs that provide a clear phenotype in the assay. Ideally we have depletions that yield mild, intermediate, and strong phenotypes in a reproducible manner. In the centrosome screen, we used dsRNAs against Polo kinase and Cnn as our controls, which both give strong phenotypes. No intermediate control was present at the start of the screen and not included.
- Controls for RNAi efficacy: Overcrowding, temperature fluctuations, and other conditions can influence the RNAi efficiency and, therefore, the results in the screen. To confirm the efficiency of the RNAi before staining and to avoid wasting precious antibodies, ensure to include dsRNAs that yield phenotypes that are easily assessed under the stereo microscope. We used String (cell cycle regulator, yields few large cells), Scar (regulator of the actin cytoskeleton, spikey cells), and Thread/Diap1 (apoptosis inhibitor, dead or no cells).

2.3 HIGH-THROUGHPUT CELL ASSAY

To obtain the best images, cells should form a continuous lawn at the end of the RNAi experiment but not be overgrown (a single monolayer only). Starting cell number and growth rate are key in obtaining this result. Having decided on the plates to use, it is important to test the starting and growing conditions. We opted for 384-well plates with a plastic base (Corning, #3712). Although more expensive options exist with better optics, these plates allowed visualization of cells with 20× and 40× lenses with a correction collar and with satisfactory resolution for our primary screen. S2R+ and S2 cells grow reproducibly on these plates. We also optimized the conditions for arrest in mitosis using the MT depolymerizing agent colchicine. As mentioned before, cells behave quite differently in multiwell plates compared to on culture dishes using coverslips. Therefore, conditions will need to be optimized during the final setup.

Outline genome-wide screening in multiwell plates (Figure 2):

1. S2R+ cells were cultured in Shield and Sang Media (Sigma S3652) with 10% fetal bovine serum (Sigma F9665) and 1% penicillin/streptomycin (Gibco 15070-063).
2. Cells were trypsinized and diluted to 7×10^5 cells/mL in serum-free Shield and Sang Medium.
3. 15 μL of cells were added to the dsRNA containing 384-well plates using a Thermo WellMate (giving a final concentration of ~10,500 cells per well).
4. Plates were gently centrifuged to allow cells to settle quickly.
5. Plates were incubated for 30–45 min at 25 °C.
6. 35 μL Shield and Sang media containing serum was added.
7. Plates were sealed with parafilm and incubated for 4 days at 25 °C.
8. Eight hours prior to fixation the media was exchanged for media containing 25 μM of the MT depolymerizing drug colchicine (Sigma #C3915). This typically resulted in 20–35% of cells arrested in mitosis at the time of fixation.

2.4 STAINING

Having a robust cellular staining is the key factor in analysis of RNAi phenotypes. Since imaging will be performed at low magnification, small objects like centrioles should be avoided. We opted for the PCM surrounding centrioles, which is around 1–3 μm in diameter in *Drosophila* cells and can easily be visualized with a 20× lens. Second, the signal-to-noise ratio should be high for easy algorithm-based analysis, and unaffected by the variability in staining within each well. In contrast to standard immunofluorescence experiments, in multiwell experiments each liquid handling step does not remove all the liquid completely, to avoid touching the bottom of the well and removing cells. Robust dyes and antibodies are needed under these conditions to obtain reliable staining. As mentioned above, antibody and dye staining should be tested as soon as possible under screening conditions to establish suitability for high-throughput screening. Below is the protocol for staining the PCM using a home-made Cnn antibody combined with phospho-histone 3 staining. Liquid was added with a liquid dispenser (Thermo WellMate) and taken off by vacuum using an eight-fingered arm, which was set up to draw off liquid ~2–3 mm above the cell layer.

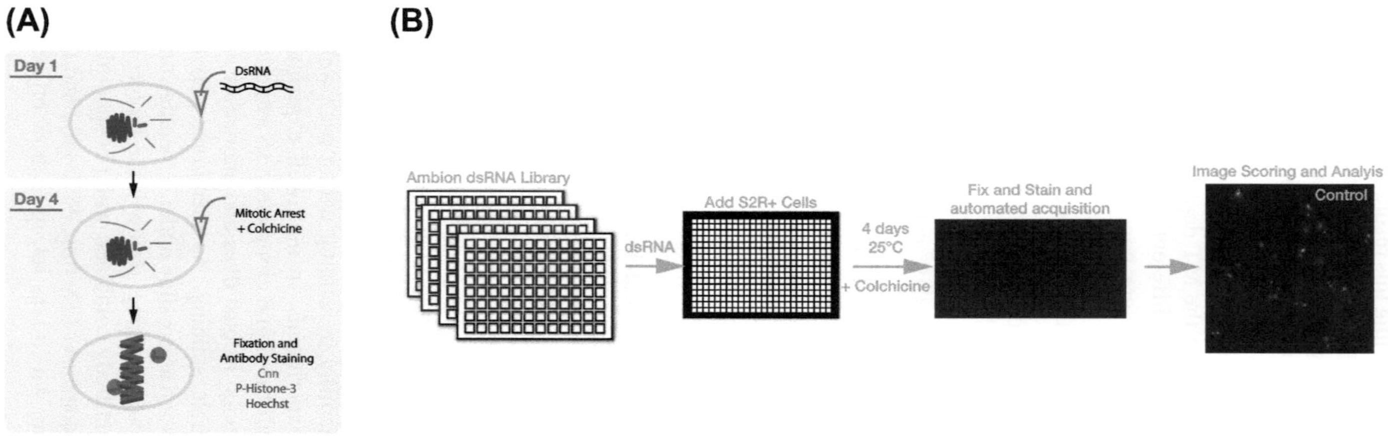

FIGURE 2 Outline of genome-wide centrosome screen.

Based on my centrosome screen in S2R+ cells (Dobbelaere et al., 2008) (A) Overview of the cell assay used to screen centrosome defects. Cells were treated with dsRNA on day one and grown for four days. Eight hours prior to staining, cells were treated with colchicine to enhance the number of mitotic cells to allow easy scoring. DNA (Hoechst), mitotic cells (phospho-H3), and centrosomes (Cnn) were stained to allow counting the number of centrosomes per mitotic cell. (B) Summary of genome-wide RNAi screen. A dsRNA library was dispensed in 384-well plates suitable for high-throughput microscopy. S2R+ cells were aliquoted to each well and incubated for 4 days. Cells were arrested and stained. Plates were examined on a microscope, with pictures acquired automatically from four fields per well.

1. Cells were washed once with phosphate-buffered saline (PBS) and fixed with 4% formaldehyde (in PBS) (Sigma #F8775) for 12 min.
2. Cells were then permeabilized with 0.5% sodium dodecyl sulfate in PBS for 10 min.
3. Cells were blocked with 5% goat serum (Sigma #G9023) in PBS-T (0.1% Triton) for 20 min and stained overnight at 4 °C with anti-Cnn antibodies (1:1000, Rabbit) to stain centrosomes (Lucas & Raff) and anti-pH3 Ser10 antibodies to label mitotic cells (1:2000, Mouse, Abcam #14955).
4. Antibodies were diluted in PBS-T with 5% goat serum.
5. The following day, cells were washed three times with PBS-T for 5 min.
6. Secondary antibodies, anti-rabbit Alexa 488 (1:1500, Molecular Probes A21206) and anti-mouse Alexa 567 (1:1500, Molecular Probes #A11004) in 5% goat serum in PBS-T were added for 2 h at room temperature.
7. Cells were washed once with PBS-T, incubated with Hoechst 33258 (final concentration of 0.2 μg/mL, Sigma #861405) in PBS for 10 min, and washed once more with PBS-T.
8. Finally, 20 μL PBS was added to each well, and plates were sealed with aluminum sealing tape (Corning #6569).

3. IMAGE ACQUISITION

To permit high-content image-based screens, a microscope is needed to acquire the images for further analysis. The microscope and accompanying software must able to accommodate multiwell plates and handle big datasets. Several commercial systems have been developed to allow for quick and reliable screening of multiwell plates in a cell-based assay, such as the ImageXpress Micro XLS System from Molecular Devices and the plate/explorer uHTS System from Perkin-Elmer, for example. The main advantage is that these systems come with laser-based autofocus and robotic arms that can load and unload plates, thereby allowing quick scanning of multiple plates. Although available at various screening centers, a simple inverted microscope can also easily be converted into a screening microscope. Besides the standard requirements, an automated stage, a lens with correction collar (to correct for the plastic bottom dishes used), and a software package that can handle multiwell plates are needed. In addition to cost savings, another advantage of building a microscope is that wells can be analyzed manually by looking through the eye-pieces, which is crucial in the setup phase. Interestingly, many cell-based screens combined with automated image acquisition and manual scoring to score all hits, highlighting the importance of manual inspection (Dobbelaere et al., 2008; Goshima et al., 2007; Kiger et al., 2003; Rohn et al., 2011). Here is the outline of our automated imaging using a Nikon TE2000E microscope, with an automated Prior stage controlled with Metamorph software (Molecular Devices) using a 20×, 0.45NA, Plan Fluor Air objective.

1. The stage was programmed to visit four different sites per well. This allowed us to examine an average of 500–2000 cells, approximately 150–400 of which were in mitosis.
2. Automated focusing was performed for each well using DNA staining.
3. Images were acquired in all three channels (Hoechst, Alexa 488, and Alexa 567) before moving to the next position.

4. SCORING PHENOTYPES AND IMAGE ANALYSIS

A genome-wide screen is a one-time experiment. Therefore, the analysis should be performed in a way that avoids missing any potential hits. False positives can be easily identified in the secondary analysis (see below). Both genome-wide *Drosophila* screens (Dobbelaere et al., 2008; Goshima et al., 2007) and their counterpart in human cells (Balestra et al., 2013) focusing on centrosomes relied on a combination of automated and manual scoring (Figure 3). In our screen (Dobbelaere et al., 2008), eight genes out of a total of 32 were missed by automated scoring, thus highlighting the importance of manual screening. The Baum Lab even scored a whole screen manually because of the diversity of phenotypes for which no easy algorithm could be designed (Rohn et al., 2011). In the following, I illustrate one possible way of image analysis. It should be noted that many other examples are available (see http://www.genomernai.org for more details).

1. Manual scoring of cell number, mitotic index, centrosome number, and centrosome size. All of these analyses were performed blind to avoid any bias.
2. Second, pictures taken with the automated microscope were manually scored using the same criteria.
3. Finally, pictures were analyzed with CellProfiler (www.cellprofiler.org) (Carpenter et al., 2006; Kamentsky et al., 2011) using a self-made pipeline. The CellProfiler software package has many predefined features for cell analysis and can be easily programmed to be used on big datasets. Cells were counted using Hoechst staining; Cnn dots were counted; mitotic cells were identified using phospho-histone staining; mitotic cell boundaries were identified starting from the phospho-histone 3 signal and grown to the background from the Cnn channel; finally, the number of Cnn foci per mitotic cell was counted.
4. The entire statistical analysis was done using the software package CellHTS developed for Bioconductor (Boutros, Bras, & Huber, 2006). This resulted in a numerical value for the number of Cnn foci per mitotic cell. The inverse of this numerical dataset was normalized and corrected for plate-by-plate variation and possible edge effects using the CellHTS software (Boutros et al., 2006, using the B-score method). The Z'-score (the degree of difference between positive and negative values) was calculated using all positive controls and all negative controls.

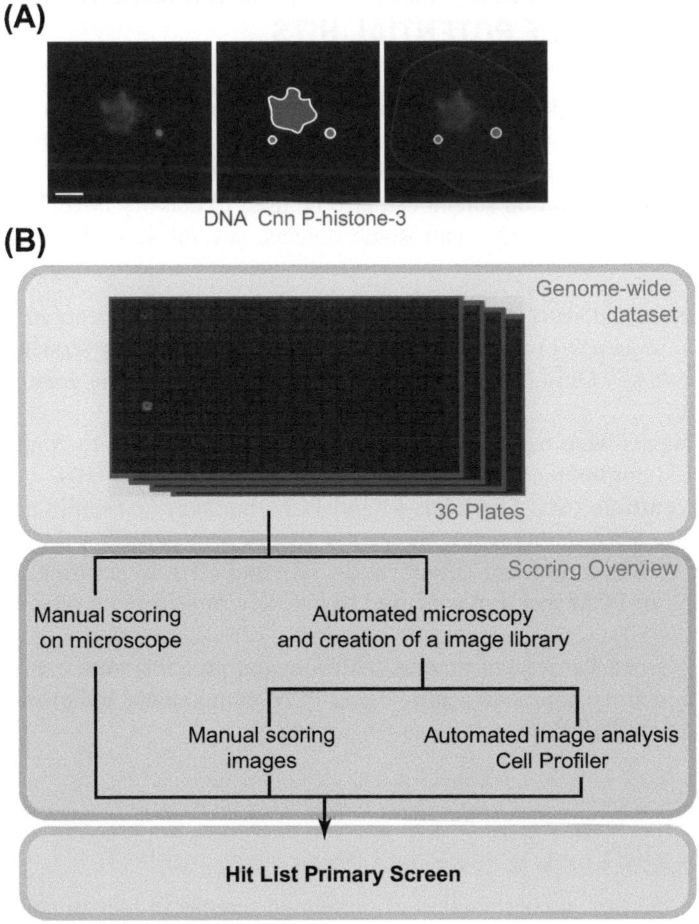

FIGURE 3 Overview of automated image analysis.

(A) Graphical representation of how objects were detected and related to each other. First, cell number was assessed using Hoechst staining. Second, centrosome (Cnn) foci and mitotic chromosomes (phospho-H3) were counted. Mitotic DNA signal was grown at the cell boundary to identify the complete mitotic cell. Cnn foci within this boundary were counted and served as the readout of the primary screen. Scale bar 5 μm. (B) Overview of the combination of manual and automated screening in the centrosome screen. Positives in at least two of the three methods were scored as hits in the screen.

5. Genes were selected for detailed secondary analysis if identified as hits in two out of the three methods. Genes eliminated included those commonly identified in previous screens (indicating they are likely false positives); those known as components of the ribosome or transcription machinery; and those that were clearly the result of off-target effects.

5. VALIDATION OF POTENTIAL HITS

Potential hits from the primary screen need validation to exclude false positives due to experimental variation, differences in cell growth, or other perturbations. In addition, off-target effects may occur due to sequence homology with other genes although this is less of an issue than in vertebrates due to the use of long dsRNAs (~500 bp). The validation screen can also be used to classify novel hits.

To identify and classify centrosome defects, several secondary screens were set up.

- Two additional dsRNAs were generated to confirm the phenotype, using the original sequence in the screen library and a new alternative sequence (Figure 4(A)). Only if both had the same phenotypes was the gene retained as a hit.
- Experiments were repeated with and without mitotic arrest to confirm that the arrest did not influence the centrosome phenotype (Figure 4(B)).
- Other centriole (Sas-4 and Sas-6) and PCM markers (γ-tubulin and DSpd-2) were tested to distinguish a molecular function in centriole duplication and/or PCM recruitment (Figure 4(B), (C), and (D)). In addition, intensity profiling of PCM markers was used to quantify intermediate phenotypes (Figure 4(E)).
- Finally, green fluorescent protein (GFP)-tagged proteins were expressed in S2 cells to identify bona fide centriole and PCM components and examine their localization during the cell cycle (Figure 4(F)).

DISCUSSION

In the last 15 years, a combination of proteomic screens in human cells with both forward and reverse genetic screens in *C. elegans* and *Drosophila* has helped define the composition and function of the centrosome as we understand it today. Although each of these methods provided novel components, only their combined application enabled us to obtain the entire set of structural components and regulators. Examining the function of centrosomes in mitosis as well as in other cellular processes will be key to understanding how centrosome dysfunction can cause cancer and give rise to genetic disorders like microcephaly and dwarfism.

Although the centrosome has been studied for more than 100 years, very few of its molecular components and regulators were known just 15 years ago. As described above, a combination of mutant analysis and later RNAi in the developing *C. elegans* embryo led to the delineation of the complete centriole duplication pathway (Fu et al., 2015), which was later found to be conserved across evolution. In many cases the low sequence conservation of centrosomal proteins and the evolutionary divergence of nematodes has precluded simple identification of homologs from sequence homology. RNAi screens in *Drosophila* and other organisms helped

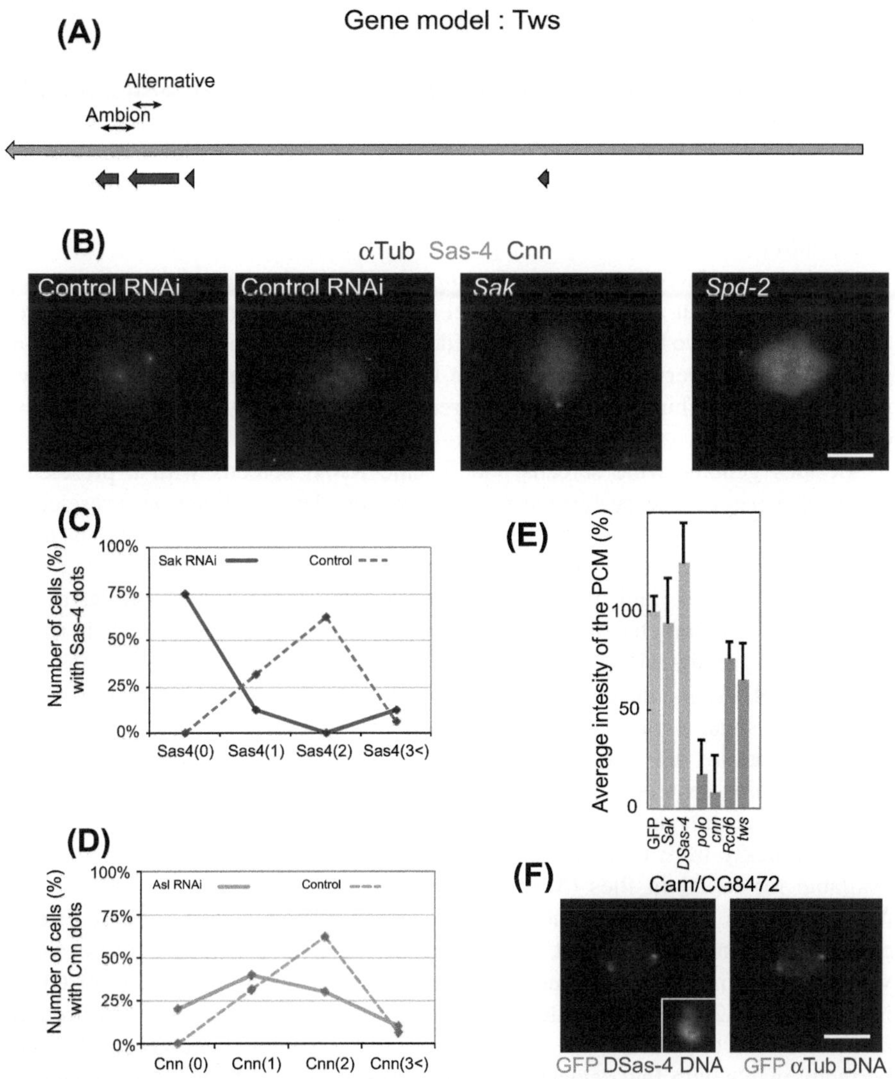

FIGURE 4 Screen validation of primary hits.

Results shown were obtained during my centrosome screen (Dobbelaere et al., 2008). (A) All hits from the primary screen were screened with a different nonoverlapping dsRNA as depicted in the schematic of the PP2A subunit Tws. (B) Centriole, PCM, and spindle defects were assessed using multiple markers in nonarrested cells. Analysis was done by immunofluorescence at higher magnification than in the original screen (63×, scale bar 5 μm). (C) Centriole duplication phenotype of Sak RNAi. Measurement of centriole number was performed by Sas-4 staining. Dotted line represents the control. (D) PCM phenotype after Asl RNAi. Measurement of PCM number was performed by Cnn staining. Dotted line represents the control. (E) Measurement of PCM size after RNAi (Cnn staining). Centriole markers in orange, PCM components and regulators in blue (dark gray in print versions). (F) Localization of centrosome component CG8472/Cam. S2-cells stably expressing GFP-construct stained with centriole and tubulin marker to analyze precise sub-cellular localization (63×, scale bar 5 μm).

identify functional homologs, as in the case of Plk4 and SAS-5/Ana2/Stil (Bettencourt-Dias et al., 2004; Delattre et al., 2004; Dobbelaere et al., 2008; Goshima et al., 2007; Stevens et al., 2009), but also novel complexes not found in *C. elegans*, such as the HAUS complex (Goshima & Kimura, 2010), which promotes MT branching by promoting nucleation from existing MTs (Goshima et al., 2007).

The success of RNAi screens in *Drosophila* tissue culture cells showed that even with a solid readout and planning, novel candidates still remain to be discovered. Similar screens have been performed to identify candidates in other pathways or organelles (for an overview of all genome-wide screens performed, see http://www.genomernai.org). Thus, genome-wide screening remains a key tool to identify novel components.

Besides genome-wide screens, small-scale RNAi screens with a preselected group of genes can be performed to identify their detailed function. Preselection based on previous screens or other datasets introduces some bias, but its major advantage is the ability to analyze phenotypes in more detail. For example, all *Drosophila* kinases and phosphatases were analyzed using multiple assays, including flow cytometry, microscopy, and other tools to understand their role in cell cycle progression (Bettencourt-Dias et al., 2004; Chen et al., 2007). dsRNA production for RNAi and the construction of small-scale libraries is simple. In addition, setting up a small library can be a worthwhile intermediate step as a pilot screen. Many sublibraries are also now commercially available or present at *Drosophila* RNAi screening centers.

Another advantage of *Drosophila* as an experimental model is that they permit moving quickly from cells to the whole organism, taking advantage of the tools available for RNAi in flies (Mohr, Hu, Kim, Housden, & Perrimon, 2014) and mutant libraries. Studying a gene product in an organism allows analysis in different cell types, during development and under physiological conditions. These can be very different from a cell culture environment. For example, depletion of CP110 in Drosophila S2R+ cells resulted in strong centriole duplication and PCM recruitment defects (Dobbelaere et al., 2008). However, a complete knock-out in flies was perfectly viable with few detectable defects (Franz, Roque, Saurya, Dobbelaere, & Raff, 2013). Analysis of genes in the fly should always be considered and the development of clustered regularly interspaced short palindromic repeats and other gene-editing techniques makes it possible to test candidates through mutant analysis within a reasonable amount of time (Bassett & Liu, 2014).

ACKNOWLEDGMENTS

I would like to thank members of the Dammermann Lab for comments on the manuscript and discussions. J.D. is supported by MFPL and a VIPS fellowship funded by the Austrian Federal Ministry of Science and Research and the City of Vienna.

REFERENCES

Andersen, J. S., Wilkinson, C. J., Mayor, T., Mortensen, P., Nigg, E. A., & Mann, M. (2003). Proteomic characterization of the human centrosome by protein correlation profiling. *Nature, 426*(6966), 570–574.

Arquint, C., Gabryjonczyk, A. M., & Nigg, E. A. (2014). Centrosomes as signalling centres. *Philosophical Transactions of the Royal Society of London B Biological Sciences, 369*(1650).

Bai, J., Binari, R., Ni, J. Q., Vijayakanthan, M., Li, H. S., & Perrimon, N. (2008). RNA interference screening in *Drosophila* primary cells for genes involved in muscle assembly and maintenance. *Development, 135*(8), 1439–1449.

Bai, J., Sepp, K. J., & Perrimon, N. (2009). Culture of *Drosophila* primary cells dissociated from gastrula embryos and their use in RNAi screening. *Nature Protocols, 4*(10), 1502–1512.

Balestra, F. R., Strnad, P., Fluckiger, I., & Gonczy, P. (2013). Discovering regulators of centriole biogenesis through siRNA-based functional genomics in human cells. *Developmental Cell, 25*(6), 555–571.

Bassett, A. R., & Liu, J. L. (2014). CRISPR/Cas9 and genome editing in *Drosophila*. *Journal of Genetics and Genomics, 41*(1), 7–19.

Bettencourt-Dias, M., Giet, R., Sinka, R., Mazumdar, A., Lock, W. G., Balloux, F., et al. (2004). Genome-wide survey of protein kinases required for cell cycle progression. *Nature, 432*(7020), 980–987.

Bettencourt-Dias, M., Rodrigues-Martins, A., Carpenter, L., Riparbelli, M., Lehmann, L., Gatt, M. K., et al. (2005). SAK/PLK4 is required for centriole duplication and flagella development. *Current Biology, 15*(24), 2199–2207.

Boutros, M., Bras, L. P., & Huber, W. (2006). Analysis of cell-based RNAi screens. *Genome Biology, 7*(7), R66.

Boutros, M., Kiger, A. A., Armknecht, S., Kerr, K., Hild, M., Koch, B., et al. (2004). Genome-wide RNAi analysis of growth and viability in *Drosophila* cells. *Science, 303*(5659), 832–835.

Carpenter, A. E., Jones, T. R., Lamprecht, M. R., Clarke, C., Kang, I. H., Friman, O., et al. (2006). CellProfiler: image analysis software for identifying and quantifying cell phenotypes. *Genome Biology, 7*(10), R100.

Chen, F., Archambault, V., Kar, A., Lio, P., D'Avino, P. P., Sinka, R., et al. (2007). Multiple protein phosphatases are required for mitosis in *Drosophila*. *Current Biology, 17*(4), 293–303.

Clemens, J. C., Worby, C. A., Simonson-Leff, N., Muda, M., Maehama, T., Hemmings, B. A., et al. (2000). Use of double-stranded RNA interference in *Drosophila* cell lines to dissect signal transduction pathways. *Proceedings of the National Academy of Sciences of the United States of America, 97*(12), 6499–6503.

Consortium, C. e. S. (1998). Genome sequence of the nematode *C. elegans*: a platform for investigating biology. *Science, 282*(5396), 2012–2018.

Dammermann, A., Maddox, P. S., Desai, A., & Oegema, K. (2008). SAS-4 is recruited to a dynamic structure in newly forming centrioles that is stabilized by the gamma-tubulin-mediated addition of centriolar microtubules. *Journal of Cell Biology, 180*(4), 771–785.

Delattre, M., Leidel, S., Wani, K., Baumer, K., Bamat, J., Schnabel, H., et al. (2004). Centriolar SAS-5 is required for centrosome duplication in *C. elegans*. *Nature Cell Biology, 6*(7), 656–664.

Dobbelaere, J., Josue, F., Suijkerbuijk, S., Baum, B., Tapon, N., & Raff, J. (2008). A genome-wide RNAi screen to dissect centriole duplication and centrosome maturation in *Drosophila*. *PLoS Biology, 6*(9), e224.

Doxsey, S. J., Stein, P., Evans, L., Calarco, P. D., & Kirschner, M. (1994). Pericentrin, a highly conserved centrosome protein involved in microtubule organization. *Cell, 76*(4), 639–650.

Franz, A., Roque, H., Saurya, S., Dobbelaere, J., & Raff, J. W. (2013). CP110 exhibits novel regulatory activities during centriole assembly in *Drosophila*. *Journal of Cell Biology, 203*(5), 785–799.

Fu, J., Hagan, I. M., & Glover, D. M. (2015). The centrosome and its duplication cycle. *Cold Spring Harbor Perspectives in Biology, 7*(2).

Glover, D. M., Leibowitz, M. H., McLean, D. A., & Parry, H. (1995). Mutations in aurora prevent centrosome separation leading to the formation of monopolar spindles. *Cell, 81*(1), 95–105.

Gonczy, P., Echeverri, C., Oegema, K., Coulson, A., Jones, S. J., Copley, R. R., et al. (2000). Functional genomic analysis of cell division in *C. elegans* using RNAi of genes on chromosome III. *Nature, 408*(6810), 331–336.

Gonczy, P., Schnabel, H., Kaletta, T., Amores, A. D., Hyman, T., & Schnabel, R. (1999). Dissection of cell division processes in the one cell stage *Caenorhabditis elegans* embryo by mutational analysis. *Journal of Cell Biology, 144*(5), 927–946.

Gonzalez, C., Saunders, R. D., Casal, J., Molina, I., Carmena, M., Ripoll, P., et al. (1990). Mutations at the asp locus of *Drosophila* lead to multiple free centrosomes in syncytial embryos, but restrict centrosome duplication in larval neuroblasts. *Journal of Cell Science, 96*(Pt 4), 605–616.

Goshima, G., & Kimura, A. (2010). New look inside the spindle: microtubule-dependent microtubule generation within the spindle. *Current Opinion in Cell Biology, 22*(1), 44–49.

Goshima, G., & Vale, R. D. (2003). The roles of microtubule-based motor proteins in mitosis: comprehensive RNAi analysis in the *Drosophila* S2 cell line. *Journal of Cell Biology, 162*(6), 1003–1016.

Goshima, G., Wollman, R., Goodwin, S. S., Zhang, N., Scholey, J. M., Vale, R. D., et al. (2007). Genes required for mitotic spindle assembly in *Drosophila* S2 cells. *Science, 316*(5823), 417–421.

Green, R. A., Kao, H. L., Audhya, A., Arur, S., Mayers, J. R., Fridolfsson, H. N., et al. (2011). A high-resolution *C. elegans* essential gene network based on phenotypic profiling of a complex tissue. *Cell, 145*(3), 470–482.

Hamaguchi, Y., & Kuriyama, R. (1982). Aster formation in sand dollar eggs by microinjection of calcium buffers and centriolar complexes isolated from starfish sperm. *Experimental Cell Research, 141*(2), 450–454.

Hamill, D. R., Severson, A. F., Carter, J. C., & Bowerman, B. (2002). Centrosome maturation and mitotic spindle assembly in *C. elegans* require SPD-5, a protein with multiple coiled-coil domains. *Developmental Cell, 3*(5), 673–684.

Horn, T., & Boutros, M. (2010). E-RNAi: a web application for the multi-species design of RNAi reagents—2010 update. *Nucleic Acids Research, 38*(Web Server issue), W332–W339.

Jakobsen, L., Vanselow, K., Skogs, M., Toyoda, Y., Lundberg, E., Poser, I., et al. (2011). Novel asymmetrically localizing components of human centrosomes identified by complementary proteomics methods. *EMBO Journal, 30*(8), 1520–1535.

Jones, T. R., Carpenter, A. E., Lamprecht, M. R., Moffat, J., Silver, S. J., Grenier, J. K., et al. (2009). Scoring diverse cellular morphologies in image-based screens with iterative feedback and machine learning. *Proceedings of the National Academy of Sciences of the United States of America, 106*(6), 1826–1831.

Kamentsky, L., Jones, T. R., Fraser, A., Bray, M. A., Logan, D. J., Madden, K. L., et al. (2011). Improved structure, function and compatibility for CellProfiler: modular high-throughput image analysis software. *Bioinformatics, 27*(8), 1179–1180.

Keller, L. C., Romijn, E. P., Zamora, I., Yates, J. R., III, & Marshall, W. F. (2005). Proteomic analysis of isolated *Chlamydomonas* centrioles reveals orthologs of ciliary-disease genes. *Current Biology, 15*(12), 1090–1098.

Kellogg, D. R., Field, C. M., & Alberts, B. M. (1989). Identification of microtubule-associated proteins in the centrosome, spindle, and kinetochore of the early *Drosophila* embryo. *Journal of Cell Biology, 109*(6 Pt 1), 2977–2991.

Kiger, A. A., Baum, B., Jones, S., Jones, M. R., Coulson, A., Echeverri, C., et al. (2003). A functional genomic analysis of cell morphology using RNA interference. *Journal of Biology, 2*(4), 27.

Kim, S., & Dynlacht, B. D. (2013). Assembling a primary cilium. *Current Opinion in Cell Biology, 25*(4), 506–511.

Kleylein-Sohn, J., Westendorf, J., Le Clech, M., Habedanck, R., Stierhof, Y. D., & Nigg, E. A. (2007). Plk4-induced centriole biogenesis in human cells. *Developmental Cell, 13*(2), 190–202.

Kuriyama, R., & Borisy, G. G. (1981). Microtubule-nucleating activity of centrosomes in Chinese hamster ovary cells is independent of the centriole cycle but coupled to the mitotic cycle. *Journal of Cell Biology, 91*(3 Pt 1), 822–826.

Li, K., Xu, E. Y., Cecil, J. K., Turner, F. R., Megraw, T. L., & Kaufman, T. C. (1998). *Drosophila* centrosomin protein is required for male meiosis and assembly of the flagellar axoneme. *Journal of Cell Biology, 141*(2), 455–467.

Lucas, E. P., & Raff, J. W. (2007). Maintaining the proper connection between the centrioles and the pericentriolar matrix requires *Drosophila* centrosomin. *Journal of Cell Biology, 178*(5), 725–732.

Megraw, T. L., Kao, L. R., & Kaufman, T. C. (2001). Zygotic development without functional mitotic centrosomes. *Current Biology, 11*(2), 116–120.

Megraw, T. L., Li, K., Kao, L. R., & Kaufman, T. C. (1999). The centrosomin protein is required for centrosome assembly and function during cleavage in *Drosophila. Development, 126*(13), 2829–2839.

Mennella, V., Keszthelyi, B., McDonald, K. L., Chhun, B., Kan, F., Rogers, G. C., et al. (2012). Subdiffraction-resolution fluorescence microscopy reveals a domain of the centrosome critical for pericentriolar material organization. *Nature Cell Biology, 14*(11), 1159–1168.

Mitchison, T., & Kirschner, M. (1984). Microtubule assembly nucleated by isolated centrosomes. *Nature, 312*(5991), 232–237.

Mohr, S. E., Hu, Y., Kim, K., Housden, B. E., & Perrimon, N. (2014). Resources for functional genomics studies in *Drosophila melanogaster. Genetics, 197*(1), 1–18.

Moritz, M., & Agard, D. A. (2001). Gamma-tubulin complexes and microtubule nucleation. *Current Opinion in Structural Biology, 11*(2), 174–181.

Moritz, M., & Alberts, B. M. (1999). Isolation of centrosomes from *Drosophila* embryos. *Methods in Cell Biology, 61*, 1–12.

Muller, H., Schmidt, D., Steinbrink, S., Mirgorodskaya, E., Lehmann, V., Habermann, K., et al. (2010). Proteomic and functional analysis of the mitotic *Drosophila* centrosome. *EMBO Journal, 29*(19), 3344–3357.

Neumann, B., Walter, T., Heriche, J. K., Bulkescher, J., Erfle, H., Conrad, C., et al. (2010). Phenotypic profiling of the human genome by time-lapse microscopy reveals cell division genes. *Nature, 464*(7289), 721–727.

Neumuller, R. A., Richter, C., Fischer, A., Novatchkova, M., Neumuller, K. G., & Knoblich, J. A. (2011). Genome-wide analysis of self-renewal in *Drosophila* neural stem cells by transgenic RNAi. *Cell Stem Cell, 8*(5), 580–593.

Nigg, E. A., Cajanek, L., & Arquint, C. (2014). The centrosome duplication cycle in health and disease. *FEBS Letters, 588*(15), 2366–2372.

Nigg, E. A., & Raff, J. W. (2009). Centrioles, centrosomes, and cilia in health and disease. *Cell, 139*(4), 663–678.

O'Connell, K. F., Caron, C., Kopish, K. R., Hurd, D. D., Kemphues, K. J., Li, Y., et al. (2001). The *C. elegans* zyg-1 gene encodes a regulator of centrosome duplication with distinct maternal and paternal roles in the embryo. *Cell, 105*(4), 547–558.

O'Connell, K. F., Leys, C. M., & White, J. G. (1998). A genetic screen for temperature-sensitive cell-division mutants of *Caenorhabditis elegans*. *Genetics, 149*(3), 1303–1321.

O'Connell, K. F., Maxwell, K. N., & White, J. G. (2000). The spd-2 gene is required for polarization of the anteroposterior axis and formation of the sperm asters in the *Caenorhabditis elegans* zygote. *Developmental Biology, 222*(1), 55–70.

Oakley, C. E., & Oakley, B. R. (1989). Identification of gamma-tubulin, a new member of the tubulin superfamily encoded by mipA gene of *Aspergillus nidulans*. *Nature, 338*(6217), 662–664.

Rodrigues-Martins, A., Bettencourt-Dias, M., Riparbelli, M., Ferreira, C., Ferreira, I., Callaini, G., et al. (2007). DSAS-6 organizes a tube-like centriole precursor, and its absence suggests modularity in centriole assembly. *Current Biology, 17*(17), 1465–1472.

Rohn, J. L., Sims, D., Liu, T., Fedorova, M., Schock, F., Dopie, J., et al. (2011). Comparative RNAi screening identifies a conserved core metazoan actinome by phenotype. *Journal of Cell Biology, 194*(5), 789–805.

Singh, S., Carpenter, A. E., & Genovesio, A. (2014). Increasing the content of high-content screening: an overview. *Journal of Biomolecular Screening, 19*(5), 640–650.

Stevens, N. R., Dobbelaere, J., Wainman, A., Gergely, F., & Raff, J. W. (2009). Ana3 is a conserved protein required for the structural integrity of centrioles and basal bodies. *Journal of Cell Biology, 187*(3), 355–363.

Strnad, P., & Gonczy, P. (2008). Mechanisms of procentriole formation. *Trends in Cell Biology, 18*(8), 389–396.

Sunkel, C. E., & Glover, D. M. (1988). polo, a mitotic mutant of *Drosophila* displaying abnormal spindle poles. *Journal of Cell Science, 89*(Pt 1), 25–38.

Tsou, M. F., & Stearns, T. (2006). Mechanism limiting centrosome duplication to once per cell cycle. *Nature, 442*(7105), 947–951.

Vincensini, L., Blisnick, T., & Bastin, P. (2011). 1001 model organisms to study cilia and flagella. *Biology of the Cell, 103*(3), 109–130.

Zwolinski, L., Kozak, M., & Kozak, K. (2013). 1Click1View: interactive visualization methodology for RNAi cell-based microscopic screening. *BioMed Research International, 2013*, 156932.

CHAPTER

Time-lapse recording of centrosomes and other organelles in *Drosophila* neuroblasts

16

Judit Pampalona*, Jens Januschke§, Paula Sampaio¶,||, Cayetano Gonzalez*,#,1

**Institute for Research in Biomedicine (IRB-Barcelona), Barcelona, Spain*
§Division of Cell and Developmental Biology, College of Life Sciences, University of Dundee, Dundee, UK
¶Instituto de Investigação e Inovação em Saúde, Universidade do Porto, Portugal
||IBMC—Instituto de Biologia Molecular e Celular, Universidade do Porto, Portugal
#Institució Catalana de Recerca i Estudis Avançats (ICREA), Barcelona, Spain
[1]Corresponding author: E-mail: gonzalez@irbbarcelona.org

CHAPTER OUTLINE

Introduction 302
1. Primary Cultures 303
2. Ex vivo Whole Mount 309
3. Fluorescent Reporters 311
4. Remarks and Troubleshooting 311
Acknowledgments 313
Supplementary data 313
References 313

Abstract

Drosophila larval neuroblasts (NBs) are an excellent model for asymmetric division and cell cycle studies in general. For decades, visualizing relevant structures like centrosomes, chromosomes, or the mitotic spindle relied exclusively on immunofluorescence on fix samples. More recently, improvements on sensitivity and acquisition speed of different confocal systems have made it possible to acquire time-resolved images of combined fluorescent reporters from single larval NBs. Here, we provide protocols to visualize centrosomes and other organelles from both primary cultures of isolated single NBs and ex vivo, whole-mounted larval brains.

Methods in Cell Biology, Volume 129, ISSN 0091-679X, http://dx.doi.org/10.1016/bs.mcb.2015.03.003

INTRODUCTION

The importance of live imaging cannot be overemphasized; a full understanding of biological processes can hardly be achieved without the time dimension. Microscopy on fixed samples is, and will remain, the technique of choice for numerous applications. It is easier and faster than time-lapse microscopy, and having no time constrains it usually affords better spatial resolution. However, because biological processes occur in time and space, time-resolved microscopy can reveal relevant features unsuspected than when using fix microscopy data; in some cases, it has even proved fix-microscopy derived interpretations wrong.

There are three different approaches for live imaging of *Drosophila* neuroblasts (NBs): primary cultures of dissociated NBs, developing embryos, and ex vivo whole mounts of larval brains. Primary cultures were the first to be developed, and are the simplest from a technical point of view. Pioneer work on primary cultures of *Drosophila* NBs was carried out using either larval (Wu, Suzuki, & Poo, 1983) or embryonic NBs (Furst & Mahowald, 1985a, 1985b). In both cases, cultured NBs were observed to give rise to ganglion-like clusters from which neurites extended. Cultured cells also presented enzymatic activities characteristic of neural tissue. NBs are only a small fraction of the cell types that populate the central nervous system, but fortunately these methods achieve a relatively high yield of NBs. This is largely due to the convenient fact that NBs are better than other neural cells at surviving the harsh dissociation procedure and at adhering to a glass substrate. To increase NB yield, Furst and Mahowald (Furst & Mahowald, 1985b) introduced an elutriation step through which they purified the fraction of larger cells. Neither Wu (Wu et al., 1983) nor more recently published protocols found this step necessary. A far more sophisticated, and very efficient, procedure to this end is to plate out nearly pure NB populations isolated by fluorescence-activated cell sorting (Harzer, Berger, Conder, Schmauss, & Knoblich, 2013).

The advent of confocal microscopy made it possible to observe living NBs in situ. This can be done in developing embryos, which requires a relatively simple procedure, or in ex vivo cultures of whole-mounted larval brains. A protocol for recording newly delaminated NBs in developing embryos was first published by Kaltschmidt (Kaltschmidt, Davidson, Brown, & Brand, 2000). Embryonic NBs at later cell cycles have been recorded using the same approach with the help of two-photon confocal microscopy (Rebollo, Roldan, & Gonzalez, 2009). Details on these protocols have been published and will not be included here (Rebollo & Gonzalez, 2010).

Unlike embryos, which are largely self-contained and only require proper gas exchange to develop normally under the microscope, ex vivo whole mounted larval brains are severed away from the individual and demand an elaborated culture medium with a certain renewal rate, as well as proper gas exchange. These cultures are more difficult to set than primary cultures. Moreover, unlike primary cultures, ex vivo whole mounts remain useful for a few hours only. These drawbacks are nonetheless compensated by the more physiological conditions that whole mount preparations afford, leaving NBs undisturbed within the tissue and within the largely maintained anatomy of the larval brain.

1. PRIMARY CULTURES

The following includes the most common current protocols. A graphic view of our preferred protocol is shown in Figure 1. Examples of recordings carried out with our protocol are shown in Figure 3 (Centrobin-Yellow Fuorescent Protein (CNB-YFP)/bright field and Asterless-Yellow Fluorescent Protein (ASL-YFP)/Tubulin-Green Fluorescent Protein (TUB-GFP)/Centrosomin-Red Fluorescent Protein (CNN-RFP) panels) and Movies 1 and 2.

1.1 Dissection media. The composition of the media referred to in this section is shown in Table 1.

1.1.1 Larval brain

1.1.1.1 Wu (Wu et al., 1983) defines a modified *Drosophila* medium that contains 27% Leibovitz L-15 medium (Gibco) and 73% of a saline solution similar to that of Schneider's medium (Table 1, column 1) supplemented with 10% heat inactivated (56 °C, 30 min) fetal calf serum (FCS), penicillin G (50 units/mL), and streptomycin sulfate (50 μg/mL).

1.1.1.2 Feiguin (Feiguin, Llamazares, & Gonzalez, 1998) uses a Ca^{2+} and Mg^{2+} free solution published by Schneider and Blumenthal (1978) (Table 1, column 2). This is the solution we currently use.

1.1.1.3 A more recent variation uses Schneider's medium (Gibco) supplemented with 10% fetal bovine serum (FBS) and 2% penicillin and streptomycin (Homem et al., 2013).

1.1.2 Embryo

1.1.2.1 Furst and Mahowald (1985b) use Chan and Gehring's medium* (Table 1, column 3) (Chan & Gehring, 1971).

1.1.2.2 Dissection of embryonic NBs has also been done in Schneider's insect medium supplemented with 2% FCS (Broadus & Doe, 1997).

1.2 Dissociation into single cells

1.2.1 Larval brain

1.2.1.1 Wu (Wu et al., 1983) first tears the brains into fine pieces with dissecting needles and incubates the fragments for 1 h in a Ca^{2+} and Mg^{2+} free solution (Schneider & Blumenthal, 1978) (Table 1, column 2) containing 0.5 mg/mL collagenase (type I, cat# C-0130, Sigma). The brain fragments are then rinsed 3× 5 min in culture media supplemented with 10% FCS, and finally dispersed into single cells by flushing the tissue through a siliconized glass pipette. Yellow tips siliconized with Sigma

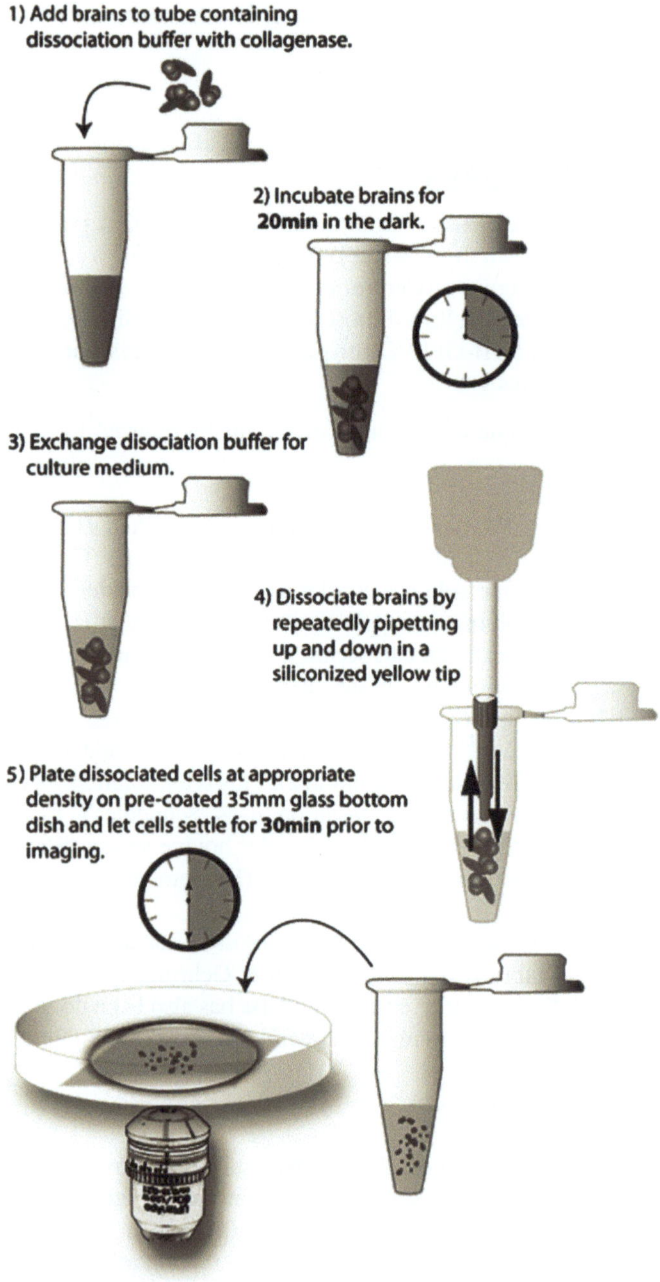

FIGURE 1 Primary culture of larval NBs.

Graphical view of the main steps of the protocol that we currently use for primary cultures of *Drosophila* larval NBs.

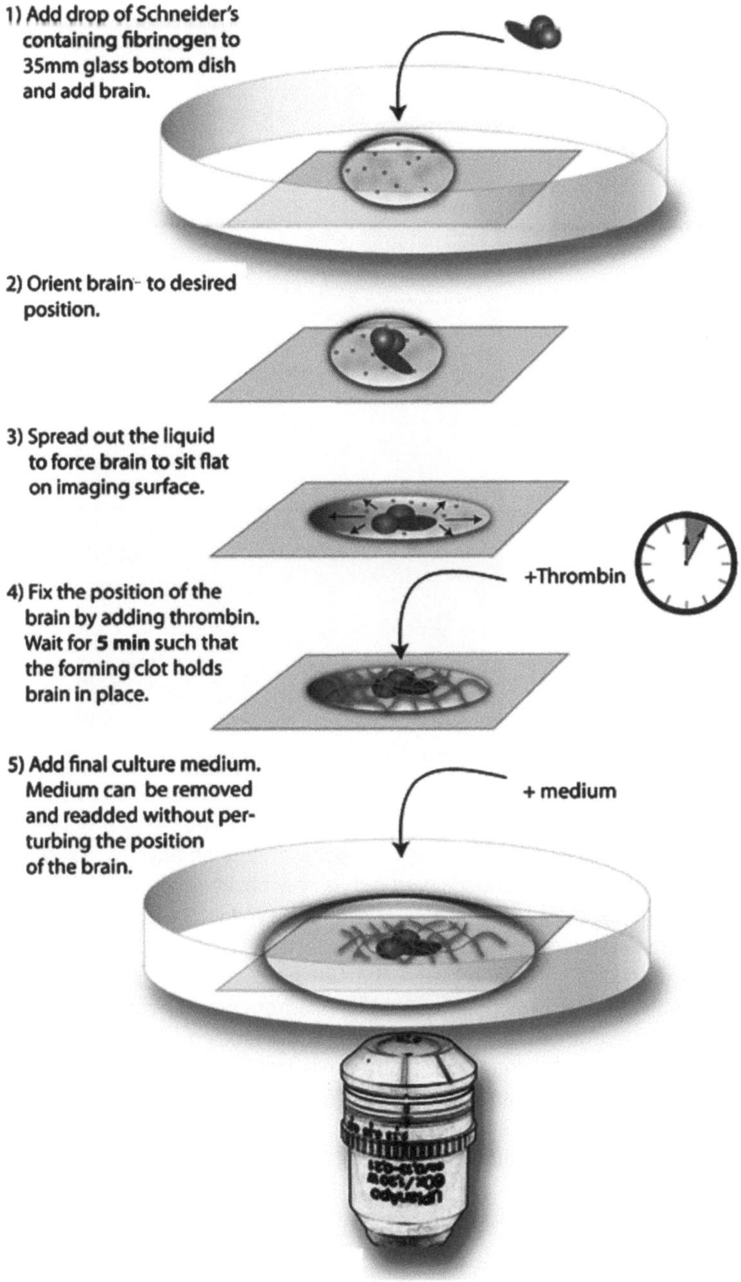

FIGURE 2 Ex vivo whole mount of larval brains.

Graphical view of the main steps of the protocol that we currently use for ex vivo, whole-mounted *Drosophila* larval brains.

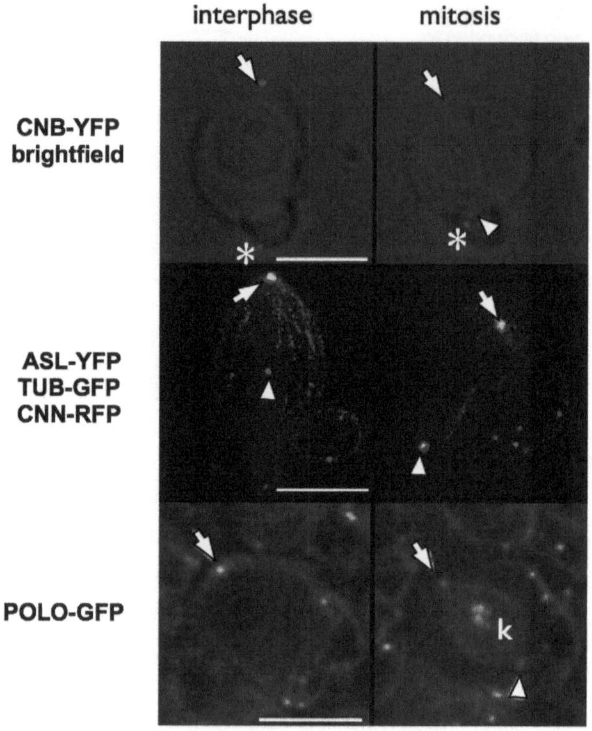

FIGURE 3 Imaging centrosomes in *Drosophila* third instar larvae NB.

Selected frames corresponding to interphase and mitosis taken from Movies 1–3. CNB-YFP, bright field) primary culture. Arrows and arrowheads point at the NB's apical and basal centrosomes, respectively. Asterisks label the centrosome of an adjacent ganglion mother cell. In the NB, CNB signal (pseudocolored red) is only observed on the daughter centriole, which during interphase is in the apical centrosome. At mitosis, both centrosomes contain newly assembled daughter centrioles that show a faint CNB label (ASL-YFP, TUB-GFP, CNN-RFP). Primary culture. In interphase, the pericentriolar material (PCM) labeled with CNN (red) is concentrated on the apical centrosome from which microtubules (green) radiate. The mother centriole (ASL; green), which is not labeled with CNN, can also be observed. CNN prominently labels both centrosomes in mitosis (POLO-GFP). Ex vivo whole-mount brain preparation. The large cell in the center is an NB located in the central brain. POLO accumulates on the apical centrosome in interphase. In mitosis POLO labels both centrosomes as well as the kinetochores (k). POLO-GFP also labels the midbody (shown in Movie 3; not shown here). Scale bar = 10 μm. (See color plate)

Table 1 Composition of Most Common Media Used for larval **brain dissection and isolation** of *Drosophila* NBs

	1 (Wu et al., 1983)	2 (Schneider & Blumenthal, 1978)	3 (Chan & Gehring, 1971)
$CaCl_2$	5.4 mM		5 mM
KCl	21.4 mM	2.7 mM	40 mM
MgCl	12.3 mM		
$MgSO_4$			15 mM
NaCl	36.0 mM	136.9 mM	55 mM
NaH_2PO_4		362.3 mM	
$NaHCO_3$	4.8 mM	11.9 mM	
KH_2PO_4	1.5 mM		
Glucose	11.1 mM	5 mM	20 mM
Sucrose			50 mM
Tris pH7.1	24.8 mM		10 mM
BSA (fraction V)			1 mg/mL

Cote (Sigma, cat#SL2-25 ML) following the manufacturer's instructions work just as well.

1.2.1.2 Feiguin (Feiguin et al., 1998) cut down the incubation time in collagenase to 30 min. We now use 0.2 mg/mL collagenase and incubate for 15–20 min. This needs to be adjusted depending on the collagenase batch. They remove collagenase by rinsing in their own culture medium (see below).

1.2.1.3 Homem (Homem et al., 2013) sticks to the 60 min incubation time. Moreover, they increase the concentration of collagenase to 1 mg/mL and add 1 mg/mL papain. They use Schneider and Blumenthal's medium (Schneider & Blumenthal, 1978) (Table 1, column 2) for rinsing 2× followed by Schneider's medium supplemented with 10% FCS and 2% penicillin and streptomycin.

1.2.2 Embryo

In the case of embryonic NBs, the first, mandatory step is dechorionation by agitation in 2.5% NaCl in 50% EtOH, followed by washing in 70% EtOH (Furst & Mahowald, 1985a, 1985b) or 95% EtOH (Broadus & Doe, 1995) and equilibration in culture medium.

1.2.2.1 Fürst and Mahowald (Furst & Mahowald, 1985a, 1985b) homogenize dechorionated embryos using a Dounce homogenizer. They then centrifuge for 5 min at 1500 rpm and resuspend the pellet in balanced saline solution (Chan & Gehring, 1971). This procedure is repeated three times before the final resuspension in Seecof's modification of Schneider's *Drosophila* medium (Seecof, Teplitz, Gerson, Ikeda, & Donady, 1972).

1.2.2.2 Broadus and Doe (Broadus & Doe, 1997) homogenize by six to eight strokes of a loose-fitting Dounce and filter the homogenate through a 30 μm Nitex mesh to retain unwanted debris. Cells are then pelleted down in a clinical centrifuge at 4 °C (setting 5, IEC) and washed twice dispersing the pellet gently in culture medium (Schneider's supplemented with 2% FCS, Sigma). Grosskortenhaus (Grosskortenhaus, Pearson, Marusich, & Doe, 2005) followed the same method, but used Chan and Gehring's medium (Chan & Gehring, 1971) with 2% FCS, instead.

1.3 Plating out

1.3.1 Larval brain. In all the following protocols cells are plated on round glass coverslips precoated with poly-L-lysine (Mw about 10K; Sigma, cat#P4707-50ML).

1.3.1.1 Wu (Wu et al., 1983) prepares a monolayer culture by adding 80–200 μL of cell suspension on poly-L-lysine coated round glass coverslips. Dispersed cells are allowed to settle for 2 h. Typically, one larval brain yields about 10^5 single cells in suspension of which a maximum of 20% adheres to the coverslip in 2 h.

1.3.1.2 Feiguin cultures in Schneider's medium supplemented with 1 mg/mL glucose, 5% FCS, 5% fly extract, and insulin 5 g/mL (Sigma, cat# 0516-5ML). 5% fly extract is prepared as follows. 1.3 g of female flies are mixed with 7 mL of Schneiders medium containing 1 mg/mL glucose and ground on ice through a glass homogenizer for about 20 min. The resulting homogenate is centrifuged for 30 min at 4 °C in an SW 55 TI rotor at 20K and the supernatant incubated for 5 min at 60 °C and centrifuged for 90 min at 4 °C. The supernatant is sterilized by filtering, aliquoted, and stored at −20 °C. We follow this procedure allowing cells to settle for 30 min.

1.3.1.3 Rebollo (Rebollo et al., 2007) implements and adapts the clot method (Forer & Pickett-Heaps, 1998). They add 4 mg/ml fibrinogen in plating medium (dissolved at 37 °C for 30 min). After plating out, cells are allowed to settle for 10 min and thrombin (0.5 U/mL, Amersham, cat# 27-0846-01) is added at a ratio of 1.5:100 (v:v). This method efficiently holds the cells down on the coverslip, thus minimizing both cell movement during time-lapse acquisition and cell loss when media is replaced. It also helps in the case of mutant cells that attach poorly to the surface.

1.3.1.4 Homem (Homem et al., 2013) uses Schneider's supplemented with 10% FCS, 2% penicillin and streptomycin, 20 mM L-Glutamine, 16.3 nM L-Glutathione, 10.8 nM α-ecdysone, and 3.5 nM insulin.

1.3.2 Embryo

1.3.2.1 Broadus and Doe (Broadus & Doe, 1997) plate 0.5 mL of cell suspension on clean glass coverslips and allow cells to settle for 30 min before adding and 1.5 mL of dissecting media. Grosskortenhaus (Grosskortenhaus et al., 2005) adds 2 mL of media, which is replaced every 30 min.

2. EX VIVO WHOLE MOUNT

The following includes the most common current protocols. A graphic view of our preferred protocol is shown in Figure 2. Examples of recordings carried out with our protocol are shown in Figure 3 (POLO-GFP panels) and Movie 3.

2.1 Dissection media

2.1.1 The most widely used is Schneider's *Drosophila* medium (Sigma-Aldrich S0146) supplemented in different ways including 10% FCS (Rusan & Peifer, 2007), Gibco Antibiotic-Antimycotic (Invitrogen) (Lerit, Plevock, & Rusan, 2014), or 1 mg/mL glucose (our own protocol). Cabernard and Doe (Cabernard & Doe, 2013) supplement sterile-filtered Schneider's *Drosophila* medium with 5 μL of 0.5 M ascorbic acid (Sigma-Aldrich A4034) and 50 μL of HyClone bovine growth serum (Thermo Scientific SH3054102) per 5 mL of Schneider's.

2.1.2 Siller (Siller, Serr, Steward, Hays, & Doe, 2005) uses D-22 (pH 6.75) insect medium (US Biological, Swampscott, MA) supplemented with 7.5% FBS.

2.2 Mounting media

2.2.1 Most protocols use their own dissection media that may be supplemented with fat bodies from third instar larvae (Cabernard & Doe, 2013) and 0.5 mM ascorbic acid (Siller et al., 2005).

2.2.2 In our hands Schneider's medium without modification works just fine. Glucose supplement and fat body are not essential.

2.3 Mounting procedure

Most of the mitotic activity in larval brains is found on the ventral side of the central brain and thoracic segments. Therefore, if recording cycling NBs is the goal, brains must be mounted such that the ventral side is the closest to the objective.

2.3.1 Cabernand and Doe (Cabernard & Doe, 2013), like Siller and colleagues previously (Siller et al., 2005) follow the procedure described by Kiehart and colleagues (Kiehart, Montague, Rickoll, Foard, & Thomas, 1994). The brain is placed on a standard gas-permeable membrane (Yellow Springs Instruments, Yellow Springs, YSI, OH Life Sciences 5793) held on a stainless steel slide and covered with a coverslip sealed with petroleum jelly. This procedure is good for both inverted and upright microscopes.

2.3.2 Rusan and Peifer (Rusan & Peifer, 2007) place the brain in the center of a pool of media in a 35 mm glass-bottomed dish (MatTek). The media is surrounded by Halocarbon oil 700 (HC), which supports a glass coverslip used to seal the chamber. Only inverted microscopes can be used.

2.3.3 Lerit et al. (Lerit et al., 2014) transfer the brains to a 30 μL drop of medium placed on the center of a gas-permeable culture dish. They then put four HC oil drops (about 30 μL each), one at each corner of the gas permeable membrane, cover with a 22 mm coverslip, and wait about 3 min as the oil and medium spread under the weight of the coverslip. This approach requires a fine adjustment of the position of the coverslip, which is controlled by removing excess media (using a tissue paper wick), until it gets close to the brain. If the tissue does not end up close enough to the coverslip, optical quality is compromised, but removing too much liquid might damage the brain. The chamber is sealed by encircling the coverslip with HC oil. This procedure is good for both inverted and upright microscopes.

2.3.4 Our own protocol takes advantage of the clot method (Forer & Pickett-Heaps, 1998), that provides a reasonable level of brain immobilization. Only inverted microscopes can be used, but it has the significant advantage that medium, which may be supplemented with drugs, can easily be replaced. It is as follows:

2.3.4.1 Prepare the fibrinogen solution: 10 mg/mL fibrinogen in Schneiders' medium (Gibco) incubated for 30 min at 37 °C until fibrinogen is dissolved.

2.3.4.2 Januschke (Januschke & Gonzalez, 2010) places a 2 μL drop of fibrinogen solution onto the bottom of a 35 mm glass bottom dish (World precision instruments). A wet paper is placed at the border of the dish to make a humid chamber.

2.3.4.3 The brain is then placed in the drop of fibrinogen and oriented with tungsten needles as desired. When multipoint visiting on automated XY stages is available, several drops/brains can be prepared in the same 35 mm dish in this way.

2.3.4.4 Spread out the drop with a tungsten needle. This is the *critical step* to place the brains as close as possible to the bottom.

2.3.4.5 Add 0.5 μL of thrombin (0.5 U/mL) on top of each brain, and let the clot form during 5 min. The clot becomes visible as a fibrillar mesh.

2.3.4.6 Carefully add 200 μL of Schneider's insect medium supplemented with 1 mg/mL glucose. The brain should stay close to the glass bottom. Rinse gently 3× with 200 μL Schneider's insect medium (may be supplemented with 1 mg/mL glucose) and cover with 1 mL of Schneider's insect medium supplemented with 1 mg/mL glucose prior to imaging. Fresh medium, which may be supplemented with drugs, can be easily added and replaced.

3. FLUORESCENT REPORTERS

A selection of the fluorescent reporters most commonly used to track centrioles, pericentriolar material, microtubules, and chromatin is shown in Table 2. Both upstream activation sequence (UAS) and ubiquitous promoters have been used with excellent results. The former allow for specific cell type and timing of expression, but having two components it is often more cumbersome to introduce in a specific mutant background; the latter is simpler, only one component, but constitutive expression may in certain cases be poisonous or result in aggregates of the fusion protein that interfere with microscopy.

4. REMARKS AND TROUBLESHOOTING

4.1 Primary cultures

- **4.1.1** NBs do not attach efficiently to the bottom glass. This is normally due to defective poly-lysine coating.
- **4.1.2** Few NBs; isolated NBs (NBs without associated ganglion mother cell (GMCs)) overstretched and elongated; poor mitotic index. These phenotypes can be caused by a variety of reasons. FCS that has gone bad is not an infrequent cause, and it is easy to amend. Readjusting collagenase incubation time and concentration might also help.
- **4.1.3** In general, for best results, record only "healthy" NBs. A variable fraction of NBs that make it through the procedure and adhere to the surface of the coverslip will never enter mitosis. Select cells that present a smooth and homogenous granular cytoplasm, are well attached to the imaging surface, and show no dramatic blebbing. NBs associated to one GMC are often good. In general, avoid cells with cell cycles longer than 2 h or mitosis longer than 15 min. It is most advisable to derive data only from cells that divide several times.

4.2 Ex vivo whole mount

- **4.2.1** Dissection per se is critical. Pinching off the optic lobes or tearing apart the ventral ganglion is a common error. These injuries produce a disruption in the whole architecture of the brain and impair imaging.
- **4.2.2** The fibrinogen clot density is critical. Too high concentrations affect NB performance while too low concentrations result in a soft clot that is not strong enough to keep the brain in place.
- **4.2.3** If the brain shifts after adding of medium to a less ideal position, it is possible to carefully adjust its position by manipulating the clot fibers and brain with tungsten needles.

Table 2 Fluorescent Fusion Proteins Commonly Used for Live Recording of Centrioles, Pericentriolar Material, Microtubules, and Chromatin in *Drosophila* NBs

	Protein	Tag	References
Centrioles	SAS4	GFP	Peel, Stevens, Basto, and Raff (2007)
	SAS6	GFP	Blachon et al. (2008)
	ANA2	GFP	Stevens, Dobbelaere, Brunk, Franz, and Raff (2010)
	ASL	YFP, mKate tomato	Januschke, Llamazares, Reina, and Gonzalez (2011), Varmark et al. (2007), Blachon et al. (2008)
	CNB Daughter centriole	YFP	Januschke et al. (2011)
	PACT Stronger in mother centriole	GFP,YFP, d2Eos	Martinez-Campos, Basto, Baker, Kernan, and Raff (2004), Conduit and Raff (2010), Januschke et al. (2011)
Pericentriolar material (PCM)	CNN	EGFP, RFP, PA-GFP	Conduit and Raff (2010), Megraw, Kilaru, Turner, and Kaufman (2002)
	POLO	GFP	Buszczak et al. (2007), Moutinho-Santos, Sampaio, Amorim, Costa, and Sunkel (1999)
	SPD-2	GFP	Dix and Raff (2007)
	γ-TUBULIN	EGFP	Hallen, Ho, Yankel, and Endow (2008)
Microtubules	Tubulin	GFP, Cherry, tdEOS	Lu, Fox, Lakonishok, Davidson, and Gelfand (2013), Rebollo and Gonzalez (2004), Rusan and Peifer (2007)
	MAP/JUPITER	Cherry; GFP	Cabernard and Doe (2013), Morin, Daneman, Zavortink, and Chia (2001)
	EB1	GFP	Rebollo et al. (2007)
Chromatin	His2AvD	GFP; mRFP1	Clarkson and Saint (1999), Schuh, Lehner, and Heidmann (2007)

ACKNOWLEDGMENTS

Work in C.G. laboratory is supported by BFU2012–32522 from the Spanish MIMECO, SGR Agaur 2009CG041413 from Generalitat de Catalunya and AdG 2011 294603 advanced grant from the European Research Council.

SUPPLEMENTARY DATA

Supplementary data related to this article can be found online at http://dx.doi.org/10.1016/bs.mcb.2015.03.003.

REFERENCES

Blachon, S., Gopalakrishnan, J., Omori, Y., Polyanovsky, A., Church, A., Nicastro, D., et al. (2008). Drosophila asterless and vertebrate Cep152 are orthologs essential for centriole duplication. *Genetics, 180*, 2081–2094.

Broadus, J., & Doe, C. Q. (1995). Evolution of neuroblast identity: seven-up and prospero expression reveal homologous and divergent neuroblast fates in *Drosophila* and *Schistocerca. Development (Cambridge, England), 121*, 3989–3996.

Broadus, J., & Doe, C. Q. (1997). Extrinsic cues, intrinsic cues and microfilaments regulate asymmetric protein localization in *Drosophila* neuroblasts. *Current Biology, 7*, 827–835.

Buszczak, M., Paterno, S., Lighthouse, D., Bachman, J., Planck, J., Owen, S., et al. (2007). The carnegie protein trap library: a versatile tool for Drosophila developmental studies. *Genetics, 175*, 1505–1531.

Cabernard, C., & Doe, C. Q. (2013). Live imaging of neuroblast lineages within intact larval brains in *Drosophila. Cold Spring Harbor Protocols, 2013*, 970–977.

Chan, L. N., & Gehring, W. (1971). Determination of blastoderm cells in *Drosophila melanogaster. Proceedings of the National Academy of Sciences of the United States of America, 68*, 2217–2221.

Clarkson, M., & Saint, R. (1999). A His2AvDGFP fusion gene complements a lethal His2AvD mutant allele and provides an in vivo marker for *Drosophila* chromosome behavior. *DNA Cell Biology, 18*, 457–462.

Conduit, P. T., & Raff, J. W. (2010). Cnn dynamics drive centrosome size asymmetry to ensure daughter centriole retention in *Drosophila* neuroblasts. *Current Biology, 20*, 2187–2192.

Dix, C. I., & Raff, J. W. (2007). *Drosophila* Spd-2 recruits PCM to the sperm centriole, but is dispensable for centriole duplication. *Current Biology, 17*, 1759–1764.

Feiguin, F., Llamazares, S., & Gonzalez, C. (1998). Methods in *Drosophila* cell cycle biology. *Current Topics in Developmental Biology, 36*, 279–291.

Forer, A., & Pickett-Heaps, J. D. (1998). Cytochalasin D and latrunculin affect chromosome behaviour during meiosis in crane-fly spermatocytes. *Chromosome Research, 6*, 533–549.

Furst, A., & Mahowald, A. P. (1985a). Cell division cycle of cultured neural precursor cells from *Drosophila. Developmental Biology, 112*, 467–476.

Furst, A., & Mahowald, A. P. (1985b). Differentiation of primary embryonic neuroblasts in purified neural cell cultures from *Drosophila. Developmental Biology, 109*, 184–192.

Grosskortenhaus, R., Pearson, B. J., Marusich, A., & Doe, C. Q. (2005). Regulation of temporal identity transitions in *Drosophila* neuroblasts. *Developmental Cell, 8*, 193–202.

Hallen, M. A., Ho, J., Yankel, C. D., & Endow, S. A. (2008). Fluorescence recovery kinetic analysis of gamma-tubulin binding to the mitotic spindle. *Biophysical Journal, 95*, 3048–3058.

Harzer, H., Berger, C., Conder, R., Schmauss, G., & Knoblich, J. A. (2013). FACS purification of *Drosophila* larval Neuroblasts for next generation sequencing. *Nature Protocols, 8*, 1088–1099.

Homem, C. C., Reichardt, I., Berger, C., Lendl, T., & Knoblich, J. A. (2013). Long-term live cell imaging and automated 4D analysis of drosophila neuroblast lineages. *PLoS One, 8*, e79588.

Januschke, J., & Gonzalez, C. (2010). The interphase microtubule aster is a determinant of asymmetric division orientation in *Drosophila* neuroblasts. *Journal of Cell Biology, 188*, 693–706.

Januschke, J., Llamazares, S., Reina, J., & Gonzalez, C. (2011). *Drosophila* neuroblasts retain the daughter centrosome. *Nature Communications, 2*, 243.

Kaltschmidt, J. A., Davidson, C. M., Brown, N. H., & Brand, A. H. (2000). Rotation and asymmetry of the mitotic spindle direct asymmetric cell division in the developing central nervous system. *Nature Cell Biology, 2*, 7–12.

Kiehart, D. P., Montague, R. A., Rickoll, W. L., Foard, D., & Thomas, G. H. (1994). High-resolution microscopic methods for the analysis of cellular movements in *Drosophila* embryos. *Methods in Cell Biology, 44*, 507–532.

Lerit, D. A., Plevock, K. M., & Rusan, N. M. (Jul 7 2014). Live imaging of Drosophila larval neuroblasts. *Journal of Visualized Experiments: JoVE*, (89). http://dx.doi.org/10.3791/51756.

Lu, W., Fox, P., Lakonishok, M., Davidson, M. W., & Gelfand, V. I. (2013). Initial neurite outgrowth in *Drosophila* neurons is driven by kinesin-powered microtubule sliding. *Current Biology, 23*, 1018–1023.

Martinez-Campos, M., Basto, R., Baker, J., Kernan, M., & Raff, J. W. (2004). The *Drosophila* pericentrin-like protein is essential for cilia/flagella function, but appears to be dispensable for mitosis. *Journal of Cell Biology, 165*, 673–683.

Megraw, T. L., Kilaru, S., Turner, F. R., & Kaufman, T. C. (2002). The centrosome is a dynamic structure that ejects PCM flares. *Journal of Cell Science, 115*, 4707–4718.

Morin, X., Daneman, R., Zavortink, M., & Chia, W. (2001). A protein trap strategy to detect GFP-tagged proteins expressed from their endogenous loci in *Drosophila. Proceedings of the National Academy of Sciences of the United States of America, 98*, 15050–15055.

Moutinho-Santos, T., Sampaio, P., Amorim, I., Costa, M., & Sunkel, C. E. (1999). In vivo localisation of the mitotic POLO kinase shows a highly dynamic association with the mitotic apparatus during early embryogenesis in *Drosophila. Biology of the Cell/under the Auspices of the European Cell Biology Organization, 91*, 585–596.

Peel, N., Stevens, N. R., Basto, R., & Raff, J. W. (2007). Overexpressing centriole-replication proteins in vivo induces centriole overduplication and de novo formation. *Current Biology, 17*, 834–843.

Rebollo, E., & Gonzalez, C. (2004). Time-lapse imaging of male meiosis by phase-contrast and fluorescence microscopy. *Methods in Molecular Biology (Clifton, NJ), 247*, 77–87.

Rebollo, E., & Gonzalez, C. (Jun 2010). Time-lapse imaging of embryonic neural stem cell division in *Drosophila* by two-photon microscopy. *Current Protocols in Stem Cell Biology* (Chapter 1):Unit1H 2. http://dx.doi.org/10.1002/9780470151808.sc01h02s13.

Rebollo, E., Roldan, M., & Gonzalez, C. (2009). Spindle alignment is achieved without rotation after the first cell cycle in Drosophila embryonic neuroblasts. *Development (Cambridge, England), 136*, 3393–3397.

Rebollo, E., Sampaio, P., Januschke, J., Llamazares, S., Varmark, H., & Gonzalez, C. (2007). Functionally unequal centrosomes drive spindle orientation in asymmetrically dividing *Drosophila* neural stem cells. *Developmental Cell, 12*, 467–474.

Rusan, N. M., & Peifer, M. (2007). A role for a novel centrosome cycle in asymmetric cell division. *Journal of Cell Biology, 177*, 13–20.

Schneider, I., & Blumenthal,, A. B. (1978). Drosophila cell and tissue culture. In M. Ashburner, & T. R. F. Wright (Eds.), *The Genetics and Biology of Drosophila* (Vol2a). (pp. 265–315). London: Academic Press.

Schuh, M., Lehner, C. F., & Heidmann, S. (2007). Incorporation of *Drosophila* CID/CENP-A and CENP-C into centromeres during early embryonic anaphase. *Current Biology, 17*, 237–243.

Seecof, R. L., Teplitz, R. L., Gerson, I., Ikeda, K., & Donady, J. (1972). Differentiation of neuromuscular junctions in cultures of embryonic *Drosophila* cells. *Proceedings of the National Academy of Sciences of the United States of America, 69*, 566–570.

Siller, K. H., Serr, M., Steward, R., Hays, T. S., & Doe, C. Q. (2005). Live imaging of *Drosophila* brain neuroblasts reveals a role for Lis1/dynactin in spindle assembly and mitotic checkpoint control. *Molecular Biology of the Cell, 16*, 5127–5140.

Stevens, N. R., Dobbelaere, J., Brunk, K., Franz, A., & Raff, J. W. (2010). Drosophila Ana2 is a conserved centriole duplication factor. *Journal of Cell Biology, 188*, 313–323.

Varmark, H., Llamazares, S., Rebollo, E., Lange, B., Reina, J., Schwarz, H., et al. (2007). Asterless is a centriolar protein required for centrosome function and embryo development in *Drosophila*. *Current Biology, 17*, 1735–1745.

Wu, C. F., Suzuki, N., & Poo, M. M. (1983). Dissociated neurons from normal and mutant *Drosophila* larval central nervous system in cell culture. *Journal of Neuroscience, 3*, 1888–1899.

CHAPTER

Centrosomes and spindles in ascidian embryos and eggs

17

Alex McDougall[1], Janet Chenevert, Gerard Pruliere, Vlad Costache, Celine Hebras, Gregory Salez, Remi Dumollard

Sorbonne Universités, UPMC Univ Paris 06, and CNRS, Laboratoire de Biologie du Développement de Villefranche-sur-mer, Observatoire Océanologique, Villefranche-sur-Mer, France

[1]Corresponding author: E-mail: dougall@obs-vlfr.fr

CHAPTER OUTLINE

Introduction 317
1. Methods 318
References 337

Abstract

During embryonic development and maternal meiotic maturation, positioning of the mitotic/meiotic spindle is subject to control mechanisms that meet the needs of the particular cell type. Here we review the methods, molecular tools, and the ascidian model we use to study three different ways in which centrosomes or spindles are positioned in three different cellular contexts. First, we review unequal cleavage in the ascidian germ lineage. In the germ cell precursors, a large macromolecular structure termed the centrosome-attracting body causes three successive rounds of unequal cleavage from the 8- to the 64-cell stage. Next, we discuss spindle positioning underlying the invariant cleavage pattern. Ascidian embryos display an invariant cleavage pattern whereby the mitotic spindle aligns in a predetermined orientation in every blastomere up to the gastrula stage (composed of 112 cells). Finally, we review methods and approaches to study meiotic spindle positioning in eggs.

INTRODUCTION

Ascidians have been used to study developmental mechanisms for more than a century (see Satoh, 2014 for a comprehensive review of the literature). Ascidians are marine invertebrate chordates that belong to the tunicate subphylum and are a sister

Methods in Cell Biology, Volume 129, ISSN 0091-679X, http://dx.doi.org/10.1016/bs.mcb.2015.03.006

group to the vertebrates (Delsuc, Brinkmann, Chourrout, & Philippe, 2006). Ascidians are a paraphyletic group split into three orders (the aplousobranchs, the phlebobranchs, and the stolidobranchs, which replaced enterogona and pleurogona as orders) and there is some fossil evidence indicating that ascidians have been present since the Early Cambrian (Chen et al., 2003). Based on noncoding DNA sequence, it has been estimated that two disparate ascidian species may be as different from each other as fish are from humans (Stolfi et al., 2014). Remarkably, however, despite these differences at the level of their noncoding DNA, disparate ascidian embryos are so similar that the system for naming the different blastomeres up to the gastrula stage developed by Conklin in 1905 for a stolidobranch ascidian (*Styela partita*) is currently used for phlebobranch embryos (*Ciona*, *Phallusia*). This is possible because the invariant cleavage pattern is perfectly conserved. Moreover, distantly related ascidians also display the same three rounds of unequal cleavage in germ cell precursors and the same relative asynchrony in cell cycle duration in vegetal versus animal blastomeres. One of our aims is to identify the proteins that regulate cell size, cell position, and cell number and so understand how developmental mechanisms (a combination of maternal mechanisms and zygotic gene-regulatory networks) control unequal cleavage, invariant cleavage, and cell cycle asynchrony. We also work on unequal cleavage during polar body extrusion in the ascidian egg. Here, we detail the cell biological tools available and methods for imaging early *Phallusia* embryos up to the 64-cell stage (see also Robin et al., 2011) focusing on unequal cleavage and invariant cleavage in 3D live imaging, as well as methods and tools for following meiotic spindle behavior in eggs. Methods to study cell cycle duration and the onset of cell cycle asynchrony in ascidian embryos (Dumollard, Hebras, Besnardeau, & McDougall, 2013) will not be covered in this review.

1. METHODS

1. Preparing ascidian eggs and sperm

We use the European ascidian *Phallusia mammillata* because its eggs and early embryos are completely transparent and also because microinjected mRNAs encoding fluorescent fusion protein markers are translated in the unfertilized eggs and early embryos (Prodon et al., 2010). *Ascidiella aspersa* eggs and early embryos are also transparent and also translate well microinjected mRNAs encoding proteins coupled to GFPs and could be used for this type of project when *Phallusia* is not available (Levasseur & McDougall, 2000). *Ascidiella* has a more widespread distribution than *P. mammillata*, e.g., it can be collected at Woods Hole, USA as well as in Europe, and northern Japan (Nishikawa et al., 2014) although its gametes are not as abundant as those of *Phallusia*.

1.1 Ascidians are covered by a protective tunic made of a type of cellulose. To remove the soft-bodied animal from its tunic, cut from the buccal siphon to the atrial siphon with scissors or a razor blade. Once cut, pry open the animal and place with eggs/sperm facing up on some tissue paper to absorb excess liquid (ascidians are hermaphrodites that simultaneously produce sperm and eggs, and some species such as *Phallusia* are self-fertilizing).

1.2 To a Petri dish (5 cm diameter) add 4.5 mL filtered seawater (0.2 μm filter). We use natural seawater supplemented with 5 mM TAPS pH buffer (adjust seawater pH to 8.2 with HCl and NaOH).

1.3 Puncture the oviduct with a needle and collect the eggs that flow out with a plastic pipette and transfer to the 5 mL Petri dish containing the filtered seawater. Massage the oviduct pushing out more eggs and continue collecting eggs until a sufficient quantity is present. *Phallusia* can produce up to about 1 mL (500,000) unfertilized eggs all arrested at metaphase I of meiosis.

1.4 Then collect the sperm from the same animal. First absorb most of the moisture from the animal, then puncture the sperm duct with a needle to collect the sperm. Transfer the sperm to an Eppendorf tube (1.5 mL) and store at 4 °C. Sperm stored in this manner can be used for a few days.

1.5 Dechorionation: To remove the follicular cells (the chorion) from the eggs, which improves microscopic observation yet does not perturb development, trypsin should be added to the eggs. Aliquots, 1 mL, of 1% trypsin solution in filtered TAPS-buffered (5 mM, pH 8.2) seawater are prepared and stored at −20 °C. Add 1 mL of 1% trypsin to the eggs in the 5 mL Petri dish. Place on a rocker and gently swirl the eggs/trypsin mixture for 1.5 h at 21 °C. Observe the process with a stereomicroscope, and when around 10% of the eggs have started to dechorionate, gently pipette the eggs in the Petri dish with a large bore plastic or glass Pasteur pipette which will mechanically dechorionate a large percentage of the remaining eggs. Pass the eggs gently up and down in the pipette about five times making no air bubbles which will kill the eggs.

1.6 To collect the dechorionated eggs prepare 5 mL dishes that have been coated with a 0.1% gelatin/formaldehyde mixture, which will prevent the dechorionated eggs from sticking to plastic or glassware (see Sardet et al., 2011 for details). Gently swirl the dechorionated eggs and collect those that come to the center and transfer them to a fresh GF-coated dish containing filtered TAPS seawater. Continue until all dechorionated eggs have been transferred. Then using the same swirling motion collect the eggs at the center of the GF-Petri dish and remove the majority of the seawater (this will contain dead and chorionated eggs). Do this about five times to prepare a batch of dechorionated eggs.

1.7 Store the dechorionated eggs at 16–19 °C until ready for use in the GF-Petri dish. For best embryonic development, the eggs should be used the same day, but they will still develop the next day depending on the batch quality.

1.8 To fertilize the eggs, first activate the sperm. Add sufficient NaOH to TAPS-buffered seawater to raise pH to 9.5. Transfer 5 μL sperm to 0.5 mL pH 9.5

seawater and leave on a rocker for 20 min. This will activate the sperm. Once activated add about 5 μL sperm near the eggs in a Petri dish (collect the eggs in the center first by the swirling motion) and observe. Fertilization triggers a shape change that is readily visible. Once a sufficient percentage of eggs have changed shape (contracted), wash the eggs by swirling again a few times with fresh exchanges of sea water to remove the excess sperm (this will prevent polyspermy).

2. Live cell markers for studying centrosome/spindle positioning in early embryos

Centrosomes and mitotic spindles can be observed in *Phallusia* eggs and embryos during live imaging or after fixation and antibody staining. Table 1 gives a list of the molecular tools we have created or used to image centrosomes and spindles in live embryos/eggs. We have also included in this table some molecular markers that we have tested but do not work well in *Phallusia*. Figure 1 shows examples of centrosome and spindle pole imaging in *Phallusia* eggs and embryos using these markers. Of these Girdin may be a useful marker of centrosomes in ascidians since it labels centrosomes during their entire cycle and the background labeling is low (Figure 1(C)).

Since spindle positioning is dynamic and can change significantly within 1 min it has become useful to compare live cell imaging data with fixed data to get an overall view of spindle behavior. In order to obtain a dynamic view of spindle orientation during the process of unequal cleavage, live cell imaging using many of the tools in Table 1 proved extremely useful and also helped to better understand our fixed data.

Centrosome and spindle positioning during unequal cleavage in *Phallusia* is difficult to understand using only fixed samples since the centrosomes and spindle move significantly at key moments during the cell cycle, particularly during prometaphase when the CAB attracts one spindle pole. Three successive cleavages, beginning at the 8-cell stage, create two small posterior germ cell precursors at the 64-cell stage (Nishikata, Hibino, & Nishida, 1999; Prodon et al., 2010). At the eight-cell stage, two centrosome-attracting bodies (CABs) become active, one in each B4.1 posterior vegetal blastomere. The CAB causes unequal cleavage of the blastomere in which it resides and is inherited by the smaller blastomere following cell division. Two small cells are thus created in the 16-cell stage embryo each containing one CAB. These small blastomeres divide unequally again creating two even smaller posterior vegetal blastomeres at the 32-cell stage. This process is repeated once more creating two tiny vegetal posterior germ cell precursors (B7.6 pair) at the 64-cell stage that each contains one CAB inherited from their great grandmothers (B4.1 pair at the 8-cell stage). The method for analyzing these three successive cell divisions is illustrated in Figure 2 (A, B live embryos, C fixed embryos).

2.1 First microinject the mRNA(s) encoding the protein(s) of interest into unfertilized eggs as prepared earlier (list 1.7). We often inject a mixture of two different mRNAs (see McDougall, Lee, & Dumollard, 2014 for a recent review of the microinjection procedure).

Table 1 Live Cell Fluorescent Markers Tested in *Phallusia*

	Gene Model	Localization	Phenotype	Source	Reference for Use in Ascidians
Centrosomes/spindle poles					
Venus/ mCherry::First 632 amino acids of Girdin	KH2012:KH.C11.506 (ci)	Centrosome/spindle poles only. Whole centrosome cycle.	No	Our lab	Unpublished
Ndel::Venus	KH2012:KH.C3.451 (ci); prtcontig02430 (pm)	Full centrosome cycle	Eventually blocks cytokinesis	Our lab	Unpublished
Lis1::Venus	KH2012:KH.C11.607 (ci); AHC0AAA96YA11.CONTIG (pm)	Spindle poles, also nuclear membrane	Eventually blocks cytokinesis	Our lab	Unpublished
Plk1::Venus/ mCherry	KH2012:KH.C12.238 (ci)	Spindle poles. Also labels midbody and CAB.	No	Our lab	Hebras and McDougall (2012)
Dynactin:: Venus (p150 glued)	KH2012:KH.C2.685 (ci)	Centrosome but signal to noise ratio worse than girdin. Also labels membranes.	No	Our lab	Unpublished
EB3::3GFP	AY893969 (synthetic construct *Homo sapiens*)	Centrosome if expressed at low levels. Dynamic microtubule plus ends if expressed at high levels. Also labels meiotic spindle poles.	No	Gift from P. Lenart	Unpublished

Continued

Table 1 Live Cell Fluorescent Markers Tested in *Phallusia*—cont'd

	Gene Model	Localization	Phenotype	Source	Reference for Use in Ascidians
Aurora A/B::Venus/mCherry	KH.C7.219.v1.A.SL1-1 (ci)	Centrosome, spindle midzone and midbody. Accumulates in nucleus.		P. Lemaire Lab	Roure et al. (2007), Hebras and McDougall (2012)
Venus/mCherry::TPX2	KH2012:KH.C3.181 (ci)	Centrosome then microtubules and midbody at higher concentration.	No	Our lab	Hebras and McDougall (2012)
Meiotic spindle poles					
Calmodulin::Venus/mCherry	KH2012:KH.C3.79	Labels meiotic spindle poles and less strongly spindle. Does not label mitotic spindle poles strongly.		Our lab	Unpublished
Microtubule					
MAP7::GFP/Rfp1/mCherry	BC052637 (mouse MAP7 N terminal fragment)	Microtubules and spindle poles during mitosis	No	Our lab	Prodon et al. (2010)
Ensconsin::3GFP	NP_003971.1 (human)	Microtubules throughout interphase/mitosis and centrosomes starting at prophase	No	Gift from H Yasuo	Unpublished
Rhodamine tubulin protein			No, but time consuming	Cytoskeleton inc	McDougall and Levasseur (1998)

Plasma membrane/cortex					
PH domain:: GFP/Tomato	EAW64507.1 (human PLC Delta 1, Met1-Ile 175)	Plasma membrane only (no internal membrane label or vesicles)	No in early embryos but possible phenotype at tadpole stage (L. Christiaen, personal communication).	Gift from T. Meyer (Stauffer, Ahn, & Meyer, 1998) Tomato version a gift from H. Yasuo	Carroll et al. (2003), Prodon et al. (2010)
ABD::Venus	P13466.1 (ABP-120 Dictyostelium)	Cortex, furrows	Yes; surface blebbing	A Gift from E. Houliston	Prodon et al. (2010)
LifeAct GFP or RFP protein	NP_014882 (Abp140, 1–17) *Saccharomyces cerevisiae*	Cortex, furrows	No		Riedl et al. (2008)
Cell mask orange	C10045	Plasma membrane and internalized vesicles.	No	Company Cat no. C10045	Molecular probes
FM4-64	T-13320	Plasma membrane and internalized vesicles.	No	Company Cat no. T-13320	Molecular probes; Prodon et al. (2010)
Centrosome proteins whose GFP fusions did not prove to be useful markers					
Gamma tubulin	KH2012:KH.C11.227 (ci); AHC0AAA242YG06RM1 (pm)	No centrosome label. GFP tried on N or C terminal ends.		Our lab	Unpublished Note: The gamma tubulin antibody GTU88 works well in fixed samples.

Continued

Table 1 Live Cell Fluorescent Markers Tested in *Phallusia*—cont'd

	Gene Model	Localization	Phenotype	Source	Reference for Use in Ascidians
Percicentrin subdomain (PACT)	KH2012: KH.C10.605.v1.B.ND1-1 (ci)	No centrosome label. GFP tried on N or C terminal ends.		Our lab based on Raff et al.	Unpublished
Centrobin	KH2012:KH.C7.69 (ci); C2982783 13.0 (pm)	Centrosome labeled in early embryos then ectopic structures labeled. GFP tried on N or C terminal ends.		Our lab	Unpublished
Alpha tubulin	KH2012:KH.C11.291 (ci); AHC0AAA4YE02RM1 (pm)	Cytoplasmic label and faint spindle poles. C Terminal tag.		Our lab	Unpublished
Odf2	Echinoderm	Labels centrosomes but also numerous bright spots or vesicles. C Terminal tag.		P. Lenart lab	Unpublished

The type of fluorescent marker, together with its gene model is indicated. Whether the fluorescent fusion tag is N or C terminal is indicated by the placement of the tag before or after the protein name, respectively. When localization was observed using more than one fluorescent tag (e.g., Venus or mCherry) the data are indicated. The phenotype column shows whether there were unwanted side-effects. It should be noted that one should be careful with combinations of two or more markers that on their own do not induce an unwanted phenotype. The source of each molecular tool is indicated, and whether there are published data on the molecular tool (we apologize for any omissions). Also, we have included some molecular markers that did not label the desired structures (in our experiments).

2.2 For studies of unequal cleavage, centrosome and spindle pole behavior should be analyzed at the 8–16, the 16–32, or the 32–64-cell stages. To perform experiments aimed at following events starting at the eight-cell stage, microinject the eggs at least 3 h before fertilization with mRNAs to allow sufficient time for translation of the fluorescent fusion proteins. We measured the molar concentration of the GFP protein that can be detected above background autofluorescence (by injecting known concentrations of GFP protein) and found that we could detect a minimum concentration of around 50 nM in eggs of *A. aspersa*, a species closely related to *Phallusia* (Levasseur & McDougall, 2000).

2.3 Coat a slide and coverslip on one side with 0.1% GF and leave to air dry then wash with water. Apply two strips of Dow Corning grease to the GF-coated side of a slide. Transfer four-cell stage embryos using a mouth pipette or similar device to the slide between the strips of grease in about 50 μL sea water. Apply the GF cover slip (GF side toward eggs). Gently compress so as to make a seal but not to compress the embryos. Fill up the remaining space with TAPS-filtered seawater. Apply grease to the remaining two edges and seal. If the embryos are not touching one another, normal tadpole larvae will develop in this chamber.

2.4 For imaging we use either a confocal microscope or a wide-field epifluorescence system. Typically, a new construct is first evaluated for relevant localizations by epifluorescence time-lapse acquisition (list 2.5), which allows rapid and long-term temporal sampling with minimal photo toxicity. In a second step, observations are made by confocal microscopy (list 2.6) which yields greater spatial detail but can damage the specimen after prolonged imaging.

2.5 For wide-field epifluorescence in toto 3D imaging, we use an x–y motorized stage and a Piezo-mounted z-stepper to observe the embryos. (A motorized z-stage may also be used, which is slower than the Piezo.) Typically, we can follow 10 embryos over time at 10 focal planes for several hours collecting a z-stack every minute. The number of focal planes can be increased to document precise movements of small objects such as centrosomes and midbodies. Exposure at each wavelength depends on the strength of the fluorescent signal, ranging from 50 to 300 ms using 2 × 2 binning (Cool Snap 2 cooled CCD camera). The imaging system is controlled by MetaMorph software.

2.6 For confocal imaging we use a Leica SP5. Because of bleaching it is advisable to monitor one unequal cleavage only at higher resolution to prevent damaging the embryo. Similar to the wide-field method, we typically image eight focal planes centered around the centrosome or spindle poles collecting z-stacks every minute.

2.7 To calculate distances and velocities, centrosome movements are tracked using acquisition software (Leica LAS or MetaMorph) or the MTrackJ plugin of ImageJ.

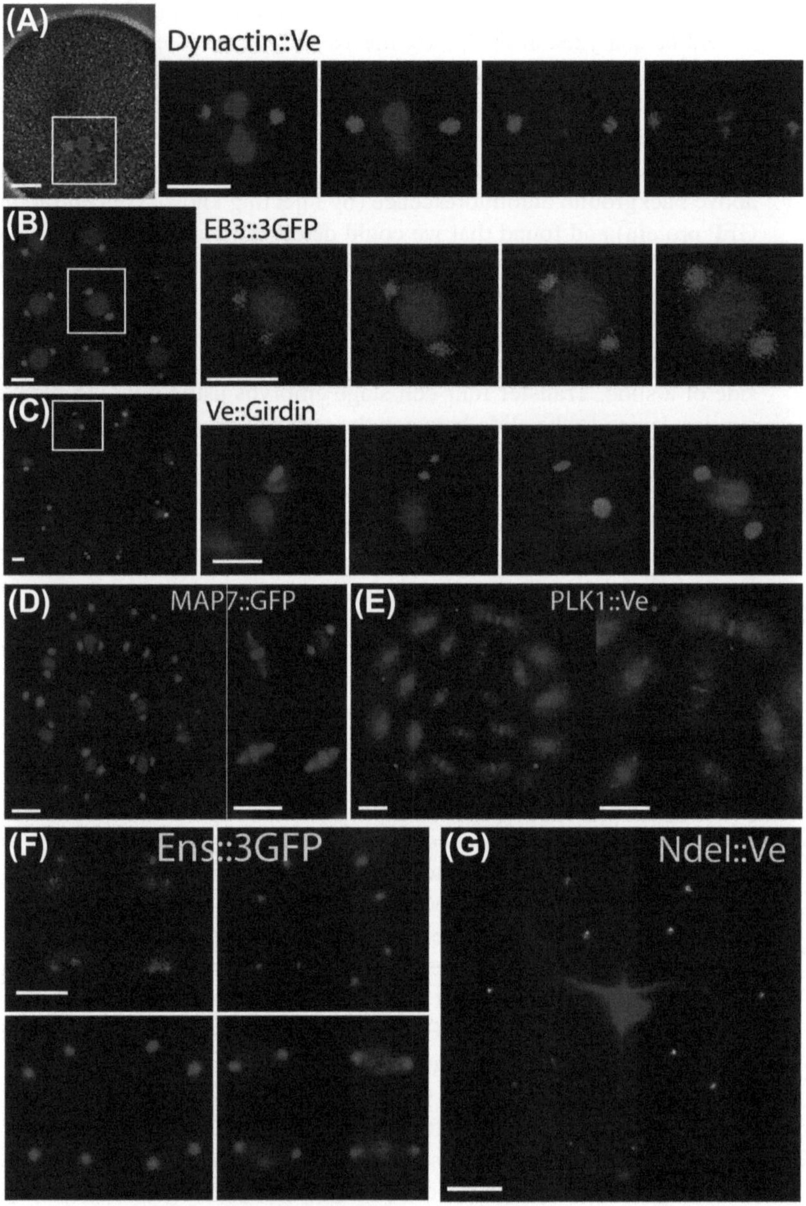

FIGURE 1 Centrosome live labels in *Phallusia*.

We observe that some markers will label centrosomes well during the entire cell cycle (girdin, EB3, Ndel) whereas others such as MAP7 label spindle poles only during mitosis (which can be useful for detecting the onset of mitosis). (A) Fertilized egg derived from an unfertilized egg injected with mRNA encoding Dynactin::Venus (green) and HH2B::Rfp1 (red). The paternal centrosomes are labeled with Dynactin::Venus and the male and female pronuclei

2.8 For all imaging experiments, the room housing the microscopy system is cooled to 19 °C.

3. Studying invariant cleavage pattern

Ascidian embryos display an invariant cleavage pattern that has been described up to the time of gastrulation for several species when the embryo has either 110 cells (stolidobranchs such as *S. partita* or *Halocynthia roretzi*) or 112 cells (phlebobranchs such as *Ciona intestinalis* and *P. mammillata*). The first cell division creates the right and left sides of the embryo which thus forth display bilateral symmetry. Bilateral symmetry is advantageous for experiments since it permits, among other things, molecular tools to be expressed in only one half of the embryo (by injecting mRNA at the two-cell stage) to determine whether they create unwanted side-effects or interesting dominant-negative phenotypes. The nomenclature system devised by Conklin (1905) for a stolidobranch (*S. partita*) is used today for stolidobranchs and phlebobranchs. This indicates that the cleavage pattern has been perfectly conserved in these two distant orders of ascidians. We do not currently know how developmental mechanisms and the physical constraints encountered by cells (see review by Thompson, 1917) operate to generate this cleavage pattern. Our current working hypothesis is that cell shape is important and Hertwig's "long-axis rule" of cell division with a planar constraint (see Wilson, 1900 for a review) underlies the cleavage pattern. The long axis rule predicts that cells divide orthogonal to their longest axis. In pre-gastrula stage ascidian embryos all blastomeres divide such that the daughter blastomeres always lie at the surface of the embryo: the first two internal blastomeres (A8.12 pair) are generated just before gastrulation (in phlebobranchs). Thus, if Hertwig's long axis rule operates then it operates parallel to the outside (or apical) surface of the embryo.

In animal cells, centrosomes duplicate following mitosis and migrate to opposite sides of the nucleus causing daughter cells to divide at right angles to their mothers. Following two successive cell divisions, a cell from a 16-cell stage

with HH2B::Rfp1. A time lapse of the same fertilized egg is shown from pronuclear apposition to metaphase. Scale bar = 20 μm. (B) An embryo previously injected with EB3::3GFP and HH2B::Rfp1. Both centrosomes are labeled around each nucleus. Scale bar = 10 μm. (C) A 16-cell stage embryo previously injected with Venus::Girdin and HH2B::Rfp1. The inset shows a time lapse. One centrosome (or perhaps two just before they separate) is shown in the first image followed by centrosome movements around the nucleus. Scale bar = 10 μm. (D) A 32-cell stage embryo previously injected with MAP7::GFP and HH2B::Rfp1. Scale bar = 20 μm. (E) A 32-cell stage embryo previously injected with Plk1::GFP. Plk1 labels the spindle poles as well as the spindle midzone and midbodies. Scale bar = 20 μm. (F) A 32-cell stage embryo previously injected with Ens::3GFP. Scale bar = 20 μm. (G) A 16-cell stage embryo previously injected with Ndel::Venus and labeled with Cell Mask (red). Scale bar = 20 μm. (See color plate)

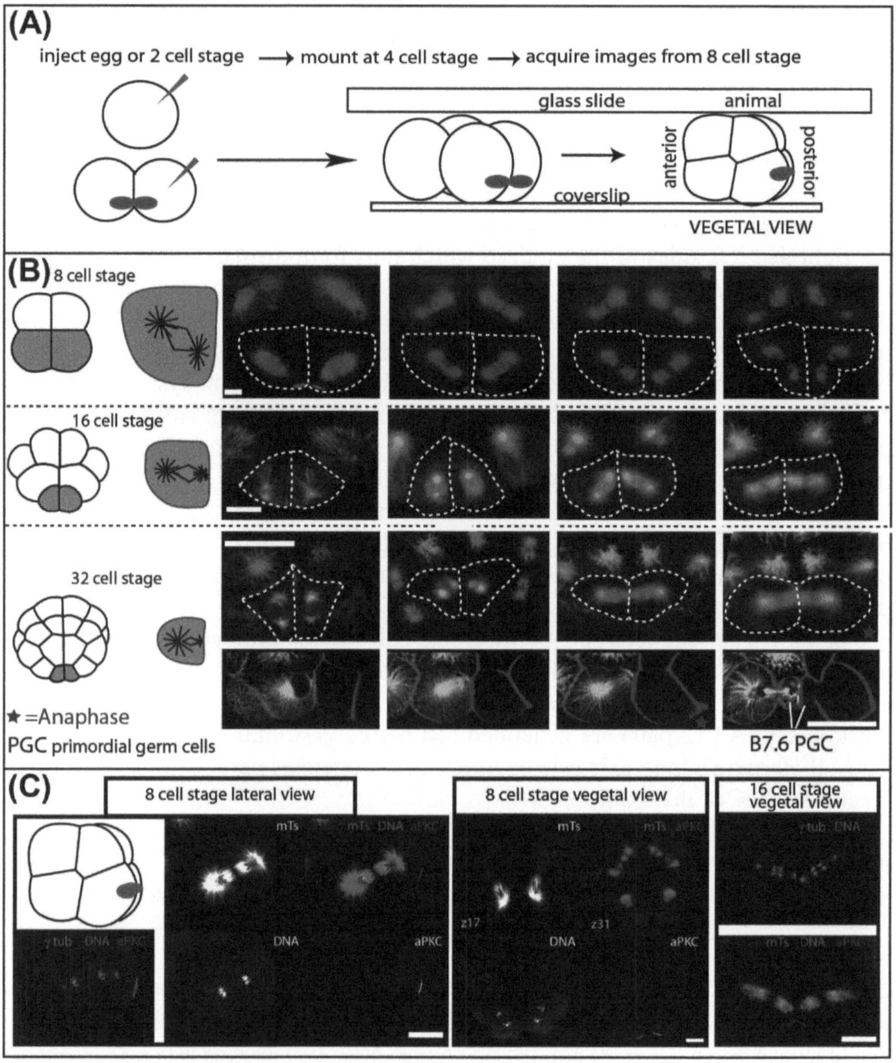

FIGURE 2 Three unequal cleavages from 8- to 64-cell stage.

(A) Method to obtain a standardized vegetal view for live imaging. Orange discs represent the pair of CABs, a structure responsible for unequal cleavage localized in the posterior vegetal cortex of the germ cell precursors. (B) Portions of confocal time lapse sequences covering the indicated stages. Green cells in the drawings (left) indicate the CAB-containing cells which are filmed (right). First, second, and third rows: Ens::3GFP label recorded by confocal microscopy. Fourth row: MAP7::GFP and FM4-64 membrane label recorded by confocal microscopy. At the eight-cell stage, the spindle forms roughly in the center of the B4.1 cell, then without changing alignment elongates toward the posterior animal cortex. The aster nearest the CAB (the midline aster) becomes progressively smaller over the three divisions. At the 16-cell stage, the spindle forms on the animal–vegetal axis of the B5.2 cell, then embryo that divides planar to the outside surface of the embryo will thus create four cells in a square pattern (by default) at the 64-cell stage (top schematic in Figure 3). In the absence of cell migration, displacement, or death (none of which occur at this stage of development), it is clear that any deviation from the square pattern would be due to centrosome migration or mitotic spindle rotation. There are two possible extreme deviations from the default pattern: one daughter cell divides in the same orientation as its mother or both do. When one daughter divides as its mother a T shape composed of four cells is formed (see schematic 2 in Figure 3). When both divide as their mother a line of four cells is formed (see schematic in Figure 3(A)). Cells from the same grandmother arranged in squares, Ts, and lines all exist in actual *Ciona* embryos (Hotta et al., 2007 images taken from http://chordate.bpni.bio.keio.ac.jp/faba/1.4/top.html). We therefore used this type of analysis to predict that mitotic spindles (or centrosomes) would rotate (through approximately 90°) at the 32-cell stage in the three specific blastomeres highlighted by white stars to give rise to the patterns observed at the 64-cell stage. As already stated, the same pattern of cell division occurs in *Phallusia* as in *Ciona* and we found that the whole mitotic spindles rotated in all three highlighted blastomeres but not appreciably in neighboring blastomeres (Figure 3(C)). We performed a similar analysis for all vegetal cells (not shown). By analyzing the shape of the cell in the planar axis where the spindle orients, we sought to determine if the spindles rotated into the longest planar cellular axis. In those cells where the spindles do not rotate, we also measured the longest planar axis of the cell to determine whether spindles did not rotate because they were already aligned in the longest axis (Dumollard, Besnardeau, Hebras, Salez, & McDougall, in preparation).

Here we provide a protocol for labeling the cell membranes and metaphase spindle poles of ascidian embryos, as well as our protocol for segmenting blastomeres to

rotates such that the vegetal centrosome migrates rapidly toward the midline and the animal centrosome shifts laterally toward the B5.1 sister cell (Prodon et al., 2010). At the 32-cell stage, one centrosome is positioned near the midline during interphase, such that the spindle forms angled toward the midline then elongates. In summary, over the three divisions the CAB centrosome is captured earlier in the cell cycle, and travels closer to the cortex as the cells become smaller, ultimately generating the tiny B7.6 pair of germ cell precursors. The red star represents anaphase when the position of the cleavage plane is defined. Scale bars = 20 μm. (C) Confocal images of CAB-containing cells fixed at the indicated stages and immunolabeled to detect endogenous microtubules or centrosomes. Gamma tubulin (green) = antibody GTU88 which labels centrosomes; aPKC (red) = antibody SC216 which labels the cortical face of the CAB (Patalano et al., 2006); mTs = rat monoclonal YL1/2 which labels asters and spindles (we also routinely use mouse monoclonal DM1a for this purpose). In all panels fixative was cold methanol/EGTA. This treatment is best for microtubule-based structures, but often results in cortical damage as can be observed in aPKC label lateral view. DNA = labeled with 1 μg/mL Hoechst. Scale bars = 20 μm. (See color plate)

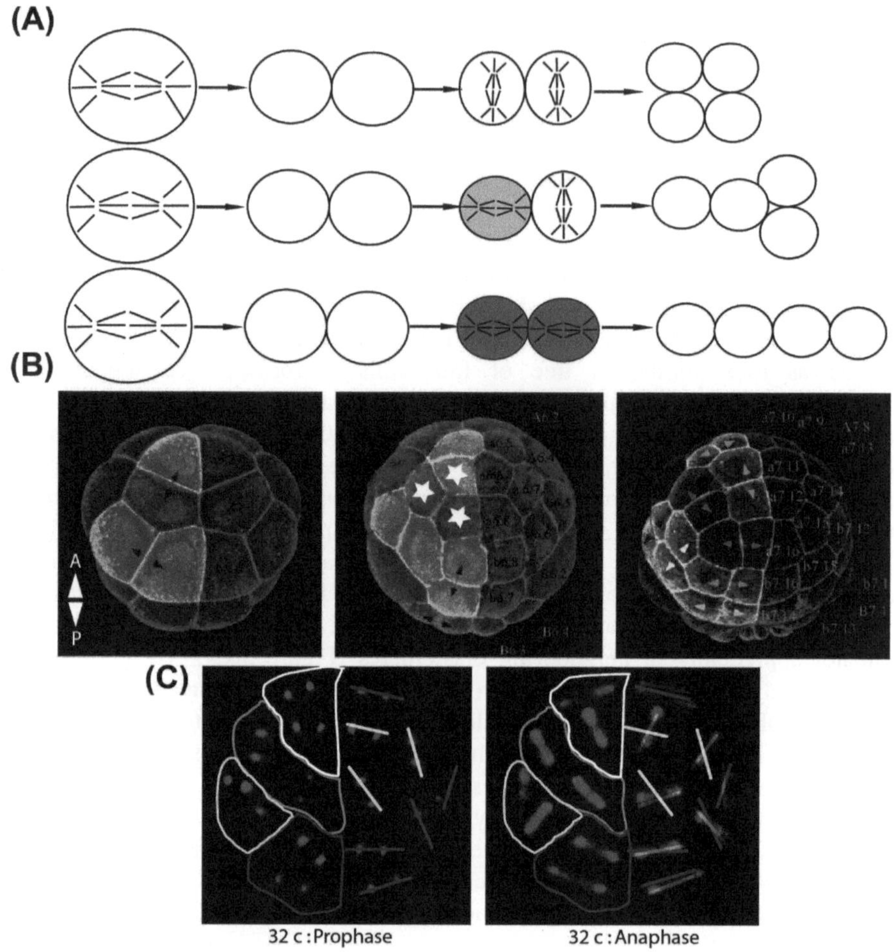

FIGURE 3 Invariant cleavage.

(A) The upper schematic represents cells dividing by default at 90° relative to their mothers, a phenomenon based on centrosome migration following mitosis. The second and third illustrations represent the two different outcomes obtained if one (colored yellow) or both (colored red) daughter cell(s) divide in the same orientation as their mother giving rise to a T configuration or to four cells in a line, respectively. (B) *Ciona intestinalis* embryos (Hotta et al., 2007; http://chordate.bpni.bio.keio.ac.jp/faba/1.4/top.html;http://chordate.bpni.bio.keio.ac.jp/faba2/2.2/top.html) where the blastomeres have been color-coded and followed over two cell divisions. The daughters and granddaughters are thus shown at the 32- and 64-cell stages, respectively. As in the schematics, blastomeres at the 64-cell stage are arranged in squares, Ts, and lines, suggesting that spindles or centrosomes likely rotated in the blastomeres indicated with white stars at the 32-cell stage (only animal blastomeres are shown). A/P indicates anterior/posterior axis. (C) Spindle position at the 32-cell stage in live *Phallusia* embryos derived from eggs injected with MAP7::GFP. Spindle position is indicated at prophase by the red and yellow lines. The gray lines (right image) indicate the new spindle orientation at anaphase superimposed on the previous prophase position. The spindles that rotate appreciably in *Phallusia* are indicated by the yellow lines, corresponding to the three blastomeres at the 32-cell stage in *Ciona* labeled with white stars in (B). (See color plate) extract the plane normal to the apical surface (the planar axis) and the plane orthogonal to the apical surface (the apicobasal axis). With these extracted planes, we can assess whether the spindle is aligned in the long axis of the cell using mathematical modeling (Minc, Burgess, & Chang, 2011).

3.1 Cell mask is a lipophilic red fluorescent marker that works well to label rapidly the cell membranes (Figure 4(A1)). In addition, following internalization cell mask vesicles accumulate around the spindle poles during mitosis (arrow in Figure 4(A1)), allowing the measurement of spindle position.

3.2 Prepare a 1 mg/mL solution of Cell Mask in DMSO and store at −20 °C. On the day of use dissolve 1 μL Cell mask stock solution in 1 mL TAPS-filtered seawater. Incubate the embryos at the eight-cell stage in this mixture for 2–4 min and then quickly transfer the labeled embryos to fresh TAPS-filtered seawater and wash three times.

3.3 Mount the embryos for confocal imaging (see list 2.3).

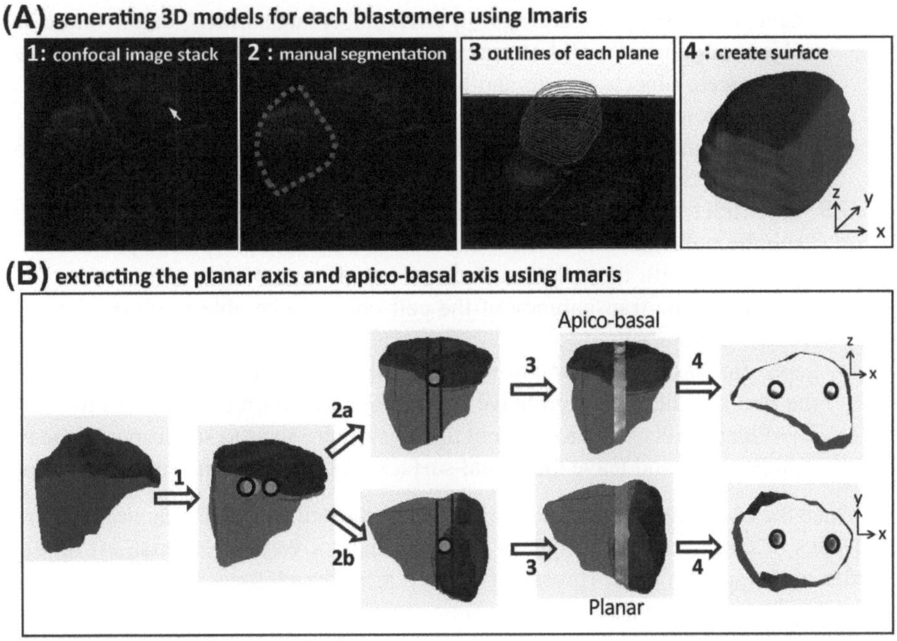

FIGURE 4 3D rendering using Imaris.

(A) Generating 3D models for each blastomere. Embryos are first labeled with Cell Mask at the eight-cell stage for 3 min (see list 3.2) and then imaged using conventional CLSM (1). The outline of cell edges is drawn for each plane (2). After all cell outlines in each plane are complete (3), the function "create surface" creates a 3D object (4). Full protocol described in list 3.5. (B) Extracting specific planes of the cell. Illustration of the protocol described in list 3.6. (See color plate)

3.4 Acquire images of one half (animal or vegetal half) of the embryo obtaining 30 z-planes. Alternatively, if using long-distance 40X objectives, the whole embryo can be imaged and then 3D rendered. In order to obtain 30 z-planes every minute, the Leica confocal microscopes (SP2, SP5) are set at 800 Hz bidirectional scanning and line-averaging of two in one color. Thirty z-sections obtained on single time points or during time-lapse 3D imaging are then imported into Imaris for 3D rendering and manual segmentation.

3.5 Open the z-stack in Imaris (we use Imaris x64 6.7.7 on a 64-bit computer in order to process the large amount of data generated by 3D time lapse imaging). The 3D rendering generated by Imaris can be viewed when the surpass icon is clicked. Before performing any segmentation, the size of the voxel must be entered in the Image properties window (click Image properties in the "Edit" Menu or press Ctrl + I). To segment a blastomere, click on the "create surface" icon. In the "create surface" window, click on the "manual segmentation" button then set the manual resolution to 205 $\times$ 205 and draw the outline of the cell on each z-plane (Figure 4(A2)) (z-plane is changed by moving the "slice position" cursor). Once all the cell contours have been drawn (Figure 4(A3)), click on "create surface" button and a 3D object of your cell will be created (Figure 4(A4)). Spindle poles are created as separate objects: create another object and draw the outlines of the spindle poles (observed as dark areas surrounded by Cell Mask positive vesicles) and select "create surface." At this point, each cell is composed of two objects: one for the cell perimeter and one for the spindle poles.

3.6 To extract the planar and the apicobasal axes of the cell (containing the spindle poles), select a blastomere and its associated spindle poles and position it with the apical surface on top (Figure 4(B)).

1—Increase the transparency of the cell object to be able to observe the spindle poles.

2a—Apicobasal axis extraction: orient the blastomere so as to superimpose the two spindle poles and leave the apical surface on top of the view.

2b—Planar axis extraction: orient the blastomere so as to superimpose the two spindle poles and leave the apical surface on the right-hand side of the view.

Once the blastomere is properly oriented use the "cut surface" function and draw two lines of cutting (the cutting line can only have a vertical orientation) on each border of the spindle poles (press cut surface after drawing each line).

3—Select the cut surface (by clicking on the cell object overlying the spindle pole) and click duplicate to create a new object (shown in yellow in Figure 4(B3)).

4—Deselect the cell object to view only the newly created cell slice and rotate it to show the spindle poles in the extracted slice. We use the snapshot icon to make a snapshot of the view and export a tiff file that can be used for spindle orientation analysis using the mathematical model (Minc et al., 2011).

4. Fixation protocols

Live imaging methodology is crucial to understand dynamic events, however, it must be borne in mind that injected GFP constructs may not reflect the true localization of endogenous unmodified proteins. Ascidian eggs and embryos are also well adapted for the study of endogenous proteins using fixation techniques (see Sardet et al., 2011 for a detailed review).

4.1 Choice of fixatives

The fixation procedures depend on the structure of interest. To visualize microtubules and spindles, oocytes or embryos are plunged into a minimum of 10 volumes of −20 °C methanol, which causes fixation by dehydration as well as good permeabilization of the cells. Methanol is often supplemented with 50 mM EGTA pH 7.5 to prevent tubulin degradation by Ca^{2+}-activated proteases (Prulière, Cosson, Chevalier, Sardet, & Chenevert, 2011). However, methanol fixation is almost always detrimental to cortical structures (such as meiotic outpocketing) in which case one should favor formaldehyde-based fixatives. For *Phallusia*, one volume of eggs or embryos is gently mixed with nine volumes of 4% formaldehyde dissolved in PBS supplemented with 500 mM NaCl for 2 h at RT or overnight at 4 °C. To visualize structures deep within the embryo, a further permeabilization step is necessary to ensure efficient antibody penetration. This is done by rinsing the eggs or embryos in 80% EtOH several times, a step which also allows long-term storage at −20 °C. This permeabilization procedure is not compatible with detection of actin filaments using phalloïdin probes. However, total actin may be visualized using an anti-actin antibody (Patalano et al., 2006).

4.2 Methanol or formaldehyde/ethanol fixed embryos have to be rehydrated before immunolabeling. This is achieved by several washes of PBS containing 3% BSA and 0.05% triton X-100 (PBT) for blocking of nonspecific sites.

4.3 Incubate with primary antibodies for 2 h at RT or overnight at 4 °C. For a previously uncharacterized antibody, try a range of concentration between 1/10 and 1/1000 in PBS or PBT.

4.4 Wash three to four times for 15 min with PBT and incubate with a fluorescent probe conjugated secondary antibody (typically 1:200 dilution).

4.5 Repeat the washing procedure. In order to visualize DNA, Hoechst at 1 μg/mL can be added to the first wash.

4.6 After mounting in an antibleach medium (we use Citifluor/AF1), samples can be observed by fluorescence microscopy. (*Phallusia* embryos do not require clearing.) For confocal microscopy, it can be useful to mount the sample between two coverslips so that either face of the thick embryo may be imaged.

5. Meiotic spindle behavior

Meiotic spindles are specialized structures adapted to forming tiny polar bodies. Meiotic spindles are small and located at the egg cortex where they orchestrate an extreme form of unequal cleavage in order to ensure that eggs lose little of their cytoplasm during meiosis I and II (approximately 0.2% total volume in each polar body in *Phallusia*). This allows eggs to retain most of their energy-laden cytoplasm to fuel embryonic development until the larvae (or juveniles in the case of ascidians since larvae are nonfeeding) are capable of feeding. Another feature of egg meiosis is that centrioles are often absent before the metaphase I spindle forms (or in some cases lost during the process of polar body extrusion). One hypothesis is that this may be important for preventing parthenogenesis; however, since mouse oocytes can display parthenogenesis (and also lack centrioles) upon loss of Mos this is not universal (Colledge, Carlton, Udy, & Evans, 1994). Mice lose their centrioles before formation of the first meiotic spindle (Dumont et al., 2007). Indirect evidence also suggests that ascidians too can form a meiotic spindle without centrioles (Dumollard et al., 2011). In ascidians, the sperm introduces a centrosome which duplicates shortly after fertilization and from which all embryonic centrosomes are derived.

Like many invertebrate species, the mature ascidian egg is arrested at metaphase I of meiosis before fertilization with a small meiotic spindle located parallel to the overlying cortex (Figure 5(A) and (B)). As in mouse eggs, a concentration of cortical F actin overlays the meiotic spindle (Figure 5(C)). Following fertilization, the meiotic spindle rotates (Figure 5(B)) and one pole moves into the outpocket of membrane/cortex which then pinches off eliminating half of the egg's homologous chromosomes in the first polar body. This occurs about 7 min after fertilization in *Phallusia*. During the following 15 min a second meiotic spindle forms, again parallel to the egg surface and rotates a second time with one pole entering the outpocket of membrane/cortex. Half of the remaining chromatids are partitioned into the second polar body. Here, we describe the methods we use to follow microtubule and spindle pole behavior during polar body extrusion, and also the protein markers we have developed to study precisely the cell cycle phases and polarization of the cortex overlying the meiotic spindle.

The meiotic divisions can be studied in vivo thanks to the fact that unfertilized *Phallusia* eggs translate exogenous mRNAs (Prodon et al., 2010), albeit less efficiently than fertilized eggs. Following injection, it is necessary to incubate overnight to allow accumulation of the fluorescent proteins. We have found that GFP fusions to MAP7, Ensconsin, or Aurora label the entire meiotic spindle, whereas calmodulin, TPX2, and EB3 localize more strongly on the meiotic spindle poles (see Table 1 and Figure 5(A)). Calmodulin::Venus labels meiotic spindle poles (more than it does mitotic spindle poles), which is similar to *Caenorhabditis elegans* where calmodulin was also detected at meiotic spindle poles and is involved in spindle rotation (Van der Voet et al., 2009).

To re-initiate meiosis in a population of unfertilized eggs, eggs prepared as in list 1.6 are either fertilized as in list 1.8 or activated by adding the calcium ionophore

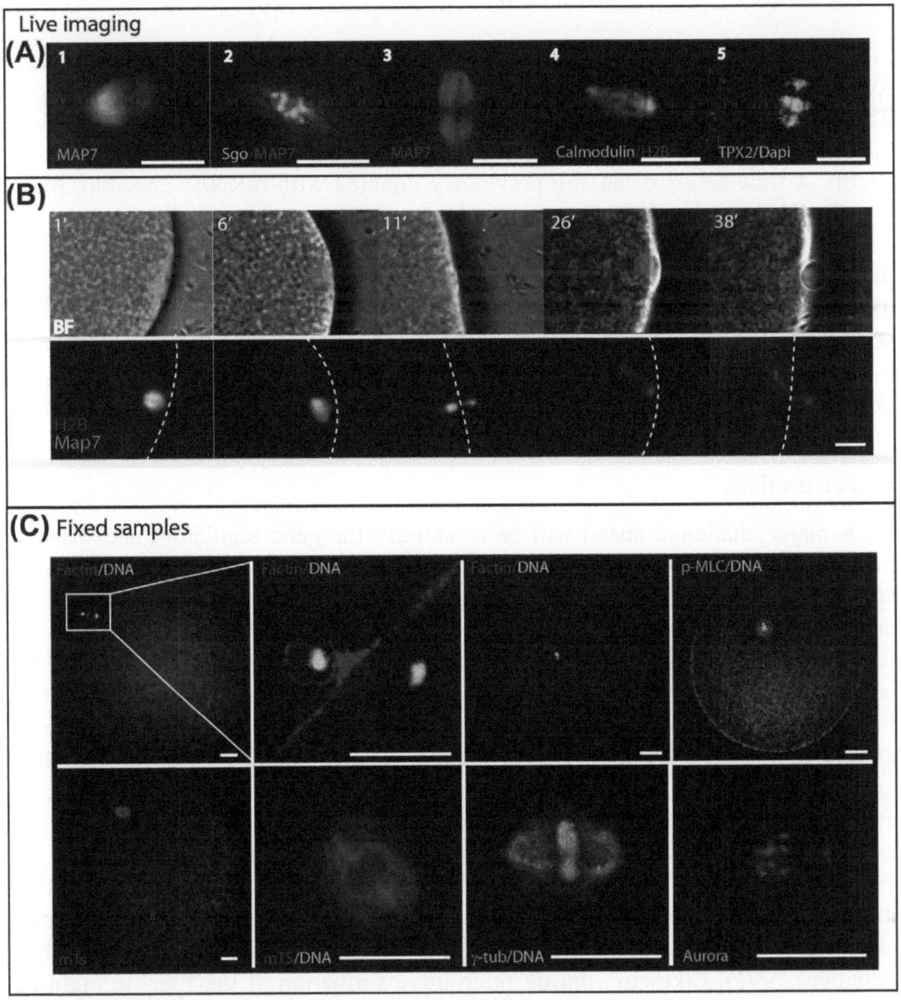

FIGURE 5 Imaging of meiotic spindles and overlying cortex.

(A) Live cell markers of meiotic spindles. (1) Metaphase I spindle labeled with MAP7::GFP. (2) Metaphase I chromosomes labeled with Shugoshin::Venus plus MAP7::Rfp1 to visualize the spindle. (3) Meiotic spindle labeled with MAP7::Rfp1 during first polar body extrusion. (4) Meiotic spindle labeled with Calmodulin::Venus (enriched at the poles) and chromosomes labeled with HH2B::Rfp1. (5) Meiotic spindle poles labeled with Venus::TPX2 and the chromosomes with DAPI. Scale bars = 10 μm. (B) Meiotic spindle behavior during extrusion of the first and second polar bodies. Unfertilized eggs were injected with mRNA encoding MAP7::GFP (green) and HH2B::Rfp1 (red), incubated overnight to allow translation, then imaged using wide-field epifluorescence time-lapse microscopy. Dotted line represents the surface of the egg. Scale bar = 10 μm. (C) Detection of meiotic structures in fixed eggs. Top row, red: phalloidin label of F actin. Left: the tiny first polar body has received half the chromatin (blue label, Hoechst). Right: in an unfertilized egg, the cortex (actin cap) overlying the meiotic spindle is rich with F-actin and phosphorylated myosin II (pMLC, green). The bottom row shows examples of meiotic spindles immunolabeled with the anti-tubulin YL1/2, anti-γ tubulin GTU88, and anti-Aurora A/B antibodies. Scale bars = 10 μm. (See color plate) ionomycin (1 mg/mL stock solution in EtOH) to a final concentration of 1 μg/mL. The use of ionomycin limits the observation to the extrusion of the first polar body but the meiotic divisions are in general more synchronous than those induced by sperm. A time series of an egg previously injected with mRNAs encoding MAP7 and the DNA marker H2B is shown in Figure 5(B).

Details of the small (approximately 10 μm) meiotic spindle in this egg (120–140 μm diameter) are best evaluated using confocal analysis of fixed samples (Figure 5(C)). The cortical actin patch overlying the meiotic spindle is detected with rhodamine phalloidin after formaldehyde fixation. To observe the spindle and associated structures, the eggs are fixed in −20 °C methanol (see list 4) and immunolabeled. Antibodies which label well the meiotic spindle include anti-tubulin YL1/2, anti-tubulin DM1A, or the anti-γ-tubulin antibody GTU88 from Sigma–Aldrich.

6. Perspectives

A major challenge ahead will be to identify the gene regulatory networks and maternal mechanisms that together determine cell shape, which in turn controls spindle orientation creating the invariant cleavage pattern that has been so well conserved during the course of evolution among the ascidians.

Many puzzles concerning the mechanism of unequal cleavage also remain. One is that the spindle poles do not appear to migrate precisely toward the CAB but instead approach the midline adjacent to the CAB (especially obvious at the 16- and 32-cell stages). Another issue is centrosome asymmetry during unequal cleavage. Many examples of centrosome asymmetry exist in different organisms. For example, in ascidians the centrosome close to the CAB becomes reduced in size as it approaches the CAB/midline (this is most extreme at the 32-cell stage, see Figure 2(B)). We do not know what causes this asymmetry. During micromere formation in sea urchin embryos from the 8- to 16-cell stage, centrosome asymmetry also occurs as one centrosome/spindle pole associates with the cortex (Holy & Schatten, 1991). Likewise, during micromere formation at the four- to eight-cell stage in the mollusk *Ilyanassa*, one centrosome/spindle pole appears on or very close to the cortex (Rabinowitz & Lambert, 2010). To conclude, centrosome asymmetry is widespread as is its association with the cortex during unequal cleavage; it would be interesting to know what causes centrosome asymmetry and its association with the cortex. We know that in the ascidian γ-tubulin is present on both spindle poles so centrosome asymmetry is not caused by transient γ-tubulin loss from one of the centrosomes as in the zygotes of the leech *Helodbella* (Ren & Weisblat, 2006). Unfortunately, our Centrobin GFP (Table 1) did not enlighten us, given that it is the absence of Centrobin from the smaller mother centrosome that causes centrosome asymmetry in *Drosophila* neuroblasts (Januschke et al., 2013).

Also, it is clear that the second meiotic spindle forms exactly where the first polar body was extruded (indeed, polar body one and two are attached to each other). This suggests that some cortical structure (e.g., the midbody that formed between the egg and the first polar body) helps direct the placement of the second polar body in *Phallusia*.

Finally, the solitary ascidian models (*C. intestinalis*, *H. roretzi*) have made significant contributions for studying many aspects of development in chordates (Satoh, 2014). For those researchers interested in performing dynamic live cell imaging studies during early embryonic development (from the unfertilized egg up to about the 64-cell stage or 112-cell early gastrula stage) using fluorescence, *Phallusia* may be useful because it is transparent and translates well exogenous mRNAs in the egg and early embryo. The methods and tools outlined here permit the analysis at the cellular and subcellular levels using *Phallusia* early embryos and eggs, where additional functional approaches can be combined (knockdown, microsurgery, dominant negative/positive constructs etc.) to better understand the cell biology of early embryonic development and egg meiosis.

REFERENCES

Carroll, M., Levasseur, M., Wood, C., Whitaker, M., Jones, K. T., & McDougall, A. (2003). Exploring the mechanism of action of the sperm-triggered calcium-wave pacemaker in ascidian zygotes. *Journal of Cell Science, 116*, 4997–5004. http://dx.doi.org/10.1242/jcs.00846.

Chen, J.-Y., Huang, D.-Y., Peng, Q.-Q., Chi, H.-M., Wang, X.-Q., & Feng, M. (2003). The first tunicate from the Early Cambrian of South China. *Proceedings of the National Academy of Sciences USA, 100*, 8314–8318. http://dx.doi.org/10.1073/pnas.1431177100.

Colledge, W. H., Carlton, M. B., Udy, G. B., & Evans, M. J. (1994). Disruption of c-mos causes parthenogenetic development of unfertilized mouse eggs. *Nature, 370*, 65–68. http://dx.doi.org/10.1038/370065a0.

Conklin, E. G. (1905). The organization and cell lineage of the ascidian egg. *J Acad Sci Philadelphia, 13*, 1–119.

Delsuc, F., Brinkmann, H., Chourrout, D., & Philippe, H. (2006). Tunicates and not cephalochordates are the closest living relatives of vertebrates. *Nature, 439*, 965–968. http://dx.doi.org/10.1038/nature04336.

Dumollard, R., Besnardeau, L., Hebras, C., Salez, G. and McDougall, A. (in preparation). Mathematical modelling of planar divisions oriented according to the Hertwig long axis rule predicts the invariant cleavage pattern of ascidian embryos.

Dumollard, R., Hebras, C., Besnardeau, L., & McDougall, A. (2013). Beta-catenin patterns the cell cycle during maternal-to-zygotic transition in urochordate embryos. *Developmental Biology, 384*, 331–342. http://dx.doi.org/10.1016/j.ydbio.2013.10.007.

Dumollard, R., Levasseur, M., Hebras, C., Huitorel, P., Carroll, M., Chambon, J.-P., et al. (2011). Mos limits the number of meiotic divisions in urochordate eggs. *Development (Cambridge, England), 138*, 885–895. http://dx.doi.org/10.1242/dev.057133.

Dumont, J., Petri, S., Pellegrin, F., Terret, M.-E., Bohnsack, M. T., Rassinier, P., et al. (2007). A centriole- and RanGTP-independent spindle assembly pathway in meiosis I of vertebrate oocytes. *The Journal of Cell Biology, 176*, 295–305. http://dx.doi.org/10.1083/jcb.200605199.

Hebras, C., & McDougall, A. (2012). Urochordate ascidians possess a single isoform of Aurora kinase that localizes to the midbody via TPX2 in eggs and cleavage stage embryos. *PloS One, 7*, e45431. http://dx.doi.org/10.1371/journal.pone.0045431.

Holy, J., & Schatten, G. (1991). Differential behavior of centrosomes in unequally dividing blastomeres during fourth cleavage of sea urchin embryos. *Journal of Cell Science, 98*(Pt 3), 423–431.

Hotta, K., Mitsuhara, K., Takahashi, H., Inaba, K., Oka, K., Gojobori, T., et al. (2007). A web-based interactive developmental table for the ascidian *Ciona intestinalis*, including 3D real-image embryo reconstructions: I. From fertilized egg to hatching larva. *Developmental Dynamics: An Official Publication of the American Association of Anatomists, 236*, 1790–1805. http://dx.doi.org/10.1002/dvdy.21188.

Januschke, J., Reina, J., Llamazares, S., Bertran, T., Rossi, F., Roig, J., et al. (2013). Centrobin controls mother-daughter centriole asymmetry in *Drosophila* neuroblasts. *Nature Cell Biology, 15*, 241–248. http://dx.doi.org/10.1038/ncb2671.

Levasseur, M., & McDougall, A. (2000). Sperm-induced calcium oscillations at fertilisation in ascidians are controlled by cyclin B1-dependent kinase activity. *Development (Cambridge, England), 127*, 631–641.

McDougall, A., Lee, K. W.-M., & Dumollard, R. (2014). Microinjection and 4D fluorescence imaging in the eggs and embryos of the ascidian *Phallusia mammillata. Methods in Molecular Biology (Clifton, NJ), 1128*, 175–185. http://dx.doi.org/10.1007/978-1-62703-974-1_11.

McDougall, A., & Levasseur, M. (1998). Sperm-triggered calcium oscillations during meiosis in ascidian oocytes first pause, restart, then stop: correlations with cell cycle kinase activity. *Development (Cambridge, England), 125*, 4451–4459.

Minc, N., Burgess, D., & Chang, F. (2011). Influence of cell geometry on division-plane positioning. *Cell, 144*, 414–426. http://dx.doi.org/10.1016/j.cell.2011.01.016.

Nishikata, T., Hibino, T., & Nishida, H. (1999). The centrosome-attracting body, microtubule system, and posterior egg cytoplasm are involved in positioning of cleavage planes in the ascidian embryo. *Developmental Biology, 209*, 72–85. http://dx.doi.org/10.1006/dbio.1999.9244.

Nishikawa, T., Oohara, I., Saitoh, K., Shigenobu, Y., Hasegawa, N., Kanamori, M., et al. (2014). Molecular and morphological discrimination between an invasive ascidian, *Ascidiella aspersa*, and its congener *A. scabra* (Urochordata: Ascidiacea). *Zoological Science, 31*, 180–185. http://dx.doi.org/10.2108/zsj.31.180.

Patalano, S., Prulière, G., Prodon, F., Paix, A., Dru, P., Sardet, C., et al. (2006). The aPKC-PAR-6-PAR-3 cell polarity complex localizes to the centrosome attracting body, a macroscopic cortical structure responsible for asymmetric divisions in the early ascidian embryo. *Journal of Cell Science, 119*, 1592–1603. http://dx.doi.org/10.1242/jcs.02873.

Prodon, F., Chenevert, J., Hébras, C., Dumollard, R., Faure, E., Gonzalez-Garcia, J., et al. (2010). Dual mechanism controls asymmetric spindle position in ascidian germ cell precursors. *Development (Cambridge, England), 137*, 2011–2021. http://dx.doi.org/10.1242/dev.047845.

Prulière, G., Cosson, J., Chevalier, S., Sardet, C., & Chenevert, J. (2011). Atypical protein kinase C controls sea urchin ciliogenesis. *Molecular Biology of the Cell, 22*, 2042–2053. http://dx.doi.org/10.1091/mbc.E10-10-0844.

Rabinowitz, J. S., & Lambert, J. D. (2010). Spiralian quartet developmental potential is regulated by specific localization elements that mediate asymmetric RNA segregation. *Development (Cambridge, England), 137*, 4039–4049. http://dx.doi.org/10.1242/dev.055269.

Ren, X., & Weisblat, D. A. (2006). Asymmetrization of first cleavage by transient disassembly of one spindle pole aster in the leech *Helobdella robusta. Developmental Biology, 292*, 103–115. http://dx.doi.org/10.1016/j.ydbio.2005.12.049.

Riedl, J., Crevenna, A. H., Kessenbrock, K., Yu, J. H., Neukirchen, D., Bista, M., et al. (2008). Lifeact: a versatile marker to visualize F-actin. *Nature Methods, 5*, 605–607. http://dx.doi.org/10.1038/nmeth.1220.

Robin, F. B., Dauga, D., Tassy, O., Sobral, D., Daian, F., & Lemaire, P. (2011). Time-lapse imaging of live *Phallusia* embryos for creating 3D digital replicas. *Cold Spring Harbor Protocols, 2011*, 1244–1246. http://dx.doi.org/10.1101/pdb.prot065847.

Roure, A., Rothbächer, U., Robin, F., Kalmar, E., Ferone, G., Lamy, C., et al. (2007). A multicassette Gateway vector set for high throughput and comparative analyses in ciona and vertebrate embryos. *PloS One, 2*, e916. http://dx.doi.org/10.1371/journal.pone.0000916.

Sardet, C., McDougall, A., Yasuo, H., Chenevert, J., Pruliere, G., Dumollard, R., et al. (2011). Embryological methods in ascidians: the Villefranche-sur-Mer protocols. *Methods in Molecular Biology (Clifton, NJ), 770*, 365–400. http://dx.doi.org/10.1007/978-1-61779-210-6_14.

Satoh, N. (2014). *Developmental genomics of ascidians*. Wiley-Blackwell. ISBN:978-1-118-65618-1.

Stauffer, T. P., Ahn, S., & Meyer, T. (1998). Receptor-induced transient reduction in plasma membrane PtdIns(4,5)P2 concentration monitored in living cells. *Current Biology: CB, 8*, 343–346.

Stolfi, A., Lowe, E. K., Racioppi, C., Ristoratore, F., Brown, C. T., Swalla, B. J., et al. (2014). Divergent mechanisms regulate conserved cardiopharyngeal development and gene expression in distantly related ascidians. *eLife, 3*. http://dx.doi.org/10.7554/eLife.03728.

Thompson, D. W. (1917). *On growth and form*. Cambridge University Press.

Van der Voet, M., Berends, C. W. H., Perreault, A., Nguyen-Ngoc, T., Gönczy, P., Vidal, M., et al. (2009). NuMA-related LIN-5, ASPM-1, calmodulin and dynein promote meiotic spindle rotation independently of cortical LIN-5/GPR/Galpha. *Nature Cell Biology, 11*, 269–277. http://dx.doi.org/10.1038/ncb1834.

Wilson, E. B. (1900). *The cell in development and inheritance*. Macmillan.

CHAPTER

Ultrastructural analysis of *Caenorhabditis elegans* cilia

18

Daniel Serwas, Alexander Dammermann[1]

Max F. Perutz Laboratories, University of Vienna, Vienna Biocenter (VBC), Vienna, Austria

[1]*Corresponding author: E-mail: alex.dammermann@univie.ac.at*

CHAPTER OUTLINE

Introduction **342**

1. Methods **344**

1.1 Specimen Fixation 344

1.1.1 Chemical fixation of C. elegans *larvae and adult worms* 345

1.1.2 High-pressure freezing–freeze substitution of C. elegans *embryos* 346

1.2 Embedding 348

1.2.1 Infiltration and pre-embedding 349

1.2.2 Embedding 350

1.3 Serial Sections 351

1.3.1 Block trimming 351

1.3.2 Cutting serial sections 353

1.3.3 Post-staining 355

1.4 Transmission Electron Microscopy 356

1.5 Electron Tomography 357

1.5.1 Adding fiducial markers 359

1.5.2 Data recording 360

1.5.3 Tilt series reconstruction and model generation 361

Discussion **363**

Acknowledgments **364**

References **364**

Abstract

In addition to organizing centrosomes, centrioles have an evolutionarily conserved role as basal bodies in the formation of cilia. The discovery of a range of human genetic disorders linked to cilia dysfunction has led to a resurgence of interest in this cellular organelle. The nematode *Caenorhabditis elegans* possesses several unique features (highly stereotypical morphology, dispensability of cilia for organismal viability and

Methods in Cell Biology, Volume 129, ISSN 0091-679X, http://dx.doi.org/10.1016/bs.mcb.2015.03.014

fertility) that make it an attractive model to study cilia assembly and function. However, both the adult worm and the embryo present particular challenges for electron microscopy (EM), which remains the gold standard for high-resolution morphological studies. Here, we present a step-by-step guide for the ultrastructural analysis of *C. elegans* cilia, including optimized protocols for standard chemical fixation as well as high-pressure freezing/freeze substitution and further processing for serial-section transmission electron microscopy and electron tomography.

INTRODUCTION

Cilia are microtubule-based cellular projections found in many eukaryotes. Like their bacterial counterparts, cilia are often motile, enabling the movement of cells, as in the case of ciliate protozoa or the flagellate sperm of many species, or fluid over the surface of the cell, as in the case of the multiciliate epithelia of the lung and oviduct. While motile cilia can also be sensory, nonmotile primary cilia are exclusive to metazoa and are specialized to detect physical and chemical stimuli. Primary cilia are widely distributed in vertebrate cells and play important roles in sensory perception, developmental signaling, and tissue homeostasis (Berbari, O'Connor, Haycraft, & Yoder, 2009). The biomedical significance of cilia is underscored by the discovery of a large number of human developmental and adult disorders, collectively referred to as ciliopathies, associated with ciliary defects (Badano, Mitsuma, Beales, & Katsanis, 2006). Ciliopathies range from disorders that affect a single type of cilium and result in a specific pathology, such as progressive blindness, infertility, or polycystic kidney disease, to broad-based disorders such as Bardet–Biedl syndrome affecting multiple tissues and organs. The most severe ciliopathies identified to date, Meckel–Gruber syndrome and hydrolethalus syndrome, result in perinatal lethality.

Cilia have been a favorite subject for ultrastructural studies since the early days of electron microscopy (Fawcett & Porter, 1954), in part because of their fascinating architectural complexity (Fisch & Dupuis-Williams, 2011; O'Toole, Giddings, McIntosh, & Dutcher, 2003). At their most basic level, cilia are built around an axoneme composed of a ninefold symmetric array of doublet microtubules that extend from the centriole-derived basal body. The basal body itself is anchored to the plasma membrane via so-called transition fibers, which in vertebrates are derived from the distal appendages present on mature centrioles. Between basal body and axoneme proper lies the transition zone, an elaborate structure distinguished by Y-links connecting axonemal microtubules to the ciliary membrane. Basal body and transition zone together act to create a ciliary gate that limits free diffusion into the cilium, defining it as a distinct cell compartment with a specialized protein composition (Reiter, Blacque, & Leroux, 2012).

The nematode *Caenorhabditis elegans* is commonly used as a model to study cilia (Inglis, Ou, Leroux, & Scholey, 2007). Cilia in *C. elegans* are exclusively found at the dendritic tips of postmitotic sensory neurons located in the head and tail of the animal, where they mediate the perception of chemosensory and mechanosensory stimuli. Compromised cilia function leads to defects in easily assayed behaviors such as chemotaxis or male mating. Uniquely among model organisms, however, cilia are not required for viability or fertility (sperm are amoeboid, not flagellate), whereas loss of cilia results in embryonic lethality in mice (Murcia et al., 2000) and severely impairs viability in flies (Basto et al., 2006). Studies in *C. elegans* are also aided by the highly stereotypic pattern of development of the worm, resulting in a nearly invariant morphology between two animals of the same genotype. Over the course of the past 30 years using both forward and reverse genetics researchers have taken advantage of these unique features to examine various aspects of cilia biology, including the molecular mechanisms underlying intraflagellar transport (Scholey, Ou, Snow, & Gunnarson, 2004) and assembly of the transition zone (Williams et al., 2011). Most of these studies have focused on the amphid channel cilia, two sets of 10 stereotypical rod-like cilia that extend through a channel created by two glial cells, the sheath, and socket cell to become partially exposed to the external environment at the tip of the nose of the animal. However, worms possess a diverse range of other cilia morphologies that are beginning to be explored (Doroquez, Berciu, Anderson, Sengupta, & Nicastro, 2014). Interestingly, all *C. elegans* cilia lack centriolar microtubule structures at their base (Perkins, Hedgecock, Thomson, & Culotti, 1986). Centriole degeneration appears to occur shortly after the onset of ciliogenesis during late embryogenesis (Dammermann et al., 2009; Sulston, Schierenberg, White, & Thomson, 1983). However, this process has so far not been studied in detail.

Transmission electron microscopy (TEM) has been extensively used in *C. elegans* from the beginning. An early milestone was the reconstruction of the entire nervous system of the worm, including its sensory anatomy of which cilia are an integral part (Hall & Russell, 1991; Perkins et al., 1986; Ward, Thomson, White, & Brenner, 1975; White, Southgate, Thomson, & Brenner, 1986). Samples have traditionally been prepared by chemical fixation with glutaraldehyde and osmium tetroxide, followed by dehydration, plastic embedding, and serial sectioning (Hall, 1995). Fixation is complicated by the largely impermeable cuticle of the larva and adult worm, which, together with the thickness of the animal, makes it difficult to achieve adequate penetration of the fixative. The egg shell of the embryo presents an even more substantial barrier, requiring pretreatment with harsh agents such as sodium hypochlorite in order to be permeabilized (Olson, Greenan, Desai, Müller-Reichert, & Oegema, 2012; Sulston et al., 1983). An alternative method of fixation is high-pressure freezing (HPF) followed by freeze substitution (FS). HPF is a rapid freezing method, which allows immobilization of biological samples in a near-native state with no or minimal ice crystal formation (reviewed in McDonald & Auer, 2006). During freeze substitution, intracellular water is exchanged for an organic solvent (e.g., acetone-based freeze substitution medium containing fixatives like glutaraldehyde and osmium tetroxide). This yields generally improved ultrastructural preservation and also eliminates the need for embryo pretreatment (Doroquez et al., 2014; Müller-Reichert, Hohenberg, O'Toole, & McDonald, 2003). HPF-FS does, however, require expensive equipment, whereas chemical fixation can be carried out in any electron microscopy facility.

The advent of electron tomography has opened up new avenues for the three-dimensional ultrastructural analysis of subcellular compartments. For electron tomography reconstructions, the specimen is imaged in a transmission electron microscope at different tilt angles and the obtained 2D images are computationally aligned into a 3D volume (Lucic, Forster, & Baumeister, 2005; McIntosh, Nicastro, & Mastronarde, 2005). Such 3D reconstructions are ideally suited for complex structures like centrioles or cilia. In combination with RNAi-mediated depletion electron tomography on *C. elegans* early embryos helped delineate the molecular pathway underlying centriole assembly and define the contribution of individual components (Pelletier, O'Toole, Schwager, Hyman, & Muller-Reichert, 2006). *C. elegans* cilia have so far been studied primarily by traditional serial-section methods. However, the application of electron tomography to cilia in organisms where they are more easily accessed such as *Chlamydomonas* and *Tetrahymena* (Heuser, Dymek, Lin, Smith, & Nicastro, 2012; O'Toole et al., 2003; Pigino et al., 2012) reveals the potential power of this technology (see also Hall, Hartwieg, & Nguyen, 2012; Müller-Reichert, Mancuso, Lich, & McDonald, 2010).

In this chapter we aim to guide the reader through the steps involved in setting up a TEM and electron tomography experiment, with detailed protocols for the preparation of both embryos and larval/adult-stage worms.

1. METHODS

1.1 SPECIMEN FIXATION

Due to the limited penetration depth of the electron beam, TEM investigations are restricted to specimens with a thickness of less than 500 nm. Ultrastructural studies of *C. elegans* therefore require mechanical sectioning, most commonly of plastic-embedded specimens. Prior to resin embedding, samples must be fixed and dehydrated. Specimen fixation is the most important step in the preparation process. This can be done by "classical" chemical fixation followed by dehydration in a graded series of an organic solvent (generally ethanol or acetone) or by HPF-FS. Chemical fixation does not require any specialized equipment. It is also the more straightforward of the two methods. In terms of image quality, classical methods tend to yield higher contrast, more visually appealing, images, whereas ultrastructural preservation is improved with HPF-FS (Doroquez et al., 2014; Müller-Reichert et al., 2003). HPF-FS also permits rapid fixation at a defined time point and is therefore ideally suited for correlative light and electron microscopy (CLEM) studies. Depending on the biological question, both methods therefore have their rationale.

In general, chemicals used for fixation and dehydration should be chosen with care and be compatible with each other. For example, phosphate buffers can form precipitates when used in conjunction with ethanol for dehydration (Hayat, 1981).

We prefer acetone for dehydration since it works best with the buffers we use. Common fixatives for EM are glutaraldehyde and osmium tetroxide, which also serves to enhance contrast. In addition to osmium, uranyl acetate is also commonly used as a contrasting agent. High concentrations of free calcium ions are known to be harmful for microtubule stability. Chelation of calcium with EGTA may therefore improve preservation of microtubule-based structures such as cilia (Osborn & Weber, 1982). We found addition of taxol to the fixative to likewise improve preservation of ciliary microtubules. Protofilament substructure is best visualized by fixation with tannic acid (Chalfie & Thomson, 1982). No one protocol suits all structures and samples. Researchers may therefore wish to experiment with a range of fixation and staining conditions.

In the following, we present protocols for chemical fixation of *C. elegans* larvae and adults (Section 1.1.1) and for fixation of embryos by HPF-FS (Section 1.1.2). The latter protocol can also be adapted for larvae and adults.

1.1.1 Chemical fixation of C. elegans *larvae and adult worms*

We have found some fixation protocols to yield variable contrast depending on the ciliary mutant strain examined, potentially due to altered permeability of the worm. We have not had this issue with the described method. Our chemical fixation protocol uses the cytoskeleton buffer (CB) as described by Small (1981). Some researchers recommend cutting of the animal to enhance penetration of the fixative (Hall, 1995). We have not found this to be necessary for amphids, although it may improve ultrastructural preservation for other cilia types.

Instrumentation:

Microcentrifuge, rotating wheel or shaker, stereomicroscope with light source suitable for *C. elegans* microdissection (we use a Nikon SMZ800 equipped with a C-DSS230 Diascopic Stand[1]).

Material:

Worms of the appropriate stage, 1.5 mL microcentrifuge tubes, Parafilm®, 1 and 3.5 mL Pasteur pipettes (Sarstedt #86.1180, 86.1171).

Reagents:

10× CB (100 mM methyl ester sulfonate, 1.5 M NaCl, 50 mM EGTA, 50 mM $MgCl_2$, 50 mM glucose in ddH_2O, pH adjusted to 6.1 with HCl, store at 4 °C, check pH and prepare 1× buffer from 10× stock immediately before use); dried acetone (Merck #100299), 40%, 60%, 80%, and 95% acetone in ddH_2O (can be stored in sealed glass bottles at 4 °C for several weeks), ddH_2O, 25% EM grade glutaraldehyde in ddH_2O (Agar Scientific #AGR1020), and 4% OsO_4 in ddH_2O (Electron Microscopy Sciences #19170). (Note: OsO_4 is extremely toxic, handle with care under a vented fume hood, and wear gloves and eye protection).

[1]Note: All instruments, materials, and reagents listed in this chapter are meant as examples and can be replaced by equivalents.

Procedure:

1. Use a worm pick or forceps to transfer 50–100 worms of the desired stage into a 1.5 mL microcentrifuge tube containing ~1 mL 2.5% glutaraldehyde in 1× CB (prepared immediately before use) and incubate overnight at 4 °C.
2. From this point all steps are carried out at 4 °C and incubations done on a shaker or rotating wheel. Tubes should be sealed with Parafilm® to prevent leakage.
3. Prepare 1× CB for washing steps and cool to 4 °C (always prepare fresh).
4. Pellet worms by centrifuging briefly (1500 g, 30 s), remove supernatant, and incubate in 1× CB for 10 min.
5. Wash two more times for 10 min with 1× CB.
6. Pellet worms and incubate for ~30 min in 0.5% OsO_4 in 1× CB (prepared immediately before use). (Do not extend incubation time beyond 45 min.)
7. Wash with 1× CB as above.
8. Pellet worms and incubate for 10 min in ddH_2O.
9. Dehydrate samples for 15 min each in 40%, 60%, 80%, twice in 95%, and finally three times in 100% acetone.
10. Proceed to Section 1.2, Embedding.

1.1.2 High-pressure freezing—freeze substitution of C. elegans *embryos*

The protocol below is based on Kolotuev, Schwab, and Labouesse (2010) and enables sorting of embryos by light microscopy, as well as rapid transfer of specimens into the HPF carrier. It can also be used for larvae and adults. Other protocols for embryo fixation are described in (Müller-Reichert, Mäntler, Srayko, & O'Toole, 2008; Sims & Hardin, 2007).

1.1.2.1 High-pressure freezing

Instrumentation:

High-pressure freezer (EM PACT, Leica), cryo tool dryer (EM CTD, Leica), stereomicroscope for *C. elegans* microdissection, microwave.

Material:

Adult worms of the desired strain, Aclar® film (199 μm thick, Electron Microscopy Sciences #50425), small binder clips, 1.8 mL cryotubes (Nunc #368632), 1.5 mm disposable biopsy punch (Miltex #33-31A-P), ESD tweezers (VOMM #3607), eyelash tool, filter paper (Whatman #1001 090), tweezers (Dumont #0108-5-PS), 10 cm glass petri dish, two liquid nitrogen Dewar flasks, HPF specimen carriers (1.5 mm in diameter, cavity depth of 200 μm, Leica #16706898), glass microscope slides (clean with ethanol before use), razor blades (degreased with acetone, uncoated, American Safety Razor #62-0179), scalpel blade, syringe needle (27 G), adhesive tape.

Reagents:

1-Hexadecene (Merck #822064), low melting point agarose (Amresco #0815), M9 buffer (22 mM KH_2PO_4, 42 mM Na_2HPO_4, 85 mM NaCl, 1 mM $MgSO_4$ in ddH_2O), liquid nitrogen.

Procedure:

1. Prepare cryotubes by punching holes in the lid with a hot syringe needle and place a metal nut into each tube to weigh down tubes in liquid nitrogen.
2. Set-up high-pressure freezer and fill liquid nitrogen Dewar flasks.
3. Place a disc of filter paper into a glass petri dish and wet with 1-hexadecene.
4. Using a razor blade cut two rectangular pieces of Aclar film to just under the size of a microscope slide.
5. Slide one of the Aclar pieces onto a glass slide in such a way that surface tension causes it to adhere to the slide.
6. Cut a rectangular hole ($\sim 3 \times 1.5$ cm) into the second Aclar piece and place it on top, creating a well. Secure both pieces at the edges of the glass slide with adhesive tape.
7. Cut a third piece of Aclar to cover the rectangular hole in the second Aclar piece, but smaller than the first two Aclar pieces and adhere it to a second slide as described above but do not secure with tape.
8. Place a 2 μL drop of M9 buffer into the well and use tweezers to transfer ~ 15 adult worms into the drop.
9. Using a syringe needle or scalpel cut worms in the middle to release embryos, separate embryos from worm parts and, if possible, remove the latter with an eyelash tool.
10. Microwave 4% low melting point agarose in M9 buffer, cool to touch and place a 30 μL drop next to the buffer droplet, intermingle drops with eyelash tool, place second Aclar-glass slide on top to spread the agarose and close sandwich with binder clamps.
11. Remove top Aclar-glass slide after agarose solidifies.
12. Select embryos of the desired stage under the dissecting microscope and excise using biopsy punch. (Note: Embryos can also be followed by live microscopy until they have developed to the desired stage.)
13. Using tweezers briefly dip HPF carrier upside down on filter paper soaked in 1-hexadecene to fill it with hexadecene.
14. Using the blade of a scalpel transfer agarose disc containing the embryo into the HPF carrier.
15. Assemble sample holder of the HPF and freeze sample.
16. From this point on only use tools cooled in liquid nitrogen and dry tools after each working step with the tool dryer.
17. Use ESD tweezers to fill a cryo tube with liquid nitrogen, let it cool down and using a second set of tweezers transfer sample into the tube. (Note: We use one tube for two to three samples.)
18. Samples can be stored in liquid nitrogen or used directly for freeze substitution.

1.1.2.2 Freeze substitution

Instrumentation:

Freeze substitution device (EM AFS2, Leica), cryo tool dryer (EM CTD, Leica).

Material:

Cryo-tubes (Electron Microscopy Sciences #34506), cryotransfer container (Leica #16701889) attached to cryotool with M4 thread (Leica #16701958), ESD tweezers (VOMM #3607), liquid nitrogen Dewar, 1 and 3.5 mL Pasteur pipettes (Sarstedt #86.1180, 86.1171).

Reagents:

Dried acetone (Merck #100299), 10% glutaraldehyde in acetone (Electron Microscopy Sciences #16530), OsO_4 (Electron Microscopy Sciences #19134), uranyl acetate (Electron Microscopy Sciences #22400), liquid nitrogen.

Procedure:

1. Only use tools cooled in liquid nitrogen, dry after each working step with tool dryer.
2. Set up AFS device and fill liquid nitrogen transfer vessel.
3. Prepare 1−1.5 mL FS medium per sample (0.5% glutaraldehyde, 2% OsO_4, and 0.25% uranyl acetate in acetone) and aliquot into cryo tubes (one per agarose disc), place tubes into AFS device chamber and let cool down for ~1 h. (Note: Do not touch tubes with hands beyond this point to prevent local heating.)
4. Cool cryotransfer container (without lid) in liquid nitrogen and place into AFS chamber.
5. Using ESD tweezers place tube containing sample into chamber.
6. Open tube in chamber, pour contents into cryotransfer container and transfer sample with ESD tweezers into cryo tube containing cooled FS medium.
7. Repeat last three steps for all samples and start freeze substitution program (see Table 1).
8. Wash three times for 15 min with acetone at 0 °C. (Note: We perform washing steps in the AFS chamber; cool Pasteur pipettes before use.)
9. Proceed to Section 1.2, Embedding.

1.2 EMBEDDING

Precise orientation of the specimen in the resin block is key to obtaining good sections. We use a two-step embedding protocol, allowing us to control specimen orientation. In the first step, the specimen is embedded in a random orientation in a thin resin layer. After polymerization, small pieces containing the specimen are cut out

Table 1 Freeze Substitution Program

Starting Temperature (°C)	Final Temperature (°C)	Time (h)	Comment
−140	−90	3	
−90	−90	61	Can be prolonged
−90	−30	20	
−30	−30	12	
−30	0	15	

and re-embedded in the desired orientation. The advantage of this method is that plastic pieces are more robust and easier to handle than the worm itself. Similar embedding approaches are described by Hall (1995) and Müller-Reichert et al. (2010).

1.2.1 Infiltration and pre-embedding

Instrumentation:

Microcentrifuge, oven, rotating wheel or shaker, stereomicroscope with light source.

Material:

Aclar® film (2 Mil, Electron Microscopy Sciences #50426), embedding mold (Ted Pella #10590), exsiccator filled with silica gel, Teflon tray (52 × 31 × 5 mm), Parafilm®, 1 and 3.5 mL Pasteur pipettes (Sarstedt #86.1180, 86.1171), pre-embedding mold (prepared by pressing a metal plate into a 1.5 mm thick PTFE sheet; see Figure 1(A)), glass scintillation vials (Wheaton #986580), toothpicks.

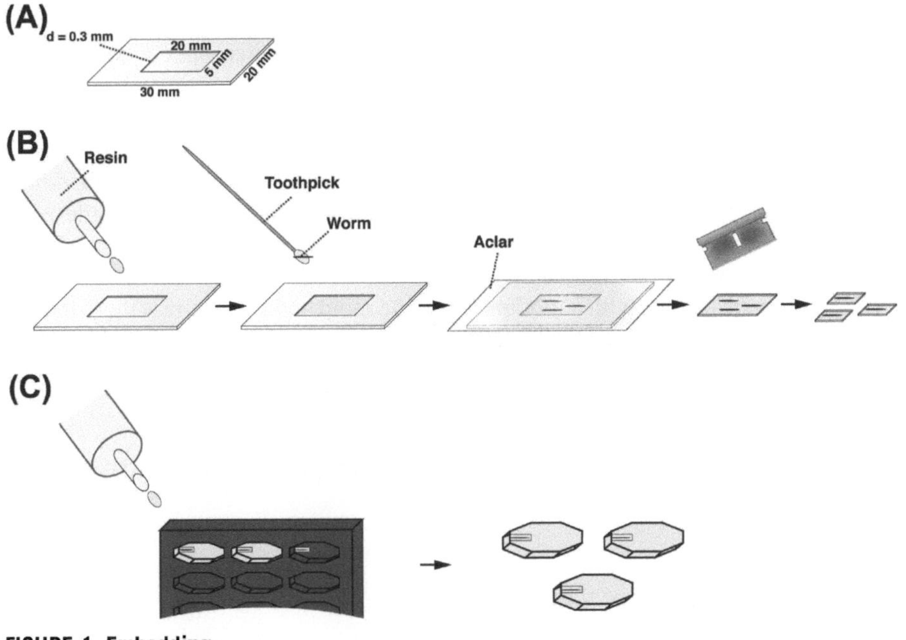

FIGURE 1 Embedding.

(A) Dimensions of pre-embedding mold prepared from 1.5-mm thick PTFE sheet. (B) Pre-embedding procedure: Specimens are embedded in a thin resin layer in the pre-embedding mold. After polymerization, individual worms are excised from the resin block for re-embedding. (C) Embedding: Resin pieces containing individual worms are carefully positioned in the embedding mold and re-embedded. After polymerization, resin blocks are obtained within which worms are oriented facing the cutting edge, enabling transverse sections of the amphids to be cut.

Reagents:

Dried acetone (Merck #100299), epoxy resin (Agar 100, Agar Scientific #AGR1031).

Procedure:

1. Pellet worms by centrifuging briefly (1500 g, 30 s) and remove acetone from fixed samples with 1 mL Pasteur pipette and incubate in a 1:2 resin/acetone mixture overnight on a rotating wheel or shaker at room temperature. (Note: Always use a 1 mL pipette for liquid removal to avoid accidentally aspirating samples. Do not attempt to remove all of the liquid to prevent drying or loss of the sample, always freshly prepare resin/acetone mixtures before use; all steps below are carried out at room temperature.)
2. Incubate for ~5 h in a 1:1 resin/acetone mixture, followed by overnight incubation in 2:1 resin/acetone mixture.
3. Transfer samples and resin/acetone mixture into teflon tray. For embryos pour samples directly into the tray. For larvae and adult worms, transfer using a 3.5 mL Pasteur pipette. (Note: Larvae and adults will be particularly fragile at this point so handle with care.)
4. Using a toothpick under a stereomicroscope transfer specimens into an embedding mold containing pure resin.
5. Incubate samples for 6–8 h in an exsiccator. (Note: It is also possible to embed agarose disks containing embryos directly without pre-embedding. If so, proceed to Section 1.2.2 Embedding. However, we have found pre-embedding to be helpful also for embryos given that agarose disks are quite fragile.)
6. Prepare several Aclar® pieces to cover each of the pre-embedding molds.
7. Fill pre-embedding molds with a thin layer of resin (Figure 1(B)).
8. Transfer specimens with a toothpick into the mold and gently press them into the resin. (Note: Several worms or embryos can be embedded in one mold but they should be sufficiently separated from each other to be cut out individually.)
9. Cover molds by gently rolling the Aclar® piece on it.
10. Gently compress sample to get a uniform resin layer.
11. Let resin polymerize at 60 °C in an oven (takes approximately 2 days).

1.2.2 Embedding

Instrumentation:

Oven, stereomicroscope with light source.

Material:

Embedding mold (Ted Pella #10590), razorblade (degreased with acetone, uncoated, American Safety Razor #62-0179), tweezers (Dumont #0108-5-PS).

Reagents:

Epoxy resin (Agar 100, Agar Scientific #AGR1031).

Procedure:

1. Peel off Aclar® film from samples, cut edges of resin layer and take resin out of the mold.

2. Under a dissecting microscope cut out rectangular pieces along the axis of the worm or embryo.
3. Place resin pieces into embedding mold (one piece per well) and position in such a way that the animal is located as shown in Figure 1(C). (Note: Orientation here is key since it defines the cutting axis for later steps. Hence, to examine cross-sections of amphids in larvae or adult animals position them perpendicular to the edge of the mold with the tip of the nose facing outward. Examination of other structures or cutting angles may require the use of differently shaped embedding molds.)
4. Fill wells with resin, if necessary re-position sample and press to the bottom.
5. Let resin polymerize at 60 °C in an oven (takes approximately 2 days).

1.3 SERIAL SECTIONS

A general problem of transmission electron micrographs is that they contain overlapping features from different z-positions, that is, they are two-dimensional projections of a three-dimensional object. While optical sectioning methods exist for light microscopy (e.g., confocal microscopy), no such methods are available for TEM. This problem is commonly addressed by mechanically sectioning the specimen after resin embedding to reduce the amount of depth information. Serial section electron micrographs can subsequently be stacked together into a 3D volume. The z-resolution which can be obtained by this method is equal to twice the section thickness (McEwen & Marko, 1999). Another approach to obtaining three-dimensional information is electron tomography (see below), which provides better z-resolution and thus more ultrastructural detail than serial section reconstructions. However, serial sectioning is a more straightforward method for tracing contiguous objects, such as ciliated neurons, over large distances (Doroqucz et al., 2014). Here we describe the steps involved in the sectioning procedure, focusing on obtaining cross sections of the amphid cilia in larvae in adult worms. Various tips and tricks on how to perform sectioning can be found in (Hagler, 2007). Sectioning is the most difficult part of the sample preparation process, requiring extensive practice and especially patience by the investigator.

1.3.1 Block trimming

Accurate trimming of the resin block is essential to obtain proper serial sections. The aim is to trim the block containing the specimen into a pyramidal shape with a flat top face (Figure 2(A)). This can be done by cutting away small slices using a razor blade, but this might not be feasible for everybody and deliver results variable in quality. We therefore prefer to use an ultramicrotome equipped with a glass knife.

Instrumentation:

Ultramicrotome (Ultracut UCT, Leica).

Material:

Glass knifes, prepared using a knifemaker (EM KMR2, Leica) from glass strips (400 × 25 × 6.4 mm, Leica #16840031), razorblade (degreased with acetone,

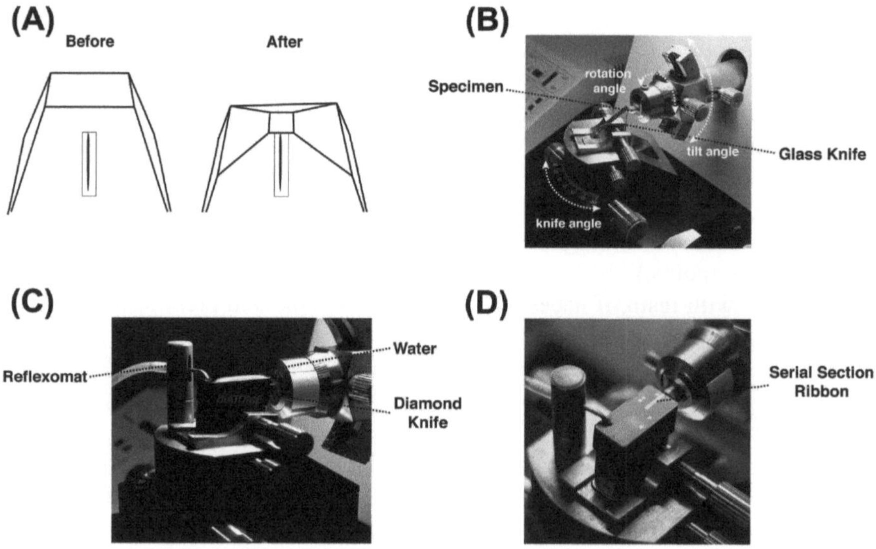

FIGURE 2 Block trimming and serial sectioning.

(A) Block trimming. Shape of the resin block before and after trimming as described in Section 1.3.1. (B) Close-up view of ultramicrotome equipped with glass knife for block trimming. (C) Ultramicrotome equipped with diamond knife for serial sectioning. Note that water bath overfilled (surface convex) as required in preparation for cutting (see Section 1.3.2, Step 16). (D) Ribbons of serial sections in water bath. Water level should be below rim (surface concave) at this point.

uncoated, American Safety Razor #62-0179), universal specimen holder (Leica #16701761).

Procedure:

1. Place resin block into the sample holder and pre-trim with a razorblade. (Note: Always remove thin slices rather than big chunks at a time.)
2. Fasten holder to the ultramicrotome arm.
3. Rotate sample such that the bottom of the block (the side with the specimen) faces upward and the mark of the holder is in line with the 90° scale of the ultramicrotome arm.
4. Adjust tilting angle to 0° and fasten glass knife to the ultramicrotome (Figure 2(B)).
5. Adjust knife angle to −20° (horizontal axis) and start to cut away ~3 μm thick slices until you reach the specimen, but make sure to leave some space between the cutting edge and the specimen. (Note: The cut surface should be shiny. If it turns opaque, it will be necessary to shift the knife laterally to an unused position or use a new knife. Otherwise edges will not be smooth and it will be impossible to obtain a ribbon of serial sections.)

6. Adjust knife angle to 20° (horizontal axis), taking care to avoid accidental contact with the resin block, and repeat the procedure.
7. Move knife back while maintaining cutting angle, rotate the sample counterclockwise 90°, and repeat cutting procedure.
8. Move knife back and rotate sample back to the original position.
9. Adjust knife angle (horizontal axis) to 0° and use motorized fine feed to carefully move knife in the direction of the sample until it nearly touches the resin block.
10. Adjust knife angle in a way that the knife edge is perfectly perpendicular to the axis of the specimen. For this use the highest possible magnification. (Note: This is the most important step during the trimming).
11. Now start to cut away 0.5–3 μm-thick slices until you arrive next to the specimen. Check from time to time how far the block has already been trimmed by taking it out and examining it under a dissecting microscope. (Note: If you plan to do a serial section reconstruction, it may be helpful to create an identifying mark, e.g., by slicing away the upper right corner. This will help in orienting and ordering sections under the electron microscope.)
12. Continue directly with Section 1.3.2, Cutting Serial Sections.

1.3.2 Cutting serial sections

Instrumentation:

Ultramicrotome (Ultracut UCT, Leica).

Material:

Diamond knife (Diatome #DU4520), dust blower (Bergeon #4657), 100 mesh copper-palladium grids (Agar Scientific #G2410PD) coated with Formvar on palladium face, eye lash tool, filter paper strips (Whatman #1001 090), cosmetic tissues (ZVG #16515), grid box (Leica #16705525), high-density polystyrene stick (Agar Scientific #C899), razorblade (degreased with acetone, uncoated, American Safety Razor #62-0179), tweezers (Dumont #0108-5-PS).

Reagents:

ddH_2O, absolute ethanol, 70% ethanol.

Procedure:

1. Clean tweezers with 70% ethanol.
2. Cut polystyrene stick tip into a wedge shape using a razor blade. Wet tip with absolute ethanol and carefully clean diamond knife edge. (Note: The diamond knife edge is quite fragile, handle with care.)
3. Rinse diamond knife and boat with ddH_2O. Dry using dust blower.
4. Replace glass knife from 3.1 with diamond knife and fasten it to the ultramicrotome. (Note: Make sure clearance angle is set to the value recommended by the manufacturer.)
5. Reset microtome arm feed.
6. Rotate sample counterclockwise 90°.
7. Use the highest possible magnification for alignment.

8. Move knife in direction of the sample, once in close proximity use electronic approach buttons (stop at a distance of ~1 mm). (Note: Take care to avoid contact with the sample during the following adjustments. If necessary, move knife back.)
9. Shift knife laterally to a position without visible imperfections where it will contact the sample surface.
10. Adjust rotation angle of sample until its bottom edge is parallel to the knife edge.
11. Adjust tilt angle of the sample such that a clear reflection of the knife is visible on the sample surface. For this make sure that bottom illumination is turned on Move microtome arm up and down; the reflection should remain the same size, otherwise tilt angle of the sample needs to be refined. If reflection broader on one side, adjust knife angle.
12. Move knife closer to the sample and refine tilt angle. Repeat until knife reflection gets nearly invisible.
13. Move knife back a short distance and define start point (distance of the bottom edge of the specimen to the knife edge should be around 10 mm) and end point (top edge of the specimen should be just below the knife edge) of the cutting window.
14. Move knife towards the block until reflection of the knife becomes nearly invisible.
15. Mount "Reflexomat" on microtome (Figure 2(C)).
16. Overfill knife boat with ddH_2O using a Pasteur pipette (water surface should appear convex, see Figure 2(C)), then reduce water level with "Reflexomat" until water surface appears reflective and slightly concave.
17. The knife edge needs to be completely wetted. If not, try to wet it with the eyelash tool. It may also be that the water level is too low, the knife edge is dirty or cracked. (Note: Make sure not to wet the back side of the knife.)
18. Set section thickness to the desired value (70 nm for standard TEM, 200 nm for electron tomography). Sectioning speeds between 0.7 and 1 mm/s are usually appropriate.
19. Start automatic cutting. (Note: If block gets wet, there is too much water in the knife boat. Stop sectioning and dry block with filter paper. It may also be necessary to dry the back side of the knife. This can sometimes be done with tissue paper by soaking away liquid, while avoiding any contact with the knife edge. If this is not possible, the knife needs to be removed and dried using the dust blower, after which the knife needs to be re-aligned (Step 7, above).)
20. During cutting, sections are compressed, but should spread once floating on the water surface. Sections will remain attached to each other, forming ribbons (Figure 2(D)). Section quality and thickness is apparent from their reflection color. Sections of 70 nm have a uniform silver color, 200 nm sections appear blue. (Note: It will take a few sections until they have a uniform color. If color still remains uneven you may have a problem with vibration sources in the

surroundings (e.g., air conditioning, heavy foot traffic). Also check if specimen and knife are properly secured. It may also be that the cutting speed is not appropriate.)

21. If you do not obtain ribbons of sections, try to re-trim top and bottom sides of the resin pyramid with a new glass knife.
22. After cutting some blank sections a small dot caused by the worm will be visible on every freshly cut section for a few seconds. Start collecting these sections. We always pick up ribbons of around 10 sections per grid. First stop the automatic sectioning, carefully nudge the second section of the ribbon counting from the knife edge with the eyelash-tool to detach the ribbon from the knife edge. (Note: Amphid cilia are located in the anterior most part of the worm, therefore start collecting as soon as the worm becomes visible. We recommend collecting every section to ensure that all regions of interest are covered. For other structures less peripherally located it is also possible to collect a few sections, examine them after drying by TEM and re-trim the sample if necessary.)
23. Carefully take a grid with cleaned tweezers without destroying the Formvar film or bending the grid (just touch it at the edges). While observing the grid at low magnification dip the grid into the water away from the ribbon and move it in the direction of the ribbon without touching it. If necessary move the ribbon into the center of the boat with an eyelash-tool. Position grid at an approximately 45° angle underneath the ribbon, palladium face up, catch one end of the ribbon, then carefully lift grid out of the water while leveling off the grid top pick up sections.
24. Dry grid by blotting from the side with a filter paper strip and place into a grid box. (Note: Do not touch section itself with the filter paper.)
25. Continue sectioning. (Note: For examination of amphids we normally prepare between 20 and 25 grids (70 nm). It is important to keep grids in the order in which they were prepared.)
26. Clean knife as described above, remove sample and store at room temperature in case more sections need to be cut later.

1.3.3 Post-staining

To enhance contrast sections need to be stained. Lead and uranium salts as aqueous or alcoholic solutions are common contrasting agents. There are special grid holder staining devices available that allow staining of multiple grids at the same time. We have obtained good results with the Pelco® Grid Staining Matrix System (Ted Pella #22510). Here we describe a staining protocol for the simultaneous staining of up to five grids, which does not require any special equipment.

Material:

Liquid waste beaker, filter paper strips (Whatman #1001 090), 14 cm glass petri dish (lid and plate covered separately with aluminum foil), Parafilm®, tweezers (Dumont #0108-5-PS).

Reagents:

ddH_2O in wash bottle, NaOH pellets, Reynolds lead citrate, prepared as described by Reynolds (1963), uranyl acetate (Electron Microscopy Sciences #22400).

Procedure:

1. Cut two 10 cm strips of Parafilm and place on lab bench.
2. Prepare 2% uranyl acetate in ddH_2O and pipette for each grid one droplet on the first Parafilm strip and place grid on top with sections facing the droplet (i.e., palladium face down). (Note: A maximum of five grids can be processed simultaneously.)
3. Cover with the lid of a petri dish and incubate for 15 min (70 nm sections) or 30–45 min (200 nm sections).
4. In the meantime, place NaOH pellets on the edges of the second Parafilm strip and pipette for each grid a droplet of lead citrate in the middle region of the parafilm and cover with the lid of a petri dish. (Note: Addition of NaOH pellets helps prevent lead citrate precipitation.)
5. Pick up grids one at a time with tweezers and rinse extensively from the edge with ddH_2O.
6. Blot grid briefly (do not let it dry) by touching the edge with a filter paper and place it face down on a droplet of lead citrate. Repeat with the other grids and cover with the petri dish.
7. Incubate each grid for no longer than 5 min.
8. Pick up grids one at a time with tweezers and rinse extensively from the edge with ddH_2O.
9. Blot grid dry from the edge with a filter paper. Place grid back into the grid box and allow to dry overnight at room temperature.

1.4 TRANSMISSION ELECTRON MICROSCOPY

The general anatomy of *C. elegans* is described in a series of publications, but it might still be difficult for beginners to orient themselves on the sample. Detailed anatomical information is provided on a number of websites including WormImage (www.wormimage.org) and WormAtlas (www.wormatlas.org). The latter also includes a "Slidable Worm" feature, which allows scanning through a series of cross-sections of the adult worm from head to tail. Especially helpful in terms of ciliary architecture are the publications by Doroquez et al. (2014), Perkins et al. (1986), and Ward et al. (1975). In this section we provide some general hints, which will hopefully make microscopy easier for the novice. For this we focus on the amphid channel cilia in the adult worm, which are most commonly used for ultrastructural studies (Figure 3(A)–(D)).

Instrumentation:

FEI Morgagni 268D microscope equipped with an 11Mpx Morada CCD camera (Olympus-SIS) or FEI Tecnai T20 microscope equipped with an FEI BM Eagle camera, both operated at 80 kV.

Procedure:

1. After microscope alignment and sample insertion, locate serial sections on the grid at low magnification.
2. At ~140× magnification screen for the worm on the section which should appear as a dark dot. (Note: Because of the larger diameter of the worm as you move from the nose towards the posterior, it is easier to locate the worm in deeper sections. It might therefore be helpful to start with grids containing more posterior sections. With knowledge of where the worm is positioned it will then be easier to locate it in more anterior sections.)
3. Increase magnification until cross-section of the worm fills the image window on the computer screen.
4. Focus with the help of the wobbler and generate an overview image. The dorsoventral body axis of the worm can be identified by the position of the amphid channels relative to the triangular-shaped pharynx (Figure 3(A)).
5. Take detailed images at higher magnification. (Note: Since amphid cilia do not run strictly parallel to each other, microtubules in some cilia will appear sharp, while others appear blurry. If all ciliary microtubules appear blurry, the cutting angle was inappropriate.)
6. Of the various ciliary substructures (Figure 3(D)), the transition fibers are particularly difficult to identify at the beginning. They are most easily located by tracking individual cilia through serial sections and examining those sections just proximal to the transition zone. (Note: It is rare to obtain cross-sections where all nine transition fibers are visible for a single cilium.)

1.5 ELECTRON TOMOGRAPHY

In electron tomography, a series of electron micrographs, the so-called tilt series, is recorded at different viewing angles of the specimen. For this, the specimen is tilted in increments of 1°–3° around an axis perpendicular to the electron beam of the microscope. From these projections a tomogram containing the 3D specimen structure is calculated in silico. Two algorithms commonly used for tomographic reconstruction are weighted back projection (WBP) and simultaneous iterative reconstruction technique (SIRT). A perfect 3D reconstruction would require a tilt series with a range of ±90°. In practice, the tilt range is limited to ±60°. This results in loss of data in form of a "missing wedge" in the tomogram's Fourier representation. Recording a double-axis tilt series reduces data loss and therefore improves the quality of the reconstruction. For this, a second tilt series perpendicular to the first one is recorded. For data recording, it is important that the sample is at eucentric height, that is, at the z-position of the sample in the electron beam where it can be tilted without lateral shifts of its projections.

A special characteristic of plastic sections is that they shrink by ~50% in the z dimension (specimen depth) upon electron beam irradiation. This shrinkage is biphasic and comprises an early rapid phase followed by a slow phase. This

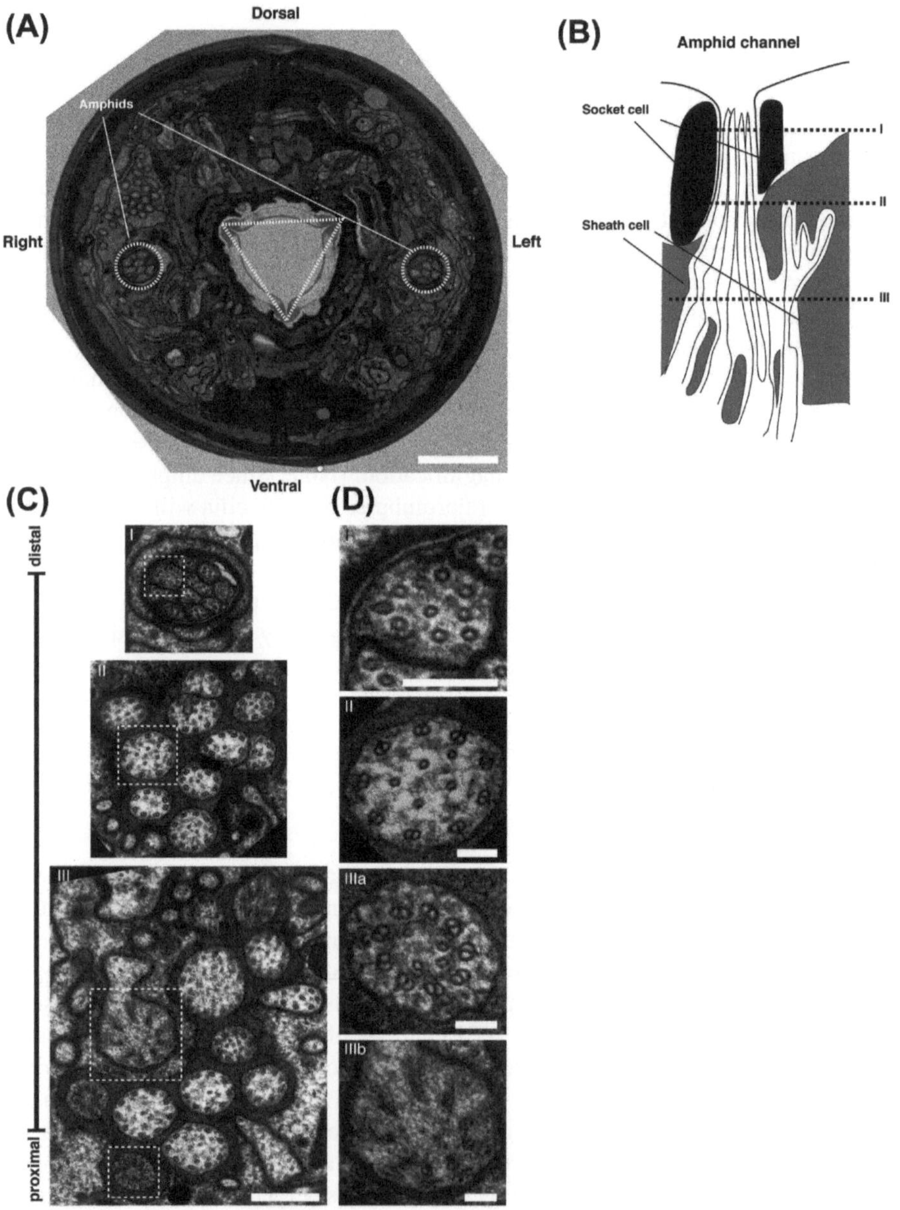

FIGURE 3 Transmission electron microscopy of *C. elegans* amphids.

(A) Transverse section through the head of an L4 larva, fixed and processed as described in Section 1.1.1. The dorsoventral body axis of the worm can be identified by the position of the amphid channels (indicated) relative to the triangular-shaped pharynx (center).
(B) Schematic of amphid sensory organ, based on Perkins et al. (1986). In amphids, 10 (channel) cilia extend into a channel formed by the sheath and socket glia cells; another

shrinkage is not a problem for conventional TEM, since the obtained image is effectively a z-projection. However, it does become an issue for electron tomography. Up to now, it has not been possible to prevent shrinkage. A solution to avoid large-scale changes during data acquisition is to pre-expose the sample and start the tilt series only after the rapid shrinking phase. In addition to shrinkage, the specimen will also shift during the tilt series, even in a perfectly aligned microscope, due to mechanical imperfections of the device. In part this can be compensated for by sample tracking during automated data collection, but precise alignment of single projections to each other has to be done after tilt series recording during tomogram reconstruction. This is facilitated by the addition of high-density markers (fiducial markers), commonly gold beads, to the specimen prior to imaging. In this section we describe the setup of a tilt series recording and later processing. For a more detailed discussion of electron tomography, we recommend the book by Frank (2006).

1.5.1 Adding fiducial markers

Instrumentation:

Microcentrifuge, sonicating water bath, vortex mixer.

Material:

Filter paper strips (Whatman #1001 090), grid support plate (Leica #16707299), 1.5 mL microcentrifuge tubes, 1000 μL adjustable micropipette, tweezers (Dumont #0108-5-PS).

Reagents:

ddH_2O in wash bottle, Gold Colloid 10 nm (BBI Solutions #EM.GC10).

Procedure:

1. Briefly vortex to resuspend gold colloid solution. For every five grids pipette 1.5 mL of gold colloid into a 1.5 mL microcentrifuge tube and pellet at maximum speed for 30 min.
2. During centrifugation place grids into grid support plate.
3. Remove half the supernatant, resuspend pellet by pipetting and sonicate in sonicating waterbath for 2 min. For more than five grids, combine resuspended pellets before sonication to ensure uniform labeling.
4. Pipette solution onto grids until they are fully submerged in gold colloid. (Note: A uniform spread of gold beads on both sides of the sample is essential for reconstruction.)

three (wing cilia) terminate in the sheath cell. (C) Serial transverse sections through amphid channel at the level of the axonemal distal segment (I), proximal segment (II), and transition zone (III), corresponding to the positions indicated in (B). Note that cilia are differently positioned along the vertical axis relative to each other such that different ciliary subdomains are found in the same section. (D) High magnification views of individual cilia at the level of the distal segment (characterized by singlet microtubules, I), proximal segment (doublet microtubules, II), transition zone (peripheral Y-links and central cylinder, (IIIa), and transition fibers (IIIb). Scale bars are 2 μm (A), 500 nm (C), and 100 nm (D).

5. Incubate for 1–2 min with gentle agitation.
6. Remove gold solution with filter paper and rinse grids with ddH_2O.
7. Dry grids with filter paper and place back into the grid box. (Note: Only touch edges of the grids.)

1.5.2 Data recording

The protocol for tilt series acquisition is based on protocols by Cindi Schwartz (http://bio3d.colorado.edu/SerialEM/#Help—Optimizing Plastic Section Tomography) and Günter Resch (http://www.nexperion.net/).

Instrumentation:

FEI Tecnai T20 microscope equipped with an FEI BM Eagle camera operated at 200 kV or FEI Tecnai F30 Helium "Polara" microscope equipped with a Gatan UltraScan 4000 CCD camera operated at 300 kV, both equipped with SerialEM (Mastronarde, 2005), available at http://bio3d.colorado.edu/SerialEM/.

Procedure:

1. Align condensor aperture, condensor astigmatism, and beam tilt pivot points on microscope before loading sample.
2. Perform a dose calibration for all spot sizes and magnifications that will be used for data recording. Center beam and spread it far enough that the entire camera field is evenly exposed and no beam edges are visible. Take an image and select "Calibration – Electron Dose". (Note: Perform a new calibration for each session.)
3. Load sample.
4. Adjust eucentric position for the grid. We do this manually by bringing an object into the center of the viewing screen at ~4200× magnification (with objective aperture inserted), activating the α-wobbler (maximum tilt angle of ±15°), and minimizing lateral movement of the object image by adjusting the z-position of the grid. We then repeat the procedure at ~20,000× magnification. (Note: If you know the approximate z-position from previous experiments, use this position as a starting point to speed up the process.)
5. We find it helpful to create an overview map of the entire grid, which makes it easier to find the region of interest. If you prefer to work without, proceed to Step 11. Otherwise set microscope to ~140× magnification, spot size 6, spread the beam fully, set record exposure time in SerialEM to 0.25 s, and take a test image. Adjust settings if necessary.
6. In SerialEM, open a new navigator window and set up a full montage. Overlap value should be 20%. Select "Align pieces in overview," unselect "Treat as very sloppy montage" in "Montage Controls" and start montage. (Note: The program will now generate a navigator map, which will take approximately 20 min.)
7. Find points of interest on the map and mark them. To do this, select "Add points" in navigator window and click on points of interest. When done, click "Stop adding." By selecting "Goto XY" in the navigator window, the stage will

move to the selected point of interest. (Note: You can also zoom in and out and adjust brightness of the map.)

8. Sometimes it is useful to create more detailed maps of regions of interest at higher magnification, especially if screening for objects that are difficult to find, like centrioles in late embryos. If you plan to work without, proceed to Step 11. Otherwise set microscope to a magnification that allows image acquisition of the entire region of interest. Take a record image and enclose region of interest with a polygon.
9. Go to the desired magnification (for centrioles ~17,000×/pixel size ~0.6 nm), focus (you may use the autofocus function, target defocus should be set to ~ −5 μm), take a test image, and adjust exposure settings if necessary.
10. Setup and start polygon map. (Note: Use binning if necessary to reduce file size. This map can be used in the same way as described for the overview map.)
11. Once you have found your feature of interest, set microscope to ~2000× magnification, spot size one, and completely spread the beam. In SerialEM, open dose meter and expose sample until dose value reaches 2000–3000 e^{-} $Å^{-2}$ (pre-exposure).
12. Go to the desired magnification and focus (defocus set to ~ −5 μm). Set exposure times for all modes in the camera parameters dialog. (Note: The field of view should contain between 20 and 30 gold beads for reconstruction and avoid regions close to grid bars. If you plan to record a double tilt series, take images at different magnifications, to make it easier to return to the original position after rotating your sample. Some microscope set-ups will also have software tools, like "Total Recall" on FEI systems, to assist with this.)
13. Reset image shift and refine eucentric height (using the "Eucentric fine" function of SerialEM) on or close to your area of interest.
14. Center area of interest and set stage to the starting tilt angle, typically −60° (use the "Walk up" function).
15. Run "Check autofocus" and set up tilt series. We use 1° angle increments, 4 s tilt delay, low mag tracking (if magnification is at 50,000×/pixel size ~0.2 nm or higher), autofocus every 4° and always above 50°. Limit image shift to 1 μm. If you plan to perform a double tilt series, append "a" to the filename. Empty buffer tank and start tilt series.
16. When tilt series is complete, if you wish to record a double tilt series, rotate sample 90° and return to your starting position. Set up and run tilt series as above, appending a "b" to the filename.

1.5.3 Tilt series reconstruction and model generation

For reconstruction of tilt series into tomograms we use the IMOD software package (Kremer, Mastronarde, & McIntosh, 1996), available as a free download from http://bio3d.colorado.edu/imod/. The comprehensive tutorials on the IMOD Website are a good way to become acquainted with the program.

While it is easy to recognize the two-dimensional shape of an object in individual tomographic slices, it may be difficult to appreciate the overall three-dimensional architecture. Visualization of an object in 3D space can be achieved by creating rendering models. We are doing this manually with the help of IMOD. Figure 4(A) shows a tomographic image of the axoneme of an amphid channel cilium, obtained from a single tilt series using the WBP algorithm and applying a smoothing filter. In IMOD, models are organized in three hierarchical levels, objects, contours, and points. The model in Figure 4(B) was obtained by tracing the outlines of different axonemal structures in the slicer window, which offers the possibility to average several slices of the tomogram, facilitating the tracing process. Contours (Figure 4(C)) were created by adding points around the structure and smoothening the resulting trace. This process was repeated over the entire z-range, creating an outline of the object (Figure 4(D)). A-microtubules and the ciliary membrane were created as closed contours, B-microtubules as open contours. Finally, contours were meshed (Figure 4(E)) and displayed for all objects. The resultant 3D model

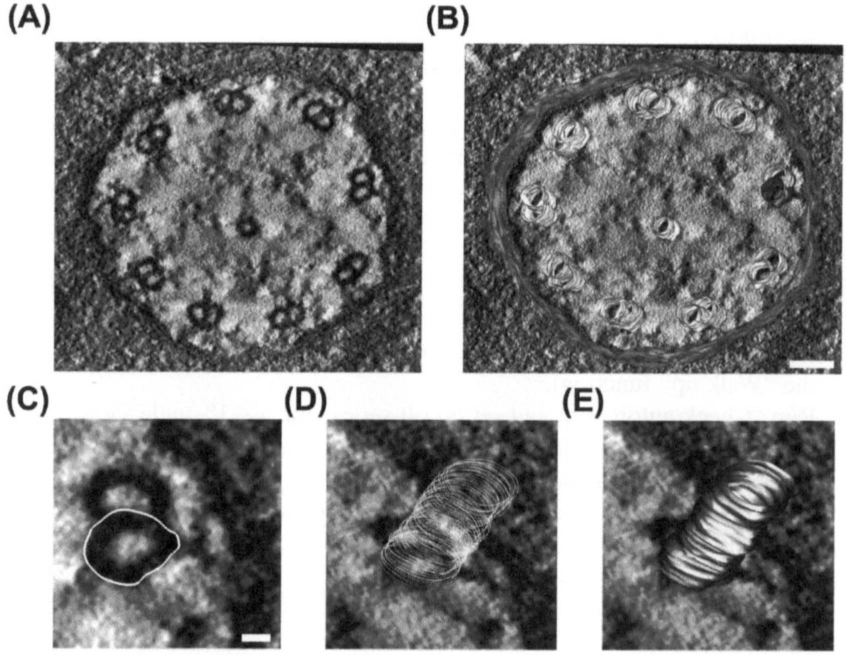

FIGURE 4 Electron tomography.

(A) A 4.5-nm thick tomographic slice of a cilium at the level of the proximal segment of the axoneme. (B) 3D reconstruction model overlaid on image in (A) highlighting axonemal microtubules (yellow) and ciliary membrane (green). (C–E) Steps in modeling a single axonemal microtubule using IMOD. First, the contour of the microtubule is traced in a single tomographic section (C), this procedure is the repeated for every second section of the tomogram (D). Finally, contours were meshed to create a surface model (E). Scale bars are 50 nm (B) and 10 nm (C). (See color plate) can be overlaid on the original tomogram and rotated in all dimensions to better appreciate the three-dimensional shape of the structure.

DISCUSSION

Although TEM has been used extensively for studies of *C. elegans* cilia, performing such studies remains a challenging task, especially for the novice. Here, we set out to guide the reader through the various steps in this process, presenting two alternative methods for sample fixation (chemical fixation and HPF-FS) and focusing on the amphid cilia most commonly used in ultrastructural studies. In addition, we provide support for the setup of an electron tomography experiment.

TEM and electron tomography are inherently morphological methods, offering insights into cellular architecture at nanometer resolution. Unlike fluorescence microscopy, processing of samples for electron microscopy do not involve labeling of specific proteins and hence require no a priori knowledge of the protein composition of particular structures. These methods are therefore ideally suited to characterize protein function by examining loss of function phenotypes in an unbiased manner. An important challenge is to relate this wealth of ultrastructural information to protein localization (Jana, Marteil, & Bettencourt-Dias, 2014). Traditionally, proteins have been localized by immuno-electron microscopy using specific antibodies to the protein or epitope tag. While such approaches have provided important insights into basal body architecture (Kilburn et al., 2007; Nakazawa, Hiraki, Kamiya, & Hirono, 2007), labeling comes at the expense of structural preservation. This is due to the fact that standard fixation conditions abolish antigenicity, while milder fixations result in poor ultrastructural preservation. One potential solution is to employ genetically encoded EM tags such as APEX or mini-SOG, which generate electron-dense signal in situ without the need for special fixation conditions (Martell et al., 2012; Shu et al., 2011). However, such approaches have so far not been extensively used. An alternative is to perform CLEM, examining protein localization by fluorescence microscopy and subsequently processing samples for EM. Of particular interest is the application of super-resolution microscopy techniques, which have greatly improved the precision of protein localization by light microscopy. Such methods have yielded important insights into the molecular architecture of the centrosome (Mennella, Agard, Huang, & Pelletier, 2014), although they have yet to be applied to cilia. In combining these techniques with electron microscopy, the biggest technical challenge lies in maintaining proper sample alignment, essential if one is to overlay protein localization on ultrastructural morphology (Kukulski et al., 2011).

Despite the increased resolution afforded by electron tomography, this technique has so far not been used extensively on *C. elegans* cilia. The only published study so far, utilizing HPF-FS prior to resin embedding and serial sectioning (Doroquez et al., 2014), reported a number of surprising findings. First, the authors could not detect any evidence for transitional fibers, structures described in numerous previous studies in *C. elegans* (see also Figure 3(D)), suggesting that these fibers are in actual fact flared axonemal microtubules. Second, the central cylinder (also known as the apical ring) found in the transition zone of many types of sensory cilia was less prominent in their tomograms, raising the possibility that it is formed by the accretion of material during conventional fixation. Finally, the authors could detect vesicles within the transition zone area, from which they were previously thought to be excluded by ciliary gating. These findings underscore the power of electron tomography to deliver important insights into cilia biology. Nevertheless, as previously discussed, there are trade-offs to the use of any EM fixation protocol. HPF-FS, for example, yields better structural preservation compared to normal chemically fixation, but might create less contrast. It will therefore be important to confirm and extend these findings using other, complementary approaches.

One such approach could be the examination of vitrified sections by cryo-electron tomography (Bouchet-Marquis & Hoenger, 2011; Chlanda & Sachse, 2014). Vitrification by rapid freezing preserves samples in their near-native state without the need for fixation or resin embedding. A general advantage of cryo-EM is that the detected electron density originates from the examined object itself (and not from stain surrounding it), increasing image resolution to near-atomic detail, but at the cost of strongly reduced signal to noise. The aforementioned shrinkage observed with resin sections upon electron irradiation also does not appear to be a problem for these kinds of samples. However, vitrified sectioning is technically highly demanding. Even in the most experienced hands it has so far not been possible to obtain vitreous sections without compression, tears, or knife marks. Vitrified sectioning has so far not been applied to *C. elegans*, but may yield important insights into its sensory anatomy.

ACKNOWLEDGMENTS

We would like to thank Thomas Heuser and Günter Resch for comments on the manuscript and discussions, Harald Kotisch for instrument photographs and other members of the CSF Electron Microscopy facility at the Vienna Biocenter for experimental protocols. Work in the Dammermann lab is supported by start-up funding from MFPL as well as grant Y597-B20 from the Austrian Science fund (FWF) to A.D.

REFERENCES

Badano, J. L., Mitsuma, N., Beales, P. L., & Katsanis, N. (2006). The ciliopathies: an emerging class of human genetic disorders. *Annual Review of Genomics and Human Genetics, 7*, 125–148.

Basto, R., Lau, J., Vinogradova, T., Gardiol, A., Woods, C. G., Khodjakov, A., et al. (2006). Flies without centrioles. *Cell, 125*(7), 1375–1386.

Berbari, N. F., O'Connor, A. K., Haycraft, C. J., & Yoder, B. K. (2009). The primary cilium as a complex signaling center. *Current Biology, 19*(13), R526–R535.

Bouchet-Marquis, C., & Hoenger, A. (2011). Cryo-electron tomography on vitrified sections: a critical analysis of benefits and limitations for structural cell biology. *Micron, 42*(2), 152–162.

Chalfie, M., & Thomson, J. N. (1982). Structural and functional diversity in the neuronal microtubules of *Caenorhabditis elegans. The Journal of Cell Biology, 93*(1), 15–23.

Chlanda, P., & Sachse, M. (2014). Cryo-electron microscopy of vitreous sections. In J. Kuo (Ed.), *Electron microscopy* (Vol. 1117, pp. 193–214). Humana Press.

Dammermann, A., Pemble, H., Mitchell, B. J., McLeod, I., Yates, J. R., 3rd, Kintner, C., et al. (2009). The hydrolethalus syndrome protein HYLS-1 links core centriole structure to cilia formation. *Genes and Development, 23*(17), 2046–2059.

Doroquez, D. B., Berciu, C., Anderson, J. R., Sengupta, P., & Nicastro, D. (2014). A high-resolution morphological and ultrastructural map of anterior sensory cilia and glia in *Caenorhabditis elegans. eLife, 3*, e01948.

Fawcett, D. W., & Porter, K. R. (1954). A study of the fine structure of ciliated epithelia. *Journal of Morphology, 94*(2), 221–281.

Fisch, C., & Dupuis-Williams, P. (2011). Ultrastructure of cilia and flagella - back to the future! *Biology of the Cell, 103*(6), 249–270.

Frank, J. (2006). *Electron tomography.* New York: Springer.

Hagler, H. K. (2007). Ultramicrotomy for biological electron microscopy. *Methods in Molecular Biology, 369*, 67–96.

Hall, D. H. (1995). Chapter 17 electron microscopy and three-dimensional image reconstruction. In F. E. Henry, & C. S. Diane (Eds.), *Methods in cell biology* (Vol. 48, pp. 395–436). Academic Press.

Hall, D. H., Hartwieg, E., & Nguyen, K. C. Q. (2012). Chapter 4-Modern electron microscopy methods for *C. elegans*. In H. R. Joel, & S. Andrew (Eds.), *Methods in cell biology* (Vol. 107, pp. 93–149). Academic Press.

Hall, D. H., & Russell, R. (1991). The posterior nervous system of the nematode *Caenorhabditis elegans*: serial reconstruction of identified neurons and complete pattern of synaptic interactions. *The Journal of Neuroscience, 11*(1), 1–22.

Hayat, M. A. (1981). *Fixation for electron microscopy.* Academic Press.

Heuser, T., Dymek, E. E., Lin, J., Smith, E. F., & Nicastro, D. (2012). The CSC connects three major axonemal complexes involved in dynein regulation. *Molecular Biology of the Cell, 23*(16), 3143–3155.

Inglis, P. N., Ou, G., Leroux, M. R., & Scholey, J. M. (2007). *The sensory cilia of* Caenorhabditis elegans. WormBook. pp. 1–22.

Jana, S. C., Marteil, G., & Bettencourt-Dias, M. (2014). Mapping molecules to structure: unveiling secrets of centriole and cilia assembly with near-atomic resolution. *Current Opinion in Cell Biology, 26*(0), 96–106.

Kilburn, C. L., Pearson, C. G., Romijn, E. P., Meehl, J. B., Giddings, T. H., Jr., Culver, B. P., et al. (2007). New tetrahymena basal body protein components identify basal body domain structure. *The Journal of Cell Biology, 178*(6), 905–912.

Kolotuev, I., Schwab, Y., & Labouesse, M. (2010). A precise and rapid mapping protocol for correlative light and electron microscopy of small invertebrate organisms. *Biology of the Cell, 102*(2), 121–132.

Kremer, J. R., Mastronarde, D. N., & McIntosh, J. R. (1996). Computer visualization of three-dimensional image data using IMOD. *Journal of Structural Biology, 116*(1), 71–76.

Kukulski, W., Schorb, M., Welsch, S., Picco, A., Kaksonen, M., & Briggs, J. A. (2011). Correlated fluorescence and 3D electron microscopy with high sensitivity and spatial precision. *The Journal of Cell Biology, 192*(1), 111–119.

Lucic, V., Forster, F., & Baumeister, W. (2005). Structural studies by electron tomography: from cells to molecules. *Annual Review of Biochemistry, 74*, 833–865.

Martell, J. D., Deerinck, T. J., Sancak, Y., Poulos, T. L., Mootha, V. K., Sosinsky, G. E., et al. (2012). Engineered ascorbate peroxidase as a genetically encoded reporter for electron microscopy. *Nature Biotechnology, 30*(11), 1143–1148.

Mastronarde, D. N. (2005). Automated electron microscope tomography using robust prediction of specimen movements. *Journal of Structural Biology, 152*(1), 36–51.

McDonald, K. L., & Auer, M. (2006). High-pressure freezing, cellular tomography, and structural cell biology. *Biotechniques, 41*(2), 137–143.

McEwen, B. F., & Marko, M. (1999). Three-dimensional transmission electron microscopy and its application to mitosis research. *Methods in Cell Biology, 61*, 81–111.

McIntosh, R., Nicastro, D., & Mastronarde, D. (2005). New views of cells in 3D: an introduction to electron tomography. *Trends in Cell Biology, 15*(1), 43–51.

Mennella, V., Agard, D. A., Huang, B., & Pelletier, L. (2014). Amorphous no more: subdiffraction view of the pericentriolar material architecture. *Trends in Cell Biology, 24*(3), 188–197.

Müller-Reichert, T., Hohenberg, H., O'Toole, E. T., & McDonald, K. (2003). Cryoimmobilization and three-dimensional visualization of *C. elegans* ultrastructure. *Journal of Microscopy, 212*(1), 71–80.

Müller-Reichert, T., Mancuso, J., Lich, B., & McDonald, K. (2010). Chapter 15-Three-Dimensional reconstruction methods for *Caenorhabditis elegans* ultrastructure. In M.-R. Thomas (Ed.), *Methods in cell biology* (Vol. 96, pp. 331–361). Academic Press.

Müller-Reichert, T., Mäntler, J., Srayko, M., & O'Toole, E. (2008). Electron microscopy of the early *Caenorhabditis elegans* embryo. *Journal of Microscopy, 230*(2), 297–307.

Murcia, N. S., Richards, W. G., Yoder, B. K., Mucenski, M. L., Dunlap, J. R., & Woychik, R. P. (2000). The Oak Ridge Polycystic Kidney (orpk) disease gene is required for left-right axis determination. *Development, 127*(11), 2347–2355.

Nakazawa, Y., Hiraki, M., Kamiya, R., & Hirono, M. (2007). SAS-6 is a cartwheel protein that establishes the 9-fold symmetry of the centriole. *Current Biology, 17*(24), 2169–2174.

Olson, S. K., Greenan, G., Desai, A., Müller-Reichert, T., & Oegema, K. (2012). Hierarchical assembly of the eggshell and permeability barrier in *C. elegans*. *The Journal of Cell Biology, 198*(4), 731–748.

Osborn, M., & Weber, K. (1982). Chapter 7-Immunofluorescence and Immunocytochemical procedures with affinity purified antibodies: Tubulin-containing structures. In W. Leslie (Ed.), *Methods in cell biology* (Vol. 24, pp. 97–132). Academic Press.

O'Toole, E. T., Giddings, T. H., McIntosh, J. R., & Dutcher, S. K. (2003). Three-dimensional organization of basal bodies from wild-type and δ-tubulin deletion strains of *Chlamydomonas reinhardtii*. *Molecular Biology of the Cell, 14*(7), 2999–3012.

Pelletier, L., O'Toole, E., Schwager, A., Hyman, A. A., & Muller-Reichert, T. (2006). Centriole assembly in *Caenorhabditis elegans*. *Nature, 444*(7119), 619–623.

Perkins, L. A., Hedgecock, E. M., Thomson, J. N., & Culotti, J. G. (1986). Mutant sensory cilia in the nematode *Caenorhabditis elegans*. *Developmental Biology, 117*(2), 456–487.

Pigino, G., Maheshwari, A., Bui, K. H., Shingyoji, C., Kamimura, S., & Ishikawa, T. (2012). Comparative structural analysis of eukaryotic flagella and cilia from Chlamydomonas, Tetrahymena, and sea urchins. *Journal of Structural Biology, 178*(2), 199–206.

Reiter, J. F., Blacque, O. E., & Leroux, M. R. (2012). The base of the cilium: roles for transition fibres and the transition zone in ciliary formation, maintenance and compartmentalization. *EMBO Reports, 13*(7), 608–618.

Reynolds, E. S. (1963). The use of lead citrate at high pH as an electron-opaque stain in electron microscopy. *The Journal of Cell Biology, 17*, 208–212.

Scholey, J. M., Ou, G., Snow, J., & Gunnarson, A. (2004). Intraflagellar transport motors in *Caenorhabditis elegans* neurons. *Biochemical Society Transactions, 32*(Pt 5), 682–684.

Shu, X., Lev-Ram, V., Deerinck, T. J., Qi, Y., Ramko, E. B., Davidson, M. W., et al. (2011). A genetically encoded tag for correlated light and electron microscopy of intact cells, tissues, and organisms. *PLoS Biology, 9*(4), e1001041.

Sims, P., & Hardin, J. (2007). Fluorescence-integrated transmission electron microscopy images. In J. Kuo (Ed.), *Electron microscopy* (Vol. 369, pp. 291–308). Humana Press.

Small, J. V. (1981). Organization of actin in the leading edge of cultured cells: influence of osmium tetroxide and dehydration on the ultrastructure of actin meshworks. *The Journal of Cell Biology, 91*(3 Pt 1), 695–705.

Sulston, J. E., Schierenberg, E., White, J. G., & Thomson, J. N. (1983). The embryonic cell lineage of the nematode *Caenorhabditis elegans. Developmental Biology, 100*(1), 64–119.

Ward, S., Thomson, N., White, J. G., & Brenner, S. (1975). Electron microscopical reconstruction of the anterior sensory anatomy of the nematode *Caenorhabditis elegans. The Journal of Comparative Neurology, 160*(3), 313–337.

White, J. G., Southgate, E., Thomson, J. N., & Brenner, S. (1986). The structure of the nervous system of the nematode *Caenorhabditis elegans. Philosophical Transactions of the Royal Society of London. B, Biological Sciences, 314*(1165), 1–340.

Williams, C. L., Li, C., Kida, K., Inglis, P. N., Mohan, S., Semenec, L., et al. (2011). MKS and NPHP modules cooperate to establish basal body/transition zone membrane associations and ciliary gate function during ciliogenesis. *The Journal of Cell Biology, 192*(6), 1023–1041.

CHAPTER

Method: In vitro analysis of pericentriolar material assembly

19

Jeffrey B. Woodruff[1], Anthony A. Hyman

Max Planck Institute of Molecular Cell Biology and Genetics, Dresden, Germany

[1]Corresponding author: E-mail: woodruff@mpi-cbg.de

CHAPTER OUTLINE

Introduction **370**

1. Methods **371**

1.1 Expression of SPD-5::GFP, SPD-2, and PLK-1 in SF9 Insect Cells 371

1.1.1 Materials 371

1.1.2 Protocol 372

1.2 Protein Purification 373

1.2.1 Purification of SPD-5::GFP 373

1.2.2 Purification of SPD-2 376

1.2.3 Purification of PLK-1 376

1.3 SPD-5::GFP Network Preparation 377

1.3.1 Materials 377

1.3.2 Protocol 377

1.4 Imaging of SPD-5::GFP Networks 380

1.5 Data Analysis 380

Conclusion **381**

Acknowledgments **381**

References **381**

Abstract

Centrosomes are major microtubule-organizing centers in eukaryotic cells and play a critical role in embryonic development and asymmetric cell division. Centrosomes comprise a pair of centrioles surrounded by an amorphous proteinaceous meshwork called the pericentriolar material (PCM). Robust deposition of PCM around the centrioles is essential for a centrosome to achieve full microtubule nucleating potential. Despite the wealth of information on PCM composition and function, the mechanism and regulation of PCM assembly have been difficult to ascertain, due in part to the lack of an in vitro system. Here, we describe methods to establish an in vitro system to study PCM assembly

Methods in Cell Biology, Volume 129, ISSN 0091-679X, http://dx.doi.org/10.1016/bs.mcb.2015.04.006

in *Caenorhabditis elegans*. Specifically, we describe (1) how to express and purify the *C. elegans* PCM proteins SPD-5, SPD-2, and PLK-1 from baculovirus-infected insect cells, (2) how to assemble these proteins into PCM-like structures in vitro, and (3) how to quantify this assembly process in a semiautomated fashion.

INTRODUCTION

Centrosomes are composed of two barrel-shaped centrioles that organize a matrix of proteins called the pericentriolar material (PCM). The PCM expands dramatically during mitosis and recruits tubulin complexes and effector proteins that regulate microtubule nucleation, anchoring, and dynamics (for a review on PCM assembly, see Woodruff, Wueseke, and Hyman (2014)). The PCM is built through the deposition of large coiled-coil domain-containing proteins around the centrioles (Dix & Raff, 2007; Hamill, Severson, Carter, & Bowerman, 2002; Kemp, Kopish, Zipperlen, Ahringer, & O'Connell, 2004; Pelletier et al., 2004; Zhu et al., 2008). When assembled, these proteins serve as scaffolds for the downstream effector proteins. The conserved Polo and Aurora A kinases regulate PCM assembly by directly phosphorylating these core coiled-coil domain-containing proteins (Barr & Gergely, 2007; Conduit et al., 2014; Haren, Stearns, & Lüders, 2009; Lee & Rhee, 2011). Despite our understanding of PCM composition and function from these in vivo studies, mechanistic questions regarding how the nanometer-scale PCM proteins organize into a micron-scale, functional organelle remain largely unanswered due in part to the lack of an in vitro *system* for PCM reconstitution.

The main obstacle to reconstituting PCM has been the expression and purification of the large coiled-coil domain-containing proteins that constitute the PCM matrix. Often, proteins that contain coiled-coil domains are insoluble, unstable, and/or express very poorly. Researchers can overcome this problem by making truncated constructs. This strategy has been useful for mapping the protein interaction domains of several centrosome proteins and, more specifically, in identifying a multimerization domain in centrosomin (Boxem et al., 2008; Conduit et al., 2014; Dzhindzhev et al., 2010). However, to reconstitute functional PCM-like assemblies in vitro, full activity of the PCM scaffold proteins is required; thus it is imperative to purify full-length proteins.

We recently developed a method for reconstituting *Caenorhabditis elegans* PCM in vitro using purified full-length SPD-5, PLK-1 (Polo-like kinase 1), and SPD-2 (Cep192) proteins (Woodruff et al., 2015). The results from this study indicate that the coiled-coil domain-containing protein SPD-5 can polymerize into supramolecular scaffolds that recruit downstream PCM proteins. Furthermore, SPD-2 and PLK-1 enhance the formation of SPD-5 assemblies in vitro. Here, we give a detailed description of the methods we used to study PCM assembly in vitro. Our protein expression protocol uses baculovirus vectors and has been optimized for enhanced expression, stability, and solubility of these three proteins inside *Spodoptera frugiperda* (SF9) cells cultured in suspension. SPD-5, PLK-1, and SPD-2 contain protease-cleavable affinity tags and are purified using tandem affinity steps. We also describe how to assemble and quantify PCM-like structures using fluorescently labeled SPD-5 and unlabeled PLK-1 and SPD-2.

1. METHODS

1.1 EXPRESSION OF SPD-5::GFP, SPD-2, AND PLK-1 IN SF9 INSECT CELLS

We engineered viruses that express (1) MBP PreScission::SPD-5::GFP-PreScission-6xHis, (2) MBP-TEV::SPD-2::TEV-6xHis, and (3) PLK-1::PreScission-6xHis (see Figure 1(A)). We found that adding an MBP tag to the N-termini of SPD-5 and SPD-2 dramatically increased their expression and stability. The N-terminal tag does not affect function. PreScission or TEV protease sites were inserted to allow cleavage and removal of all purification tags. PreScission protease is preferable, as it is more active at 4 °C. However, we found experimentally that PreScission protease cleaves SPD-2 internally at a cryptic site. Thus, TEV protease must be used to remove purification tags from SPD-2.

Our protocol is adjusted to ensure low multiplicity of infection (MOI) during transfection and amplification (MOI = 0.1−1). We also limit the number of virus amplification steps; thus, we use P1 virus stocks to initiate protein expression. These steps help prevent the formation of defective virus particles that show reduced target protein expression and/or express truncated target proteins.

1.1.1 Materials

ESF 921 Insect Cell Culture Medium, Protein-Free (expression systems)
Fetal Bovine Serum (PAA)
Protease Inhibitor Cocktail III (Calbiochem)
PMSF (Sigma)
SF9-ESF *S. frugiperda* insect cell line
SF9 cell resuspension buffer
- 50 mM Tris HCl, pH 7.4
- 150 mM NaCl
- 1% glycerol
- 0.5 mM DTT

Protease inhibitors (final concentration)
- 1 mM PMSF
- 100 μM AEBSF
- 0.08 μM Aprotinin
- 5 μM Bestatin
- 1.5 μM E-64
- 2 μM Leupeptin
- 1 μM PepstatinA

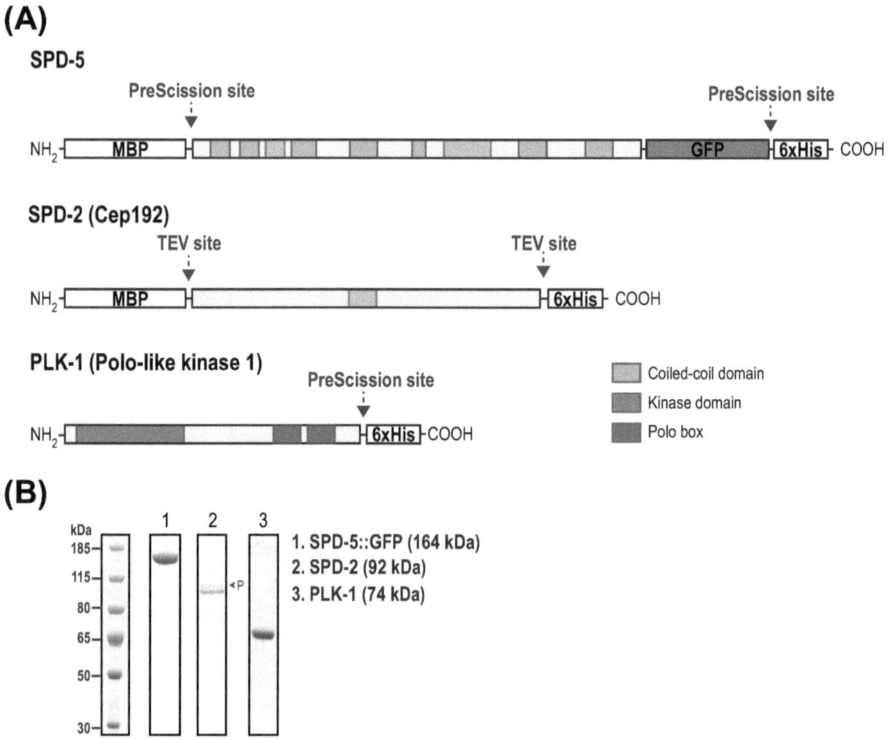

FIGURE 1 Purification of *Caenorhabditis elegans* PCM proteins.

(A) Schematic of the SPD-5, SPD-2, and PLK-1 proteins expressed using baculovirus-mediated infection of insect cells. During purification, all affinity tags can be removed by site-specific proteolysis with either PreScission or TEV protease. (B) Coomassie-stained gels showing purified SPD-5, SPD-2, and PLK-1 after removal of affinity tags. The expected molecular weights of the full-length proteins are listed on the right. The red (gray in print versions) P indicates a phosphorylated species of the full-length SPD-2. This band disappears after phosphatase treatment (data not shown).

Modified with permission from Woodruff et al. (2015).

To make final resuspension buffer + protease inhibitor solution, add 10 μL Calbiochem Protease Cocktail III and 50 μL PMSF (200 mM) to 10 mL SF9 cell resuspension buffer.

1.1.2 Protocol

1. Generate baculoviruses expressing the SPD-5, PLK-1, and SPD-2 constructs through general viral transfection protocols. Harvest virus 4–5 days after transfection. Pass cell supernatant through a 0.45 μm filter. This is the P0 stock.

2. Infect 50 mL SF9 cells in suspension (log phase, 0.5 million cells/mL) with 12.5 μL of P0 virus. Add fetal bovine serum to a final concentration of 2%.
3. Incubate cells at 27 °C with moderate shaking. Harvest cells after 72 h by centrifugation. Save the supernatant and discard the cell pellet.
4. Pass the supernatant through a 0.45 μm filter. This is the P1 stock.
5. Infect 500 mL SF9 cells (log phase, 1 million cells/mL) with 5 mL of P1 virus.
6. Incubate cells at 27 °C with moderate shaking.
7. Harvest cells after 72 h by gentle centrifugation (700× g for 5 min; slow deceleration). Discard supernatant by decanting.
8. Add protease inhibitors to resuspension buffer.
9. Resuspend cells in 5 mL resuspension buffer + protease inhibitor solution.
10. Split resuspended cells into three 15 mL Falcon tubes (there should be approximately 3 mL of cell suspension per tube).
11. Flash freeze cells in liquid N_2. Store at −80 °C.

1.2 PROTEIN PURIFICATION

1.2.1 Purification of SPD-5::GFP

SPD-5 sticks nonspecifically to a wide variety of resins commonly used for protein purification. Traditional protein purification techniques that require large volumes of resin, such as size exclusion chromatography, are not advisable. One exception to this rule is Ni-NTA. Also, we found that using batch binding and gravity elution columns produces higher protein yields than using packed columns with high pressure solvent flow (e.g., fast protein liquid chromatography (FPLC) systems).

1.2.1.1 Materials

Ni-NTA affinity resin (Qiagen)
Amylose affinity resin (New England Biolabs)
Glutathione sepharose 4B (GE Healthcare)
15 mL disposable chromatography columns (Bio-rad)
30 K Amicon Ultra centrifugal filter units (Millipore)
0.22 μm cellulose acetate spin filters (Agilent Technologies)
Superdex 200 HR 10/30 column (GE Healthcare)
PreScission protease (1 mg/mL; GST-tagged)
TEV protease (1 mg/mL; GST-tagged)
Buffer A
- 50 mM Tris HCl, pH 7.4
- 500 mM NaCl
- 30 mM imidazole
- 1% glycerol
- 0.1% CHAPS

Buffer B
- 50 mM Tris HCl, pH 7.4
- 500 mM NaCl
- 250 mM imidazole
- 1% glycerol
- 0.1% CHAPS

Buffer C
- 50 mM Tris HCl, pH 7.4
- 500 mM NaCl
- 1% glycerol
- 0.1% CHAPS
- 0.5 mM DTT

Protease inhibitors
- 1 mM PMSF
- 100 μM AEBSF
- 0.08 μM Aprotinin
- 5 μM Bestatin
- 1.5 μM E-64
- 2 μM Leupeptin
- 1 μM PepstatinA

1.2.1.2 Protocol

1. Assemble all reagents and cool down centrifuge, centrifuge tubes, and the dounce homogenizer.
   a. Put the dounce homogenizer on ice and wash once with cold water, then once with cold Buffer A.
   b. Label the ultracentrifuge tubes with tape, then put them on ice.
   c. Close the door of the Beckman Ultracentrifuge, turn on the vacuum, and cool down to 4 °C.
   d. Prepare Ni-NTA columns. Pour 5 mL of Ni-NTA beads into a plastic column. Wash twice with water (50 mL), then once with Buffer A (25 mL).
   e. Prepare amylose columns. Pour 2.5 mL of amylose beads into a plastic column. Wash twice with water, then once with Buffer C.

   Note: Steps 2–3 must be performed quickly!!!

2. Add protease inhibitors to lysis buffer.
3. Add 12 mL lysis Buffer A (with protease inhibitors) to 3 mL of frozen cell suspension (fill Falcon tube to the top). Quickly thaw the cells by incubating in a beaker with warm water.

   Note: All remaining steps must occur at 4 °C.

4. Once the cell suspension is loose (partially unfrozen), place the falcon tube on the rotator and rotate for 15 min in the cold room.

5. Check if the cells are fully thawed and resuspended. If not, then pipet up and down a few times.
6. Pour the cells into the clean, prechilled dounce homogenizer. Add 10 mL of Buffer A (+protease inhibitors) to the falcon tube and then pour into the dounce. Gently plunge up and down five times.
7. Use a pipet to remove the homogenized lysate and dispense it into a prechilled ultracentrifuge tube.
8. Balance the tubes by weighing them and adding buffer.
9. Place tubes in a 50.2 Ti rotor, then place the rotor in the Beckman Ultracentrifuge. Spin at 40,000 rpm for 25 min.
10. Remove tubes from the rotor and pour off the supernatant into a prechilled 50 mL falcon tube. This is the high-speed supernatant. Save 20 μL for later analysis on an SDS-PAGE gel.
11. Pour the high-speed supernatant over the prepared Ni-NTA column. Allow the supernatant to drip through once (about 10 min). Collect this flow through in your 50 mL falcon tube. Then, pour the flow through back into the column. Close the column and put in on the rotator and rotate for 60 min or longer.
12. Open the column and let the unbound protein flow through. Save the flow through for later analysis.
13. Wash off non-specifically bound proteins from the Ni-NTA. Fill the column to the top with Buffer A. Repeat 5 times. Check for unwanted proteins in the flow through using a Bradford assay.
14. Elute bound protein with Buffer B. Close the column and rotate it on the rotator for 30 min or longer. Open the column and collect the eluate in a 50 mL falcon tube. Save 20 μL for later analysis on an SDS-PAGE gel.
15. Load the eluate onto the prepared amylose column. Close the column and rotate on the rotator for 60 min or longer.
16. Open the column and collect the flow through. Save 20 μL for later analysis on an SDS-PAGE gel.
17. Wash the column with Buffer C. Repeat 5 times. Check for unwanted proteins in the flow through using a Bradford assay.
18. Elute your protein from the column by adding 25 mL Buffer C plus 100 μL PreScission protease (1 mg/mL). Close the column and rotate on the rotator overnight in the cold room.
19. Open the column and collect the eluate. Once the column is empty, add 20 mL of Buffer C and collect the leftover protein in the same tube used for collecting the eluate.
20. Add 500 μL of prewashed glutathione beads. Incubate for 1 h. This will remove the PreScission protease.
21. Pass eluate + glutathione beads over an empty chromatography column to remove the beads. Collect the flow through in a new 50 mL Falcon tube.
22. Concentrate the flow through using a 15 mL 30 K Amicon concentrator. Just use one. Spin at 4000 rpm for 10 min in a tabletop centrifuge with a swinging

bucket rotor. Dump the flow through. Load again. Repeat until all of the eluate has been concentrated down to about 1 mL (the time needs to be determined experimentally, but spin for at least 10 min).

23. Add 14 mL of Buffer C to the concentrator. Spin until the protein is concentrated to about 1 mL. This step will ensure that any remaining cleaved His tags are removed (the concentrator has a 30 kDa cutoff, so the His tag should easily flow through).
24. Measure the absorbance of the protein at 280 nm using a NanoDrop 1000 and determine protein concentration from the absorbance by entering its extinction coefficient (E(1%)).
25. Distribute protein in 15 μL aliquots to prechilled PCR tubes. Label them and snap-freeze them in liquid nitrogen. Plunge freezing is the best way to ensure uniform and quick freezing.
26. Store the PCR tubes in a small cardboard box and keep at −80 °C.
27. Analyze high-speed supernatant, Ni-NTA flow through, Ni-NTA eluate, amylose flow through, amylose eluate, and final product on an SDS-PAGE gel.

1.2.2 Purification of SPD-2

Follow the same protocol used for SPD-5::GFP described above. Important: use the appropriate protease for cutting (TEV instead of PreScission protease). At the very end, spin protein at 80,000 rpm for 10 min to remove any aggregates that may have formed.

1.2.3 Purification of PLK-1

1. Follow steps 1−13 in the SPD-5::GFP purification protocol.
2. Elute PLK-1 from the Ni-NTA resin by adding 25 mL Buffer B plus 100 μL PreScission protease (1 mg/mL). Incubate overnight.
3. Save the eluate and pass through a 0.22 μm cellulose acetate spin filter to remove large particles.

   Note: Subsequent steps can occur at room temperature.

4. Connect a 24 mL Superdex 200 size exclusion column to a FPLC system. Wash column with two column volumes of filtered and degassed water, followed by two column volumes of filtered and degassed Buffer C.
5. Inject the filtered protein solution into the FPLC, run, then collect 0.5 mL fractions in a multiwell plate.
6. Analyze the fractions using SDS-PAGE gel electrophoresis followed by Coomassie staining. PLK-1 should elute as a single peak in the 14−15 mL fractions.
7. Collect the fractions containing full-length PLK-1 and concentrate using an Amicon 30 K concentrator.
8. Measure protein concentration, aliquot protein into PCR tubes, flash freeze, and store as per steps 24−28 in the SPD-5::GFP purification protocol.

1.3 SPD-5::GFP NETWORK PREPARATION

SPD-5::GFP will assemble into micron-scale networks over time in a concentration-dependent manner. This means that SPD-5::GFP will form networks during the final concentration step during purification. However, performing a proper network assembly assay requires SPD-5::GFP to be in the unassembled state. Diluting SPD-5::GFP into a high salt, high detergent buffer and incubating overnight at 4 °C will ensure that all preexisting networks are dissolved. Network assembly can then be initiated by diluting SPD-5::GFP into a low salt, no detergent buffer and raising the temperature to 23 °C. Inert red fluorescent beads are added to the sample to aid in maintaining proper focus during imaging. If blocked properly, these beads should not interfere with SPD-5::GFP network assembly or morphology.

1.3.1 Materials

HEPES-KCl Buffer

- 25 mM HEPES, pH 7.4
- 150 mM KCl
- 0.5 mM DTT

Network Dilution Buffer

- 50 mM Tris HCl, pH 7.4
- 500 mM NaCl
- 1% Glycerol
- 0.1% CHAPS
- 0.5 mM DTT

0.2 μm FluoroSpheres, red fluorescent beads (Molecular Probes)

Before use, wash the beads in water, then block the beads in a solution of 5 mg/mL BSA or 5 mg/mL ovalbumin for 16 h at 4 °C. Wash the beads 3× to remove unbound blocking agent, then resuspend in HEPES-KCl buffer and store at 4 °C. You will need to adjust the final bead concentration so that you have >10 beads per field of view during microscopy. Fewer beads will result in improper autofocusing.

20× ATP·$MgCl_2$ mix

- 4 mM ATP
- 200 mM $MgCl_2$

Dilute into HEPES-KCl buffer. Aliquot and store at −20 °C.

0.5 mg/mL ovalbumin

Dissolve ovalbumin (Sigma) in HEPES-KCl buffer. Aliquot and store at −20 °C.

1.3.2 Protocol

1. Day 1: Dilute SPD-5::GFP to 250 nM using Network Dilution Buffer. Incubate overnight at 4 °C. Dilution and overnight storage will promote disassembly of preexisting SPD-5 networks.

2. Day 2: Set up a master network reaction in 1.5 mL microcentrifuge tubes.
   Example 1. 12.5 nM SPD-5 assembly reaction

1 μL	250 nM SPD-5::GFP
1 μL	Preblocked red fluorescent beads
1 μL	0.5 mg/mL ovalbumin
17 μL	HEPES-KCl buffer
20 μL	

   Example 2. 12.5 nM SPD-5 + 12.5 nM PLK-1 + 12.5 nM SPD-2 assembly reaction

1 μL	250 nM SPD-5::GFP
1 μL	250 nM PLK-1
1 μL	250 nM SPD-2
1 μL	Preblocked red fluorescent beads
1 μL	0.5 mg/mL ovalbumin
1 μL	20X ATP·$MgCl_2$ mix
14 μL	HEPES-KCl buffer
20 μL	

3. Distribute master mix in 20 μL aliquots to other 1.5 mL microcentrifuge tubes. Incubate for desired amount of time at 23 °C.
4. Place an 18 × 18 mm coverslip on the Pressing Device (illustrated in Figure 2(A)).
5. Create centering marks on a nonfrosted microscope slide using the slide guide. To do this, place the slide on top of the Slide Guide (illustrated in Figure 2(A)) and trace the black lines using a pen.
6. Pipet 2.3 μL of a reaction into the center mark area of the marked microscope slide. Important: Do not turn the slide after pipetting, instead pipette the reaction on the bottom side of the slide directly. Turning the slide will cause the drop to spread and dry out, thus affecting the morphology of the networks.
7. Place the slide with the hanging drop sample on the foam supports of the Pressing Device (see Figure 2(A)). Align the slide with the cover slip using the centering marks.

Note: The Pressing Device is needed to prevent bubbles from forming under the coverslip. Bubbles will cause denaturation and unwanted aggregation of protein that will significantly alter your results. The centering marks and cover slip holder are necessary to maintain reproducibility in later analysis steps.

8. Press down on the slide gently. Use both hands and apply pressure evenly above the foam supports. Press until the hanging drop just contacts the coverslip.

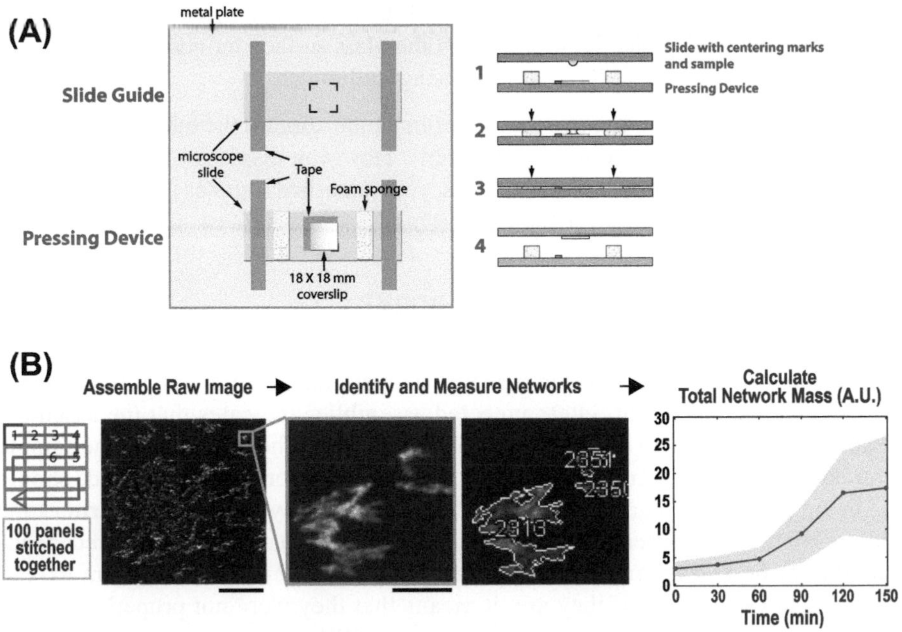

FIGURE 2 Analysis of SPD-5::GFP assemblies.

(A) Schematic of the devices used to prepare the samples for imaging. These devices ensure reproducible positioning of the cover slip and prevent the formation of bubbles in the sample. A new nonfrosted microscope slide is placed over the slide guide and the centering marks are traced with a pen. The sample is applied in the center of the centering marks on the bottom side of the marked slide (Step 1). An 18 × 18 mm coverslip is loaded into the Pressing Device. Second, the sample slide is placed on the Pressing Device and gentle pressure is evenly applied above the two sponges, bringing the sample droplet in contact with the coverslip (Step 2). The slide is then pressed down farther until the cover slip adheres to the free slide (Step 3). Finally, the free slide is lifted up and the newly adhered cover slip is sealed with nail polish (Step 4). The sample is now ready for imaging. (B) Overview of the imaging and analysis protocol. The objective scans across the sample and collects 100 adjacent panels in a square pattern. Every two panels, the focal plane is readjusted using an autofocusing routine that identifies fluorescent red beads in the sample (not shown). The image panels are stitched together to assemble the final raw image. This image is analyzed with an image processing script that identifies and calculates the integrated fluorescence intensity for each SPD-5::GFP network. The integrated fluorescence intensities for all of the networks in a sample are summed up, giving the Total Network Mass. Finally, after performing several experimental replicates, the mean Total Network Mass is determined and plotted. The graph on the right depicts the formation of SPD-5::GFP networks over time for an initial concentration of 25 nM SPD-5::GFP. The red (gray in print versions) line represents the mean Total Network Mass per time point and the pink shaded area represents the 95% confidence interval (n = 10). Scale bars, 25 μm (first panel), 2.5 μm (inset).

Modified with permission from Woodruff et al. (2015).

The coverslip should then adsorb to the slide, if not, then press down a little bit more. This will spread the networks on the glass surface for easy detection.

9. Remove slide and seal the coverslip using nail polish.

Important: Once you remove sample from a tube, discard the tube. Pipetting will disrupt the networks and initiate disassembly. However, disassembly will not occur if the networks are adsorbed to glass. Thus, your squashed samples are relatively stable and can be visualized over the next 30 min.

1.4 IMAGING OF SPD-5::GFP NETWORKS

We use automated microscopy to collect a series of images and then stitch them together into one final image for analysis (see Figure 2(B), first panel). The appearance of bright, interconnected assemblies indicates that the assembly reaction was successful (Figure 2(B), middle panels). The red beads in the sample can be used for autofocusing, as the SPD-5::GFP networks (if assembled) should be in the same focal plane as the beads. This is especially useful in the earlier time points of the experiment before network formation has begun. The red beads should not be fluorescent when illuminated in the GFP channel (475 nm excitation/523 nm emission). If they are, it means that they were not properly blocked and are sticking nonspecifically to monomeric SPD-5::GFP.

Important: SPD-5::GFP networks may not be distributed homogenously under the squashed coverslip. Therefore, to ensure adequate quantification of network formation, it is essential to sample a large area. We typically collect >100 images per sample (40× objective, 1024 × 1024 pixels per image).

1.5 DATA ANALYSIS

SPD-5::GFP networks can be identified and analyzed using Fiji or ImageJ, which are freely available. In brief, the stitched image is thresholded to generate a mask. The mask is then used to create selections that outline the SPD-5::GFP networks. Finally, the selections are applied to the original stitched image, and the total area and mean fluorescence intensity of SPD-5::GFP networks are measured. These data can be compiled from multiple replicates and timepoints to generate a network assembly curve (Figure 2(B), last panel).

1. Open stitched image in Fiji or ImageJ.
2. Subtract background from image.
3. Under "Image," click "Duplicate."
4. Click on the duplicated image. Under "Image/Adjust," select "Threshold." Choose dark background. Adjust the lower bound until only the SPD-5::GFP networks are identified. Click "Apply." This is the masked image.
5. Invert masked image. The background should now be white.
6. Under "Edit/Selection," select "Create selection." This will outline the networks with a yellow line.

7. Under "Edit/Selection," click "Add to Manager." This will store the ROI generated in step 5.
8. Click on the original image.
9. Under "Image/Overlay," click "From ROI Manager." This will overlay the outlines generated in step 5 onto the unaltered image.
10. Under "Analyze," click "Measure." With the default settings, this will measure the total area and average fluorescence intensity of the material within the ROI. Multiplying these values will generate the integrated fluorescence intensity, which can be used as an estimate of the total amount of SPD-5::GFP network existing within the stitched image.

CONCLUSION

Here, we report how to express and purify the *C. elegans* PCM proteins SPD-5, PLK-1, and SPD-2 and how to use them to form PCM-like assemblies in vitro. We also describe imaging and analysis methods to quantify the formation of these assemblies over time. Using this system, we have recently shown that SPD-5 can polymerize into supramolecular assemblies that serve as hubs for downstream PCM proteins (Woodruff et al., 2015). In the future, this system can be used for identifying proteins that influence SPD-5 polymerization and also for identifying structural features of SPD-5 that mediate self-assembly.

Recent work from Conduit et al. (2014) suggests that the *Drosophila* protein centrosomin may perform a similar function as SPD-5 in forming the foundation of the fly PCM. However, full-length centrosomin has yet to be purified; thus, it is not known if full-length centrosomin can assemble micron-scale assemblies in vitro. We are hopeful that the methods described here can be applied to the in vitro study of PCM proteins from *Drosophila*, humans, and other centrosome-containing organisms.

ACKNOWLEDGMENTS

We thank David Drechsel, Barbara Borgonovo, Regis Lemaitre, Regina Wegner, Aliona Pozniakovsky, and Andrei Pozniakovsky for help and advice with protein expression and purification; Dan White for help with imaging; Oliver Wueseke, Louise Jawerth, and Bea Ferreira Gomes for help with image analysis and for critical reading of the manuscript; Karen Oegema and the Hyman lab for helpful discussions. J.B.W. was supported by Alexander von Humboldt and EMBO fellowships.

REFERENCES

Barr, A. R., & Gergely, F. (2007). Aurora-A: the maker and breaker of spindle poles. *Journal of Cell Science, 120*, 2987–2996.

Boxem, M., Maliga, Z., Klitgord, N., Li, N., Lemmens, I., Mana, M., et al. (2008). A protein domain-based interactome network for *C. elegans* early embryogenesis. *Cell, 134*, 534–545.

Conduit, P. T., Feng, Z., Richens, J. H., Baumbach, J., Wainman, A., Bakshi, S. D., et al. (2014). The centrosome-specific phosphorylation of Cnn by Polo/Plk1 drives Cnn scaffold assembly and centrosome maturation. *Developmental Cell, 28*, 659–669.

Dix, C. I., & Raff, J. W. (2007). *Drosophila* Spd-2 recruits PCM to the sperm centriole, but is dispensable for centriole duplication. *Current Biology, 17*, 1759–1764.

Dzhindzhev, N. S., Yu, Q. D., Weiskopf, K., Tzolovsky, G., Cunha-Ferreira, I., Riparbelli, M., et al. (2010). Asterless is a scaffold for the onset of centriole assembly. *Nature, 467*, 714–718.

Hamill, D. R., Severson, A. F., Carter, J. C., & Bowerman, B. (2002). Centrosome maturation and mitotic spindle assembly in *C. elegans* require SPD-5, a protein with multiple coiled-coil domains. *Developmental Cell, 3*, 673–684.

Haren, L., Stearns, T., & Lüders, J. (2009). Plk1-dependent recruitment of γ-tubulin complexes to mitotic centrosomes involves multiple PCM components. *PLoS One, 4*, e5976.

Kemp, C. A., Kopish, K. R., Zipperlen, P., Ahringer, J., & O'Connell, K. F. (2004). Centrosome maturation and duplication in *C. elegans* require the coiled-coil protein SPD-2. *Developmental Cell, 6*, 511–523.

Lee, K., & Rhee, K. (2011). PLK1 phosphorylation of pericentrin initiates centrosome maturation at the onset of mitosis. *Journal of Cell Biology, 195*, 1093–1101.

Pelletier, L., Özlü, N., Hannak, E., Cowan, C., Habermann, B., Ruer, M., et al. (2004). The *Caenorhabditis elegans* centrosomal protein SPD-2 is required for both pericentriolar material recruitment and centriole duplication. *Current Biology, 14*, 863–873.

Woodruff, J. B., Wueseke, O., Viscardi, V., Mahamid, J., Ochoa, S. D., Bunkenborg, J., et al. (2015). Regulated assembly of a supramolecular centrosome scaffold in vitro. *Science, 348*, 808–812.

Woodruff, J. B., Wueseke, O., & Hyman, A. A. (2014). Pericentriolar material structure and dynamics. *Philosophical Transactions of the Royal Society of London Series B-Biological Sciences, 369*.

Zhu, F., Lawo, S., Bird, A., Pinchev, D., Ralph, A., Richter, C., et al. (2008). The mammalian SPD-2 ortholog Cep192 regulates centrosome biogenesis. *Current Biology, 18*, 136–141.

CHAPTER

Monitoring SPB biogenesis in fission yeast with high resolution and quantitative fluorescent microscopy

20

Imène B. Bouhlel[*,§], **Kathleen Scheffler**[*,§], **Phong T. Tran**[*,§], **Anne Paoletti**[*,§,1]

*Centre de Recherche, Institut Curie, Paris, France

§CNRS-UMR144, Paris, France

[1]Corresponding author: E-mail: Anne.paoletti@curie.fr

CHAPTER OUTLINE

Introduction ... **384**

1. Monitoring SPB Duplication in Fixed SPB-Labeled Strains ... **384**

1.1 Cell Growth and Fixation ... 386

1.2 Imaging SPBs in Fixed Cells ... 386

1.3 Analysis of SPB Status ... 386

2. Quantitative Analysis of SPB Biogenesis in Live Cells ... **387**

2.1 PDMS Chambers for Live Imaging ... 387

2.2 Live Cell Imaging ... 389

2.3 Quantitative Analysis of SPB Biogenesis ... 389

Conclusion ... **390**

Acknowledgments ... **390**

References ... **390**

Abstract

Like centrosomes, yeast spindle pole bodies (SPBs) undergo a tightly controlled duplication cycle in order to restrict their number to one or two per cell and promote the assembly of a bipolar spindle at mitotic entry. This conservative duplication cycle is tightly coordinated with cell cycle progression although the mechanisms that ensure this coordination remain largely unknown. In this chapter, we describe simple high resolution microscopy- and quantitative light microscopy-based methods that allow to monitor SPB biogenesis in fission yeast and may be useful to study the molecular pathways controlling the successive phases of the duplication cycle.

Methods in Cell Biology, Volume 129, ISSN 0091-679X, http://dx.doi.org/10.1016/bs.mcb.2015.03.005

INTRODUCTION

Spindle pole bodies or SPBs are the yeast equivalent of centrosomes and have an essential function in bipolar spindle assembly. They share key functional features with centrosomes, including microtubule nucleation and anchoring, attachment to the nucleus as well as a strict regulation of their copy number, controlled by a conservative duplication mechanism that restricts them to one copy per cell before duplication, and two copies afterward. Nevertheless, the composition and structure of SPBs is quite different from that of centrosomes. Indeed, SPB components generally share limited homology with centrosomal proteins outside of the γ-TurC involved in microtubule nucleation. Moreover, these organelles are lacking centrioles, and instead are formed of stacked layers referred to as SPB plaques, tightly associated with one another and with the nuclear envelope (Adams & Kilmartin, 2000; Ding, West, Morphew, Oakley, & McIntosh, 1997; Jaspersen & Ghosh, 2012; Jaspersen & Winey, 2004; Lim, Zhang, & Surana, 2009; Uzawa et al., 2004). (Figure 1(A)).

In fission yeast, electron microscopy analysis performed in the late-1990s revealed the ultrastructure of SPBs at different cell cycle stages (Ding et al., 1997). It showed that during interphase, the SPB is sitting on top of the nuclear envelope rather than being inserted in it like in *Saccharomyces cerevisiae*. It also identified, like in budding yeast, an appendage called the half-bridge, on the side of the SPB plaques, which duplicates first to create at its tip an assembly site for the new SPB (Bouhlel et al., in press; Elserafy et al., 2014; Kilmartin, 2003; Lee et al., 2014; Li et al., 2006; Paoletti et al., 2003). Side-by-side duplicated SPBs eventually insert in the nuclear envelope at mitotic entry and separate after cleavage of the bridge and nucleation of intranuclear microtubules to generate a bipolar spindle that can segregate chromosomes.

Determining the duplication status of SPBs by electron microscopy remains very challenging and tedious in fission yeast, and only allows the analysis of a handful of samples. New progress will be aided by faster methods to track SPB duplication. In addition, the availability of methods to monitor SPB biogenesis in live cells is important to establish how the different steps of the duplication cycle are coordinated with cell cycle progression.

In this chapter, we describe two methods to monitor SPB duplication cycle based on high resolution microscopy of fixed cells and quantitative light microscopy of live cells. These analyses can be performed on fission yeast strains expressing red and green fluorescent SPB components from their endogenous locus.

1. MONITORING SPB DUPLICATION IN FIXED SPB-LABELED STRAINS

SPB duplication can be monitored by imaging some of the key structural components of the SPB plaques such as Ppc89 (Rosenberg et al., 2006) and Sid4 that

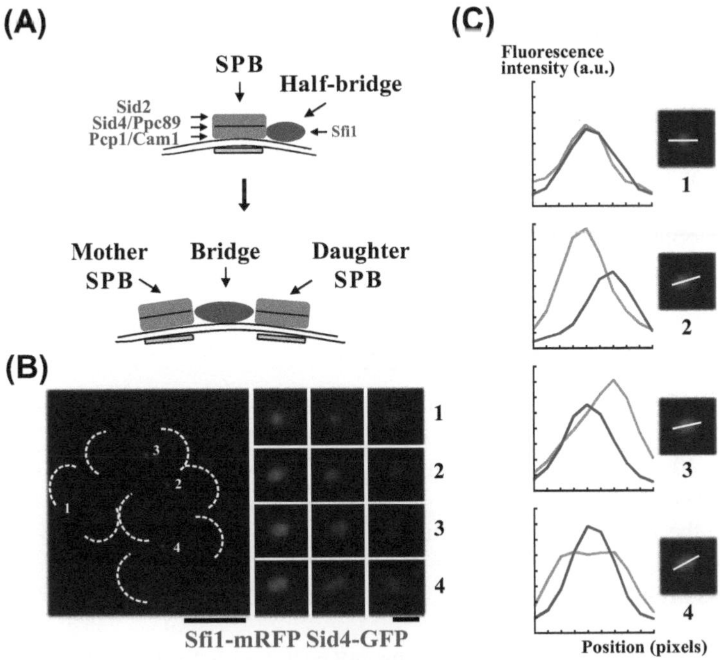

FIGURE 1

Monitoring SPB duplication in fixed SPB-labeled strains. (A) Scheme of the SPB with SPB plaques in green and half-bridge in red before duplication (top) or the duplicated SPBs linked by the bridge that maintain them side by side (bottom). The localization of the plaque components Sid2, Sid4, Ppc89, Pcp1, and Cam1 and of the half-bridge component Sfi1 is indicated. (B) Localization at SPBs of Sid4-GFP and Sfi1-mRFP. Bar: 5 μm. Insets: magnification of SPBs from cells numbered according to cell cycle progression in the left panel. Bars: 500 nm. *(Reproduced from Bouhlel et al., in press.)* (C) Fluorescence intensity profiles of Sfi1-mRFP and Sid4-GFP across SPBs shown in insets measured on a 4-pixels-wide line oriented as shown in white. These SPBs are the same as those shown in (B). Note that Sfi1 maximum intensity does not coincide with Sid4 maximum intensity (SPB 2,3), except for SPB 1 where the SPB-half-bridge axis might be oriented perpendicular to the field of view. After duplication (SPB 4), Sfi1 localizes between the two maxima of intensity of Sid4-GFP. Note the asymmetric peak of Sid4-GFP in SPB 3 suggesting that the assembly of the daughter SPB has started. (See color plate)

are C-terminally tagged with GFP (Chang & Gould, 2000) in combination with the half-bridge/bridge component Sfi1 (Bouhlel et al., in press; Lee et al., 2014; Ohta, Sato, & Yamamoto, 2012) C-terminally tagged with mRFP (Figure 1(B)). This analysis can be completed with strains expressing Sid2-GFP associated with the SPB cytoplasmic surface (Sparks, Morphew, & McCollum, 1999), and GFP-tagged calmodulin Cam1 or Pcp1 that associate with the nuclear-facing surface of the SPB (Flory, Morphew, Joseph, Means, & Davis, 2002; Fong, Sato, & Toda,

2010). Imaging is performed on methanol-fixed cells, which preserves the fluorescence of SPB components best, in order to block microtubule-dependent SPB oscillations that would blurry the fluorescent signal and lower the resolution. Imaging can be performed on any classical epifluorescence microscope or on a spinning disc confocal microscope equipped with a high aperture 100× objective, an automated z control and a high resolution CCD camera. We typically use a DM 5000 B upright microscope (Leica Microsystems), equipped with a 100×/1.4NA oil immersion PlanApo objective, a Pifoc objective stepper, and a Coolsnap HQ CCD camera (Photometrics).

1.1 CELL GROWTH AND FIXATION

1. Grow cells at 25 °C in YE5S medium until exponential phase (OD at 595 nm 0.2–0.8).
2. Filter 20–30 mL of culture on 0.45 μm Durapore membrane filter (Millipore, HVLP4700).
3. Place the filter into a 50 mL tube half-filled with cold methanol (−20 °C) and vortex.
4. Quickly remove the membrane from the tube and centrifuge cells at 4000 rpm for 2 min.
5. Remove methanol and rehydrate the cell pellet with 1 mL PEM buffer (100 mM Pipes, 1 mM EGTA, 1 mM $MgSO_4$).
6. Transfer cells to a 1.5 mL microcentrifuge tube and wash three times with 1 mL PEM.

At this stage, the cells resuspended in PEM buffer can be stored in the dark and at 4 °C for several days before imaging.

1.2 IMAGING SPBs IN FIXED CELLS

1. Mount 2 μL of cells between coverslip and slide.
2. Select a field with numerous cells, focus on a medial focal plane, and acquire an image of cells in transmitted light or DIC mode to define cell contours.
3. Acquire two successive series of green and red z-stacks of nine planes spaced by 0.5 μm (i.e., −2 μm below to +2 μm above the medial focal plane).
4. Image multichannel fluorescent beads (100 nm diameter) in the same conditions.

1.3 ANALYSIS OF SPB STATUS

1. Make maximum intensity projections of the green and red z-stacks. Check the perfect superimposition of signals between the two stacks acquired successively with the same color channel to ensure that the cells did not move during the imaging process. Check also the proper registration of red and green channels with the fluorescent beads.

2. Determine the SPB status based on the signal obtained for the SPB plaque component on high magnification views. A duplicated SPB appears as two juxtaposed points separated by the completely isotropic bridge signal. It can also appear as an elongated anisotropic signal when the two duplicated SPBs are not fully resolved. It should be noted that a minority of duplicated SPBs may be overlooked if the elongation axis is oriented perpendicular to the field of view. This is however expected to happen at low frequency only since duplicated SPBs orient along the cell long axis due to the spatial constraints imposed by cell geometry on microtubule bundles that are anchored to SPBs, and run parallel to their duplication axis (Vogel et al., 2007).
3. Trace the mean fluorescence intensity profile along a 4-pixels-wide line covering all the signal and parallel to the signal elongation axis using the linescan tool of Metamorph software, or equivalent, to confirm the SPB status. A broad signal with two maxima is observed upon duplication compared to the central single peak obtained for the half-bridge or bridge throughout the duplication cycle (Figure 1(C)).
4. Correlate the SPB duplication status with cell cycle progression as judged by cell length during G2 phase that can be measured from the recorded DIC or bright field image or presence of two separated SPBs during mitosis, or of two separated SPBs as well as a septum in postmitotic cells.

2. QUANTITATIVE ANALYSIS OF SPB BIOGENESIS IN LIVE CELLS

The SPB duplication cycle is accompanied by modifications of the content of individual components of SPB plaques and bridge, which can be tracked by quantitative live cell imaging. This method can complement the qualitative analysis of SPB duplication status proposed in the first part of the chapter.

2.1 PDMS CHAMBERS FOR LIVE IMAGING

Long-term imaging for over a cell cycle can be performed on agar pads as described earlier (Tran, Paoletti, & Chang, 2004) or in dedicated 4–5 μm thick poly-dimethylsiloxane (PDMS) microfluidic chambers sticked on glass coverslips (Figure 2(A)). In our hands, such chambers are better suited than agar pads for time-lapse movies with red fluorescent proteins. Production methods for such chambers have been described previously (Terenna et al., 2008; Velve-Casquillas, Le Berre, Piel, & Tran, 2010). Briefly:

1. Spin-coat Su-8 negative photoresist onto a silicon wafer (Su-8 2005 Microchem).
2. Transfer the features of a photomask (Microtronics Engineering GmbH) produced according to chamber design in L-edit software (Tanner EDA) by

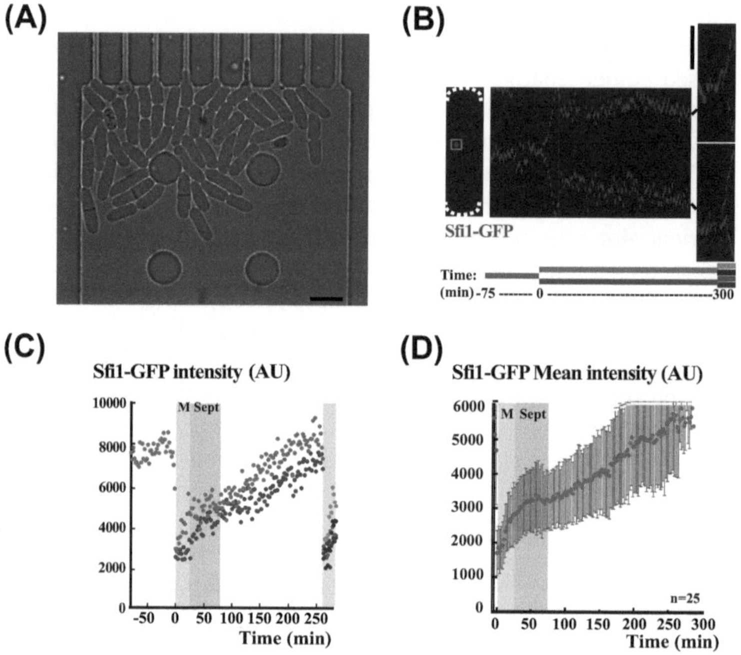

FIGURE 2

Quantitative analysis of SPB biogenesis by live cell imaging. (A) Transmitted light image of fission yeast cells growing in a PDMS chamber for live imaging. Bar: 10 μm. (B–C) Kymograph of SPBs in a wild-type cell expressing Sfi1-GFP and quantitative analysis of Sfi1-GFP intensity at SPB during cell cycle progression. Bars in (B) (bottom) represent SPBs tracked in (C) with a similar color code. Bar: 5 μm. t = 0: SPB segregation. M: Mitosis (blue (very light gray in print versions) zone). Sept: septation (gray zone). *Reproduced from Bouhlel et al., in press.* (D) Mean Sfi1-GFP intensity at SPB before (orange (light gray in print versions)) and after segregation (green (dark gray in print versions)) at t = 0. n = 25. Bars: SD. Reproduced from Bouhlel et al., in press.

laser-etching into a thin layer of chromium on a quartz plate onto the photoresist layer, by exposure and cross-linking with UV light (365 nm).

3. Develop the photoresist with the developer (Su-8 developer, Microchem) and clean it with isopropyl alcohol and nitrogen gas.
4. Use the wafer as a master mold to repeatedly cast PDMS chambers (Sylgard 184, Dow Corning).
5. Assemble chambers by peeling off a PDMS replica from a mold, introducing inlet and outlet holes, and bonding the replica to a glass coverslip after surface activation with a plasma cleaner (Harrick Scientific).
6. Insert metallic connectors in inlet and outlet holes and connect them to appropriate Tygon tubing to allow cell injection into the chamber with a syringe and needle after filling the chamber with YE5S medium.

2.2 LIVE CELL IMAGING

Time-lapse imaging can be performed on an automated spinning disc confocal microscope equipped with a temperature control box to maintain cells at a constant temperature of 25 °C, a high aperture 60× or 100× objective, and a sensitive CCD camera or an EMCCD camera. We typically use Nikon Eclipse Ti-E microscope equipped with the Perfect Focus System, a 100×/1.45 NA PlanApo oil immersion objective, a Mad City Lab piezo stage, a Yokogawa CSU×1 confocal unit, a Photometrics HQ2 CCD camera, and a laser bench (Errol) with 491 and 561 nm diode lasers of 100 mW each (Cobolt). Lasers should be set at a very low power to limit fluorescence bleaching during time-lapse movies. With a 100× objective, camera binning can be set to 2 to increase signals. Acquisition is made in nine-plane z-stacks (+2 μm to −2 μm every 0.5 μm) to always capture SPBs even if their z position changes over time. Acquisitions are performed every 2−5 min depending on the brightness of the signal, to allow proper capture of intensity variations at different phases of the duplication cycle while limiting bleaching that would alter intensity measurements.

2.3 QUANTITATIVE ANALYSIS OF SPB BIOGENESIS

Simple image analysis allows extracting quantitative data from time-lapse movies.

1. Perform quantitative analysis of SPB intensity over time manually on sum or maximum projections of z-stacks in Metamorph (Molecular devices) or ImageJ software by measuring the integrated fluorescence intensity within a small rectangle of a few pixels wide (5−10) centered on the SPB at each time point (Figure 2(B)). Of note, maximum projections are well suited for SPBs that do not span over several planes due to their limited size. Alternatively, automatic tracking of SPBs can be implemented to speedup the analysis using dedicated softwares such as MIA (Racine et al., 2006).
2. Measure background intensity in a square of similar size placed in the cell cytoplasm and deduce it from the SPB intensities measured in 1.
3. Plot SPB intensity over time to follow the SPB duplication cycle. Mitotic entry can be recognized by the sudden drop in SPB intensity when the two SPBs segregate from one another. Intensity increase during the subsequent cell cycle reflects the behavior of specific SPB components during SPB biogenesis (Figure 2(C)).
4. Generate mean curves for several SPBs after time registration between cells selecting an appropriate time reference such as mitotic entry or anaphase completion as evidenced by SPB separation and maximum distance (Figure 2(D)).
5. Estimate overall bleaching rate during the movie by measuring the mean intensity of all SPBs at the first and last time points. Bleaching corrections can be introduced if necessary assuming an exponential decay of fluorescence over time.
6. Complete the analysis by ratio measurement between different time points along the cell cycle, for instance, between mitotic entry and end of anaphase.

CONCLUSION

The simple linear geometry of side-to-side duplicated SPBs allows for their direct visualization by classical high resolution light microscopy in strains expressing SPB plaque components tagged with fluorescent proteins. This property permits an easy monitoring of SPB duplication process that can be coupled to quantitative analysis of fluorescent signals to describe the dynamics of their accumulation on SPBs as cells progress along the cell cycle. Such methods were recently used to describe the behavior of the SPB component Sfi1 (Bouhlel et al., in press; Lee et al., 2014). In this particular case, based on fluorescence intensity profiles, we managed to distinguish Sfi1 localization on the half-bridge, from that of components of SPB plaques, before or after SPB duplication. Quantitative analysis of fluorescent signals allowed us to identify two phases of Sfi1 accumulation. The first one, at mitotic exit, could be attributed to half-bridge duplication, an event that precedes the assembly of the new SPB. It also revealed that the bridge is destabilized at mitotic entry and loses about a third of its Sfi1 molecules when it splits into two half-bridges. We found that this event is under the control of Cdc31 phosphorylation on a Cdk1 consensus site and controls the timely separation of the two SPBs necessary for bipolar spindle assembly. Quantitative analysis of fluorescence was also used recently to study the nuclear side SPB component Pcp1. This revealed a late phase of accumulation, shortly before mitosis entry (Walde & King, 2014). These examples illustrate how the systematic use of quantitative microscopy can reveal the differential behavior of various SPB components during SPB biogenesis and help deciphering how this complex process is controlled molecularly.

ACKNOWLEDGMENTS

The authors wish to thank Vincent Fraisier from the PICT-IBiSA Lhomond Imaging facility of Institut Curie. IBB received a PhD fellowship from Université Paris-Sud. KS received a PhD fellowships from Université Pierre et Marie Curie and ARC. This work was funded by ANR, LNCC Comité de Paris, and ARC. AP and PT are members of Labex CelTisPhyBio, part of Idex PSL*.

REFERENCES

Adams, I. R., & Kilmartin, J. V. (2000). Spindle pole body duplication: a model for centrosome duplication? *Trends in Cell Biology, 10*, 329–335.

Bouhlel, I. B., Ohta, M., Mayeux, A., Bordes, N., Dingli, F., Boulanger, J., et al. Cell cycle control of spindle poles bodies duplication and splitting by Sfi1 and Cdc31in fission yeast. *Journal of Cell Science*, in press.

Chang, L., & Gould, K. L. (2000). Sid4p is required to localize components of the septation initiation pathway to the spindle pole body in fission yeast. *Proceedings of the National Academy of Sciences of the United States of America, 97*, 5249–5254.

Ding, R., West, R. R., Morphew, D. M., Oakley, B. R., & McIntosh, J. R. (1997). The spindle pole body of *Schizosaccharomyces pombe* enters and leaves the nuclear envelope as the cell cycle proceeds. *Molecular Biology of the Cell, 8*, 1461–1479.

Elserafy, M., Saric, M., Neuner, A., Lin, T. C., Zhang, W., Seybold, C., et al. (2014). Molecular mechanisms that restrict yeast centrosome duplication to one event per cell cycle. *Current Biology, 24*, 1456–1466.

Flory, M. R., Morphew, M., Joseph, J. D., Means, A. R., & Davis, T. N. (2002). Pcp1p, an Spc110p-related calmodulin target at the centrosome of the fission yeast *Schizosaccharomyces pombe*. *Cell Growth & Differentiation, 13*, 47–58.

Fong, C. S., Sato, M., & Toda, T. (2010). Fission yeast Pcp1 links polo kinase-mediated mitotic entry to gamma-tubulin-dependent spindle formation. *EMBO Journal, 29*, 120–130.

Jaspersen, S. L., & Ghosh, S. (2012). Nuclear envelope insertion of spindle pole bodies and nuclear pore complexes. *Nucleus, 3*, 226–236.

Jaspersen, S. L., & Winey, M. (2004). The budding yeast spindle pole body: structure, duplication, and function. *Annual Review of Cell and Developmental Biology, 20*, 1–28.

Kilmartin, J. V. (2003). Sfi1p has conserved centrin-binding sites and an essential function in budding yeast spindle pole body duplication. *Journal of Cell Biology, 162*, 1211–1221.

Lee, I. J., Wang, N., Hu, W., Schott, K., Bahler, J., Giddings, T. H., Jr., et al. (2014). Regulation of spindle pole body assembly and cytokinesis by the centrin-binding protein Sfi1 in fission yeast. *Molecular Biology of the Cell, 25*, 2735–2749.

Lim, H. H., Zhang, T., & Surana, U. (2009). Regulation of centrosome separation in yeast and vertebrates: common threads. *Trends in Cell Biology, 19*, 325–333.

Li, S., Sandercock, A. M., Conduit, P., Robinson, C. V., Williams, R. L., & Kilmartin, J. V. (2006). Structural role of Sfi1p-centrin filaments in budding yeast spindle pole body duplication. *Journal of Cell Biology, 173*, 867–877.

Ohta, M., Sato, M., & Yamamoto, M. (2012). Spindle pole body components are reorganized during fission yeast meiosis. *Molecular Biology of the Cell, 23*, 1799–1811.

Paoletti, A., Bordes, N., Haddad, R., Schwartz, C. L., Chang, F., & Bornens, M. (2003). Fission yeast cdc31p is a component of the half-bridge and controls SPB duplication. *Molecular Biology of the Cell, 14*, 2793–2808.

Racine, V., Hertzog, A., Jouanneau, J., Salamero, J., Kervrann, C., & Sibarita, J.-B. (2006). Multiple-target tracking of 3D fluorescent objects based on simulated annealing. In *ISBI* (pp. 1020–1023).

Rosenberg, J. A., Tomlin, G. C., McDonald, W. H., Snydsman, B. E., Muller, E. G., Yates, J. R., III, et al. (2006). Ppc89 links multiple proteins, including the septation initiation network, to the core of the fission yeast spindle-pole body. *Molecular Biology of the Cell, 17*, 3793–3805.

Sparks, C. A., Morphew, M., & McCollum, D. (1999). Sid2p, a spindle pole body kinase that regulates the onset of cytokinesis. *Journal of Cell Biology, 146*, 777–790.

Terenna, C. R., Makushok, T., Velve-Casquillas, G., Baigl, D., Chen, Y., Bornens, M., et al. (2008). Physical mechanisms redirecting cell polarity and cell shape in fission yeast. *Current Biology, 18*, 1748–1753.

Tran, P. T., Paoletti, A., & Chang, F. (2004). Imaging green fluorescent protein fusions in living fission yeast cells. *Methods, 33*, 220–225.

Uzawa, S., Li, F., Jin, Y., McDonald, K. L., Braunfeld, M. B., Agard, D. A., et al. (2004). Spindle pole body duplication in fission yeast occurs at the G1/S boundary but maturation is blocked until exit from S by an event downstream of cdc10+. *Molecular Biology of the Cell, 15*, 5219–5230.

Velve-Casquillas, G., Le Berre, M., Piel, M., & Tran, P. T. (2010). Microfluidic tools for cell biological research. *Nano Today, 5*, 28–47.

Vogel, S. K., Raabe, I., Dereli, A., Maghelli, N., & Tolić-Nørrelykke, I. (2007). Interphase microtubules determine the initial alignment of the mitotic spindle. *Current Biology, 17*, 438–444.

Walde, S., & King, M. C. (2014). The KASH protein Kms2 coordinates mitotic remodeling of the spindle pole body. *Journal of Cell Science, 127*, 3625–3640.

Volumes in Series

Founding Series Editor
DAVID M. PRESCOTT

Volume 1 (1964)
Methods in Cell Physiology
Edited by David M. Prescott

Volume 2 (1966)
Methods in Cell Physiology
Edited by David M. Prescott

Volume 3 (1968)
Methods in Cell Physiology
Edited by David M. Prescott

Volume 4 (1970)
Methods in Cell Physiology
Edited by David M. Prescott

Volume 5 (1972)
Methods in Cell Physiology
Edited by David M. Prescott

Volume 6 (1973)
Methods in Cell Physiology
Edited by David M. Prescott

Volume 7 (1973)
Methods in Cell Biology
Edited by David M. Prescott

Volume 8 (1974)
Methods in Cell Biology
Edited by David M. Prescott

Volume 9 (1975)
Methods in Cell Biology
Edited by David M. Prescott

Volume 10 (1975)
Methods in Cell Biology
Edited by David M. Prescott

Volume 11 (1975)
Yeast Cells
Edited by David M. Prescott

Volume 12 (1975)
Yeast Cells
Edited by David M. Prescott

Volume 13 (1976)
Methods in Cell Biology
Edited by David M. Prescott

Volume 14 (1976)
Methods in Cell Biology
Edited by David M. Prescott

Volume 15 (1977)
Methods in Cell Biology
Edited by David M. Prescott

Volume 16 (1977)
Chromatin and Chromosomal Protein Research I
Edited by Gary Stein, Janet Stein, and Lewis J. Kleinsmith

Volume 17 (1978)
Chromatin and Chromosomal Protein Research II
Edited by Gary Stein, Janet Stein, and Lewis J. Kleinsmith

Volume 18 (1978)
Chromatin and Chromosomal Protein Research III
Edited by Gary Stein, Janet Stein, and Lewis J. Kleinsmith

Volume 19 (1978)
Chromatin and Chromosomal Protein Research IV
Edited by Gary Stein, Janet Stein, and Lewis J. Kleinsmith

Volume 20 (1978)
Methods in Cell Biology
Edited by David M. Prescott

Advisory Board Chairman
KEITH R. PORTER

Volume 21A (1980)
Normal Human Tissue and Cell Culture, Part A: Respiratory, Cardiovascular, and Integumentary Systems
Edited by Curtis C. Harris, Benjamin F. Trump, and Gary D. Stoner

Volume 21B (1980)
Normal Human Tissue and Cell Culture, Part B: Endocrine, Urogenital, and Gastrointestinal Systems
Edited by Curtis C. Harris, Benjamin F. Trump, and Gray D. Stoner

Volume 22 (1981)
Three-Dimensional Ultrastructure in Biology
Edited by James N. Turner

Volume 23 (1981)
Basic Mechanisms of Cellular Secretion
Edited by Arthur R. Hand and Constance Oliver

Volume 24 (1982)
The Cytoskeleton, Part A: Cytoskeletal Proteins, Isolation and Characterization
Edited by Leslie Wilson

Volume 25 (1982)
The Cytoskeleton, Part B: Biological Systems and In Vitro Models
Edited by Leslie Wilson

Volume 26 (1982)
Prenatal Diagnosis: Cell Biological Approaches
Edited by Samuel A. Latt and Gretchen J. Darlington

Series Editor
LESLIE WILSON

Volume 27 (1986)
Echinoderm Gametes and Embryos
Edited by Thomas E. Schroeder

Volume 28 (1987)
***Dictyostelium discoideum*: Molecular Approaches to Cell Biology**
Edited by James A. Spudich

Volume 29 (1989)
Fluorescence Microscopy of Living Cells in Culture, Part A: Fluorescent Analogs, Labeling Cells, and Basic Microscopy
Edited by Yu-Li Wang and D. Lansing Taylor

Volume 30 (1989)
Fluorescence Microscopy of Living Cells in Culture, Part B: Quantitative Fluorescence Microscopy—Imaging and Spectroscopy
Edited by D. Lansing Taylor and Yu-Li Wang

Volume 31 (1989)
Vesicular Transport, Part A
Edited by Alan M. Tartakoff

Volume 32 (1989)
Vesicular Transport, Part B
Edited by Alan M. Tartakoff

Volume 33 (1990)
Flow Cytometry
Edited by Zbigniew Darzynkiewicz and Harry A. Crissman

Volume 34 (1991)
Vectorial Transport of Proteins into and across Membranes
Edited by Alan M. Tartakoff

Selected from Volumes 31, 32, and 34 (1991)
Laboratory Methods for Vesicular and Vectorial Transport
Edited by Alan M. Tartakoff

Volume 35 (1991)
Functional Organization of the Nucleus: A Laboratory Guide
Edited by Barbara A. Hamkalo and Sarah C. R. Elgin

Volume 36 (1991)
***Xenopus laevis*: Practical Uses in Cell and Molecular Biology**
Edited by Brian K. Kay and H. Benjamin Peng

Series Editors
LESLIE WILSON AND PAUL MATSUDAIRA

Volume 37 (1993)
Antibodies in Cell Biology
Edited by David J. Asai

Volume 38 (1993)
Cell Biological Applications of Confocal Microscopy
Edited by Brian Matsumoto

Volume 39 (1993)
Motility Assays for Motor Proteins
Edited by Jonathan M. Scholey

Volume 40 (1994)
A Practical Guide to the Study of Calcium in Living Cells
Edited by Richard Nuccitelli

Volume 41 (1994)
Flow Cytometry, Second Edition, Part A
Edited by Zbigniew Darzynkiewicz, J. Paul Robinson, and Harry A. Crissman

Volume 42 (1994)
Flow Cytometry, Second Edition, Part B
Edited by Zbigniew Darzynkiewicz, J. Paul Robinson, and Harry A. Crissman

Volume 43 (1994)
Protein Expression in Animal Cells
Edited by Michael G. Roth

Volume 44 (1994)
***Drosophila melanogaster*: Practical Uses in Cell and Molecular Biology**
Edited by Lawrence S. B. Goldstein and Eric A. Fyrberg

Volume 45 (1994)
Microbes as Tools for Cell Biology
Edited by David G. Russell

Volume 46 (1995)
Cell Death
Edited by Lawrence M. Schwartz and Barbara A. Osborne

Volume 47 (1995)
Cilia and Flagella
Edited by William Dentler and George Witman

Volume 48 (1995)
Caenorhabditis elegans: Modern Biological Analysis of an Organism
Edited by Henry F. Epstein and Diane C. Shakes

Volume 49 (1995)
Methods in Plant Cell Biology, Part A
Edited by David W. Galbraith, Hans J. Bohnert, and Don P. Bourque

Volume 50 (1995)
Methods in Plant Cell Biology, Part B
Edited by David W. Galbraith, Don P. Bourque, and Hans J. Bohnert

Volume 51 (1996)
Methods in Avian Embryology
Edited by Marianne Bronner-Fraser

Volume 52 (1997)
Methods in Muscle Biology
Edited by Charles P. Emerson, Jr. and H. Lee Sweeney

Volume 53 (1997)
Nuclear Structure and Function
Edited by Miguel Berrios

Volume 54 (1997)
Cumulative Index

Volume 55 (1997)
Laser Tweezers in Cell Biology
Edited by Michael P. Sheetz

Volume 56 (1998)
Video Microscopy
Edited by Greenfield Sluder and David E. Wolf

Volume 57 (1998)
Animal Cell Culture Methods
Edited by Jennie P. Mather and David Barnes

Volume 58 (1998)
Green Fluorescent Protein
Edited by Kevin F. Sullivan and Steve A. Kay

Volume 59 (1998)
The Zebrafish: Biology
Edited by H. William Detrich III, Monte Westerfield, and Leonard I. Zon

Volume 60 (1998)
The Zebrafish: Genetics and Genomics
Edited by H. William Detrich III, Monte Westerfield, and Leonard I. Zon

Volume 61 (1998)
Mitosis and Meiosis
Edited by Conly L. Rieder

Volume 62 (1999)
Tetrahymena thermophila
Edited by David J. Asai and James D. Forney

Volume 63 (2000)
Cytometry, Third Edition, Part A
Edited by Zbigniew Darzynkiewicz, J. Paul Robinson, and Harry Crissman

Volume 64 (2000)
Cytometry, Third Edition, Part B
Edited by Zbigniew Darzynkiewicz, J. Paul Robinson, and Harry Crissman

Volume 65 (2001)
Mitochondria
Edited by Liza A. Pon and Eric A. Schon

Volume 66 (2001)
Apoptosis
Edited by Lawrence M. Schwartz and Jonathan D. Ashwell

Volume 67 (2001)
Centrosomes and Spindle Pole Bodies
Edited by Robert E. Palazzo and Trisha N. Davis

Volume 68 (2002)
Atomic Force Microscopy in Cell Biology
Edited by Bhanu P. Jena and J. K. Heinrich Hörber

Volume 69 (2002)
Methods in Cell–Matrix Adhesion
Edited by Josephine C. Adams

Volume 70 (2002)
Cell Biological Applications of Confocal Microscopy
Edited by Brian Matsumoto

Volume 71 (2003)
Neurons: Methods and Applications for Cell Biologist
Edited by Peter J. Hollenbeck and James R. Bamburg

Volume 72 (2003)
Digital Microscopy: A Second Edition of Video Microscopy
Edited by Greenfield Sluder and David E. Wolf

Volume 73 (2003)
Cumulative Index

Volume 74 (2004)
Development of Sea Urchins, Ascidians, and Other Invertebrate Deuterostomes: Experimental Approaches
Edited by Charles A. Ettensohn, Gary M. Wessel, and Gregory A. Wray

Volume 75 (2004)
Cytometry, 4th Edition: New Developments
Edited by Zbigniew Darzynkiewicz, Mario Roederer, and Hans Tanke

Volume 76 (2004)
The Zebrafish: Cellular and Developmental Biology
Edited by H. William Detrich, III, Monte Westerfield, and Leonard I. Zon

Volume 77 (2004)
The Zebrafish: Genetics, Genomics, and Informatics
Edited by William H. Detrich, III, Monte Westerfield, and Leonard I. Zon

Volume 78 (2004)
Intermediate Filament Cytoskeleton
Edited by M. Bishr Omary and Pierre A. Coulombe

Volume 79 (2007)
Cellular Electron Microscopy
Edited by J. Richard McIntosh

Volume 80 (2007)
Mitochondria, 2nd Edition
Edited by Liza A. Pon and Eric A. Schon

Volume 81 (2007)
Digital Microscopy, 3rd Edition
Edited by Greenfield Sluder and David E. Wolf

Volume 82 (2007)
Laser Manipulation of Cells and Tissues
Edited by Michael W. Berns and Karl Otto Greulich

Volume 83 (2007)
Cell Mechanics
Edited by Yu-Li Wang and Dennis E. Discher

Volume 84 (2007)
Biophysical Tools for Biologists, Volume One: In Vitro Techniques
Edited by John J. Correia and H. William Detrich, III

Volume 85 (2008)
Fluorescent Proteins
Edited by Kevin F. Sullivan

Volume 86 (2008)
Stem Cell Culture
Edited by Dr. Jennie P. Mather

Volume 87 (2008)
Avian Embryology, 2nd Edition
Edited by Dr. Marianne Bronner-Fraser

Volume 88 (2008)
Introduction to Electron Microscopy for Biologists
Edited by Prof. Terence D. Allen

Volume 89 (2008)
Biophysical Tools for Biologists, Volume Two: In Vivo Techniques
Edited by Dr. John J. Correia and Dr. H. William Detrich, III

Volume 90 (2008)
Methods in Nano Cell Biology
Edited by Bhanu P. Jena

Volume 91 (2009)
Cilia: Structure and Motility
Edited by Stephen M. King and Gregory J. Pazour

Volume 92 (2009)
Cilia: Motors and Regulation
Edited by Stephen M. King and Gregory J. Pazour

Volume 93 (2009)
Cilia: Model Organisms and Intraflagellar Transport
Edited by Stephen M. King and Gregory J. Pazour

Volume 94 (2009)
Primary Cilia
Edited by Roger D. Sloboda

Volume 95 (2010)
Microtubules, in vitro
Edited by Leslie Wilson and John J. Correia

Volume 96 (2010)
Electron Microscopy of Model Systems
Edited by Thomas Müeller-Reichert

Volume 97 (2010)
Microtubules: In Vivo
Edited by Lynne Cassimeris and Phong Tran

Volume 98 (2010)
Nuclear Mechanics & Genome Regulation
Edited by G.V. Shivashankar

Volume 99 (2010)
Calcium in Living Cells
Edited by Michael Whitaker

Volume 100 (2010)
The Zebrafish: Cellular and Developmental Biology, Part A
Edited by: H. William Detrich III, Monte Westerfield and Leonard I. Zon

Volume 101 (2011)
The Zebrafish: Cellular and Developmental Biology, Part B
Edited by: H. William Detrich III, Monte Westerfield and Leonard I. Zon

Volume 102 (2011)
Recent Advances in Cytometry, Part A: Instrumentation, Methods
Edited by Zbigniew Darzynkiewicz, Elena Holden, Alberto Orfao, William Telford and Donald Wlodkowic

Volume 103 (2011)
Recent Advances in Cytometry, Part B: Advances in Applications
Edited by Zbigniew Darzynkiewicz, Elena Holden, Alberto Orfao, Alberto Orfao and Donald Wlodkowic

Volume 104 (2011)
The Zebrafish: Genetics, Genomics and Informatics
3rd Edition
Edited by H. William Detrich III, Monte Westerfield, and Leonard I. Zon

Volume 105 (2011)
The Zebrafish: Disease Models and Chemical Screens 3rd Edition
Edited by H. William Detrich III, Monte Westerfield, and Leonard I. Zon

Volume 106 (2011)
Caenorhabditis elegans: Molecular Genetics and Development 2nd Edition
Edited by Joel H. Rothman and Andrew Singson

Volume 107 (2011)
Caenorhabditis elegans: Cell Biology and Physiology 2nd Edition
Edited by Joel H. Rothman and Andrew Singson

Volume 108 (2012)
Lipids
Edited by Gilbert Di Paolo and Markus R Wenk

Volume 109 (2012)
Tetrahymena thermophila
Edited by Kathleen Collins

Volume 110 (2012)
Methods in Cell Biology
Edited by Anand R. Asthagiri and Adam P. Arkin

Volume 111 (2012)
Methods in Cell Biology
Edited by Thomas Müler Reichart and Paul Verkade

Volume 112 (2012)
Laboratory Methods in Cell Biology
Edited by P. Michael Conn

Volume 113 (2013)
Laboratory Methods in Cell Biology
Edited by P. Michael Conn

Volume 114 (2013)
Digital Microscopy, 4th Edition
Edited by Greenfield Sluder and David E. Wolf

Volume 115 (2013)
Microtubules, in Vitro, 2nd Edition
Edited by John J. Correia and Leslie Wilson

Volume 116 (2013)
Lipid Droplets
Edited by H. Robert Yang and Peng Li

Volume 117 (2013)
Receptor-Receptor Interactions
Edited by P. Michael Conn

Volume 118 (2013)
Methods for Analysis of Golgi Complex Function
Edited by Franck Perez and David J. Stephens

Volume 119 (2014)
Micropatterning in Cell Biology Part A
Edited by Matthieu Piel and Manuel Théry

Volume 120 (2014)
Micropatterning in Cell Biology Part B
Edited by Matthieu Piel and Manuel Théry

Volume 121 (2014)
Micropatterning in Cell Biology Part C
Edited by Matthieu Piel and Manuel Théry

Volume 122 (2014)
Nuclear Pore Complexes and Nucleocytoplasmic Transport - Methods
Edited by Valérie Doye

Volume 123 (2014)
Quantitative Imaging in Cell Biology
Edited by Jennifer C. Waters and Torsten Wittmann

Volume 124 (2014)
Correlative Light and Electron Microscopy II
Edited by Thomas Müller-Reichert and Paul Verkade

Volume 125 (2015)
Biophysical Methods in Cell Biology
Edited by Ewa K. Paluch

Volume 126 (2015)
Lysosomes and Lysosomal Diseases
Edited by Frances Platt and Nick Platt

Volume 127 (2015)
Methods in Cilia & Flagella
Edited by Renata Basto and Wallace F. Marshall

Volume 128 (2015)
Building a Cell from its Component Parts
Edited by Jennifer Ross and Wallace F. Marshall

Index

Note: Page numbers followed by "b", "f" and "t" indicate boxes, figures and tables respectively.

A

Activation domain (AD), 256
Adenosine triphosphate (ATP), 21–22
Adherent cells, adaptation for, 180–182
  quality control/structural integrity of purified centrosome, 181–182
Analog-sensitive kinases (AS kinases), 21–22
Antibiotic resistance genes excision, 91–92
Antigen-binding fragments (Fabs), 136
Ascidian(s), 317–318
  3D rendering using Imaris, 331f
  embryos, 317–318, 327, 329–332
  invariant cleavage, 330f
  methods, 318–337
  unequal cleavages, 328f–329f
Autoactivation, 262
Automatic centriole counting, 118–119

B

Bardet–Biedl syndrome, 342
Basal body (BB), 62
Binding domain (BD), 256
Biotin identification (Bio-ID), 155
  advantages, 155–157
  biotinylation, 155
  to centrosome, 156f
  choice of cell line and expression method, 160
  data analysis, 166–167
  materials, 157–159
Biotin stock, 158
BirA*-fusion protein expression
  large-scale BioID pulldown, 163–166
  plasmid generation, 159
  validation, 160–163
Block trimming, 351–353
Bovine serum albumin (BSA), 145
Buffer recipes, 47–48

C

CABs, *see* Centrosome-attracting bodies
Caenorhabditis elegans (*C. elegans*), 280–281, 342–343, 370–371
  centriole components genetic identification in, 281–282
  electron tomography, 344
  electron tomography, 357–363
  embedding, 348–351
  purification, 372f
  serial sections, 351–356
  specimen fixation, 344
    chemical fixation, 345–346
    HPF–freeze substitution, 346–348
  TEM, 356–357
Calf intestinal phosphatase (CIP), 25
Cancer, 52
CB, *see* Cytoskeleton buffer
Cell staining, 286–287
Cellular staining, 289–291
Centrioles, 192, 212, *see also Trichonympha* centriole
  analysis, 193f
  biogenesis, 105f
  centriole isolation, 194–195
  cryotomography, 200, 201f
  electron microscopy, 195
  isolation, 194–195
  material and reagents, 194
  position, 5–6
  software, 196
  subtomogram averaging, 203–206, 204f–205f
    sample preparation and transfer, 200–202
    tomogram acquisition, 202–203
    tomogram reconstruction using IMOD, 203
  Trichonympha, 194–195
Centrosome protein interactions, array based screen for, 259
  autoactivation, 262
  dividing proteins, 260–261
  false positive rate identification, 262
  interpreting screening results, 267
  prescreen planning, 262–263
  screening for interactions, 263–267
  Y2H
    library generation, 261–262
    system selection, 259–260
Centrosome-attracting bodies (CABs), 320–329, *see also* Spindle pole bodies (SPBs)
Centrosome/centriole amplification, 52
Centrosome(s), 40–41, 52, 131–132, 160, 172–173, 230, 252–253, 370, *see also* Three-dimensional culture; Yeast-two hybrid system (Y2H system)
  aberrations and mitotic outcomes, 53–54

Centrosome(s) (*Continued*)
aberrations in human cancer
counting, 54–55
evaluation, 55
immunostaining, 55–56
tissue, 54, 55f
accumulation, 54
analysis, 56–57
in benign and malignant human tissue, 57f
and cancer, 52
centriole components genetic identification, 281–282
characterization of molecular composition, 281
direct protein interaction identification, 253–255
Drosophila advantage, 296
enrichment, 164–165
functions and structure, 280–281
genome-wide
RNAi screen, 282
screen, 288–291
image acquisition, 291–292
image analysis, 292–293, 293f
isolation, 177–179
maturation, 280
MTOC, 252f
overduplication, 54
potential hits validation, 294
protein composition, 253
protein–protein interactions, 255–256
proteome identification of purified centrosomes, 282–283
purification procedure, 174f–175f, 175
adherent cells, adaptation for, 180–182
effect of $CaCl_2$ on centrosome, 182–184, 183f
immunolabeling of centrosomal fractions, 179–180, 180f
materials, 175–176
optimization of $MgCl_2$ concentration, 177, 178f
solutions and reagents, 176
RNAi in *Drosophila* cells, 283–284
scoring phenotypes, 292–293
screen design, 284–288, 285f
in *C. elegans* and *Drosophila*, 294
primary hits validation, 295f
structure, 2
Centrosomin (Cnn), 283
Chimera software, 143
Cilia, 62, 342
Ciliopathies, 342
CIP, *see* Calf intestinal phosphatase
CLEM, *see* Correlative light and electron microscopy
Clustered, regularly interspaced, short palindromic repeats systems (CRISPR systems), 21
CrispR-Cas9 technology, 86–88
CRISPR/Cas system, 21
genome editing using, 23f
CRISPR/Cas9-mediated genome engineering
cloning oligonucleotides into PX459 vector, 24–27
CRISPR/Cas system, 21
functional analysis, 33
gRNA for sequence-specific DNA cleavage, 22–24
AS kinases, 21–22
repair template design, 27–28
screening, 28–33
transfection, 28–33
Cnn, *see* Centrosomin
Conventional electron microscopy, 195
Correlative light and electron microscopy (CLEM), 2–3, 344, *see also* Transmission electron microscopy (TEM)
cell
culture, 3
fixation and postfixation recording of cell, 5–6
preparation for microscopy, 4–5
centriole position, 5–6
chemicals, buffers, and media, 15–16
dehydration, 6–7
electron microscopy, 12–13
embedding, 6–7
glass coverslip removal, 8
instrumentation, 16
light microscopy, 5
marking position of target cell on polymerized resin, 7–8
picking up serial sections, 12
preparation of formvar-coated slot grids, 13–14
prestaining, 6–7
staining of sections, 12
trimming, 8–9, 9f
ultrathin serial sectioning, 10
Coverslips, 236
CRE recombinase, antibiotic resistance gene excision by, 91–92
CRISPR systems, *see* Clustered, regularly interspaced, short palindromic repeats systems
CRISPR-RNA (crRNA), 21
Cryo-electron microscopy (cryo-EM), 173, 182f
Cryomicroscopy, 195
sample preparation and transfer, 200–202
Cryotomography of centrioles, 200, 201f

sample preparation and transfer, 200–202
tomogram acquisition, 202–203
tomogram reconstruction using IMOD, 203
Cytoskeleton buffer (CB), 345

D

Data
analysis, 166–167
processing, 79
alignment and reconstruction of tilt series, 71–75
filtering of tomographic volume, 75–76
modeling/segmentation of features from tomograms, 76–78
recording, 360–361
Dehydration, 6–7
Deoxyribonucleic acid (DNA), 84–85
Differential interference contrast (DIC), 6
Dimethyl sulfoxide (DMSO), 158
Direct protein interaction identification, 253–255
Dishevelled (Dvl), 63
DMSO, *see* Dimethyl sulfoxide
dominant negative Mastermind (dnMM), 115–117
Dorsal telencephalon explants preparation, 222–223
Double-immunofluorescence analysis, 57
Double-strand break (DSB), 20
double-stranded DNA (dsDNA), 27, 286, *see also* guide RNA (gRNA)
Drosophila, 280–281
centriole components genetic identification in, 281–282
Drosophila neuroblast (*Drosophila* NB), 302
ex vivo whole mount, 305f, 309–311
fluorescent reporters, 311, 312t
imaging centrosomes in, 306f
primary cultures, 303–309, 304f, 307t
Drosophila neuroepithelial cells, imaging centrosomes in, 214, 216f–217f
materials
equipment, 215–217
reagents, 215
methods
dissection, 217–218
immunolabeling of fixed brains, 218–219
larval staging, 217
live imaging, 219–220
Drosophila syncytial embryos, 230
collection and preparation, 232
fruit juice agar plates preparation, 232
setting up fly collection chambers, 233
yeast paste preparation, 232
imaging and data analysis, 242–246
microinjection of embryos, 238
breaking needle tip, 239
embryo injection, 240–241
loading sample in needle, 239
pulling glass microinjection needles, 238–239
mounting embryos for spinning disc confocal microscopy, 235–238
coverslips, 236
heptane-glue solution preparation, 235
MatTek dishes, 236
perturbation via cold treatment, 241
assay, 241–242
assay with microinjection, 242
protocol, 231
samples preparation for microinjection, 233
drugs or lyophilized peptides, 233
fluorescently coupling affinity-purified antibodies, proteins, or peptides, 234
unlabeled affinity-purified antibodies or other proteins, 233–234
in vitro transcribed mRNA, 235
DSB, *see* Double-strand break
dsDNA, *see* double-stranded DNA
DT40 cells, 84–86, 96f, *see also* MCF-10A cells
analysis, 93–99
immunofluorescence, 95–97
time-lapse imaging, 97–98
centrosome purification from, 98–99
design and preparation of targeting constructs, 88–89
excision of antibiotic resistance genes by CRE recombinase, 91–92
genetic manipulation, 86–88
heterozygous mutant cells generation, 89–91
materials, 90–91
notes, 90
homozygous mutant cells generation, 91
outline of gene targeting in, 87f
in situ tagging of one or both alleles, 92–93, 94f
Dual-axis tomography, 65
Dvl, *see* Dishevelled

E

EDTA, *see* Ethylene diamine tetraacetic acid
Eggs, 317–318, 320
EHS murine tumor, *see* Engelbreth–Holm–Swarm murine tumor
"Electrical tape" method, 108–111

Electron microscopy (EM), 2, 12–13, 65, 132, 195, *see also* Correlative light and electron microscopy (CLEM)
Electron Microscopy Sciences (EMS), 15
Electron tomography, 69–71, 79, 344, 357–363
  sectioning and preparation for, 68–69, 79
Embedding, 6–7, 348–351, 349f
  infiltration, 349–350
  pre-embedding, 349–350
Embryonic NSC, 214
Embryonic stem cells (ES cells), 84–85
"Empty stains", 262
Engelbreth–Holm–Swarm murine tumor (EHS murine tumor), 38
Ethylene diamine tetraacetic acid (EDTA), 158
Ex vivo whole mount, 305f, 309–311

F

Fabs, *see* Antigen-binding fragments
False positive rate identification, 262
Fast protein liquid chromatography (FPLC), 373
Fetal bovine serum (FBS), 88, 303
Fetal calf serum (FCS), 303
FFPE tissue, *see* Formalin-fixed, paraffin-embedded tissue
Fiducial markers, 359–360
Field emission gun (FEG), 195
Flip-in recombination, 160
Fluorescein isothiocyanate (FITC), 113–114
Fluorescence labeling method, 135–136
Fluorescent antibody staining, 145
Fluorescent reporters, 311, 312t
Formalin-fixed, paraffin-embedded tissue (FFPE tissue), 55–56
FPLC, *see* Fast protein liquid chromatography
Freeze substitution (FS), 343–344

G

Genome engineering, 20, *see also* CRISPR/Cas9-mediated genome engineering
Genome-wide screen, 292–293
  cellular staining, 289–291
  controls, 288
  genome-wide centrosome screen, 290f
  high-throughput cell assay, 288–289
  library selection, 288
Genomic DNA extraction, 30–31
Glass coverslip removal, 8
Glycine rinse, 46
Green fluorescent protein (GFP), 3, 106, 159, 294
guide RNA (gRNA), 21, *see also* double-stranded DNA (dsDNA)
  for sequence-specific DNA cleavage, 22–24

H

HDR, *see* Homology-directed repair
Heterozygous mutant cells generation, 89–91
  materials, 90–91
  notes, 90
High spatial resolution imaging, 220, *see also* Centrosomes
  materials
    equipment, 222
    reagents, 220–222
  methods
    dorsal telencephalon explants preparation, 222–223
    imaging of centrosomes and mitotic spindles, 225–226
    immunolabeling of whole-mount cortical explants, 223–225
High-pressure freezing (HPF), 343–344, 346–347
High-throughput cell assay, 288–289
Homologous recombination (HR), 20, 160
Homology-directed repair (HDR), 20
Homozygous mutant cells generation, 91
Horse-radish peroxidase (HRP), 157

I

IM, *see* Infrahyoid musculature
Image acquisition, 291–292
  guidelines for, 48–49
Image analysis, 292–293, 293f
Image processing, 196
Imaging buffers, 145–146
Immunofluorescence technique (IF technique), 45–47, 95–97
  experiment, 199–200
  trichonympha centrioles detection by, 195
Immunolabeling
  of centrosomal fractions, 179–180, 180f
  of whole-mount cortical explants, 223–225
IMOD, tomogram reconstruction using, 203
In situ tagging, 92–93
In vitro transcribed mRNA, 235
Infrahyoid musculature (IM), 66
Insertions and deletion (InDel), 22
Interphase centrosomes, mitotic centrosomes *vs.*, 284–286

K

Known Interacting Protein (KIP), 268–270

L

Leica microsystem, 10
Lentivirus, 160
Ligation and transformation, 26–27
Light microscopy, 5
Live-cell imaging, 4f, 48
Live-cell microscopy, 5
Loss-of-function techniques, 84–85
Lymphoblastic cell line KE-37, 95

M

Mass spectrometry (MS), 155
  preparation of sample for, 166
Matrigel
  indirect IF staining of MCF-10A acini cultured, 45–48
MatTek dishes, 236
MCC, *see* Multiciliated cell
MCF-10A cells, *see also* DT40 cells
  events in acinar morphogenesis, 40f
  indirect IF staining, 45–48
  propagation in monolayer cultures, 41
Meckel–Gruber syndrome, 342
messenger RNA (mRNA), 106–107
Microinjection experiments, 231
Microtubule (MT), 62, 230, 252–253, 280
Microtubule-organizing center (MTOC), 104, 172–173, 252–253
Mitotic centrosomes, interphase centrosomes *vs.*, 284–286
Monoethanolamine (MEA), 139–140
Morpholino oligos (MOs), 107
Mosaic *xenopus* embryos generation, 113–114
Motile cilia, 62–63
Mouse dissection and tissue preparation
  materials, 78
  methods, 65–67
mRNA, *see* messenger RNA
MS, *see* Mass spectrometry
MT, *see* Microtubule
MTOC, *see* Microtubule-organizing center
Multiciliated cell (MCC), 62–63, 104
  cell cycle-regulated centriole duplication, 104–106
  ciliated epithelial explants, 111f
  ciliated epithelial spheroid culture, 114–115
  code for centriole number automated quantification, 119b
  data processing
    alignment and reconstruction of tilt series, 71–75
    filtering of tomographic volume, 75–76
    modeling/segmentation of features from tomograms, 76–78
  electron tomography, 69–71
  fully-automatic, program-assisted counting of centrioles, 118–119
  manipulating MCC formation, 115–117, 117t
  materials
    data processing, 79
    electron tomography, 79
    mouse dissection, 78
    sample sectioning and preparation for electron tomography, 79
    TEM sample preparation, 78
  methods
    mouse dissection and tissue preparation, 65–67
    sectioning and preparation for electron tomography, 68–69
    TEM tissue sample preparation, 67–68
  modulation of multiciliated cell fate, 116f
  molecular markers for, 109t–110t
  mosaic *xenopus* embryos generation, 113–114
  rationale, 63–65
  Xenopus, 104f
    driving specific expression in, 112–113
    visualization of centrioles and structures in, 106–112
Multicolor imaging, 136–138
Mutant library generation, 268–271

N

NA, *see* Numerical aperture
NAD, *see* Nonlinear anisotropic diffusion
NAP, *see* Note associated to point
Neural stem cell (NSC), 212–214
Nonhomologous end-joining pathway (NHEJ pathway), 22
Nonlinear anisotropic diffusion (NAD), 75–76
Normalized spectral abundance factor analysis (NSAF analysis), 167
Note associated to point (NAP), 66
Nucleotide (nt), 22
Numerical aperture (NA), 5

O

Oligonucleotide
  annealing, 26
  phosphorylation, 26
Osmium tetroxide (OsO_4), 7

P

PALM, *see* Photoactivated localization microscopy

PAM, *see* Protospacer adjacent motif
Paraformaldehyde (PFA), 112
PBS, *see* Phosphate-buffered saline
PCR, *see* Polymerase chain reaction
PDMS, *see* Poly-dimethyl-siloxane
PEI, *see* Polyethylenimine
Pericentriolar material (PCM), 2, 132, 172–173, 181–182, 212, 230, 252–253, 280, 370, *see also* SPD-5::GFP network
 C. elegans purification, 372f
 expression of SPD-5::GFP, SPD-2, and PLK-1 in SF9 insect cells, 371–373
 protein purification, 373–376
Permeabilization, 46
PFA, *see* Paraformaldehyde
Phallusia mammillata (*P. mammillata*), 318–320
 centrosome live labels in, 326f–327f
 live cell fluorescent markers testing in, 321t–324t
Phosphate-buffered saline (PBS), 7, 41, 56, 112, 139–140
Photoactivated localization microscopy (PALM), 131
PNK, *see* Polynucleotide kinase
Polo-like kinase 1 (PLK-1), 370–371
 expression of SPD-5::GFP, SPD-2, and PLK-1 in SF9 insect cells, 371–373
 purification, 376
Polo-like kinase 4 (Plk4), 22
Poly-dimethyl-siloxane (PDMS), 387–388
Polyethylenimine (PEI), 158
Polymerase chain reaction (PCR), 26, 86–88
 amplification, 32
Polymerized resin, marking position of target cell on, 7–8
Polynucleotide kinase (PNK), 26
Post-staining, 355–356
Potential hits validation, 294
Prescreen planning, 262–263
Prestaining, 6–7
Primary antibody selection, 136
Primary cultures, 303–309, 304f, 307t
Protein–protein interactions, 253
 in centrosome, 255–256
Protospacer adjacent motif (PAM), 22
PX459 vector, cloning oligonucleotides into, 24–27

R

Radial glial cells, See Embryonic NSCs
Red fluorescent protein (RFP), 108
Reference markers, 136
Refractive index (RI), 133
Region of interest (ROI), 65, 118
Repair template design, 27–28
Restriction enzyme digest, 32–33
Rib cage (RC), 66
Ribonucleic acid interference (RNAi), 84–85, 282
 in drosophila cells, 283–284
 first genome-wide RNAi screen for centrosomes, 282
Room temperature (RT), 46, 56

S

SAF, *see* Spectral abundance factor
Sample drift, 141
Sample mounting, 138–140
scFv, *see* single chain variable fragments
Screen design, centrosomes, 284, 285f, *see also* Genome-wide screen
 analysis, 287–288
 in *C. elegans* and *Drosophila*, 294
 cell line selection and growing conditions, 286
 cell staining to visualize centrosomes, 286–287
 interphase *vs.* mitotic centrosomes, 284–286
 primary hits validation, 295f
Screening, 28–33
 for interactions, 263–267
 for loss of interaction, 268–271
SDS, *see* Sodium dodecyl sulfate
SDS-PAGE, *see* Sodium dodecyl sulfate polyacrylamide gel electrophoresis
Secondary antibodies, 136–138
Sequencing clones, 33
Serial sections, 351
 block trimming, 351–353
 cutting serial sections, 353–355
 post-staining, 355–356
SerialEM software, 69–70
SF9, *see Spodoptera frugiperda*
SG, *see* Submandibular glands
single chain variable fragments (scFv), 136
single-stranded DNA (ssDNA), 27
Slot grids, preparation of formvar-coated, 13–14
Sodium dodecyl sulfate (SDS), 97
Sodium dodecyl sulfate polyacrylamide gel electrophoresis (SDS-PAGE), 157, 159
Sp, *see Streptococcus pyogenes*
SPBs, *see* Spindle pole bodies
SpCas9
 gRNA for sequence-specific DNA cleavage by, 22–24
SPD-5::GFP network
 data analysis, 380–381
 imaging, 380

preparation, 377–380
Specimen fixation, 344
chemical fixation, 345–346
HPF–freeze substitution, 346–348
Spectral abundance factor (SAF), 167
Spindle pole bodies (SPBs), 384, *see also* Centrosome-attracting bodies (CABs)
duplication status, 258, 384
monitoring SPB duplication in fixed SPB-labeled strains, 384–386, 385f
analysis of SPB status, 386–387
cell growth and fixation, 386
imaging SPBs in fixed cells, 386
plaques, 384
quantitative analysis of SPB biogenesis in live cells, 387, 388f, 389
imaging, 389
PDMS chambers for live imaging, 387–388
Spindle positioning, 320–329
Spodoptera frugiperda (SF9), 370–371
expression of SPD-5::GFP, SPD-2, and PLK-1 in SF9 insect cells, 371–373
ssDNA, *see* single-stranded DNA
Stable cell lines, 163
Staining of sections, 12
Stochastic optical reconstruction microscopy (STORM), 131
microscopy image acquisition, 140–141
Stolidobranch ascidian (*Styela partita*), 317–318
Streptavidin affinity purification experiments, 165–166
Streptococcus pyogenes (Sp), 22
Styela partita, see Stolidobranch ascidian
Subdiffraction resolution fluorescence microscopies, 131
image acquisition and data analysis, 141
3D volume alignment and averaging, 141–143, 142f
3DSIM microscope image acquisition, 140
intensity profiles, 143–144
STORM microscopy image acquisition, 140–141
reagents, protocols
fixation, 144–145
fluorescent antibody staining, 145
imaging buffers, 145–146
sample preparation, 144
sample choice and preparation, 133
cells, embryos, isolated centrosomes, 133–134
fixation, 135
fluorescence labeling method, 135–136
primary antibody selection, 136
reference markers, 136
sample mounting, 138–140
secondary antibodies and multicolor imaging, 136–138
Submandibular glands (SG), 66
Subtomogram averaging, 196, 203–206, 204f–205f, *see also* Tomogram
Superresolution microscopies, 131

T

TALEN, *see* Transcription activator-like effector nuclease
TEM, *see* Transmission electron microscopy
Three-dimensional culture, 38
centrosome, 40–41
comparative structure of mammary gland and acini formation, 39f
image acquisition, guidelines for, 48–49
live-cell imaging, 48
MCF-10A
events in acinar morphogenesis, 40f
indirect IF staining, 45–48
propagation in monolayer cultures, 41
plating cells in, 45
preparation of overlay 3D culture, 42–45
Three-dimensional structured illumination microscopy (3DSIM), 131
microscope image acquisition, 140
3D volume alignment and averaging, 141–143, 142f
Tilt series, 357
reconstruction and model generation, 361–363
Time-lapse imaging, 97–98
Tomogram
acquisition, 196, 202–203
reconstruction using IMOD, 196, 203
Transcription activator-like effector nuclease (TALEN), 20–21
Transfection, 28–33
Transient expression, 162–163
Transmission electron microscopy (TEM), 343–344, 356–357, *see also* Correlative light and electron microscopy (CLEM); Electron microscopy (EM)
analysis by, 200
C. elegans amphids, 358f–359f
tissue sample preparation
materials, 67–68
methods, 67–68
Trichonympha centriole
centrioles detection, 195

Trichonympha centriole (*Continued*)
isolation, 194–197, 198f–199f
analysis, 197–200
Trimming, 8–9, 9f

U

Ultrathin serial sectioning, 10
Untranslated region (UTR), 93

V

Ventricular surface, 214
Visualization, 196

W

Wash buffer, 158–159
Weighted back projection (WBP), 357
Whole-cell lysis, 165

X

Xenopus MCC, 104f
centrioles visualization, 106–108
via antibody staining and immunofluorescence, 112
in fixed embryos, 108
in live embryos, 108–112
ciliated epithelial explants, 111f
driving specific expression in, 112–113

Y

Yeast-two hybrid system (Y2H system), 253, 257f
array-based Y2H screen for centrosomeeprotein interactions, 265f
dissection of complex, multicomponent protein machines using, 256
advantages of, 257–258
principle of, 256
limitations of, 258–259
separation of function mutations by reverse, 267–268
mutant library generation, 268–271
rationale, 268
screening for loss of interaction, 268–271
yeast media recipes, 263t
Your Favorite Gene in prey plasmid (YFG-prey plasmid), 263–264

Z

Zinc finger nuclease (ZFN), 20–21

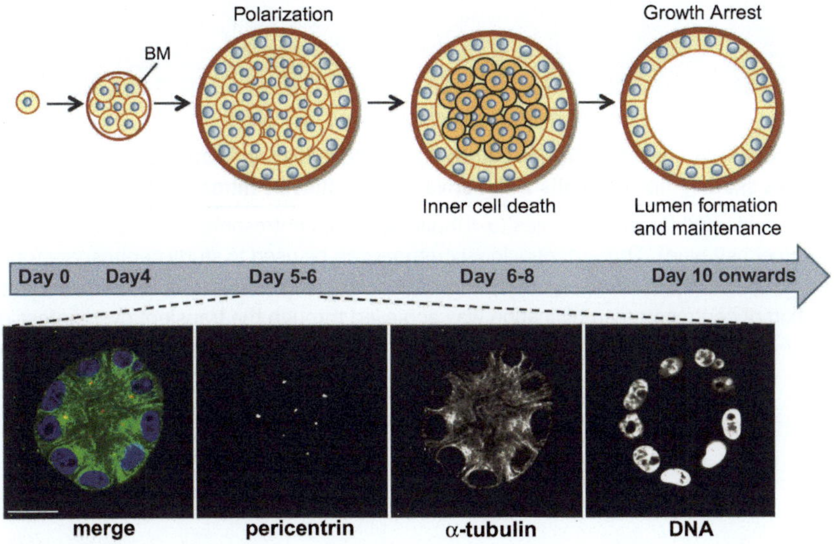

TERESA ARNANDIS AND SUSANA A. GODINHO, FIGURE 2 Events in MCF-10A acinar morphogenesis.

During the development of MCF-10A mammary epithelial acini in 3D cultures, a sequence of biological events is observed. In the early stages of 3D culture, apicobasal polarization becomes evident within cell clusters (day 5–6), before the appearance of bona fide lumen formation. This can be observed in the depicted IF images where the centrosome is localized apically. At day 6–8, two distinct populations of cells differentiate within each acini: an outer layer of cells in direct contact with the matrix and an inner subset of cells lacking matrix contact. Throughout morphogenesis, this outer cell layer remains polarized. From day 8 onwards, the centrally located cells undergo cell death, which contributes to the formation of a hollow lumen. The acini remain hollow thereafter. Cells were stained for microtubules (α-tubulin; green), centrosomes (pericentrin; red), and DNA (Hoechst; blue). Scale bar: 20 μm.

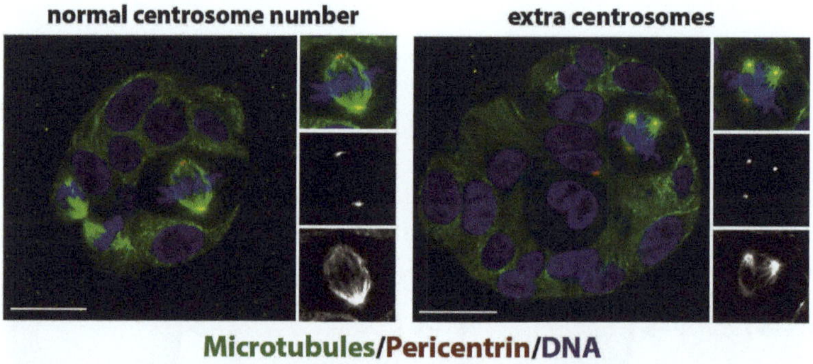

TERESA ARNANDIS AND SUSANA A. GODINHO, FIGURE 4 Immunofluorescence of MCF-10A 3D structures showing dividing cells with normal and amplified centrosome numbers.

Cells were stained for microtubules (α-tubulin; green), centrosomes (pericentrin; red), and DNA (Hoechst; blue). Three-dimensional cultures can be used to study centrosome function and mitosis. Examples of bipolar (left) and multipolar spindles (right) in 3D acini are shown. Induction of centrosome amplification was achieved through the transient overexpression of PLK4 (Polo-like kinase 4) as previously shown (Godinho et al., 2014). Scale bar: 20 μm.

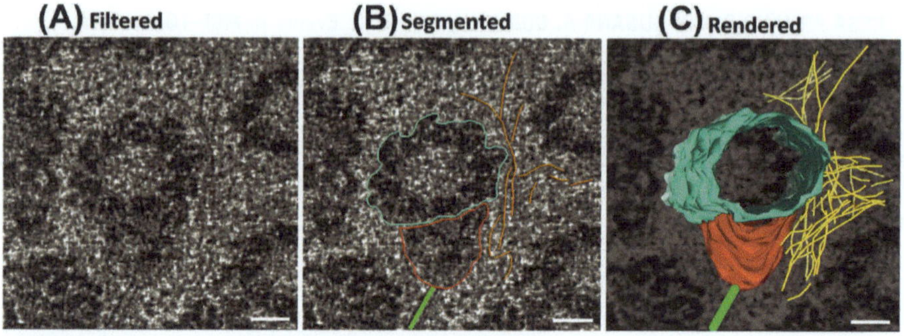

DANIEL K. CLARE *ET AL.*, FIGURE 5 Segmenting tomograms.

Z section shown in the xy plane of a volume reconstructed from a dual-axis tilt series after nonlinear anisotropic diffusion filtering (A), then segmented (B), and rendered (C). The microtubules (MTs) are shown in green, the basal body (BB) in cyan, the basal foot in red, and the actin in gold. The MT and actin filaments were modeled with open contours, while the BB and basal foot cap were modeled with closed contours. The MT and actin filaments were rendered as tubes (20 pixel and 6 pixels, respectively), the BB and the basal foot were rendered as surfaces. Scale bar: 100 nm.

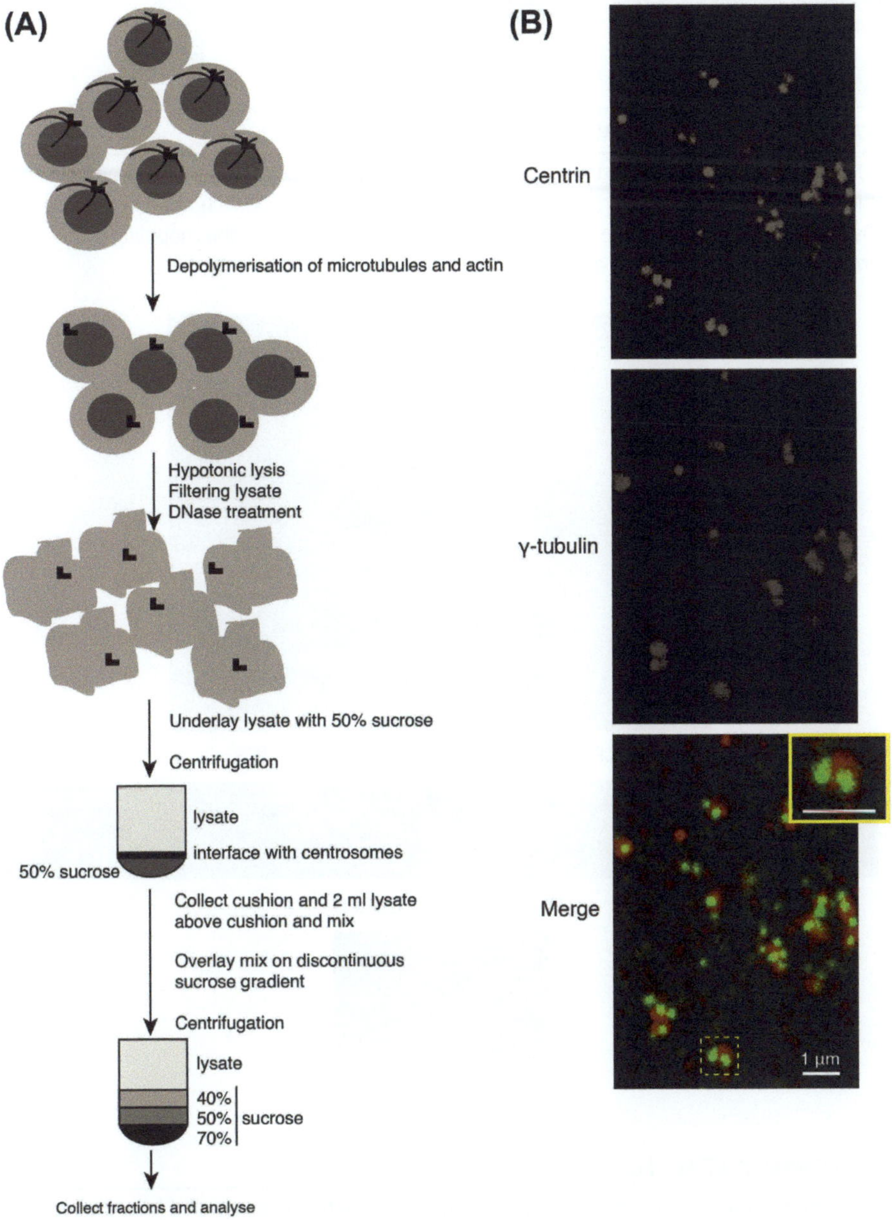

PAVITHRA L. CHAVALI AND FANNI GERGELY, FIGURE 3

(A) Protocol for centrosome isolation from DT40 cells. (B) A field of centrosomes is shown after sedimentation onto coverslips. Centrioles are labeled by anticentrin antibodies (green in merge), whereas the pericentriolar matrix (PCM) is labeled with anti-γ-tubulin antibodies (red in merge). Scale bar = 1 μm.

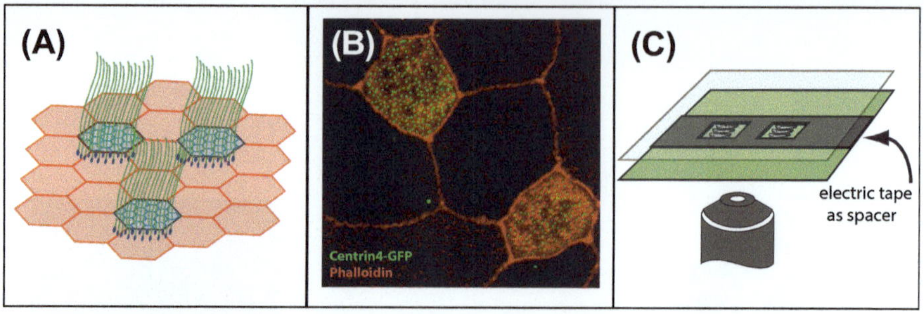

SIWEI ZHANG AND BRIAN J. MITCHELL, FIGURE 1 ***Xenopus*** **MCCs.**

(A) Illustration of the ciliated epithelia that covers the skin of *Xenopus* embryos. (B) Confocal image of multiciliated cells with centrin4-GFP marking centrioles/basal bodies (green) and rhodamine-phalloidin marking actin in the cell periphery (red). (C) Illustration of the mounting of *Xenopus* embryos using electrical tape as a spacer.

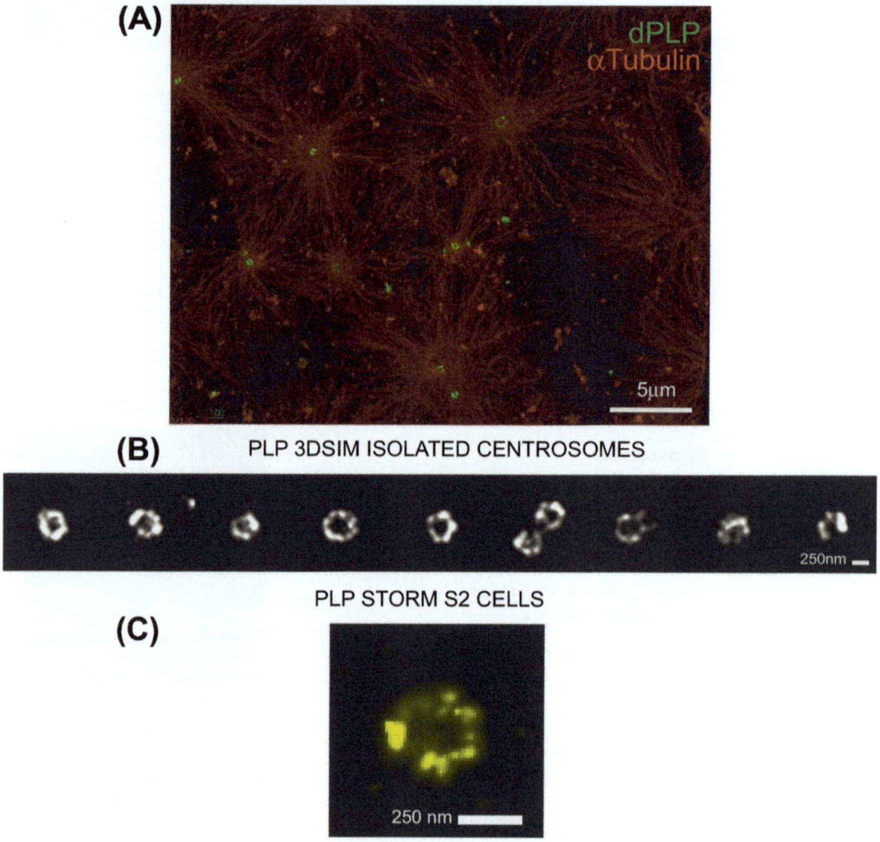

VITO MENNELLA ***ET AL.*****, FIGURE 1**

(A) 2D maximum intensity projections of SIM images of centrosomes isolated from *Drosophila* syncytial blastoderm embryos. After purification, centrosomes are allowed to regrow microtubules with rhodamine—tubulin and are labeled with rabbit antipericentrin like protein (Plp) antibodies recognized with antirabbit Alexa Fluor 488. Scale bars, 500 nm. (B) Examples of biochemically purified centrosomes labeled with rabbit anti-Plp recognized with antirabbit Alexa Fluor 488. Notice the variability in centrosome orientation and cell cycle stage. (C) STORM image of a centrosome in an S2 cell labeled with rabbit anti-Plp antibodies recognized with antirabbit Alexa Fluor 647. The secondary antibodies are labeled with a ratio of 0.3647 molecules per antibody.

Adapted from Mennella et al. (2012).

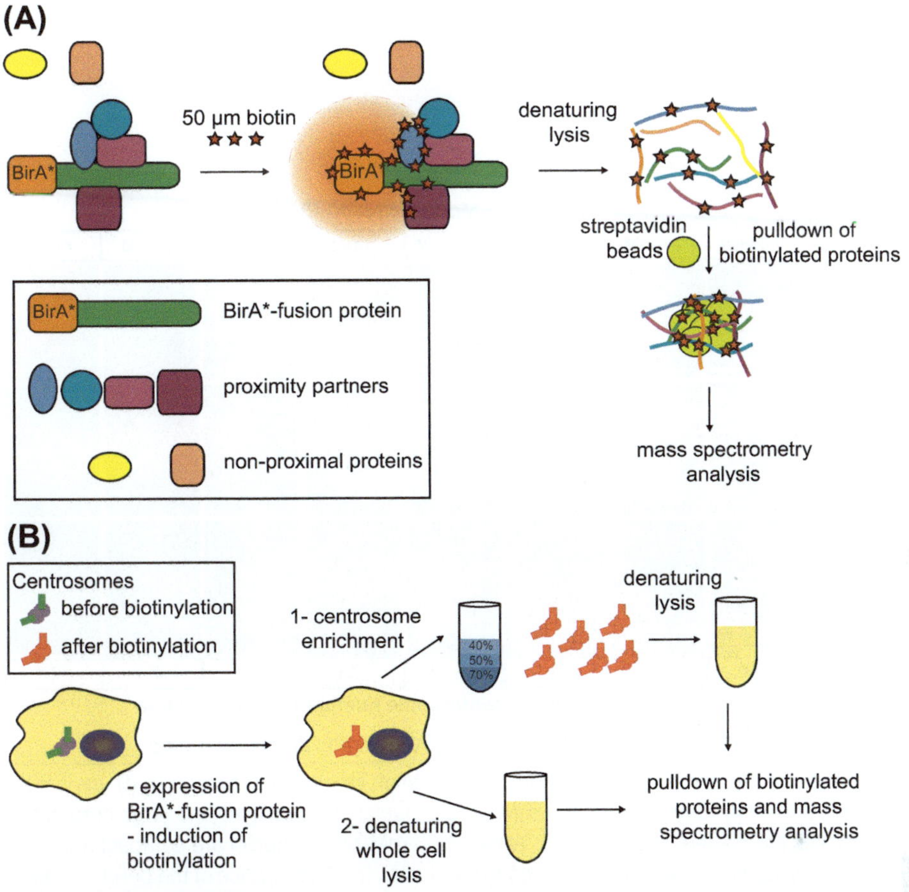

ELIF NUR FIRAT-KARALAR AND TIM STEARNS, FIGURE 1 Work flow for the application of the BioID approach to the centrosome.

(A) In the BioID approach, the protein of interest is tagged with a mutant form of *E. coli* biotin ligase BirA (R118G), denoted as BirA*. Following incubation of cells with 50 μM of biotin for 18–24 h, BirA* biotinylates proteins that are in the proximity of the BirA*-fusion protein and we refer to such proteins as "proximity interactors." The labeling radius of BirA* is estimated to be around 10 nm. The proteins outside the proximity radius are denoted as "nonproximal proteins." Following incubation of cells with biotin, proteins are solubilized and biotinylated proteins are precipitated by streptavidin beads for subsequent mass spectrometry analysis. (B) Cell line of choice is transfected with the expression plasmid for the BirA*-fusion protein. Cells transiently or stably expressing the BirA*-fusion protein are incubated with excess biotin to induce biotinylation of the proximity interactors. These cells are then processed for identification of the biotinylated proteins by the following two methods: (1) Centrosome-enriched fractions are prepared from these cells by sucrose gradient fractionation, and centrosomes are solubilized under denaturing conditions. (2) Cells are lysed under denaturing conditions. Following lysis, the biotinylated proteins are affinity captured and are identified by mass spectrometry.

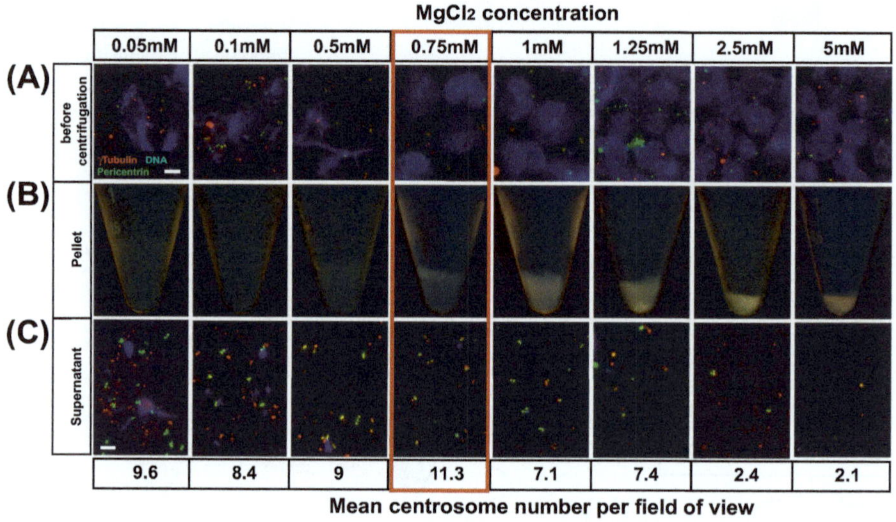

DELPHINE GOGENDEAU *ET AL.*, FIGURE 2 Influence of $MgCl_2$ during lysis step.

(A) Lysates of IIA1.6 murine cells before centrifugation labeled by γ-tubulin (red), pericentrin (green), and DNA (blue). Below 0.5 mM $MgCl_2$, nuclei are damaged and large chromatin aggregates are present in the lysates. Above 1.25 mM $MgCl_2$, nuclei are compact and more importantly, centrosomes are associated to the nuclei. (B) Appearance of the pellets following 10 min centrifugation at 2500 rpm. The volume of the nuclei pellet decreases as the $MgCl_2$ concentration increases (from 0.4 mL at 0.02 mM $MgCl_2$ to 0.1 mL at 10 mM $MgCl_2$). At low $MgCl_2$ concentrations, from 0.05 to 0.1 mM, the nuclei pellet appears translucent and highly viscous, whereas at higher concentrations, from 2.5 to 5 mM $MgCl_2$, the nuclei pellet is white and very compact. At the optimal concentration, the pellets should still appear viscous and whitish, but not too white. (C) Supernatants labeled by γ-tubulin (red), pericentrin (green), and DNA (blue). Under 0.1 mM $MgCl_2$, DNA is still found associated with centrosomes. Above 2.5 mM $MgCl_2$, supernatants contain very few centrosomes, still associated with the nuclei. Concentrations between 0.5 and 1.25 mM give a good purity of centrosomes. The number given under each panel gives the average number of centrosomes per field of view (at least 10 fields were counted for each condition). For IIA.6 cells, the optimal $MgCl_2$ concentration is 0.75 mM. Bars = 10 μm.

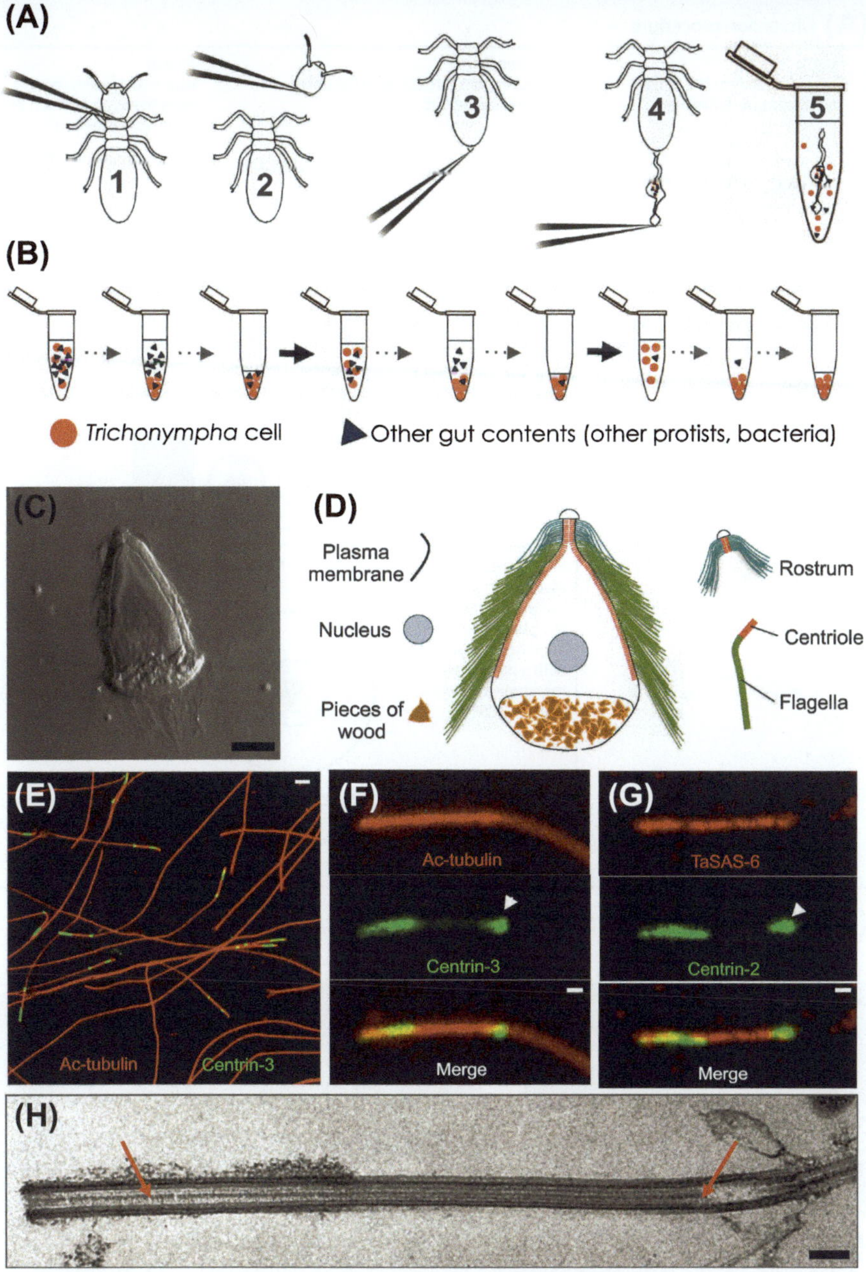

PAUL GUICHARD *ET AL.*, FIGURE 2 Isolation of *Trichonympha* centrioles.

(A) Dissection of *Zootermopsis nevadensis* termite and extraction of its gut. (B) *Trichonympha* cells are isolated from the rest of the gut contents by three rounds of washes and sedimentation. (C) Intact isolated *Trichonympha* cells under DIC microscope. Scale bar: 20 μm. (D) Schematic representation of a *Trichonympha* cell. Note the rostrum, which dissociates later than the cell body. (E–G) Isolated *Trichonympha* centrioles connected to flagella stained with antibodies against centrin (green) (E–G) and acetylated tubulin (red, E, F) or TaSAS-6 (red, G). Note that in the *Trichonympha* centriole, the pattern of centrin (centrin-2 or centrin-3) staining is different from that observed for *Chlamydomonas* or human centrioles. Indeed, beside the evolutionary conserved localization of the proteins at the distal region of centrioles (arrowheads, F, G), centrin antibodies also decorate the proximal-most third of the proximal region. Scale bars represent 2 μm in (E) and 500 nm in (F–G). (H) Conventional transmission electron microscopy of the exceptionally long *Trichonympha* centriole. Note that the central hub of the cartwheel is visible along the centriole (red arrow). Scale bar: 200 nm.

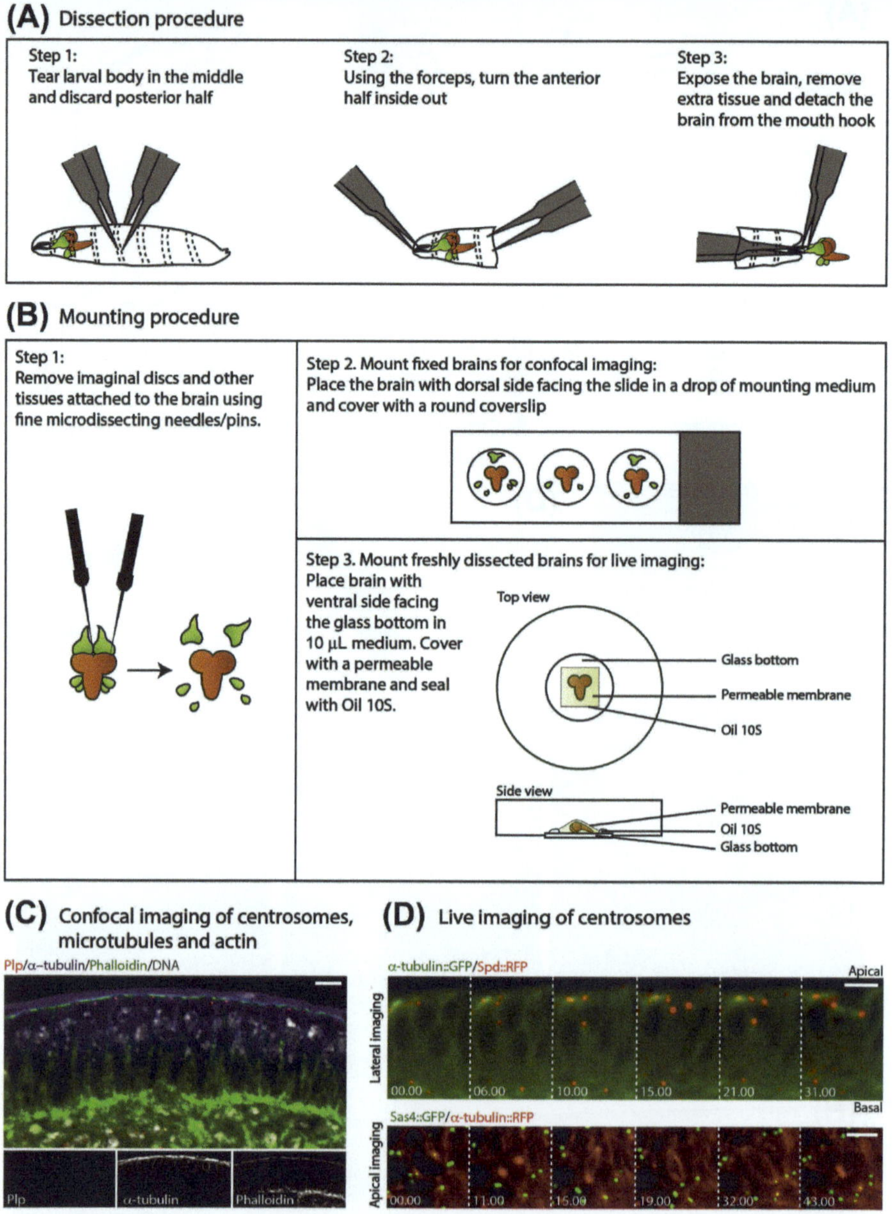

MARIA A. RUJANO *ET AL.*, FIGURE 2 Imaging of *Drosophila* neuroepithelial cells.

(A) Schematic representations of the successive steps of dissection required for isolating the *Drosophila* larval brain. (B) Larval brain mounting for imaging. The left-side panel shows a scheme for cleaning the brain before imaging. Right panel shows schemes of mounted samples for confocal imaging of fixed brains (top) and live imaging of fresh samples (bottom). (C) Example of confocal acquisition for centrosomes, microtubules, and actin staining. Centrosomes are depicted in red and immunostained with *Drosophila* pericentrin-like protein (Plp), microtubules are in blue and immunostained with α-tubulin, actin is in red and stained with phalloidin, and DNA is in gray labeled with Hoechst. (D) Example of spinning disc confocal live imaging of centrosomes and microtubules. Centrosomes are visualized with SPD::RFP (top panel) or Sas4::GFP (bottom panel) and microtubules are visualized with α-tubulin::GFP (top panel) or α-tubulin::RFP (bottom panel). Time is given in minutes. The neuroepithelium is imaged from the lateral side (top panel) and from the apical side (bottom panel). Four-dimensional z-stacks of 5–10 μm at 0.3 μm intervals were acquired every 30 or 60 s using an X60, NA 1.4 oil immersion objective. Scale bars: 5 μm.

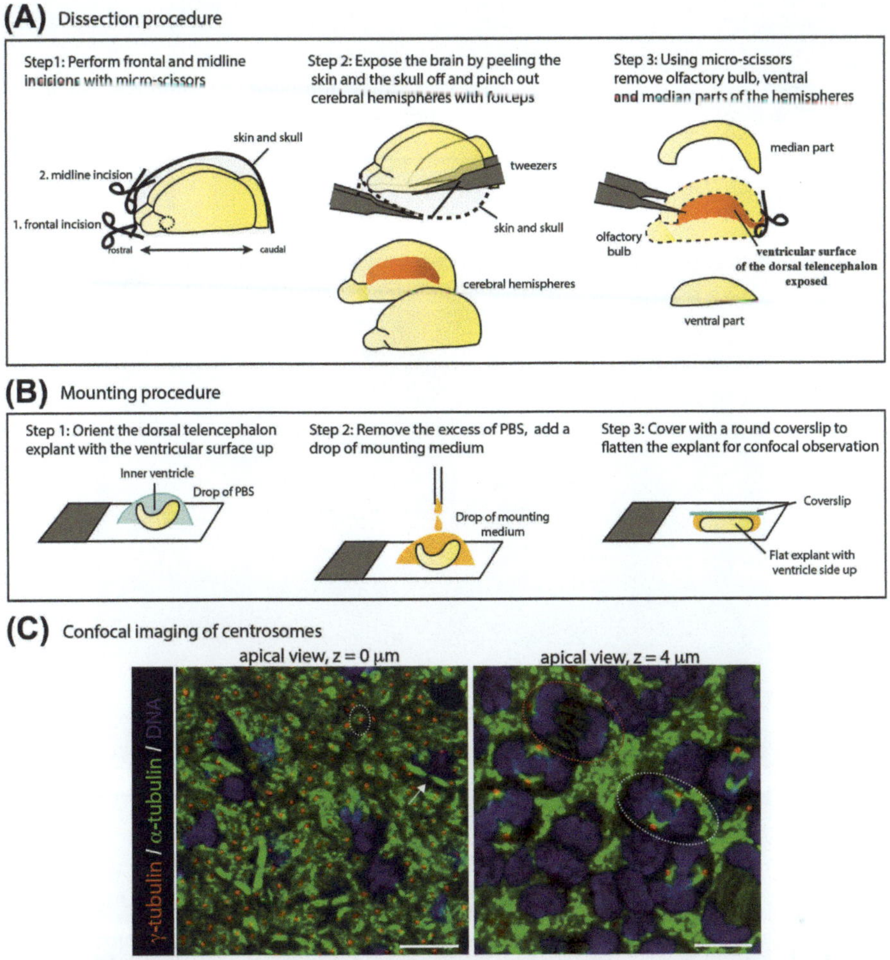

MARIA A. RUJANO *ET AL.*, FIGURE 3 En face views imaging of centrosomes in the mouse neocortex.

(A) Schematic representations of the successive steps of dissection required for isolating dorsal telencephalon explants from mouse embryonic brains before fixation and immunolabeling procedures. (B) Drawings explaining the mounting procedure to obtain apical side up dorsal telencephalon explants to image centrosomes in interphase and mitotic figures from an en face point of view. (C) Examples of en face view confocal acquisitions showing centrosomes (γ-tubulin in red), microtubules (α-tubulin in green), and DNA in blue. The same region was imaged at different z-positions along the apicobasal axis. At the uttermost apical position ($z = 0$ µm) (left-hand panel), the interphase centrosomes within the apical processes can be visualized (dashed white circle). In addition, the cleavage furrows are also visualized (white arrow). Just 4 µm above ($z = 4$ µm) (right-hand panel) mitotic cells are visualized in different stages (white dashed circle surrounds one metaphase and red dashed circle surrounds one anaphase). In the left-hand panels, note that centrosomes in interphase belong to cells at different stages of the cell cycle, i.e., one or two centrosomes can be visualized per apical process. In the right-hand panels, note that the parallel orientation of the mitotic spindle in relation to the ventricular lining enables the visualization of the two mitotic centrosomes at the same z-position. Scale bars: 8 µm.

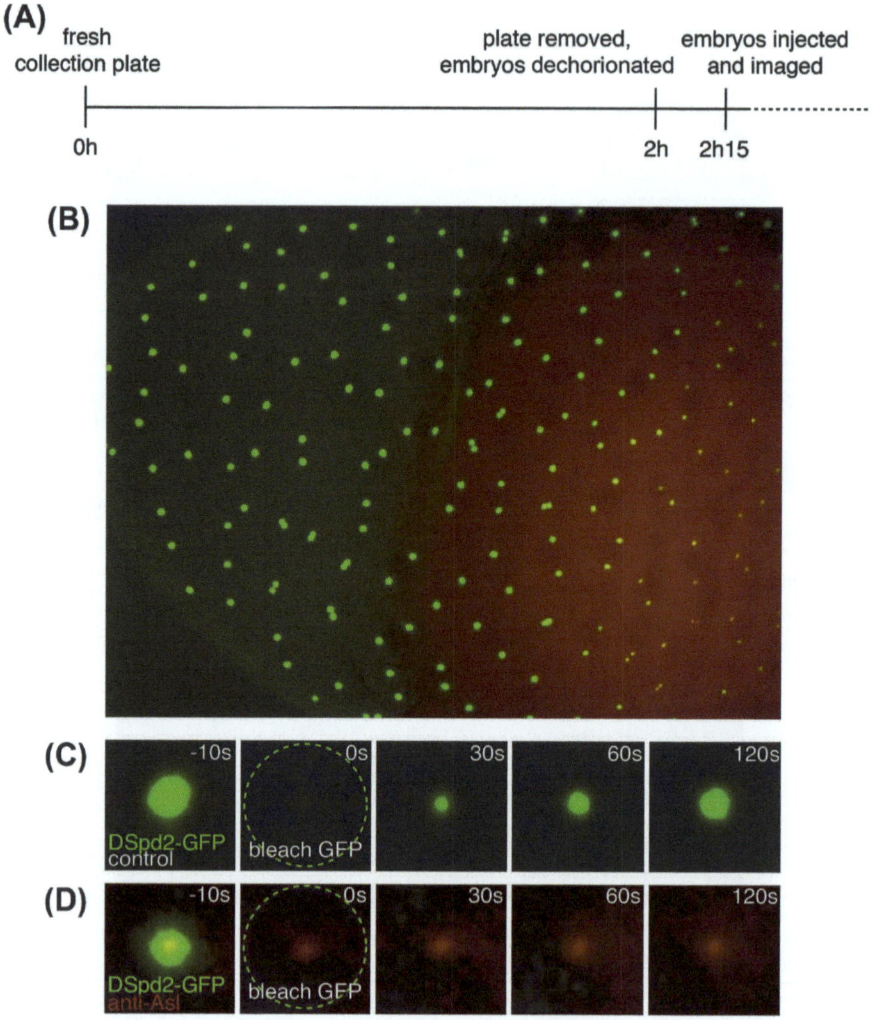

PAUL T. CONDUIT *ET AL.*, FIGURE 1 Antibody-mediated inhibition combined with FRAP analysis.

(A) A timeline of the experiment. (B) An image of a syncytial embryo expressing Spd-2-GFP (green) shortly after having been injected with Texas-red-labeled anti-Asl antibodies (red). Note how the antibody is concentrated on one side of the field of view, allowing a comparison between centrosomes that have been bound by the antibody (bound centrosomes) and centrosomes that have not been bound by the antibody (control centrosomes). (C, D) Examples from an FRAP experiment where the Spd-2-GFP signal was photobleached from a control centrosome (C) or from a bound centrosome (D). Time before and after photobleaching (t = 0 s) is indicated in the top right of each panel. Note how the Spd-2-GFP signal recovers robustly at the control centrosome, but not at the bound centrosome, showing that Asl inhibition perturbs the incorporation of new Spd-2-GFP molecules into the centrosome.

Modified from Conduit, et al. (2014) – eLife.

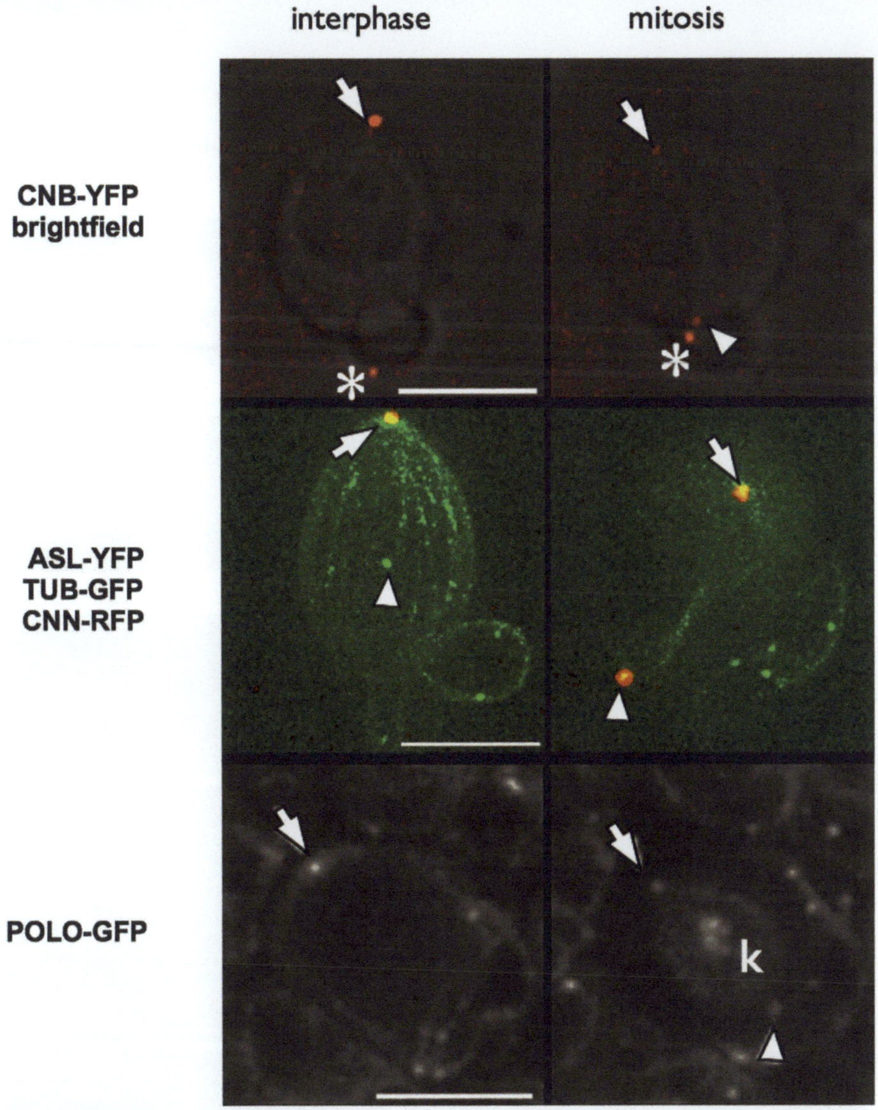

JUDIT PAMPALONA *ET AL.*, FIGURE 3 Imaging centrosomes in *Drosophila* third instar larvae NB.

Selected frames corresponding to interphase and mitosis taken from Movies 1–3. CNB-YFP, bright field) primary culture. Arrows and arrowheads point at the NB's apical and basal centrosomes, respectively. Asterisks label the centrosome of an adjacent ganglion mother cell. In the NB, CNB signal (pseudocolored red) is only observed on the daughter centriole, which during interphase is in the apical centrosome. At mitosis, both centrosomes contain newly assembled daughter centrioles that show a faint CNB label (ASL-YFP, TUB-GFP, CNN-RFP). Primary culture. In interphase, the pericentriolar material (PCM) labeled with CNN (red) is concentrated on the apical centrosome from which microtubules (green) radiate. The mother centriole (ASL; green), which is not labeled with CNN, can also be observed. CNN prominently labels both centrosomes in mitosis (POLO-GFP). Ex vivo whole-mount brain preparation. The large cell in the center is an NB located in the central brain. POLO accumulates on the apical centrosome in interphase. In mitosis POLO labels both centrosomes as well as the kinetochores (k). POLO-GFP also labels the midbody (shown in Movie 3; not shown here). Scale bar = 10 μm.

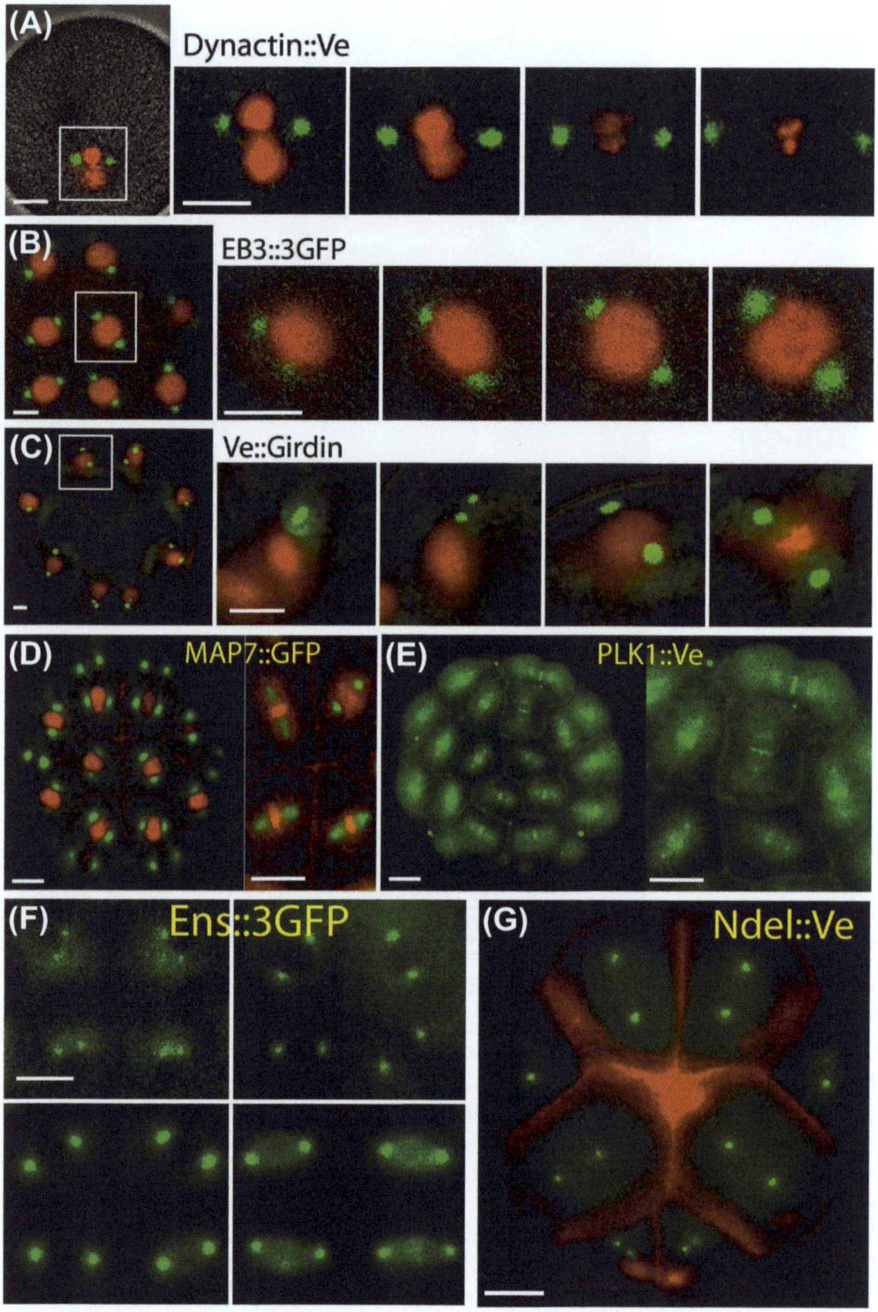

ALEX McDOUGALL *ET AL.*, FIGURE 1 Centrosome live labels in *Phallusia*.

We observe that some markers will label centrosomes well during the entire cell cycle (girdin, EB3, Ndel) whereas others such as MAP7 label spindle poles only during mitosis (which can be useful for detecting the onset of mitosis). (A) Fertilized egg derived from an unfertilized egg injected with mRNA encoding Dynactin::Venus (green) and HH2B::Rfp1 (red). The paternal centrosomes are labeled with Dynactin::Venus and the male and female pronuclei with HH2B::Rfp1. A time lapse of the same fertilized egg is shown from pronuclear apposition to metaphase. Scale bar = 20 μm. (B) An embryo previously injected with EB3::3GFP and HH2B::Rfp1. Both centrosomes are labeled around each nucleus. Scale bar = 10 μm. (C) A 16-cell stage embryo previously injected with Venus::Girdin and HH2B::Rfp1. The inset shows a time lapse. One centrosome (or perhaps two just before they separate) is shown in the first image followed by centrosome movements around the nucleus. Scale bar = 10 μm. (D) A 32-cell stage embryo previously injected with MAP7::GFP and HH2B::Rfp1. Scale bar = 20 μm. (E) A 32-cell stage embryo previously injected with Plk1::GFP. Plk1 labels the spindle poles as well as the spindle midzone and midbodies. Scale bar = 20 μm. (F) A 32-cell stage embryo previously injected with Ens::3GFP. Scale bar = 20 μm. (G) A 16-cell stage embryo previously injected with Ndel::Venus and labeled with Cell Mask (red). Scale bar = 20 μm.

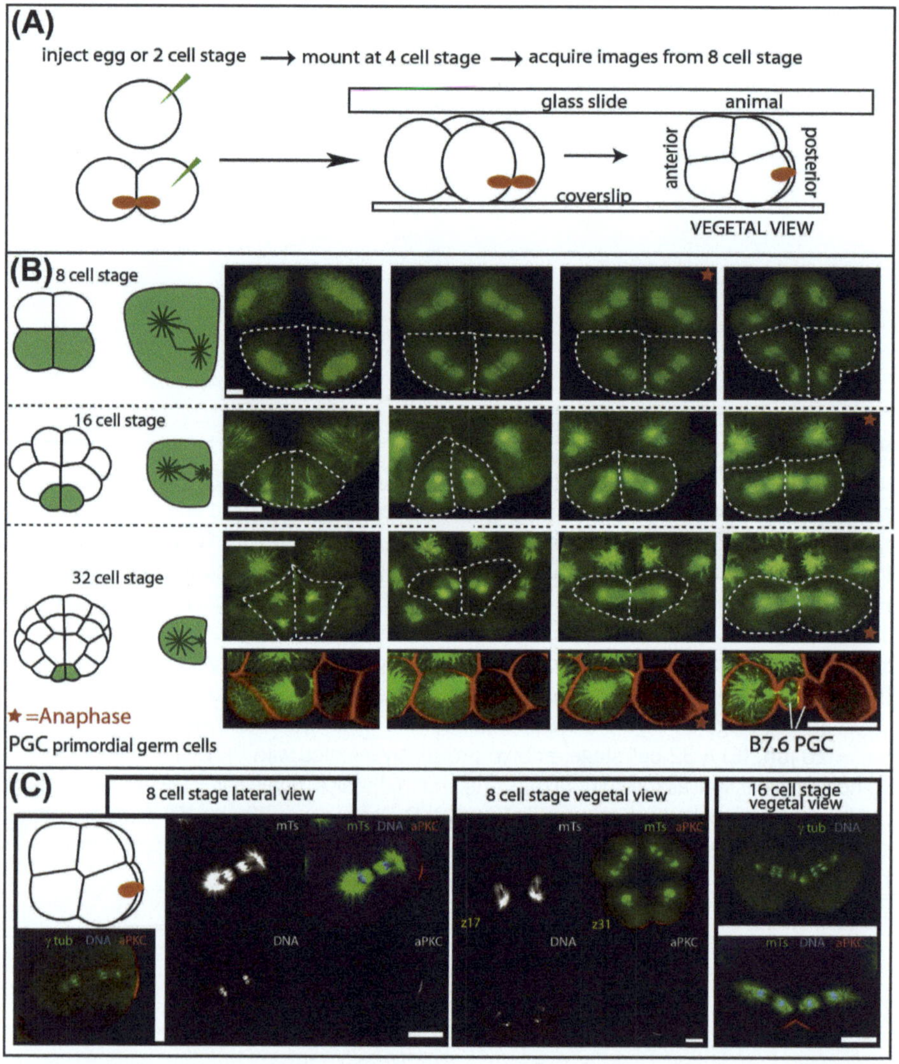

ALEX McDOUGALL *ET AL.*, FIGURE 2 Three unequal cleavages from 8- to 64-cell stage.

(A) Method to obtain a standardized vegetal view for live imaging. Orange discs represent the pair of CABs, a structure responsible for unequal cleavage localized in the posterior vegetal cortex of the germ cell precursors. (B) Portions of confocal time lapse sequences covering the indicated stages. Green cells in the drawings (left) indicate the CAB-containing cells which are filmed (right). First, second, and third rows: Ens::3GFP label recorded by confocal microscopy. Fourth row: MAP7::GFP and FM4-64 membrane label recorded by confocal microscopy. At the eight-cell stage, the spindle forms roughly in the center of the B4.1 cell, then without changing alignment elongates toward the posterior animal cortex. The aster nearest the CAB (the midline aster) becomes progressively smaller over the three divisions. At the 16-cell stage, the spindle forms on the animal–vegetal axis of the B5.2 cell, then rotates such that the vegetal centrosome migrates rapidly toward the midline and the animal centrosome shifts laterally toward the B5.1 sister cell (Prodon et al., 2010). At the 32-cell stage, one centrosome is positioned near the midline during interphase, such that the spindle forms angled toward the midline then elongates. In summary, over the three divisions the CAB centrosome is captured earlier in the cell cycle, and travels closer to the cortex as the cells become smaller, ultimately generating the tiny B7.6 pair of germ cell precursors. The red star represents anaphase when the position of the cleavage plane is defined. Scale bars = 20 μm. (C) Confocal images of CAB-containing cells fixed at the indicated stages and immunolabeled to detect endogenous microtubules or centrosomes. Gamma tubulin (green) = antibody GTU88 which labels centrosomes; aPKC (red) = antibody SC216 which labels the cortical face of the CAB (Patalano et al., 2006); mTs = rat monoclonal YL1/2 which labels asters and spindles (we also routinely use mouse monoclonal DM1a for this purpose). In all panels fixative was cold methanol/EGTA. This treatment is best for microtubule-based structures, but often results in cortical damage as can be observed in aPKC label lateral view. DNA = labeled with 1 μg/mL Hoechst. Scale bars = 20 μm.

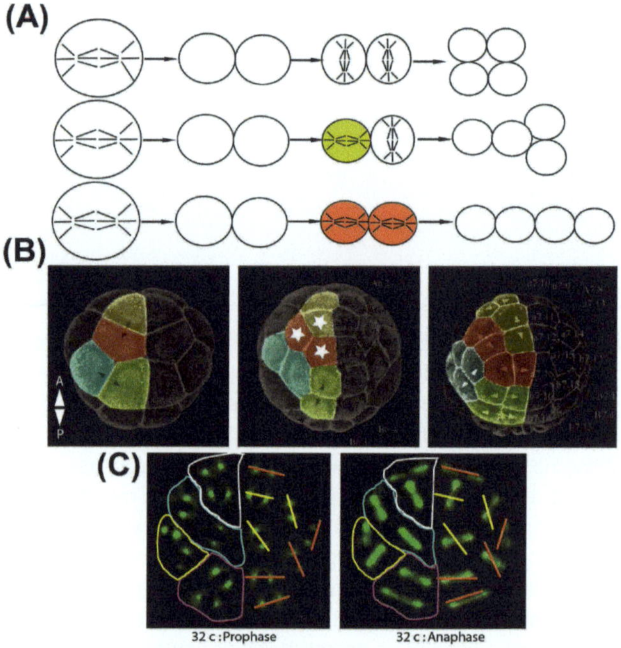

ALEX McDOUGALL *ET AL.*, FIGURE 3 Invariant cleavage.

(A) The upper schematic represents cells dividing by default at 90° relative to their mothers, a phenomenon based on centrosome migration following mitosis. The second and third illustrations represent the two different outcomes obtained if one (colored yellow) or both (colored red) daughter cell(s) divide in the same orientation as their mother giving rise to a T configuration or to four cells in a line, respectively. (B) *Ciona intestinalis* embryos (Hotta et al., 2007; http://chordate.bpni.bio.keio.ac.jp/faba/1.4/top.html;http://chordate.bpni.bio.keio.ac.jp/faba2/2.2/top.html) where the blastomeres have been color-coded and followed over two cell divisions. The daughters and granddaughters are thus shown at the 32- and 64-cell stages, respectively. As in the schematics, blastomeres at the 64-cell stage are arranged in squares, Ts, and lines, suggesting that spindles or centrosomes likely rotated in the blastomeres indicated with white stars at the 32-cell stage (only animal blastomeres are shown). A/P indicates anterior/posterior axis. (C) Spindle position at the 32-cell stage in live *Phallusia* embryos derived from eggs injected with MAP7::GFP. Spindle position is indicated at prophase by the red and yellow lines. The gray lines (right image) indicate the new spindle orientation at anaphase superimposed on the previous prophase position. The spindles that rotate appreciably in *Phallusia* are indicated by the yellow lines, corresponding to the three blastomeres at the 32-cell stage in *Ciona* labeled with white stars in (B).

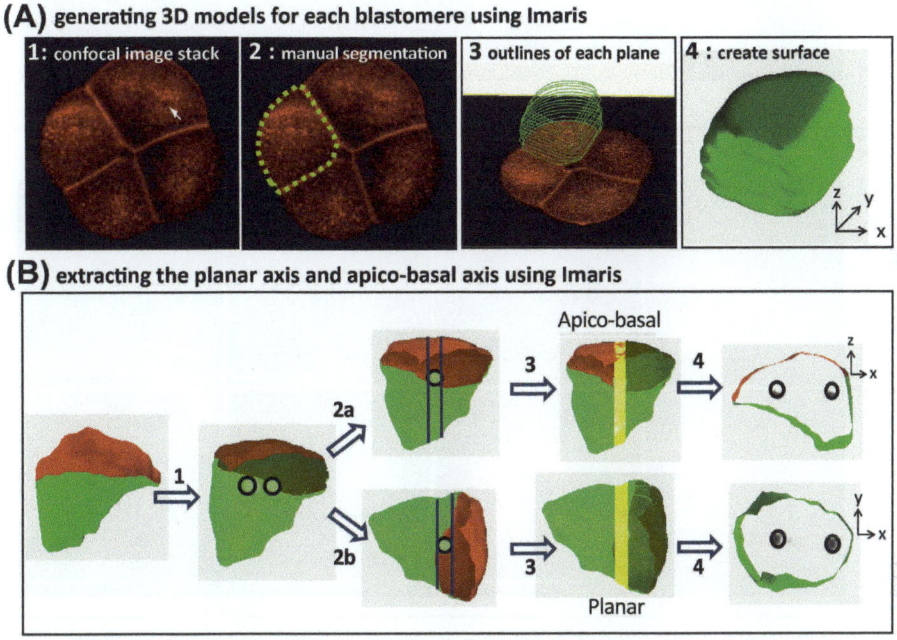

ALEX McDOUGALL *ET AL.*, FIGURE 4 3D rendering using Imaris.

(A) Generating 3D models for each blastomere. Embryos are first labeled with Cell Mask at the eight-cell stage for 3 min (see list 3.2) and then imaged using conventional CLSM (1). The outline of cell edges is drawn for each plane (2). After all cell outlines in each plane are complete (3), the function "create surface" creates a 3D object (4). Full protocol described in list 3.5. (B) Extracting specific planes of the cell. Illustration of the protocol described in list 3.6.

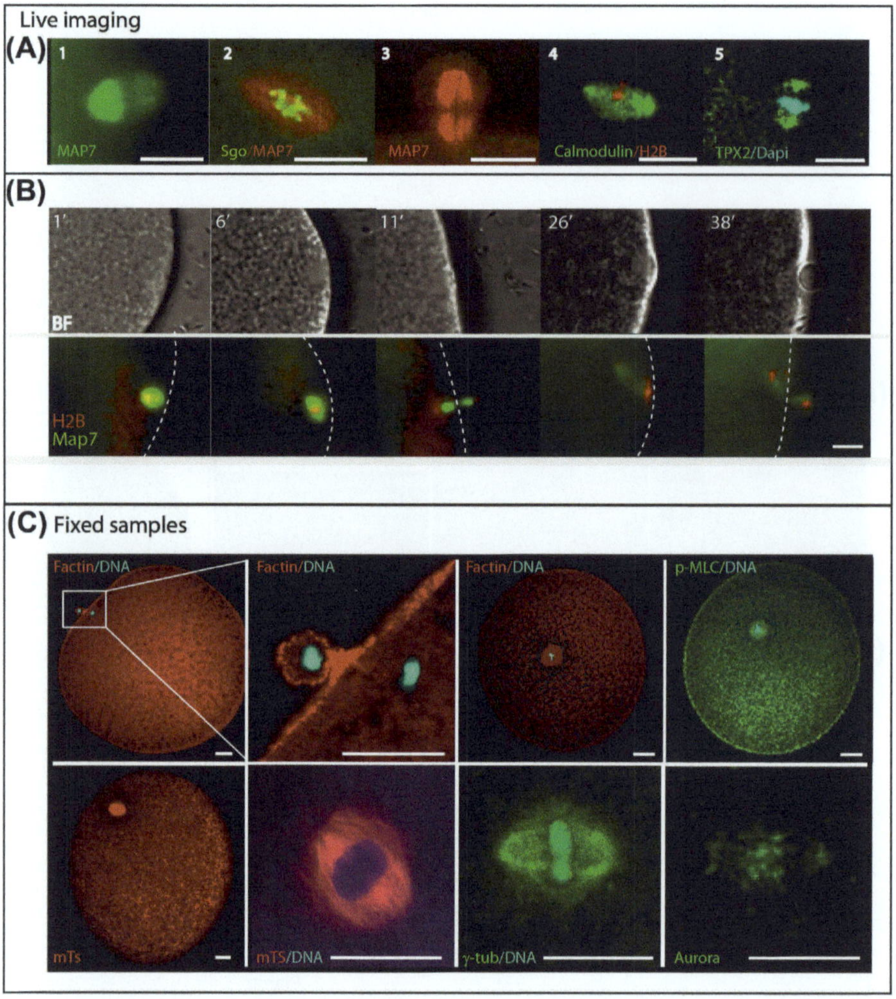

ALEX McDOUGALL *ET AL.*, FIGURE 5 Imaging of meiotic spindles and overlying cortex.

(A) Live cell markers of meiotic spindles. (1) Metaphase I spindle labeled with MAP7::GFP. (2) Metaphase I chromosomes labeled with Shugoshin::Venus plus MAP7::Rfp1 to visualize the spindle. (3) Meiotic spindle labeled with MAP7::Rfp1 during first polar body extrusion. (4) Meiotic spindle labeled with Calmodulin::Venus (enriched at the poles) and chromosomes labeled with HH2B::Rfp1. (5) Meiotic spindle poles labeled with Venus::TPX2 and the chromosomes with DAPI. Scale bars = 10 μm. (B) Meiotic spindle behavior during extrusion of the first and second polar bodies. Unfertilized eggs were injected with mRNA encoding MAP7::GFP (green) and HH2B::Rfp1 (red), incubated overnight to allow translation, then imaged using wide-field epifluorescence time-lapse microscopy. Dotted line represents the surface of the egg. Scale bar = 10 μm. (C) Detection of meiotic structures in fixed eggs. Top row, red: phalloidin label of F actin. Left: the tiny first polar body has received half the chromatin (blue label, Hoechst). Right: in an unfertilized egg, the cortex (actin cap) overlying the meiotic spindle is rich with F-actin and phosphorylated myosin II (pMLC, green). The bottom row shows examples of meiotic spindles immunolabeled with the anti-tubulin YL1/2, anti-γ tubulin GTU88, and anti-Aurora A/B antibodies. Scale bars = 10 μm.

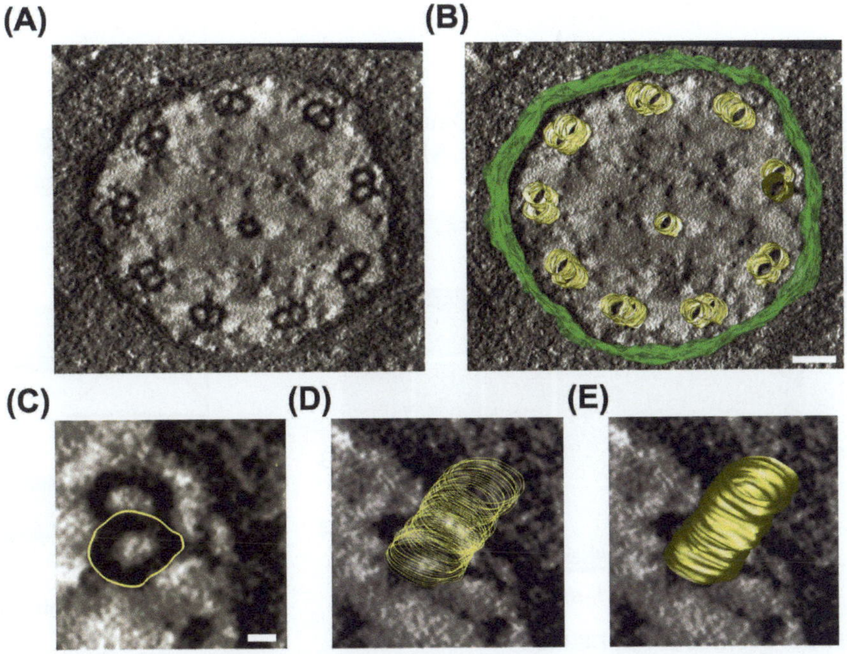

DANIEL SERWAS AND ALEXANDER DAMMERMANN, FIGURE 4 Electron tomography.

(A) A 4.5-nm thick tomographic slice of a cilium at the level of the proximal segment of the axoneme. (B) 3D reconstruction model overlaid on image in (A) highlighting axonemal microtubules (yellow) and ciliary membrane (green). (C–E) Steps in modeling a single axonemal microtubule using IMOD. First, the contour of the microtubule is traced in a single tomographic section (C), this procedure is the repeated for every second section of the tomogram (D). Finally, contours were meshed to create a surface model (E). Scale bars are 50 nm (B) and 10 nm (C).

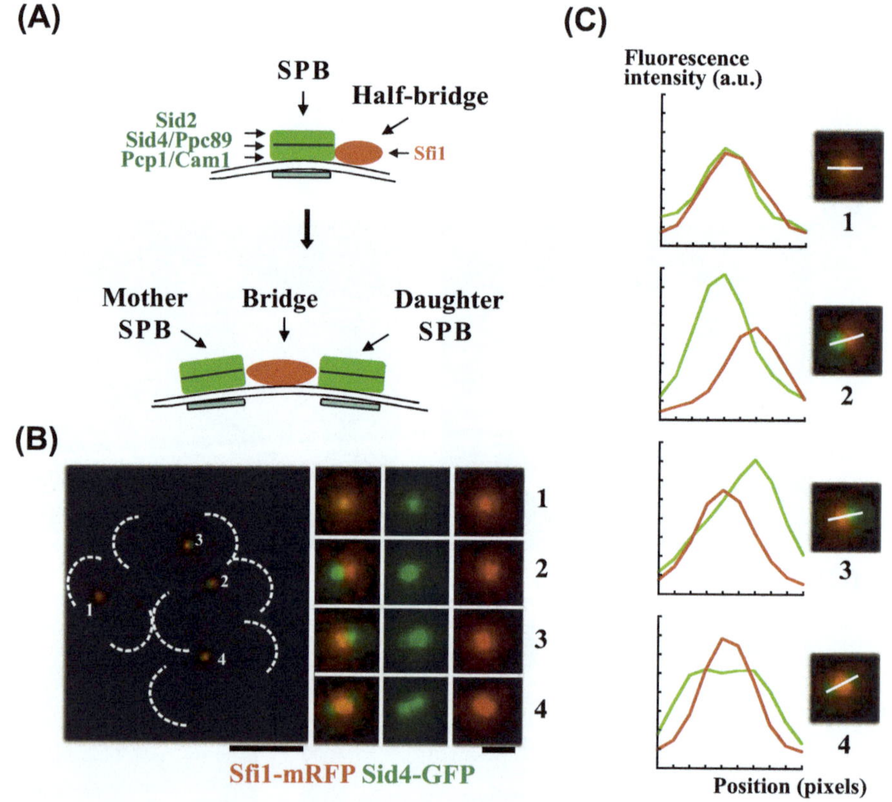

IMÈNE B. BOUHLEL *ET AL.*, FIGURE 1

Monitoring SPB duplication in fixed SPB-labeled strains. (A) Scheme of the SPB with SPB plaques in green and half-bridge in red before duplication (top) or the duplicated SPBs linked by the bridge that maintain them side by side (bottom). The localization of the plaque components Sid2, Sid4, Ppc89, Pcp1, and Cam1 and of the half-bridge component Sfi1 is indicated. (B) Localization at SPBs of Sid4-GFP and Sfi1-mRFP. Bar: 5 μm. Insets: magnification of SPBs from cells numbered according to cell cycle progression in the left panel. Bars: 500 nm. *(Reproduced from Bouhlel et al., in press.)* (C) Fluorescence intensity profiles of Sfi1-mRFP and Sid4-GFP across SPBs shown in insets measured on a 4-pixels-wide line oriented as shown in white. These SPBs are the same as those shown in (B). Note that Sfi1 maximum intensity does not coincide with Sid4 maximum intensity (SPB 2,3), except for SPB 1 where the SPB-half-bridge axis might be oriented perpendicular to the field of view. After duplication (SPB 4), Sfi1 localizes between the two maxima of intensity of Sid4-GFP. Note the asymmetric peak of Sid4-GFP in SPB 3 suggesting that the assembly of the daughter SPB has started.